Axoplasmic Transport

Editor
Zafar Iqbal, Ph.D.
Assistant Professor
Department of Neurology
and
Interdepartmental Graduate Program in Neuroscience
Northwestern University
School of Medicine
Chicago, Illinois

CRC Press, Inc.
Boca Raton, Florida

Library of Congress Cataloging-in-Publication Data
Main entry under title:

Axoplasmic transport.

Includes bibliographies and index.
1. Axonal transport. I. Iqbal, Zafar, 1941—
[DNLM: 1. Axoplasmic Flow. WL 102.5 A9711]
QP363.A9425 1986 591.1'88 85-26914
ISBN 0-8493-6334-9

This book represents information obtained from authentic and highly regarded sources. Reprinted material is quoted with permission, and sources are indicated. A wide variety of references are listed. Every reasonable effort has been made to give reliable data and information, but the author and the publisher cannot assume responsibility for the validity of all materials or for the consequences of their use.

Direct all inquiries to CRC Press, Inc., 2000 Corporate Blvd., N.W., Boca Raton, Florida, 33431.

International Standard Book Number 0-8493-6334-9

Library of Congress Card Number 85-26914
Printed in the United States

With affection to my children
Jameel and Shareen

PREFACE

Axoplasmic transport is the intracellular movement of cellular components required for the maintenance and normal physiological functioning of neuronal cells. This book provides an up-to-date reference for both established investigators as well as for those entering in the field.

The chapters contributed by the acknowledged experts in the area of axoplasmic transport discuss state of the art information about the subject. The first six chapters describe the history and methods of the study of transport and the involvement of energy, ions, calmodulin, microtubules and other cellular components in transport. In the next five chapters the transport of polypeptides, lipids, nucleic acids, neurotransmitter containing components and various other particles in nerve fibers, is discussed. A significant portion of this book is devoted to axoplasmic transport studies related to metabolic and toxic neuropathies. Areas dealing with retrograde transport, regeneration and the relevance of transport in neurotropic functions are described in the last four chapters, followed by a discussion on the mechanism of axoplasmic transport.

As editor, I found that the collaborative efforts of my fellow neuroscientists gave me a renewed education in the field of axoplasmic transport. I am thankful to all the contributors for their cooperation in completing this book. I am especially grateful to Drs. S. Ochs, M. I. Sabri, and R. Hammerschlag for their advice and help. I am also indebted to my father, Professor S. A. Sandilvi and to my late mother, Saleha, who have been a constant inspiration to me.

THE EDITOR

Zafar Iqbal, Ph.D., is associated with the Chicago Medical School, North Chicago as an associate professor of Pharmacology. After receiving his Ph.D. degree in biochemistry in 1971 from the All-India Institute of Medical Sciences, New Delhi, Dr. Iqbal served as a research associate and later as an assistant professor in the Departments of Biochemistry and Physiology, and Biophysics at the Indiana University School of Medicine at Indianapolis from 1972 to 1982. He also served as an assistant professor at the Northwestern University School of Medicine, Chicago from 1982 to 1985 before assuming his present assignment.

Dr. Iqbal is a member of several professional societies including the American Physiological Society, International Brain Research Organization, International Society for Neurochemistry, Society for Neuroscience, American Society for Neurochemistry, Society of Biological Chemistry and Indian Neuroscience Academy, Indiana Academy of Sciences and Sigma Xi. He has been recipient of research grants at national and international level and has served as visiting scientist in different countries.

CONTRIBUTORS

Lawrence Austin, Ph.D.
Department of Biochemistry
Monash University
Clayton, Victoria
Australia

Mark A. Bisby, D.Phil.
Professor
Department of Medical Physiology
University of Calgary
Calgary, Alberta
Canada

Goutam Chakraborty, Ph.D.
Research Associate
Department of Physiology
UMDNJ—New Jersey Medical School
Newark, New Jersey

Shew Chan, Ph.D.
Senior Research Associate
CRESAP Neuroscience Lab.
Northwestern University
Evanston, Illinois

Annica Dahlström, M.D., Ph.D.
Professor
Institute of Neurobiology
University of Goteborg
Goteborg, Sweden

J. Alejandro Donoso, Ph.D.
Research Associate Professor
Department of Neurology
Veterans Administration Medical Center
Kansas University Medical Center
Kansas City, Missouri

John S. Elam, Ph.D.
Professor
Department of Biological Science
Florida State University
Tallahassee, Florida

Hugo L. Fernandez, Ph.D.
Professor of Physiology
Kansas University Medical Center
Associate Research Career Scientist
Veterans Administration Medical Center
Kansas City, Missouri

L. C. J. Gaziri
Professor of Physiology
Department of Physiological Sciences
University of Londrina
Londrina, Brazil

Richard Hammerschlag, Ph.D.
Head, Section of Cellular Neurochemistry
Division of Neurosciences
Beckman Research Institute of the City of Hope
Duarte, California

Nicholas A. Ingoglia, Ph.D.
Professor
Physiology and Neuroscience
New Jersey Medical School
Newark, New Jersey

Zafar Iqbal, Ph.D.
Associate Professor
Department of Pharmacology
Chicago Medical School
North Chicago, Illinois

Johannes Jakobsen, M.D.
Resident
Department of Neurology
Diabetes Research Laboratory
University Hospital
Aarhus, Denmark

Ralph A. Jersild, Jr., Ph.D.
Professor
Department of Anatomy
Indiana University School of Medicine
Indianapolis, Indiana

Martin Kanje, Ph.D.
Professor
Department of Zoophysiology
University of Lund
Lund, Sweden

Masanori Kurokawa, M.D. and Ph.D.
Professor
Department of Biochemistry
Institute of Brain Research
Faculty of Medicine
Tokyo University
Tokyo, Japan

Sidney Ochs, Ph.D.
Professor of Physiology
Department of Physiology and Biophysics
Indiana University Medical School
Indianapolis, Indiana

C. G. Rasool, Ph.D.
Assistant Professor
Department of Neurology
Harvard Medical School
Boston, Massachusetts
and
Associate Biochemist
Center for Neurologic Diseases
Brigham and Women Hospital
Boston, Massachusetts

Mohammad I. Sabri, Ph.D.
Assistant Professor
Department of Neuroscience and Neurology
Albert Einstein College of Medicine
Bronx, New York

Per Sidenius, M.D.
Resident
Second Clinic of Internal Medicine
Diabetes Research Laboratory
University Hospital
Aarhus, Denmark

Kenneth A. Skau
Assistant Professor of Pharmacology
College of Pharmacy
University of Cincinnati
Cincinnati, Ohio

George C. Stone, Ph.D.
Assistant Research Scientist
Division of Neurosciences
Beckman Research Institute of the City of Hope
Duarte, California

Toshifumi Takenaka, M.D.
Professor of Physiology
Department of Physiology
Yokohama City University
Yokohama, Japan

Bruce Tedeschi
Assistant Professor of Cell Biology
Department of Zoology and Physiology
Louisiana State University
Baton Rouge, Louisiana

Dieter G. Weiss
Privatdozent, Dr. rer. nat.
Institut für Zoologie
Technische Universität München
Munchen-Garching, West Germany

David L. Wilson
Professor
Departments of Biology and Physiology and Biophysics
University of Miami
Miami, Florida

Michael F. Zanakis, Ph.D.
American BioInterface Corporation
The Cooper Union
New York, New York

TABLE OF CONTENTS

Chapter 1

A BRIEF HISTORICAL INTRODUCTION TO AXOPLASMIC TRANSPORT

Sidney Ochs

TABLE OF CONTENTS

I. DEVELOPMENT OF THE CONCEPT OF TRANSPORT IN NERVE

A. The Older Conception of Nerve Function

The idea of a spirit-like influence flowing in nerve can be traced back to ancient Greek physiological speculation. As later formulated by Galen in the 2nd century, a spirit-like principle is inhaled during inspiration and carried by the blood to the brain where an animal (animating) spirit is abstracted into the ventricles of the brain. The animal spirit then moves down within the hollow nerves to effect motion and sensation. The hollow nerve concept was an extrapolation from what appeared to be the case in the bovine optic nerve.[1] This was an extraordinarily long-lasting concept with traces of it seen lingering even into the 19th century.[2] In the mechanistic twist given to Galen's theory by Descartes in the 17th century,[3] the animal spirits, were viewed as the "liveliest and subtlest" of particles. To explain reflex action the animal spirits are able to leave the ventricles when flaps within the individual nerve fibers are pulled on by sensory stimulation. Under control of the pineal, the animal spirits then flow from the ventricles of the brain down within the opened individual hollow nerve fiber into their muscles to inflate them and thus cause the body to move in an appropriate fashion.

An inflation of muscle by animal spirits or a nerve fluid was disproved by the plethysmographic studies of Swammerdam soon after Descartes' formulation. Using the isolated frog muscle preparation, Swammerdam in the 17th century found on stimulation of its nerve that there was no increase in muscle volume during contraction. Similar plethysmographic studies made by Steno and Glisson at this time using the human arm also showed the lack of an increased muscle volume on contraction.[4,5]

In the 17th and 18th centuries various theories were advanced to account for nerve and muscle action with most retaining the concept of nerves consisting of hollow tubes with a nerve fluid of some kind moving down inside them. In one theory the nerve substance was considered to bring about an explosive reaction in the muscle to account for its contraction. There were also more physical conceptions of nerve action, with the nerve fibers transmitting some kind of a vibratory influence, either like sound waves or as a vibration of ether within the fibers.[6]

When electricity came under renewed scientific study in the late 18th century, it was usually dealt with as a flow of one or more fluids. Galvani[7] conceived of the animal electricity produced in the body as a fluid of a lively and peculiar kind which he assimilated to the old idea of a nerve spirit or fluid moving down within hollow nerve tubes. He wrote that, "the electric fluid is produced by the activity of the cerebrum, that it is extracted in all probability from the blood and that it enters the nerves and circulates within them in the event that they are hollow and empty or, as seems more likely, they are carriers for a very fine lymph or other similarly subtle fluid which is secreted from the cortical substance of the brain, as many believe. If this be the case, perhaps at last the nature of animal spirits, which has been hidden and vainly sought after for so long will be brought to light with clarity".

B. Nerve Form and Material Transport

The concept of a nerve fluid had been given substantiation by the early microscopic studies of Van Leeuwenhoek[8] as described in a letter written in 1717. When he placed freshly cut cross-sections of bovine optic nerve under his simple microscopes the nerve fibers were seen to be circular. Within a minute or so, a pearly fluid exudate appeared over the center of each of the circular cross-sections to form a little mound. This soon passed off as a vapor with the fibers collapsing to form flattened bands. Van Leeuwenhoek concluded that normally the nerve fluid within the fibers distends the walls of the nerve tubes, causing them to assume a cylindrical form and on cutting the nerves the escape of the nerve fluid allows the pressure on the walls to decrease and the fibers to collapse.

Chromatic and other optical distortions inherent in the lenses then used for microscopy made studies at higher magnification difficult until the achromatic compound lens microscope became available in the early part of the 19th century. With such a microscope Ehrenberg[9] in the 1830s clearly pictured the individual fibers. However, because he separated nerve fibers by compressing the entire nerve under microscopic control,[10] the fibers assumed an abnormal beaded shape which led to their confusion with nerve cell bodies. Purkinje and Valentin soon thereafter identified the nerve cell body by the presence of the nucleus.[11,12] Unfortunately, from their microscopical studies, these pioneers considered the nerve cell bodies and fibers to be two separate and distinct neural entities.[2,13] The cell body was considered to be the active element which in some fashion causes the nerve fluid contained in the apposed nerve fibers, the passive element, to circulate.[14,15] The fibers were viewed as closed loops with their central ends apposed to the cell bodies in the ganglia or in the central nervous system, their peripheral ends acting as sensory endings or contacting muscles to cause them to contract. This concept of a circulation of nerve fluid in looped fibers could have been influenced by some earlier hypotheses made of a circulation of nerve fluid in nerve fibers[16] inspired by Harvey's demonstration of the circulation of blood.

C. Cell Body-Fiber Dependence

Remak in an early paper in 1838 and definitively in his thesis 2 years later,[17] was the first to advance the idea that the nerve fibers and cell bodies constitute a single neural entity, i.e., to espouse an "early neuron doctrine".[2,18] Remak wrote that, "the organic fibers originate from the substance of the nucleated globules itself. In spite of the fact that this observation is very difficult and requires great dexterity in preparation as well as in observation, it is so well founded that it already would not be possible to doubt (it)".[19] This early conception of the neuron as an entity was in direct opposition to the dual neural element concept of Valentin and Purkinje and led to a long and dogged dispute between Remak and Valentin.[19] Only late in the 19th century, with the work of His,[20] Forel,[21] and Ramon y Cajal[22,23] among others, did the evidence for a single neuron element become massive and convincing and in Waldeyer's review,[24] the concept became formally baptized as the neuron doctrine. Finally, the outgrowth of fibers from cells shown directly in the tissue culture studies of Harrison[25] served to remove lingering reservations even as late as 1910 concerning the neuron doctrine.[26]

In spite of the uncertain status of the true nature of the neuron in the mid-19th century, Waller[27] inferred from his experiments on nerve degeneration that the viability of the nerve fiber depends on its connection with the cell body. He considered that some trophic influence passes from the cell body into the fiber to maintain its viability. This was clearly indicated in a classic study where the roots and nerves of the 2nd cervical ganglia of kittens were separately transected and those portions of the fibers of the nerves or roots connected to the dorsal root ganglion were seen to retain their normal form, while the part of the fiber separated from the ganglion underwent degeneration. Waller[28] concluded that, "the ganglion corpuscles (cell bodies) present in the dorsal root ganglion exert a trophic influence necessary to maintain the form and function of the fibers ascending in the dorsal root fibers as well as on the sensory fibers descending in the peripheral nerve". And, "as long as the influence of the ganglion over the nerve fiber occurs, this equilibrium (forces of renewal as opposed to those of degeneration) is maintained, but as soon as the connection of the ganglion corpuscle with the nerve fiber is destroyed, its peripheral (severed) end ... is subjected to forces of destruction (degeneration)". Waller[29] later figuratively put it that "a ganglion therefore was to the fibers connected to it what a river was to the rivulet that trickled from it, a source of nutritive energy".

D. Fibrillary Elements Within the Fibers — Cytoskeleton

The presence of a nerve fluid capable of moving in hollow nerve fibers appeared to conflict with another of the pioneer observations made by Remak,[30] namely that a fibrillary, somewhat solid material, was contained in the primitive nerve fiber, or nerve band, as he called it. This view of the axon was verified by Purkinje[31] who renamed it the nerve axis or axon, the term by which it is referred to today. There was some doubt as to the reality of the presence of fibrillary material in nerve fibers. Schultze[32] later in the 19th century using more advanced staining methods than previously available, clearly established its presence in the nerve fiber. He considered that the function of the neurofibrils was to conduct the nerve impulse.

By the early part of this century it was generally recognized that the axon membrane is the seat of the nerve action potential[33] and that some other role had to be assigned to the neurofibrils. Parker[34] proposed that the fibrillary elements are the means by which the cell body exerts a metabolic control over its fiber. Considering the T-shaped neurons of the dorsal root ganglion earlier demonstrated by Ranvier, Parker wrote that,

> "in the ordinary sensory neurons, nerve impulses originate at the peripheral end, make their way centrally over the neurite, and without entering the body of the cell, pass on to discharge at the central end of the neuron. The metabolic influences on the other hand originate in the region of the nucleus of the cell body, pass down its neck to the tract of nervous transmission where they separate into two streams, one flowing peripherally over the neurite and the other centrally over the central nerve fiber process. And thus, the course of the neurofibrils does not follow that of the nerve impulses but does duplicate exactly that of the metabolic influences. I conclude therefore that the neurofibrillar system in the neuron is concerned specifically with the distribution of the metabolic influences and not with the conduction of nerve impulses. These influences start in the region of the neuronic nucleus and spread over the lines of neurofibrils throughout the whole neurone."

Parker's view was indeed very much in line with our modern one of transport, but he could not say what the nature of the materials transported might be. Gerard,[35] summarizing the evidence that the nerve impulse depends on oxidative metabolism, proposed that the enzymes necessary to maintain the metabolism of the fibers were being continually transported down within the nerve fibers from their source in the nerve cell bodies. This concept was similar to one earlier expressed by Goldscheider, who suggested that a transport of ferment-like substances from the cell body moves down along the whole course of the axon to their extremities as is required for the nutrition of the axons.[36]

Scott,[37] struck by the similarity of the staining of the neurosomes in the neuron cells to zymogen granules in secretory cells, considered that "nerve cells are true secretory cells". He also wrote that, "in the body of the nerve cell a substance is formed from the nucleus and Nissl bodies which gradually passes into the nerve fibers; and also that (the) stimulation of other cells by a nerve fiber is brought about by the passage of some of this substance into the cells on which the fiber acts".[38] He could thus be considered one of the forerunners of the concept of transport and neurotransmission as well as noting the similarity of the neuron to that of a secretory cell.

Torrey,[39] a student of Parker, proposed that the substance supplied by the cell body to its nerve fibers was a hormone-like material necessary for the maintenance of the nerve fiber and he also considered that "the substance involved may be enzymatic in character, produced by the nucleus of the neuron, and transported peripherally by the way of the neurofibrils." Again, we are close to the modern concept based on molecular biology showing the role of DNA and RNA to synthesize proteins in the cell body and the revelations of fine structure in the fibers by the electron microscope, with a differentiation of the neurofilaments, the cytoskeleton, into microtubules, neurofilaments, and microfilaments.

E. Damming and Fluid Outflow as Evidence of Transport

Throughout most of the 19th century the dominant belief was not only that nerve fibers and nerve cell bodies were separate elements, but that a divided nerve could heal by a reunion rather than by a regeneration of new fibers from the proximal part of the divided fibers. This was the case in spite of the early evidence of Waller[27] showing regeneration to be due to an outgrowth of new fibers. Only with the acceptance of the neuron doctrine could the concept of regeneration as an outgrowth of neurites from the proximal stump become established.[26,40,41]

The effect of a partial constriction of nerve is of interest in that in some sense the old concept of a fluid flow in nerve fibers was carried over in explaining the phenomena seen. Ramon y Cajal[41] in his studies of regeneration using partially compressed nerves described a series of bulges and narrowings in the fibers above the compressed region which he interpreted as due to a starting and stopping of regenerative nerve fiber growth. A new growth was identified by the fine neurites seen in and beyond the region of partial constriction with growth cones at their distal ends. Weiss and Hiscoe[42] who studied the effects of partial compressions of nerve trunks by means of contracted arterial segments, saw a somewhat similar picture. However, the bulges and other form changes in the fibers above the constriction were differently interpreted. They considered the axonal contents to be flowing or constantly growing down within their sheaths at a rate of 1 to 3 mm/day as a column of semisolid axoplasm. On meeting with the obstruction due to the partial constriction, the column becomes dammed up, throwing the fibers into the beads and the other tortuosities seen.

A somewhat similar concept of a continual outflow of axoplasm within the nerve fibers had also been advanced by Young[43] at about the same time. In an amputated portion of nerve the fibers appeared to contract to form folds and spirals while "the central portion swells as if under the influence of an increased turgor pressure". The normally oriented molecules in the fiber (the neurofibrils of Parker) were considered to become disorganized in the amputated nerve stumps. On the second day after cutting, the central ends of the nerve fibers are swollen and "during the succeeding days the material of the central end flows out from the tube, forming one or several regenerating strands". Young[44] had earlier observed an outflow of axoplasm from the cut end of squid giant axons and from this he had inferred that the continued production of axoplasm by the cell body acts to force axoplasm down within the nerve fibers. The pressure exerted by the downflowing axoplasm is opposed by the lateral pressure in the walls of the fibers. Those concepts were carried over by Young to the myelinated nerve fibers.[45] The classical beads and ovoids, and other form changes of Wallerian degeneration which appear in amputated nerves were explained by Young as due to the unopposed surface tension of the myelin wall giving rise to the similar appearing beads shown in model experiments and attributed to surface tension by Plateau.[46] The oozing of axoplasm from the cut ends of myelinated nerve fibers described by Lubińska[47] seemed to be in accord with Young's concept. However, the assumption of a surface tension produced by a fluid-like myelin sheath giving rise to Plateau figures was not reconcilable with the laminar nature of the myelin sheath deduced by Schmitt from X-ray diffraction studies and later shown by electron microscopy to be produced by a wrapping of Schwann cell membranes around the axons.[48]

F. Tracer Techniques, Enzymes, and Optical Methods

While the morphological studies of Weiss and Young published in the 1940s suggesting a fluid downflow in nerve fibers did much to stimulate further study of the phenomenon of material transport in nerve, the concept that a semisolid fluid slowly moves down the axons at a rate of several millimeter per day was soon to undergo revision with the introduction of new techniques. Samuels et al.[49] in 1951 first reported isotopes as tracers to the study of downflow using ^{32}P-orthophosphate as a precursor. After its uptake by spinal cord motoneurons from the blood there was a differential outflow of various labeled phosphorus components in the motor fibers, a finding which suggested a more complex outflow of axonal contents. Lubińska[47] in the 1960s discovered, after crushing nerves, that the acetylcholinesterase present in motor fibers became redistributed with time. The enzyme accumulated just above a distal crush as expected of an anterograde transport and to accumulate also just below an upper crush indicating a retrograde transport as well. This result was clearly inconsistent with a simple downflow of the axoplasm as a whole. Dahlström[50] later in 1969 found that amines in adrenergic nerve fibers accumulated at nerve crushes or ligations at a fast rate. The use of isotopes of amino acids as precursors injected near cell bodies for selective uptake revealed, after its incorporation, a still faster and regular transport of labeled proteins at a rate of 410 mm/day in addition to a slow transport.[51] Optical studies also revealed particles moving within nerve fibers at a fast rate in both the anterograde and retrograde directions.[51]

The modern era of investigation of material transport has established the cell body as the site of the synthesis of a wide range of materials supplied to the axon, components which include proteins and polypeptides, lipids and other substances which turn over in the membrane and the cytoskeletal elements and within the various other organelles of the fiber. Also synthesized are neurotransmitter and neurotransmitter components supplied to the nerve terminals, the vesicle membrane and those enzymes needed for local synthesis and turnover of neurotransmitter within the terminals. Additionally, there is a transport of trophic substances destined to be carried from the nerve terminals into post-synaptic cells to regulate some of its functions.

The mechanism of transport in the axon for anterograde transport materials is not as yet definitively established but it is recognized to be a local process within the nerve fiber, one which requires a supply of ATP and, in many nerves at least, depends on calcium.[51] The retrograde transport of materials has similar requirements. A number of models have been proposed to account for the transport mechanism[51] but here we are dealing with current problems. For their further explication we turn to the following chapters of this book.

REFERENCES

1. **Clarke, E.,** The doctrine of the hollow nerve in the seventeenth and eighteenth centuries, in *Medicine, Science and Culture: Historical Essays in Honor of Owsei Temkin*, Stevenson, L. G. and Multhauf, R. P., Eds., Johns Hopkins Press, Baltimore, 1968, 123.
2. **Ochs, S.,** Waller's concept of the trophic dependence of the nerve fiber on the cell body in the light of early neuron theory, *Clio Med.*, 10, 253, 1975.
3. **Hall, T. S.,** *Treatise of Man by Rene Descartes*, (with translation and commentary), Harvard University Press, Cambridge, 1972.
4. **Bastholm, E.,** *The History of Muscle Physiology*, Munksgaard, Copenhagen, 1950.
5. **Keele, K. D.,** Physiology in, *Medicine in Seventeenth Century England*, Debus, A. G., Ed., University of California Press, Berkeley, 1974, chap. 7.
6. **Newton, I.,** *Opticks*, 4th ed., 1730, republished, Dover, New York, 1952.
7. **Galvani, L.,** *Commentary on the Effects of Electricity on Muscular Motion*, 1791, (translated from the Latin by Margaret G. Foley with notes and a critical introduction by I. Bernard Cohn), Burndy Library, Norwalk, Conn., 1954.
8. **van Leeuwenhoek, A.,** Letter to Abraham van Bleiswyk (March 2, 1717) (translated), in *The Human Brain and Spinal Cord*, Clarke, E. and O'Malley, C. D., Eds., University of California Press, Berkeley, 1968, 32.

9. **Ehrenberg, C. G.,** Observations on the structure hitherto unknown of the nervous system in man and animals (translated from the German by Craigie, D.), in *Essays on Physiology and Hygiene,* Haswell, Philadelphia, 1838.
10. **Liddell, E. G. T.,** *The Discovery of Reflexes,* Clarendon Press, Oxford, 1960.
11. **Kruta, V. J. E.** Purkyne's contribution to the cell theory, *Clio Med.,* 6, 109, 1971.
12. **Kruta, V.,** A note on the history of Purkyne cells, in *Jan Evangelist Purkyne 1787-1869,* Kruta, V., Ed., Universita Jana Evangelisty Purkyne, Brno, Czechoslovakia, 1971, 125.
13. **Van der Loos, H.,** The history of the neuron, in *The Neuron,* Hyden, H., Ed., Elsevier, Amsterdam, 1967, chap. 1.
14. **Valentin, G.,** Uber die Scheide der Ganglienkugel und deren Fortzetsung, *Muller's Arch.,* 138, 1839.
15. **Valentin, G.,** Traite de neurologie, in *Encyclopedie Anatomique,* Vol. 4, (translated from the German by A. J. L. Jourdan), Paris, Bailliere, 1843.
16. **Clarke, E.,** The neural circulation. The use of analogy in medicine, *Med. Hist.,* 22, 291, 1978.
17. **Remak, R.,** Observationes anatomicae et microscopicae de systematis nervosi structura, Dissertation, Berolini, Reims, 1838.
18. **Ochs, S.,** The early history of material transport in nerve, *Physiologist,* 22, 16, 1979.
19. **Kisch, B.,** Forgotten leaders of medicine: Valentine, Gruby, Remak, Auerbach, *Trans. Am. Philos. Soc.,* 44, 139, 1954.
20. **His, W.,** Die Neuroblasten und deren Entstehung in embryonalen Marke, Abh. Sachs Ges. akad., *Wiss. Math. Phys. Klasse,* 15, 311, 1889.
21. **Forel, A.,** Einige Hirnanatomische Betrachtungen und Ergebnisse, *Arch. Psychiatr. Nervenkr.,* 18, 162, 1887.
22. **Ramon y Cajal, S.,** *Histologie du système nerveux de l'homme et des vertébrés (1909-1911),* Vols. 1 and 2, (translated by Azoulay, L.), republished by Consejo Superior de Investigaciones Cientificas, Instituto Ramon y Cajal, Madrid, 1954.
23. **Ramon y Cajal, S.,** *Neuron Theory or Reticular Theory?,* (translated by Purkiss, M. U. and Fox, C. A.,) Consejo Superior de Investigaciones Cientificas, Instituto Ramon y Cajal, Madrid, 1954.
24. **Waldeyer, W.,** Ueber Einige Neuere Forschungen im Gebiete der Anatomie des Central Nervensystems, *Dtsch. Med. Wochenschr.,* 17, 1213, 1891.
25. **Harrison, R. G.,** The outgrowth of the nerve fiber as a mode of protoplasmic movement, *J. Exp. Zool.,* 9, 787, 1910.
26. **Ochs, S.,** A brief history of nerve repair and regeneration, in *Nerve Repair and Regeneration: Its Clinical and Experimental Basis,* Jewett, D. L. and McCarroll, H. R., Eds., C. V. Mosby, St. Louis, 1980, 1.
27. **Waller, A. V.,** A new method for the study of the nervous system (English translation), *London J. Med.,* 43, 609, 1852.
28. **Waller, A. V.,** Huitieme Memoire sur le systeme nerveux, *C. R. Acad. Sci.,* 35, 561, 1852.
29. **Waller, A.,** *The Nutrition and Reparation of Nerves,* paper delivered at the Royal Institution of Great Britain, Read and Co., London, 1861.
30. **Remak, R.,** Weitere mikroskopische Beobachtungen uber die Primitivfasern des Nervensystem der Wirbelthiere, Neue Notiz, *Beg. Natur-u. Heilk.,* 3, 35, 1837.
31. **Purkinje, J.,** Beobachtungen Betreffend der innerste Struktur der Nerven *(Ber. Versamm. deutsch. Natur. u. Arzte. Prague,* 15, 177, 1837), in *Opera Omnia,* Vol. 1, Academia Nakladalelstvoic Ceskoslovenske Akademie, Prague, 1918, republished, V. Kruta, Ed., 1968, 88.
32. **Schultze, M.,** The general characters of the structures composing the nervous system, in *Human Comparative Histology,* Vol. 1, Stricker, S., Ed., New Sydenham Society, London, 1871, chap. 3.
33. **Bernstein, J.,** *Elektrobiologie,* Veiweg, Braunschweig, 1912.
34. **Parker, G. H.,** The neurofibril hypothesis, *Q. Rev. Biol.,* 4, 155, 1929.
35. **Gerard, R. W.,** Nerve metabolism, *Physiol. Rev.,* 12, 469, 1932.
36. **Barker, L. F.,** *The Nervous System and its Constituent Neurons,* Appleton and Co., New York, 1899.
37. **Scott, F. H.,** On the metabolism and action of nerve cells, *Brain,* 28, 506, 1905.
38. **Scott, F. H.,** On the relation of nerve cells to fatigue of their nerve fibres, *J. Physiol. (London),* 34, 145, 1906.
39. **Torrey, T. W.,** The relation of taste buds to their nerve fibers, *J. Comp. Neurol.,* 59, 203, 1934.
40. **Ochs, S.,** The early history of nerve regeneration beginning with Cruikshank's observations in 1776, *Med. Hist.,* 21, 261, 1977.
41. **Ramon y Cajal, S.,** *Studies on Degeneration and Regeneration of the Nervous System,* Vol. 1, translated by R. M. May, 1928, Oxford University Press, Oxford, reprinted, Hafner, New York, 1968, 290.
42. **Weiss, P. and Hiscoe, H. B.,** Experiments on the mechanism of nerve growth, *J. Exp. Zool.,* 107, 315, 1948.
43. **Young, J. Z.,** Contraction, turgor and the cytoskeleton of nerve fibers, *Nature (London),* 153, 333, 1944.
44. **Young, J. Z.,** Structure of nerve fibers in cephalopods and crustacea, *Proc. R. Soc. London, Ser. B,* 121, 319, 1936.

45. **Young, J. Z.,** Factors influencing the regeneration of nerves, *Adv. Surg.*, 1, 165, 1949.
46. **Plateau, J.,** *Statique experimentale et theoretique des liquides sounmis aux seules forces moleculaires*, Paris, 1873.
47. **Lubińska, L.,** Axoplasmic streaming in regenerating and in normal nerve fibres, *Prog. Brain Res.*, 13, 1, 1964.
48. **Geren, B. B.,** The formation from the Schwann cell surface of myelin in the peripheral nerves of chick embryos, *Exp. Cell Res.*, 7, 558, 1954.
49. **Samuels, A. J., Boyarsky, L. L., Gerard, R. W., Libet, B., and Brust, M.,** Distribution, exchange and migration of phosphate compounds in the nervous system, *Am. J. Physiol.*, 164, 1, 1951.
50. **Dahlström, A.,** Synthesis, transport, and life-span of Amine storage granules in sympathetic adrenergic neurons, in *Cellular Dynamics of the Neuron*, Barondes, S., Ed., Academic Press, New York, 1969, 153.
51. **Ochs, S.,** *Axoplasmic Transport and Its Relation to Other Nerve Functions*, John Wiley & Sons, New York, 1982.

Chapter 2

ANALYTICAL TECHNIQUES FOR THE STUDY OF AXOPLASMIC TRANSPORT

S. Y. Chan and Z. Iqbal

TABLE OF CONTENTS

I. INTRODUCTION

Since the discovery of the movement of intracellular materials in neuronal processes, as described in the previous chapter, a large number of experimental procedures has been employed in the study of axoplasmic transport. In this chapter, we will describe, in brief, some of the most commonly used methods. Details of some of these procedures are described in various chapters of this volume and recent reviews.[1-6]

The existing approaches for the study of the transport can arbitrarily be divided into physiological/biochemical and physical methods. There is a great deal of overlapping in these categories and very often, an investigator has to employ a combination of methodology to achieve the desired goals.

II. PHYSIOLOGICAL/BIOCHEMICAL METHODS

The majority of axoplasmic studies have been performed by utilizing a class of materials in a particular nervous system, either endogenous, such as enzymes[7] and neurotransmitters[8] or exogenous, such as horseradish peroxidase[9,10] and radiolabeled precursors for proteins,[11,12] lipids,[13] and nucleic acids.[14] The two general methods to estimate the velocity of anterograde as well as retrograde transport are the so-called segmental analysis and the accumulation or ligation technique, both of which originated from physiological procedures. However, the identification of transported materials often requires biochemical techniques such as enzyme assays, SDS gel electrophoresis, subcellular fractionation, chromatography etc. In recent years, more and more biochemical techniques have been incorporated into the traditional methods and have become a powerful tool to study material transport in nerve processes.

A. Segmental Analysis

This procedure[11,15] entails the injection of a small amount of radiolabeled precursors into the proximity of well-defined neuron cell bodies. After the uptake and incorporation of these precursors into the cellular machinery, the outflow pattern of labeled activity moving down the nerve fibers can be analyzed by cutting the nerve into small segments and counting the radioactivity in each segment. The velocity of the transported materials is determined by the downflow of radioactivity at various time intervals after incorporation. Figure 1 depicts the experimental approach used for this method. Typically, 5 $\mu\ell$ of [^{3}H]-amino acid having high specific activity is injected into the L7 dorsal root ganglion of anaesthetized cat over a period of several minutes using a micropipet which is left in place for a short time after injection to prevent leakage of the tracer through the injection site. After a period of [^{3}H]-precursor incorporation into proteins and downflow, the sciatic nerves are taken out, sliced into 2 to 5 mm segments, and the radioactivity determined. The rate of advancement of the crest of radioactivity gives the velocity of the most rapidly transported material. A number of amino acids such as [^{3}H]-leucine,[11] [^{3}H]-proline[12] and [^{35}S]-methionine[16] have been used as precursors for the study of transport of different proteins.

Besides amino acids, other precursors can be used for the analysis of specific components. As described in Chapter 9, radiolabeled fucose and glucosamine are used for the analysis of glycoproteins and glycolipids. Likewise, glycerol and choline are used to study the transport of phospholipids.[17] For the investigation of nucleic acid transport, labeled nucleosides are employed, as detailed in Chapters 8 and 17. The study of neurotransmitter transport is carried out either by injecting radiolabeled transmitter such as [^{14}C]-norepinephrine[18] or radiolabeled transmitter precursors such as [^{3}H]-5-hydroxytrytophan[19] and [^{14}C]-DOPA[20] for the study of serotonin and dopamine, respectively.

Incorporation of labeled amino acids into protein is usually rapid, generally 5 to 20 min.[21,22] However, incorporation of other precursors such as [^{32}P]-phosphate,[23,24] may require longer periods, up to 3 to 4 hr.

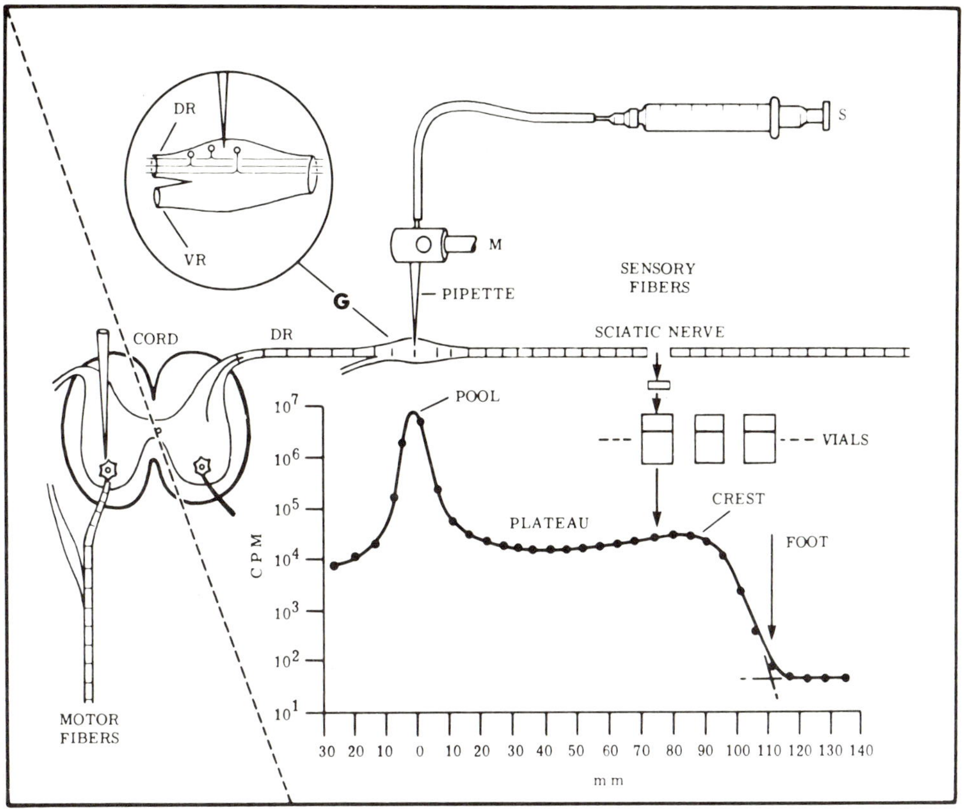

FIGURE 1. Injection and sampling technique showing transport. The L7 ganglion shown in the inset contains T-shaped neurons with one branch ascending in the dorsal root, the other descending in the sciatic nerve. A pipette containing ^{3}H-leucine is passed into the ganglion; and after its injection and the incorporation of precursor, the downflow of labeled components in the fibers is sampled at various times by sacrificing the animal and sectioning the nerve. Each segment is placed in a vial, solubilized, scintillation fluid added, and the activity counted. The outflow pattern is displayed on the ordinate log scale; the abscissa is in millimeters, taking the distance from the center of the ganglion as zero. The fastest moving activity is a crest falling abruptly to baseline levels. Behind it is a plateau with a high level of activity remaining in the ganglion. The left side of the cord shows motoneurons injected in the region of the L7 cell body. Removal and sampling of the ventral root and sciatic nerve at a later time gives a similar outflow pattern. (From Ochs, S., in *Hypothalamic Peptide Hormones and Pituitary Regulation*, Porter, J. C., Ed., Plenum Press, New York, 1977, chap. 2. With permission.)

In addition to the injection of precursors into cell body area and the determination of the transported material *in situ,* the outlined procedure can also be used to study the uptake and transport of material in vitro. Peripheral nerves together with its attached cell bodies (located in ganglion) are isolated and placed into a multicompartmental tray containing appropriate incubation media in each compartment.[25,26] Labeled precursors are then applied to cell bodies (sealed off from axons with grease) and after a predetermined time period, nerves are segmented and analyzed for the transported materials. For the studies of the effect of pharmacological agents, the in vitro chamber system is very useful. The drugs can be added into the area containing cell bodies to see their effect on the uptake of a precursor and/or the synthesis of components which are destined for the transport in the fibers[26-28] while the exposure of the nerve fibers to the compound of interest can be useful in determining its effect on the movement of materials within fibers.[29] Although the in vitro system has proved to be valuable in dissecting the overall transport process into uptake, synthesis and movement along nerve fibers, it can be applied only for the studies aimed at the components which are transported at a fast rate, as isolated nerves cannot be kept functional for a long period of time in vitro.

We have studied the effect of ions and various pharmacological agents on transport using segmental procedure by a combination of both *in situ* and in vitro systems. In performing such studies, attention should be given to the perineurial sheath of peripheral nerves since it is a permeability barrier to certain drugs and ions.[30,31] As in our studies, the cat sciatic nerves had to be desheathed before exposing to suitable incubation media for the desired effect.[31,32] The procedure for such a preparation is depicted in Figure 2. The precursor, in this case [^{3}H]-leucine, is injected into the L7 dorsal root ganglion as described and about 2 hr of *in situ* incorporation and transport is allowed. The sciatic nerve, together with the ganglion is then removed from the animal and placed into a special chamber filled with buffer. The peroneal branch of the sciatic nerve is desheathed since it has less branching, whereas the tibial branch is left intact as control since it is smaller, more branched, and harder to desheath. After desheathing, the whole nerve is then placed in temperature controlled (37°C) and oxygenated test medium for a predetermined period before segmental analysis. It should be noted that the two branches should be adjusted to the same length before desheathing since the desheathed nerve is often longer than the intact nerve, which will complicate the analysis of transport. A method of adjusting the nerve length, taking advantage of the band of Fontana,[33] is described in Chan et al.[32] By utilizing this procedure, we have shown that the transport of labeled proteins was dependent on an appropriate concentration of Ca^{2+}.[32] The fact that transport requires microtubular integrity was also supported by experiments involving the exposure of desheathed nerve fibers to a class of antimicrotubular agents, vinca alkaloids.[34] Recently, we have also shown that anticalmodulin drug trifluoperazine (TFP) blocks the transport, thus suggesting an important role for calmodulin in the transport process (see Chapter 5 for details).

The desheathing procedure may be physically damaging to axons in some cases and is not applicable to small and greatly branched nerves. An alternate procedure for permeabilization of nerve has been recently devised. The procedure requires the incubation of nerve with a mild concentration of detergents such as soponin.[35] Such permeabilized nerves which are viable physiologically have been used to study the effect of various drugs on transport.

The studies mentioned above utilizing the tracer technique, *in situ,* in chamber or a combination of both have mainly been conducted in the sciatic nerves of cat,[11] rat,[26] frog,[27,28] or chicken[36] for the ease of access and the fact that sciatic nerve offers a longer length for transport studies. The system also offers the possibility to study both the sensory axons[11] (by injection of tracer into the ganglion) and motor axons[37] (by injection into the ventral horn).

The study of axoplasmic transport by segmental analysis has also utilized a number of other nervous systems such as optic, olfactory, or vagus. The optic system entails the injection of radioactive tracer into the eye and the labeled components are subsequently analyzed by autoradiography in the optic axons.[38] The procedure, first employed in mice by Taylor and Weiss,[39] has been extended to goldfish,[38] rabbit,[14,40,41] and other optic systems.[42] The study of slower transported components has been performed mainly through the use of segmental analysis procedure in optic systems.[41,42] The vagus nerve has been popular because of long unbranched nerve and both the motoneurons and sensory fibers can be studied in this system.[43,44,45] The olfactory nerve system is suitable for the study of transport rate due to its long length, especially in garfish[15,46] and pike.[47]

Although the segmental analysis is probably the most commonly used method in the transport studies, it is not free of limitations. These include the loss of precursor from the injection site,[48] possibility of precursor diffusion down the axons,[40] and blood-borne label reaching the nerve and getting incorporated locally.[48] The system also poses a problem when the investigator wishes to identify a specific component being transported in the nerve fiber.

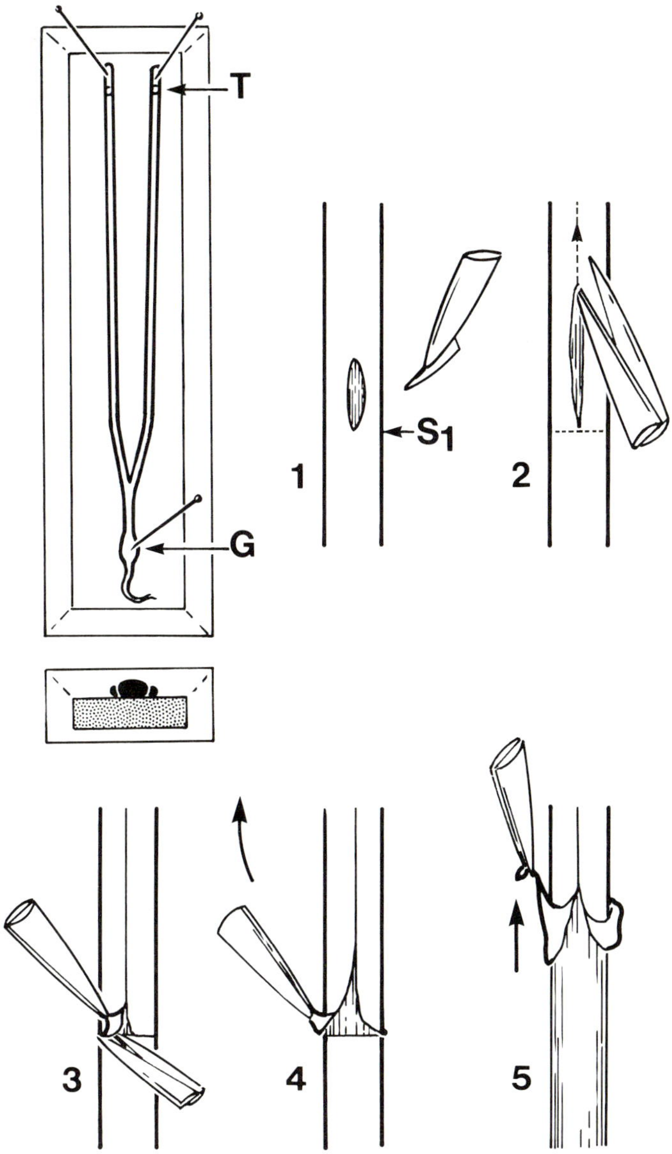

FIGURE 2. Desheathing procedure in cat sciatic nerves. The nerve was pinned down on the chamber on the ganglion (G) and the ends. A tie (T) was placed on the ends of each branch to serve as a marker after they were stretched. Step 1: a slit was made on the peroneal branch approximately 35 mm distal to the ganglion by means of a knife made from a sliver of razor blade. Step 2: the blade of a pair of iris scissors was used to slit the perineural sheath distally. Step 3: the sheath was then cut circumferentially at the proximal end of the slit. Steps 4 and 5: the perineurium was peeled back down the nerve to the point where branching occurs, usually 135 to 150 mm distal to the ganglion. (From Chan, S., et al., *J. Physiol.*, 301, 477, 1980. With permission.)

Some of the above mentioned shortcomings can be avoided by the use of single neurons. Microquantities of tracers can be injected directly into the cell body of the giant cerebral neurons of Aplysia[19,49] and the rate of transport of labeled components observed. In this system, the radioactivity remains confined to the axon of the neuron along the entire length of the nerve.

Proteins can also be labeled, in vivo, covalently by *N*-succinimidyl [2,3-^{3}H]-propionate (NSP) which can penetrate cell membranes and bind to intraneuronal proteins.[50,51] No significant change has been found for the electrophoretic mobilities in NSP labeled proteins either in SDS-PAGE or isoelectro-focusing gels. The NSP labeling was originally thought to be nonselective and could be applied to any part of the neuron as no protein synthesis was required. However, a recent report[52] indicates that the [^{3}H]-NSP has a tendency to prominantly bind to a small number of proteins, which may limit its usage as a probe to study axoplasmic transport.

B. Ligation and Accumulation Technique

The transport of endogenous components such as structural proteins, enzymes, neurotransmitters etc. have been successfully studied by the accumulation technique with which the transported components are disrupted either by ligating the nerve fibers with a fine thread[53] or by exposing a region of the nerve to cold block.[54] The accumulation or depletion of those components are determined by sectioning out a small segment at the disruption point and analyzing the accumulated material(s) with a variety of biochemical, histochemical, or electron microscopic methods.

The bidirectional movement of the materials in nerve fibers can be determined by double-ligation technique. In this procedure, initially used for the study of acetycholinesterase (AChE)[55,56] transport (also see review[57]), the materials accumulating on both the proximal and distal sides of a nerve ligation are detected. Usually, the accumulation is confined to a narrow region extending a few millimeters from the point of ligation on either side and the amount of enzyme accumulated is due to the redistribution, caused by the intra-axonal transport of the enzyme being analyzed. Such conclusion is derived from the observations that the amount of enzyme damming at both ends is balanced by a depletion of the enzyme in the middle segment of the nerve, thus the total amount of enzyme remains unchanged[57] and the inhibition of protein synthesis has no significant effect on the amount of enzyme accumulated.[58]

The cold block technique,[54,59-61] a modification of the ligation procedure requires the exposure of a region of nerve to cold temperature (4°C) at which the transport is blocked locally, causing the damming of transported material at the cooling region. Rewarming to body temperature allows the accumulated material to move along the nerve beyond the block. This technique has an advantage over the conventional ligation methods as it allows the resumption of original conditions by rewarming and thus avoids damage to the nerve fibers by ligation or crushing. Additionally, injury to nerves caused by the ligation could induce a response in local synthesis[62,63] or degradation of certain materials under investigation.

The above mentioned procedures have been used extensively in the study of transport of enzymes[53,54,56] and neurotransmitters,[63,64] as well as the transport of mitochondria.[65] Accumulation techniques offer certain advantages over the segmental analysis, as they allow the determination of bidirectional flow of endogenous materials. Further, the procedure can be used for measuring transport in short axons where profiles of activity along the nerves are difficult to obtain.

The investigators using the accumulation technique should also be aware of the few pitfalls in the method. Besides the artifacts encountered due to injury in the ligation procedure, the rate of transport obtained from the accumulated samples is a weighed average since the materials accumulated at the disruption point are also subjected to turnaround or retrograde transport.[63] For those components which are transported in small amounts, the available analysis techniques may not be sensitive enough for detection.

C. Combination of Tracer and Accumulation Technique

There are a number of intraneuronal components which are transported in small amounts, thus making their detection difficult by either the tracer or ligation method. By combining both techniques, we have been able to detect the fast transport of calmodulin and microtubular proteins.[66] The radiolabeled precursor is injected into the ganglion of cat sciatic nerve, as described in Section II.A. After 1 to 2 hr of uptake and transport of labeled components from the cell body, a ligation is placed distal to the ganglion to stop further transport of labeled components. A second ligation is placed a few centimeter distal to the first ligation down the axon. After a predetermined time interval, the labeled material associated with the fast moving crest is analyzed for calmodulin activity. This procedure is potentially valuable for the analysis of fast-transported components which are transported in small amount.

D. Biochemical Analysis of Transported Materials

The number of components being transported in either directions at any given time is large and the composition complex. The segmental analysis, in general, demonstrates that a class of labeled components is moving at a given rate. However, the identification of the various materials being transported in nerve depends heavily on biochemical techniques designed for specific purposes. Using a subcellular fractionation procedure, it has been shown that most of the particulate proteins including some transmitter synthesizing enzymes and glycoproteins are transported at a fast rate.[67,68] On the other hand, most of the soluble proteins including cytoskeletal elements are found to be moving in nerve fibers at a slow rate.[69]

The soluble components can be further separated and characterized by gel filtration or by ion-exchange chromatography. For example, after injection of ^{45}Ca into cat dorsal root ganglion, about 25% of ^{45}Ca was found present in the soluble fraction. Using gel filtration columns containing Sephadex G-100 and Bio-Gel A-5m, the radioactivity was found to be associated with a calcium-binding protein.[70] The particulate components, on the other hand, need to be solubilized before they can be separated on the chromatographic columns. With the advent of sodium dodecyl sulfate-polyacrylamide gel electrophoresis[71] (SDS-PAGE), these insoluble particulate components could be resolved into a number of different bands, separating on the gel according to their molecular weight. A single dimensional gel separation technique could identify about a dozen of fast transported components.

With the introduction of two-dimensional gels (2DG) which separates proteins according to two independent physico-chemical criteria, namely the isoelectric point and molecular weight; it has now been possible to identify over 100 individual fast transported proteins.[72] The technique, first applied by O'Farrell[73] in 1975 in the separation of *E. coli* proteins has been modified for use with neuronal tissue.[72,74-76] Essentially, the samples are mixed with urea, SDS, nonionic detergent such as NP-40 and then separated on thin gels of acrylamide or agarose containing ampholytes of different isoelectric points, by electrophoresis. The proteins migrate to their corresponding isoelectric points and thus get separated from one another. In the second dimension, the proteins are separated according to their molecular weights using conventional slab gel SDS-PAGE technique. After second dimensional electrophoresis, the gel is dried and subjected to fluorography. An alternative method is to remove the spots from the original gel using the fluorographic pattern as a guide and process for radioactivity determination by scintillation counting.[74] With this punch and count procedure, the sensitivity of detecting the protein components is greatly enhanced. In our laboratory, we have been able to demonstrate fast transport of tubulin in rat sciatic nerve.[75]

The 2DG procedure has been very helpful in the identification and determination of proteins and glycoproteins[76,77] which can be transported anterogradely or retrogradely and at fast or slow rate. The precursors employed are either [^{3}H]-labeled or [^{35}S]-labeled. By the use of double-labeling with [^{35}S]-methionine as protein precursor, along with [^{3}H]-labeled fucose or galactose for glyco-moiety precursor, identification of transported glycoproteins has been made possible with great success (see Chapter 9 for details).

III. PHYSICAL METHODS

Physical methods as a means to study axoplasmic transport are relatively new and still under development. Besides complimenting the physiological/biochemical techniques, physical methodology opens up new frontiers in obtaining informations that are off-limit to the other approaches. If current technical problems can be resolved, this method of analysis will probably contribute much informations regarding the underlying mechanism(s) governing axoplasmic transport.

In general, physical methods of studying axoplasmic transport can be divided into two broad categories: (1) the direct-monitoring techniques[3,78] and (2) the optical or microscopical methods.[3,79,80]

A. Direct-Monitoring Techniques

By injection of a radiolabeled precursor into cell bodies and allowing downflow, labeled materials can be monitored by external radiation detectors, also called position-sensitive detectors. These devices, capable of detecting ionizing radiations such as gamma or strong beta, can be placed at desired positions along the axon in order to evaluate the labeled material(s) being transported. There are three types of external radiation detectors reported so far, namely, the gamma or scintillation camera, the silicon p-n junction diode, and the multiwire proportional chamber (MWPC) or multiple proportional counter (MPC).

The gamma camera was first used to externally monitor axoplasmic transport in cat nerve using [^{75}Se]-methionine as the labeled precursor.[81] This is the first instance reported to use an external detector successfully for such purpose. However, it was later found that the detector had rather poor spatial resolution as well as few suitable radiolabeled precursors. The use of silicon p-n junction diode was pioneered by Takenaka et al. in 1978.[82] The semiconductor radiation detector is windowless and has a diameter of just 2 mm and thickness of 1 mm. [^{14}C]- and [^{32}P]-labeled[82,83] precursors can be used as radiation source and both in vivo[83] and in vitro[82] detections have been accomplished. The semiconductor detector needs to be repositioned during the course of the experiment due to the small size of the detector, which results in variations of detection efficiency. The low level of activity detected due to self-absorption by the nerve fibers also limits its use on relatively small nerves such as frog and rat sciatic nerves. The multiple proportional counter, introduced by Snyder et al.[84,85] seems to avoid some of the above-mentioned problems. Simplistically, MPC consists of a plane of parallel anode wires with an electrically common ground. β-Particles emitted from the labeled material(s) inside the axon are detected by any of the anode wires that are above the region of the radiation source and thus, the progress of the transported material along the axon continuously monitored. It is a relatively complicated device and the mechanical and technical workings are detailed by Synder.[86]

On the whole, direct-monitoring devices have certain advantages over conventional techniques. Their ability to follow temporal changes in axoplasmic transport without injury to the axon provides a more reliable procedure to analyze transported components by biochemical techniques. The devices (aside from their limitations), once setup, are simple to use in various nervous systems. However, direct-monitoring is mostly restricted to in vitro studies so far although in vivo study has been reported.[83] If in vivo studies can be done routinely, both fast- and slow-transport can potentially be monitored in one preparation and tissue assayed for different transport components. The existing devices have very limited range of detection and have not been improved to the extent that thicker nerves such as cat sciatic nerves can be included. Thicker nerves have the advantage that more tissue is available for biochemical analyses.

B. Optical Methods

Saltatory organelle movements have been observed to travel at fast rates in living axons of a variety of animals by light microscopy.[87] This new method for studying axoplasmic transport, though still under development, has had major advances recently due to the introduction of advanced optics and new microscopic techniques to the field.[79]

The most common optical method for structural visualization inside living cells is the phase contrast microscopy. It has been used to observe organelle movements in nonmyelinated crab axons[88] and nonneuronal cells.[89] However, phase optics are only useful for very thin tissues and give poor resolution of myelinated axons due to interference of out-of-focus haloes. Darkfield microscopy has been shown to be useful in the visualization of organelle movement, especially in myelinated axons,[90] however, problems related to background light scattering hinder the detection of small particles in those axons. Differential interference-contrast (DIC) or Nomarski microscopy, coupled with video-recording, can provide high-resolution optical image of relatively thicker specimens.[79] However, DIC images are difficult to record since the highest resolution can be obtained only when the light intensity is lowest. However, a new method, named Video-Enhanced Contrast, Differential Interference Contrast (AVEC-DIC) microscopy[91] has overcome this limitation and provides very high resolution of imaging. The AVEC-DIC may be capable of visualizing single microtubules[91] and it represents an exciting new way of investigating intracellular movement. Recently, Hodge and Adelman[92] reported the technique of pseudostereoscopic viewing of image pairs as an efficient means to evaluate particle movements inside living axons.

In general, light microscopic methods offer the advantage of greater spatial and temporal resolution than any other available methods and very often, reveal details that are not accessible to others. It also allows the continuous monitoring of a particular organelle movement, instead of the average velocity estimated by other methods. However, there are still many difficulties since it requires rather restricted conditions for the system to work. Besides the requirement of high resolution and magnification, the axon preparation under light microscopic investigation need to be thin so that light can penetrate.[93] In vivo study is often impossible due to light penetration and scattering problems. Rather intense illumination is also required in Darkfield and Nomarski optics and the axons under investigation may be damaged by the heat emitted. In addition, organelles have to be large and refractile enough in order to be seen in the light microscope. These problems limit a great deal the usefulness of the light microscopy techniques in the evaluation of transport. However, as these limitations are gradually overcome through innovations and advanced technology, optical techniques will emerge as an important tool in the study of axoplasmic transport.

ACKNOWLEDGMENTS

The authors thank Dr. Daniel P. Stromska for helpful suggestions in the preparation of this manuscript. Thanks are also due to Mrs. Florence S. Sales for secretarial assistance.

REFERENCES

1. **Grafstein, B. and Forman, D. S.,** Intracellular transport in neurons, *Physiol. Rev.*, 60, 1167, 1980.
2. **Wilson, D. L. and Stone, G. C.,** Axoplasmic transport of proteins, *Annu. Rev. Biophys. Bioeng.*, 8, 27, 1979.
3. **Snyder, R. E. and Smith, R. S.,** Physical methods for the study of the dynamics of axonal transport, *CRC Crit. Rev. Bioeng.*, 10(2), 89, 1984.
4. **Elam, J. S. and Cancalan, P., Eds.,** Axonal transport in neuronal growth and regeneration, in *Advances in Neurochemistry*, Vol. 6, Plenum Press, New York, 1984.
5. **Ochs, S.,** *Axoplasmic Transport and its Relation to other Nerve Functions*, John Wiley & Sons, New York, 1982.

6. **Lasek, R. J. and Hoffman, P. N.,** The neuronal cytoskeleton, axonal transport and axonal growth, in *Cell Motility*, Vol. 3, Goldman, R., Pollard, T. and Rosenbaum, J., Eds., Cold Spring Harbor, Cold Spring Harbor, N.Y., 1976, 1021.
7. **Niemierko, S. and Lubinska, L.,** Two fractions of axonal acetylcholinesterase exhibiting different behaviour in severed nerves, *J. Neurochem.*, 14, 761, 1967.
8. **Maclean, D. B. and Lewis, S. F.,** De novo synthesis and axoplasmic transport of [^{35}S] methionine-substance P in explants of nodose ganglion/vagus nerve, *Brain Res.*, 310(2), 325, 1984.
9. **LaVail, J. H. and LaVail, M. M.,** Retrograde axonal transport in the central nervous system, *Science*, 176, 1416, 1972.
10. **Leong, S. K. and Wong, W. C.,** Retrograde transport of horseradish peroxidase by regenerating preganglionic axons in the cat's cervical sympathetic trunk, *Neurosci. Lett.*, 33, 235, 1982.
11. **Ochs, S. and Ranish, N.,** Characteristics of the fast transport system in mammalian nerve fibers, *J Neurobiol.*, 1, 247, 1969.
12. **Elam, J. S. and Agranoff, B. W.,** Rapid transport of protein in the optic system of the goldfish, *J. Neurochem.*, 18, 375, 1971.
13. **Miani, N.,** Evidence of a proximo-distal movement along the axon of phospholipid synthesized in the nerve cell body, *Nature (London)*, 193, 887, 1962.
14. **Autilio-Gambetti, L., Gambetti, P., and Shafer, B.,** RNA and axonal flow. Biochemical and autoradiographic study in the rabbit optic system, *Brain Res.*, 53, 387, 1973.
15. **Gross, G. W. and Beidler, L. M.,** Fast axonal transport in the C-fibers of the garfish olfactory nerve, *J. Neurobiol.*, 4(5), 413, 1973.
16. **McLean, W. G., McKay, A. L., and Sjostrand, J.,** Electrophoretic analysis of axonally transported proteins in rabbit vagus nerve, *J. Neurobiol.*, 14(3), 227, 1983.
17. **Miani, N.,** Proximo-distal movement of phospholipid in the axoplasm of the intact and regenerating neurons, in *Progress in Brain Research: Mechanisms of Neural Regeneration*, Vol. 13, Singer, M. and Schade, J. P., Eds., Elsevier, Amsterdam, 1964, 115.
18. **Livett, B. G., Geffen, L. B., and Austin, L.,** Axoplasmic transport of ^{14}C-noradrenaline and protein in splenic nerves, *Nature (London)*, 217, 278, 1968.
19. **Goldman, J. E. and Schwartz, J. H.,** Cellular specificity of serotonin storage and axonal transport in identified neurones of Aplysia Californica, *J. Physiol. (London)*, 242, 61, 1974.
20. **Fibiger, H. C., McGeer, E. G., and Atmadja, S.,** Axoplasmic transport of dopamine in Nigro-striatal neurons, *J. Neurochem.*, 21, 373, 1973.
21. **Ochs, S., Sabri, M. I., and Ranish, N.,** Somal site of synthesis of fast transported materials in mammalian nerve fibers, *J. Neurobiol.*, 1, 329, 1970.
22. **Wettstein, F. O., Noll, H., and Penman, S.,** Effect of cycloheximide on ribosomal aggregates engaged in protein syntheses *in vitro, Biochem. Biophys. Acta*, 87, 525, 1964.
23. **Nichols, T. R., Smith, R. S., and Snyder, R. E.,** The action of puromycin and cycloheximide on the initiation of rapid axonal transport in amphibian dorsal root neurons, *J. Physiol. (London)*, 332, 441, 1982.
24. **Snyder, R. E., Nichols, T. R., and Smith, R. S.,** Rapid orthograde transport of [^{32}P]-labelled material in amphibrian sensory axons: a multiwire proportional chamber study, *Can. J. Physiol. Pharmacol.*, 58, 513, 1980.
25. **Edstrom, A. and Mattsson, H.,** Fast axonal transport *in vitro* in the sciatic system of the frog, *J. Neurochem.*, 19, 205, 1972.
26. **Paulson, J. C. and McClure, W. O.,** Microtubules and axoplasmic transport. Inhibition of transport by podophyllotoxin: an interaction with microtubule protein, *J. Cell Biol.*, 67, 461, 1975.
27. **Dravid, A. R. and Hammerschlag, R.,** Axoplasmic transport of proteins *in vitro* in primary afferent neurons of frog spinal cord: effect of Ca^{2+}-free incubation conditions, *J. Neurochem.*, 24, 711, 1975.
28. **Lavoie, P. -A.,** Importance of monovalent ions for the fast axonal transport of proteins, *Can. J. Physiol. Pharmacol.*, 59(1), 31, 1981.
29. **Ekstrom, P., Kanje, M., and Edstrom, A.,** Effects of phenothiazines and dibenzazepines on axonal transport and microtubule assembly *in vitro, Acta Physiol. Scand.*, 116, 121, 1982.
30. **Feng, T. P. and Gerard, R. W.,** Mechanisms of nerve asphyxiation: with a note on the nerve sheath as a diffusion barrier, *Proc. Soc. Exp. Biol. Med.*, 27, 1073, 1930.
31. **Ochs, S., Worth, R. M., and Chan, S. Y.,** Calcium requirement for axoplasmic transport in mammalian nerve, *Nature (London)*, 270, 748, 1977.
32. **Chan, S. Y., Ochs, S., and Worth, R. M.,** The requirement for calcium ions and the effect of other ions on axoplasmic transport in mammalian nerve, *J. Physiol. (London)*, 301, 477, 1980.
33. **Clarke, E. and Bearn, J. B.,** The spinal nerve bands of Fontana, *Brain*, 95, 1, 1972.
34. **Chan, S. Y., Worth, R., and Ochs, S.,** Block of axoplasmic transport *in vitro*, by Vinca Alkaloids, *J. Neurobiol.*, 11(3), 251, 1980.
35. **Forman, D. S., Brown, K. J., and Livengood, D. R.,** Fast axonal transport in permeabilized lobster giant axons is inhibited by Vanadate, *J. Neurosci.*, 3(6), 1279, 1983.

36. **Austin, L., Bray, J. J., and Young, R. J.,** Transport of proteins and ribonucleic acid along nerve axons, *J. Neurochem.*, 13, 1267, 1966.
37. **Ochs, S.,** Effect of maturation and aging on the rate of fast axoplasmic transport in mammlian nerve, *Prog. Brain Res.*, 40, 349, 1973.
38. **Murray, M. and Grafstein, B.,** Changes in the morphology and amino acid incorporation of regenerating goldfish optic neurons, *Exp. Neurol.*, 23, 544, 1969.
39. **Taylor, A. C. and Weiss, P.,** Demonstration of axonal flow by the movement of tritium-labeled protein in mature optic nerve fibers, *Proc. Natl. Acad. Sci. U.S.A.*, 54, 1521, 1965.
40. **Haley, J. E., Wisniewski, H. M., and Ledeen, R. W.,** Extra-axonal diffusion in the rabbit optic system: a caution in axonal transport studies, *Brain Res.*, 179, 69, 1979.
41. **Chihara, E.,** Axoplasmic and nonaxoplasmic transport along the optic pathway of albino rabbits: a theoretical pattern of distribution, *Invest. Ophthalmol. Visual Sci.*, 18, 339, 1979.
42. **Grafstein, B.,** The eyes have it: axonal transport and regeneration in the optic nerve, in *The Nervous System. The Basic Neurosciences,* Vol. 1, Tower, D. B., Ed., Raven, New York, 1975, 147.
43. **Miani, N.,** Analysis of the somato-axonal movement of phospholipids in the vagus and hypoglossal nerves, *J. Neurochem.*, 10, 859, 1963.
44. **Donoso, J. A., Illanes, J. -P., and Samson, F.,** Dimethylsulfoxide action on fast axoplasmic transport and ultrastructure of vagal axons, *Brain Res.*, 120, 287, 1977.
45. **McLean, W. G., Frizell, M., and Sjostrand, J.,** Axonal transport of labelled proteins in sensory fibers of rabbit vagus nerve *in vitro, J. Neurochem.*, 25, 695, 1975.
46. **Gross, G. W.,** The effect of temperature on the rapid axoplasmic transport in C-fibers, *Brain Res.*, 56, 359, 1973.
47. **Gross, G. W. and Kreutzberg, G. W.,** Rapid axoplasmic transport in the olfactory nerve of the pike. I. Basic transport parameters for proteins and amino acids, *Brain Res.*, 139, 65, 1978.
48. **McEwen, B. S. and Grafstein, B.,** Fast and slow components in axonal transport of protein, *J. Cell Biol.*, 38, 494, 1968.
49. **Aletta, J. M. and Goldberg, D. J.,** Rapid and precise down regulation of fast axonal transport of transmitter in an identified neuron, *Science*, 218, 913, 1982.
50. **Fink, D. J. and Gainer, H.,** The use of a labeled acylating probe for the study of fast axonal transport *in vivo, Brain Res.*, 177, 208, 1979.
51. **Fink, D. J. and Gainer, H.,** Axonal transport of proteins: a new view using *in vivo* covalent labeling, *J. Cell Biol.*, 85, 175, 1980.
52. **Logan, M. J., McLean, W. G., and Meiri, K. F.,** Limitation to the usefulness of N-succinimidyl propionate for the study of retrograde axonal transport, *Neurosci. Lett.*, 36, 203, 1983.
53. **Partlow, L. M., Ross, C. D., Motwani, R., and McDougal, D. B., Jr.,** Transport of axonal enzymes in surviving segments of frog sciatic nerve, *J. Gen. Physiol.*, 60, 388, 1972.
54. **Brimijoin, S.,** Stop-flow: a new technique for measuring axonal transport, and its application to the transport of dopamine-β-hydroxylase, *J. Neurobiol.*, 6, 379, 1975.
55. **Lubinska, L.,** Acetylcholinesterase in mammalian peripheral nerves and characteristics of its migration, *Acta Neuropathol. Suppl.*, 5, 136, 1971.
56. **Lubinska, L., Niemierko, S., Oderfeld, B., and Szwarc, L.,** Behavior of acetylcholinesterase in isolated nerve segments, *J. Neurochem.*, 11, 493, 1964.
57. **Lubinska, L.,** On axoplasmic flow, *Int. Rev. Neurobiol.*, 17, 241, 1975.
58. **Frizell, M., Hasselgren, P. O., and Sjostrand, J.,** Axoplasmic transport of acetylcholinesterase and choline acetyltransferase in the vagus and hypoglossal nerve of the rabbit, *Exp. Brain Res.*, 10, 526, 1970.
59. **Brimijoin, S. and Dyck, P. J.,** Axonal transport of dopamine-β-hydroxylase and acetylcholinesterase in human peripheral neuropathy, *Exp. Neurol.*, 66, 467, 1979.
60. **Litchy, W. J. and Brimijoin, S.,** Concentration dependence of rapid axonal transport: a study of the transport kinetics of [^{35}S] methionine-labeled protein in postganglionic sympathetic fibers of the bullfrog, *J. Neurosci.*, 3(10), 2075, 1983.
61. **Brimijoin, S. and Helland, L.,** Rapid retrograde transport of dopamine-β-hydroxylase as examined by the stop-flow technique, *Brain Res.*, 102, 217, 1976.
62. **Kapeller, K. and Mayor, D.,** An electron microscopic study of the early changes distal to a constriction in sympathetic nerves, *Proc. R. Soc. London Ser. B*, 172, 53, 1969.
63. **Wooten, G. F., Weise, V. K., and Kopin, I. J.,** Norepinephrine synthesis contributes to the accumulation of catecholamines after ligation of rat sciatic nerves, *Brain Res.*, 136, 174, 1977.
64. **Ben-Jonathan, N., Maxson, R. E., and Ochs, S.,** Fast axoplasmic transport of noradrenaline and dopamine in mammalian peripheral nerve, *J. Physiol. (London)*, 281, 315, 1978.
65. **Khan, M. A. and Ochs, S.,** Slow axoplasmic transport of mitochandria (MAO) and lactic dehydrogenase in mammalian nerve fibers, *Brain Res.*, 96, 267, 1975.
66. **Iqbal, Z.,** this volume, chap. 5.

67. **Brimijoin, S. and Wiermaa, M. J.,** Direct comparison of the rapid axonal transport of norepinephrine and dopamine-β-hydroxylase activity, *J. Neurobiol.*, 8, 239, 1977.
68. **Edstrom, A. and Mattsson, H.,** Electrophoretic characterization of leucine-, glucosamine- and fucose-labelled proteins rapidly transported in frog sciatic nerve, *J. Neurochem.*, 21, 1499, 1973.
69. **Grafstein, B., McEwen, B. S., and Shelanski, M. L.,** Axonal transport of neurotubule protein, *Nature (London)*, 227, 289, 1970.
70. **Iqbal, Z. and Ochs, S.,** Fast axoplasmic transport of a calcium-binding protein in mammalian nerve, *J. Neurochem.*, 31, 409, 1978.
71. **Ames, G. F. -L.,** Resolution of bacterial proteins by polyacrylamide gel electrophoresis on slabs, *J. Biol. Chem.*, 249, 634, 1974.
72. **Stone, G. C., Wilson, D. L., and Hall, M. E.,** Two-dimensional gel electrophoresis of proteins in rapid axoplasmic transport, *Brain Res.*, 144, 287, 1978.
73. **O'Farrell, P. H.,** High resolution two-dimensional electrophoresis of proteins, *J. Biol. Chem.*, 250, 4007, 1975.
74. **Wilson, D. L.,** Two dimensional polyacrylamide gel electrophoresis of proteins, in *Handbook of Neurochem.*, Vol. 2, Lajlha A., Ed., Plenum Press, New York, 133, 1982.
75. **Stromska, D., Iqbal, Z., and Ochs, S.,** Evidence for the fast transport of tubulin and actin in mammalian sciatic nerve, *Soc. Neurosci. Abstr.*, 9, 1191, 1983.
76. **Hammerschlag, R., Stone, G. C., Bolen, F. A., Lindsey, J. D., and Ellisman, M. H.,** Evidence that all newly synthesized proteins destined for fast axonal transport pass through the golgi apparatus, *J. Cell Biol.*, 93, 568, 1982.
77. **Brady, S. T. and Lasek, R. J.,** Nerve-specific enolase and creativer phosphokinase in axonal transport: soluble proteins and the axoplasmic matrix, *Cell*, 23, 515, 1981.
78. **Snyder, R. E. and Smith, R. S.,** Application of position-sensitive radioactivity detectors to the study of the axonal transport of β-emitting isotopes, in *Axoplasmic Transport,* Weiss, D. G., Ed, Springer-Verlag, Basel, 1982, 442.
79. **Forman, D. S.,** Microscopic methods for the observation of axonal transport in living axons, in *Axoplasmic Transport,* Weiss, D. G., Ed., Springer-Verlag, Basel, 1982, 424.
80. **Smith, R. S.,** Axonal transport of optically detectable organelles, in *Axoplasmic Transport,* Weiss, D. G., Ed., Springer-Verlag, Basel, 1982, 181.
81. **Widen, L., Greitz, T., Michelayannakis, J., Ruden, B. I., and Asard, P.,** External detection of axonal flow, *Electroenceph. Clin. Neurophysiol.*, 38, 214, 1975.
82. **Takenaka, T., Horier, H., and Sugita, T.,** New technique for measuring dynamic axonal transport and its application to temperature effects, *J. Neurobiol.*, 9, 317, 1978.
83. **Takenaka, T. and Ochs, S.,** External detection of axoplasmic transport using ATP-^{32}P as precursor, *J. Neurobiol.*, 11, 571, 1980.
84. **Snyder, R. E., Reynolds, R. A., Smith, R. S., and Kendal, W. S.,** Application of a multiwire proportional chamber to the detection of axoplasmic transport, *Can. J. Physiol. Pharmacol.*, 54, 238, 1976.
85. **Snyder, R. E., Reynolds, R. A., Smith, R. S., and Kendal, W. S.,** Erratum: application of a multiwire proportional chamber to the detection of axoplasmic transport, *Can. J. Physiol. Pharmacol.*, 54, 945, 1976.
86. **Snyder, R. E.,** The design and construction of a multiple proportional counter used to study axonal transport, *J. Neurosci. Methods*, 11, 79, 1984.
87. **Forman, D. S. and Shain, W. G., Jr.,** Ratrachotoxin blocks saltatory organelle movement in electrically excitable neuroblastoma cells, *Brain Res.*, 212, 242, 1981.
88. **Adams, R., Baker, P. F., and Bray, D.,** Particle movement in crustacean axons that have been rendered permeable by exposure to brief intense electric fields, *J. Physiol.*, 326, 7P, 1982.
89. **Forman, D. S.,** Vanadate inhibits saltatory organelle movement in a permeabilized cell model, *Exp. Cell Res.*, 141, 139, 1982.
90. **Hammond, G. R. and Smith, R. S.,** Inhibition of the rapid movement of optically detectable axonal particles by colchicine and vinblastine, *Brain Res.*, 128, 227, 1977.
91. **Allen, R. D., Allen, N. S., and Travis, J. L.,** Video-enhanced contrast, differential interference contrast (AVEC-DIC) microscopy: a new method capable of analyzing microtubule-related motility in the reticulopodial network of *Allogromia laticollaris*, *Cell Motility*, 1, 291, 1981.
92. **Hodge, A. J. and Adelman, W. J., Jr.,** Direct visualization of particle velocity distribution by pseudostereostopic viewing of time-lapsed sequential images. Application to fast axonal transport, *Cell Motility*, 4, 231, 1984.
93. **Forman, D. S., Padjen, A. L., and Siggins, G. R.,** Axonal transport of organelles visualized by light microscopy: cinemicrographic and computer analysis, *Brain Res.*, 136, 197, 1977.

Chapter 3

PRELUDE TO FAST AXONAL TRANSPORT: SEQUENCE OF EVENTS IN THE CELL BODY

Richard Hammerschlag and George C. Stone

TABLE OF CONTENTS

I. STRATEGY FOR IDENTIFYING INTRASOMAL PATHWAY

The fast anterograde transport of newly synthesized proteins can be conveniently subdivided into three phases. *Initiation* comprises the recognition, sorting, and routing to proximal regions of axon (or dendrites) of a select population of proteins. *Translocation* involves the longitudinal migration of these proteins within the axon. *Termination* is the arrival of proteins at their distal sites of function. Our studies have focused on defining the initiation phase — the events that interface synthesis in the soma with the onset of transport in the axon.

Since fast axonal transport is predominantly a process for delivering materials to the neuronal membrane, a fruitful approach has been to compare it to schemes employed by nonneuronal cells for transporting membrane and secretory proteins to their cell surface.[1] The well-characterized intracellular pathways of such proteins as digestive enzymes in pancreas,[2] acetylcholinesterase and acetylcholine receptor in muscle,[3] and immunoglobulins in plasma cells,[4] have served as models for examining the initiation or somal phase of fast axonal transport.

Our main strategy for probing these events has involved in vitro pharmacological manipulations of dorsal root ganglion neurons from bullfrog.[5,6] Preparations are incubated with a test drug after cell bodies within the ganglia have been exposed to a radioactive precursor of either protein, glycoprotein, or lipid. After various time periods, the profile of labeled fast-transported material within the axon is determined by assessing the acid-insoluble radioactivity in successive segments of the nerve trunk. A decreased amount of transport is taken to indicate that a step in the soma has been affected and that less material was exported to the axon. A somal location of drug action is verified if after exposing only the desheathed nerve trunk to the drug, the amount of transported material passing the exposed region is unaffected. Indications of the subcellular site of blockade in the cell bodies have been achieved by ultrastructural examination, and by utilizing different radioactive precursors known to be incorporated into fast-transported proteins at distinct regions of the soma.

This general approach, in combination with subcellular fractionation studies, has brought the outlines of the intrasomal pathway into focus (Figure 1).[1] Newly synthesized proteins destined for fast axonal transport are rapidly assembled into, or sequestered within, membrane. They are then sorted into subpopulations and routed through the Golgi apparatus where many acquire carbohydrate side chains and/or sulfate residues. Their exit from the Golgi occurs in association with coated vesicles and may involve entry into a storage pool from which they are delivered to the fast transport system and routed to distinct membrane domains. The evidence for each step of this itinerary will be presented in the body of this review.

II. ASSOCIATION WITH *DE NOVO* SYNTHESIZED MEMBRANE

Numerous studies have demonstrated an interrelation between fast-transported protein and lipid. Phospholipids,[7,8] cholesterol,[9] and gangliosides[10] undergo transport at rates comparable to the fast rate of protein transport. In addition, labels associated with fast-transported lipid and protein show parallel subcellular fractionation patterns, and local injection of colchicine blocks transport of both labels.[7] Observations that inhibition of protein synthesis reduces the amount of fast-transported lipid as well as protein exported from the soma[11,12] first suggested that fast-transport occurs as a concerted movement of assembled membrane rather than as an assembly line movement of membrane components.

Complementary support for this concept has come from our findings that inhibiting the synthesis of either phospholipid or cholesterol in the cell body leads to a proportional decrease in the amount of fast-transported protein exported to the axon.[13] The drug fenfluramine affects phospholipid biosynthesis by inhibiting the enzyme phosphatidic acid phosphohy-

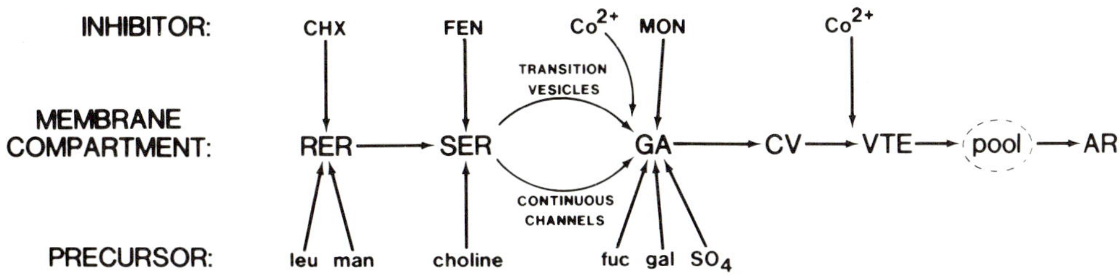

FIGURE 1. Flow diagram describing routes of fast-transported proteins through subcellular compartments in the neuron. The protein itineraries have been ascertained by taking advantage of site specific incorporation of various radiolabeled precursors in combination with several inhibitors also known to act at specific subcellular locations. Details are described in the text.

Membrane compartments	Inhibitors	Precursors
RER — Rough endoplasmic reticulum	CHX — Cycloheximide	leu — Leucine
SER — Smooth endoplasmic reticulum	FEN — Fenfluramine	man — Mannose
GA — Golgi apparatus	Co^{2+} — Cobalt	fuc — Fucose
CV — Coated vesicles	MON — Monensin	gal — Galactose
VTE — Vesiculo-tubular elements		SO_4 — Sulfate
AR — Axonal reticulum		

drolase, thereby diverting synthesis away from phosphatidyl choline and phosphatidyl ethanolamine production. Fenfluramine was found to inhibit fast-transport of ^{3}H-protein when ganglia were exposed to the drug but not when desheathed nerve trunks were selectively exposed. The extent of depression of protein transport was directly proportional to the inhibition of labeled choline incorporation into phosphatidyl choline (Figure 2). These findings suggest that phospholipid synthesis is required for the initiation of fast-transport in the soma but not to maintain ongoing transport in the axon. In similar studies, diazacholesterol, which blocks the final step in cholesterol biosynthesis, was found to inhibit fast-transport of both protein and phospholipid. Thus, it appears that inhibition of somal synthesis of any one of the three major components of membrane: protein, phospholipid, or cholesterol leads to a depression of fast axonal transport.

Association of individual fast-transported proteins with somal membranes was demonstrated by preparing a microsomal fraction from pulse-labeled dorsal root ganglia.[14] Labeled proteins of this fraction were separated by two-dimensional gel electrophoresis and the gel pattern compared with the pattern of fast-transported proteins generated by accumulation at a nerve ligature. This analysis revealed that nearly all of 48 fast-transported species examined were present among the proteins of the microsomal fraction. This observation regarding proteins within the soma is compatible with the long-established finding that fast-transported proteins are predominantly membrane associated in the axon.[15]

III. SORTING INTO SUBPOPULATIONS

While fast-transported proteins apparently remain membrane-associated during their passage through the neuron, the question arises as to whether they all travel a common pathway. Our first indication that they do not came from studies in which transport was inhibited by selectively bathing dorsal root ganglia in a medium either lacking Ca^{2+} or supplemented with the Ca^{2+} antagonist, Co^{2+}.[16,17] Under these conditions, protein synthesis and the rate of transport were unaffected, but the *amount* of protein exported from the soma into the axon was markedly decreased.

Next, we asked whether this Ca^{2+}-mediated pathway was common for all fast-transported

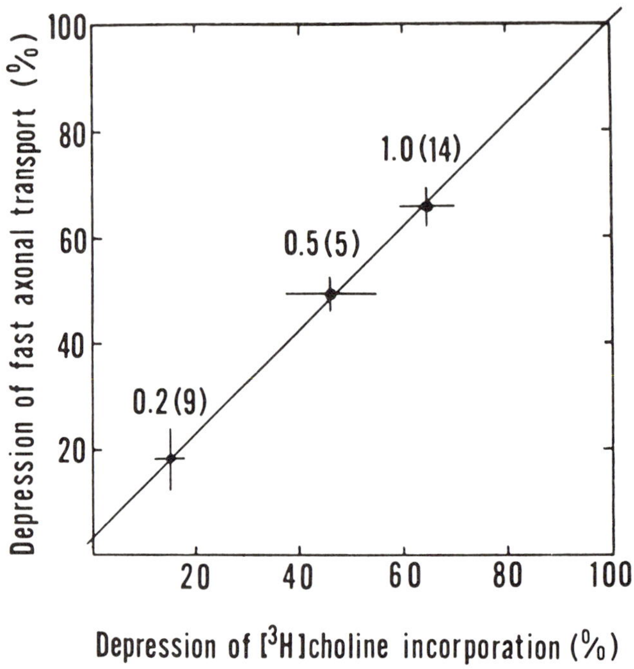

FIGURE 2. Fast axonal transport depends on newly synthesized protein and lipid. Reducing incorporation of ^{3}H-choline into dorsal root ganglion phospholipids with fenfluramine inhibits the fast axonal transport of ^{3}H-protein. The correlation is described by values for depression of incorporation and transport ($\pm$ SEM) at three concentrations of fenfluramine (m*M*). The number of trials at each concentration is shown in parentheses.[13]

proteins. When the effect of Co^{2+} on the passage of each of 48 individual fast-transported proteins was examined via two-dimensional gel electrophoresis, the export of all proteins was inhibited.[18] Quantitative analysis, however, revealed a differential effect of Co^{2+} in that fast-transported proteins of molecular weight above approximately 35,000 daltons were inhibited to a significantly greater extent than were proteins below this molecular weight (Figure 3).

In seeking an explanation for these results, we also had to consider data suggesting the existence of two separate Ca^{2+}-dependent steps. This likelihood arose from studies examining the effects of Co^{2+} on fast-transport of ^{3}H-galactose-labeled glycoproteins.[19] Incorporation of galactose (a late Golgi event) was depressed under conditions where leucine incorporation was unaffected. Also, the decrease in galactose-labeled proteins entering the axon was considerably less than the decrease in transport of leucine-labeled protein. These data are consistent with the existence of two Ca^{2+}-dependent steps, one involved in delivery of proteins to the Golgi complex and a second in transferring these proteins from the Golgi to the axonal transport system. In this scheme (Figure 1), leucine incorporation is unaffected by Co^{2+}, but leucine-labeled proteins pass through two Ca^{2+}-dependent steps. By contrast, since a reduced number of proteins reach the Golgi complex, galactose incorporation appears depressed. However, those proteins that do become galactosylated then pass through only one Ca^{2+} step.

Perhaps the simplest model to explain both sets of data is for one population of proteins to pass through a single Ca^{2+}-dependent step while the other population passes through both steps. Ultrastructural studies have revealed a possible basis for this model.[20] Passage of proteins from the endoplasmic reticulum to the Golgi complex may occur either by vesicles that bud from the former membrane system and fuse with the latter, or via membrane

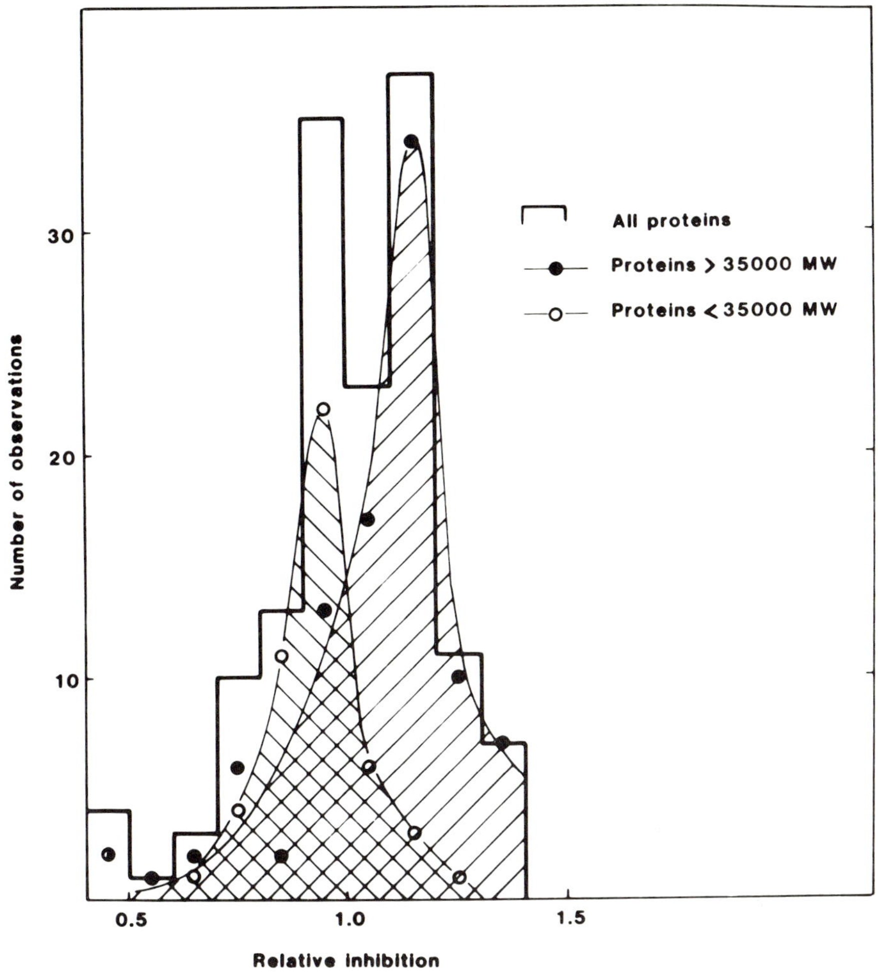

FIGURE 3. Subpopulations of fast-transported proteins are differentially affected by exposure of dorsal root ganglia to 0.8 m*M* Co^{2+}. The initially observed bimodal distribution was explained by grouping proteins according to molecular weight. Individual fast-transported proteins of > 35,000 mol wt were affected to a greater extent than proteins below this cutoff. The two subpopulations (cross-hatched regions) are significantly different ($p < 0.01$).[18]

channels that appear to connect the two systems.[21] While vesicle fusion is likely to be Ca^{2+}-dependent, passage via continuous channels has no obvious Ca^{2+} requirement. (In fact, these channels are morphologically unaltered in Co^{2+} treated ganglia.) The post-Golgi Ca^{2+} step, which would then be common for all fast-transported proteins, may be the fusion of Golgi-derived vesicles occurring during delivery of proteins to the transport system of the axon (See Section V).

The subpopulations defined on the basis of their differential sensitivity to Co^{2+} were found to have additional identifying characteristics. On the basis of labeling experiments with the ^3H-sugars fucose, galactose, and mannose, greater than 85% of the fast-transported proteins identified as glycoproteins had molecular weights over 35,000 (see Section IV).[22,23] Further, in pulse/chase studies designed to examine the duration of the initiation phase of fast-transport, the lower molecular weight proteins were cleared from the cell bodies with little delay relative to the higher molecular weight species which appeared to accumulate in the cell bodies prior to their export to the axons.[24]

The subpopulations may be a reflection of the longer transit time required for the formation of oligosaccharide side chains (comprising en bloc addition of core sugars in the rough endoplasmic reticulum, trimming of distal core sugars, and sequential addition of terminal sugars in the Golgi complex).

IV. PASSAGE THROUGH THE GOLGI APPARATUS

The role of the Golgi apparatus (GA) in the initiation of fast axonal transport was examined since this structure appears to be an obligatory way station in the routing of membrane-associated proteins to the cell surface.[25] Electron microscope autoradiographic studies provided the first indication that neuronal proteins synthesized in the rough endoplasmic reticulum pass through the GA before undergoing fast-transport in the axon.[26] We sought to extend this observation by determining whether passage through the GA is restricted to those proteins requiring post-translational modification (such as glycosylation or sulfation) or represents a common step for all proteins destined for fast axonal transport. Recent confirmation of a central role for the GA in intracellular transport has come from studies exploiting the affinity of the sodium ionophore, monensin, for this structure.[27,28] Biochemical and ultrastructural evidence suggest that monensin acts by preventing the formation of Golgi-derived vesicles that normally ferry materials to the plasma membrane.

We have found monensin to be a potent inhibitor of fast axonal transport.[29] At less than micromolar concentrations, transport of protein was depressed in parallel with that of lipid under conditions where protein and lipid synthesis were unaffected. With labeled galactose as precursor (a sugar added to glycoproteins by enzymes localized to distal Golgi cisternae) transport was again depressed, but only to the extent that incorporation was reduced. Since monensin has no direct effect on galactosyltransferase activity, the findings suggest that monensin acts to reduce the number of glycoprotein molecules reaching the sites of transferase action. Morphological examination has confirmed that monensin disrupts the Golgi cisternae of dorsal root ganglion neurons.

Next, studies were designed to determine if fast-transported proteins from both subpopulations (see Section II.) were affected by monensin.[29] This would indicate whether non-glycosylated proteins may bypass the GA en route to the fast-transport system in the axon. Quantitative analyses of the 48 most abundant individual fast-transported species, selected after separation on two-dimensional gels, revealed that monensin depressed all the proteins to a similar extent (Table 1). Since only half of the 48 species examined had been previously identified as glycoproteins (see Section IV.), it is likely that the GA has a critical role beyond post-translational glycosylation, perhaps involving the sorting and packaging of proteins prior to their entry into the axon.

V. POST-TRANSLATION MODIFICATION

Post-translational modifications can play an important role in determining the fate of a newly synthesized protein.[30] Addition of carbohydrate residues, for example, has been shown to determine intracellular destination of proteins,[31] to protect against proteolysis, allowing proteins to reach their prescribed destination,[30] and to be required for function once proteins have reached their final membrane domain.[32] Recent studies have sought to identify fast-transported proteins that become glycosylated with the aim of exploring whether molecular recognition processes involved in intracellular transport are in part dependent upon such post-translational modification.

Early studies of axonal transport using labeled sugars as precursors indicated that glycoproteins are found only in the fast phase of transport and that they are predominantly among the higher molecular weight species.[33] When ^{3}H-fucose or ^{3}H-galactose-labeled fast-

Table 1
COMPARATIVE EFFECTS OF MONENSIN ON FAST-TRANSPORTED PROTEINS AND GLYCOPROTEINS

Relative inhibition[a]	
Glycoproteins	**Nonglycoproteins**
1.01	1.05
1.25	1.07
1.10	1.27
0.79	0.64
0.84	0.82
0.81	0.81
1.11	1.06
0.95	0.98
0.87	1.09
1.14	0.60
1.20	1.16
1.12	1.08
1.01 ± 0.15	0.97 ± 0.20

[a] Fast axonal transport of ^{35}S-met-labeled proteins was carried out in the presence or absence of 1×10^{-7} *M* monensin. Following two-dimensional gel electrophoresis of nerve segments from control and experimental preparations, the relative abundance[18] of 24 protein spots, 12 each previously identified as glycoproteins or nonglycoproteins,[22] was determined. Individual values represent the mean of three control/experimental gel pairs. Overall means (± SD) are not significantly different ($p > 0.1$).

transported proteins were separated via two-dimensional gel electrophoresis, and individual gel spots quantitatively analyzed, over 60% of the 92 most abundantly labeled spots were found to be glycoproteins (Figure 4); approximately 85% of the identified glycoproteins migrated to positions corresponding to molecular weights greater than 35,000.[22]

Moreover, most of the carbohydrate-containing species appeared on the gel as families of spots, each member having a similar molecular weight but different isoelectric point. This effect is apparently due in large part to biological variations in the number of charged sugar residues present on a given polypeptide. Support for this explanation was obtained by treatment of fast-transported proteins with neuraminidase which, by removing sialic acid residues, alters isoelectric point.[22] While several glycoprotein families collapsed to a single major species, other families of protein spots were either unaffected by the enzyme or were shifted in register to more basic isoelectric points. Seeking other origins for this charge microheterogeneity, the presence of sulfate moieties associated with fast-transported proteins was examined.[23,34] A precedent for sulfation of neural proteins is that as many as 40% of the hexose and hexosamine residues of brain glycoproteins were found to contain O-linked sulfate.[35] In addition, sulfated macromolecules have been shown to undergo fast axonal transport, with $^{35}SO_4$ associated with both glycoproteins and proteoglycans.[33]

When visualized by two-dimensional gel electrophoresis, fast-transported $^{35}SO_4$-labeled macromolecules could be seen as poorly resolved high molecular weight proteoglycan-like material and as families of well-resolved spots corresponding in migration to previously identified glycoproteins.[23] The most heavily labeled sulfate-containing species, however, corresponded to fast-transported proteins that had been found not to be glycosylated (Figure 4). Base hydrolysis of these individual species followed by high pressure liquid chromatography demonstrated the presence of tyrosine sulfate residues.[34] Included among several

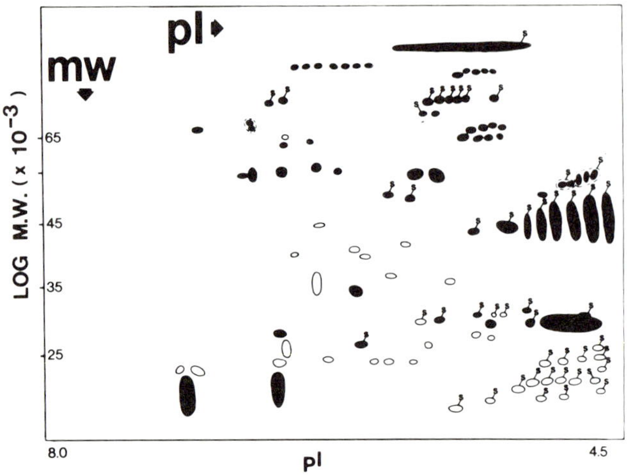

FIGURE 4. Fast-transported proteins are post-translationally modified in several ways. This reconstruction of a two-dimensional gel pattern identifies individual fast-transported proteins that possess only carbohydrate side chains (●), side chains containing sulfated sugars (●^{S}), and non-glycosylated proteins containing sulfate linked to tyrosine (○^{S}). Proteins that have not undergone detectable glycosylation or sulfation are also shown (○).[22,23]

previous reports of polypeptides containing sulfated tyrosine residues[36-38] are indications that the sulfate plays a key role in the expression of physiological activity.[39]

Although the existance of sulfated amino acid and carbohydrate residues in fast-transported proteins is now apparent, understanding the biological role(s) of this modification will require establishing the destinations and functions of the sulfated molecules (see Section VII.).

VI. TRANSFER FROM THE GOLGI APPARATUS VIA COATED VESICLES

Intracellular transport of proteins between discontinuous regions of membrane occurs via populations of vesicles that bud-off from one region and fuse with the next.[40] One particular vesicular structure, the so-called coated vesicle, has received considerable attention as a vehicle mediating exocytotic transfers from the Golgi apparatus, as well as endocytotic transfers from the plasma membrane.[41] Thus, a strategy was devised to determine whether fast-transported proteins could be detected in association with Golgi-derived coated vesicles in dorsal root ganglion (DRG) neurons.[42] Since it was impractical to isolate coated vesicles directly from DRG of bullfrog (wet weight $\simeq$ 2 mg/ganglion), vesicles were co-purified from ^{3}H-leucine-labeled DRG and unlabeled frog brain. Purified vesicles isolated from a crude microsomal fraction subjected to D_2O/Ficoll density gradient centrifugation were observed to have the same electrophoretic protein profile and the same characteristic surface lattice as coated vesicles purified from calf brain. ^{3}H-proteins associated with the coated vesicle fraction were co-electrophoresed on two-dimensional gels with ^{35}S-methionine-labeled fast-transported proteins generated in separate preparations of spinal nerve. When protein spots were removed from the gel using the ^{35}S fluorographic pattern as a guide, approximately two thirds of the 67 individual fast-transported proteins analyzed were found to contain significant levels of ^{3}H.

Co-migration implies that an array of coated vesicle proteins are similar in both net charge and molecular weight to mature fast-transported proteins in the axon. Thus, such proteins associated with the vesicle fraction must have already undergone Golgi-level post-translational modifications, including terminal glycosylation and sulfation prior to their incorporation into coated vesicles. We infer from these findings that coated vesicles play a major, although not necessarily exclusive, role in the exit of fast-transported proteins from the GA. Further, since clathrin, the major coat protein, is not fast-transported it appears that the post-Golgi coated vesicles shed their coats prior to their possible fusion to become vesiculo-tubular elements (see Figure 1). Studies are in progress to determine if the noncomigrating proteins can be detected by increasing their specific activity, if they exist in a different molecular form in the vesicle than in the axon, or if they exit the Golgi cisternae by a different route.

VII. ENTRY INTO A STORAGE POOL

Several studies have indicated that all fast-transported materials do not directly enter the axon following their passage through the cell body. Observations of a prolonged release of lipid,[43,44] protein,[45] and glycoprotein[46] have been interpreted to suggest that some materials may enter a storage pool from which they are slowly released but fast-transported once they reach the axon.

A somal pool has also been invoked to explain the seemingly contradictory findings that exposure of cell bodies to Co^{2+} or monensin strongly inhibits the amount of radiolabeled protein accumulating at a nerve ligature but has relatively little effect on the accumulation of the fast-transported enzymes, acetylcholinesterase (AChE), and adenylate cyclase.[47,48] Such a pool would exist distal to the sites where Co^{2+} and monensin block intracellular passage of newly synthesized proteins. The pool explanation seems more plausible in light of follow-up studies showing that AChE and the cyclase still accumulate to a level approaching that in control nerve when protein synthesis is shutdown by exposing DRG to cycloheximide in addition to Co^{2+} or monensin (Figure 5). Additional studies are required to localize the pool more precisely, to determine whether all fast-transported proteins pass through the pool, and to examine the kinetics of protein depletion from the pool. The existance of a somal pool of material committed to fast axonal transport suggests that the amount of material delivered for transport can be regulated independently of protein synthesis.

VIII. DELIVERY TO DISCRETE MEMBRANE DOMAINS

Once post-Golgi vesicles have shed their coats and given rise to vesiculo-tubular elements (see Figure 1 and Section V.), the fast-transported proteins contained within such elements have only covered a portion of their itinerary. While as yet unidentified molecular recognition mechanisms enable appropriate proteins to find their way out of the soma, the proteins still must reach distal regions of the neuron.[49-51] In fact, since specific proteins have been detected at discrete membrane domains, it may well be that recognition mechanisms exist for the routing of fast-transported proteins to distinct destinations.[52] One of the first approaches to attempt to document such a selective delivery process has been to compare the proteins undergoing fast-transport in the bifurcating axons of primary afferent neurons.

Early in vivo studies comparing proteins transported toward synaptic terminals in the spinal cord with those destined for sensory endings in the periphery yielded intuitively comforting results in that one-dimensional gel patterns of fast-transported proteins did suggest different proteins were being delivered to the two functionally distinct regions of the neuron.[53] Several attempts to reproduce these findings have been unsuccessful, however, with local incorporation suspected as the basis for differences in transport seen in the orginal study.[54] The most detailed analysis attempting to demonstrate differential delivery in this system has involved the in vitro labeling of bullfrog dorsal root ganglia followed by analysis of fast-transported proteins via two-dimensional gel electrophoresis.[55] Using this preparation, in

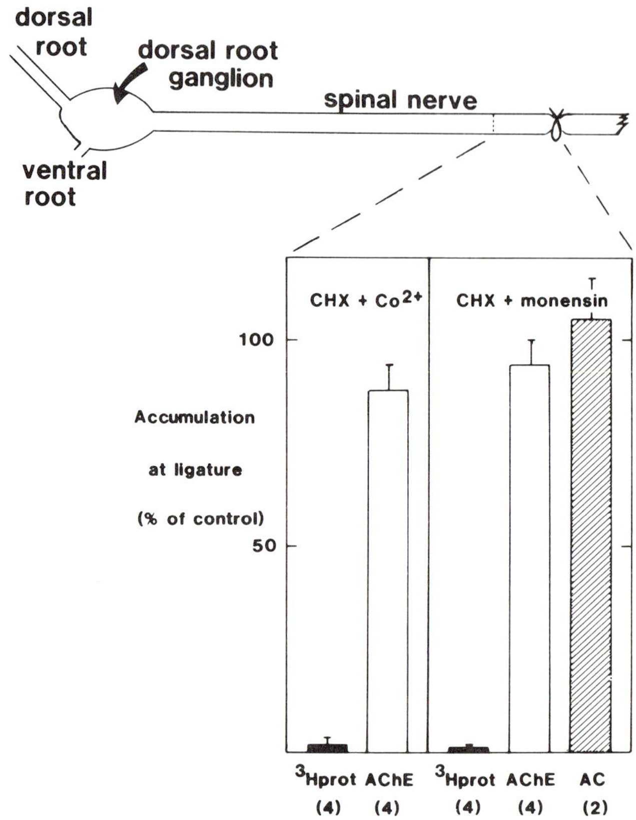

FIGURE 5. Exposure of cycloheximide (CHX), plus Co^{2+} or monensin, to the dorsal root ganglion inhibits the fast axonal transport of ^3H-proteins but not the transport of endogenous acetylcholinesterase (AChE) or adenylate cyclase (AC). Differential accumulation of endogenous versus newly synthesized proteins in the presence of these agents is consistent with a pool occurring after fast-transported proteins exit the Golgi. Values for accumulation are ± SD for (n) preparations.

which local incorporation can be virtually ruled out, no qualitative differences were detected when approximately 150 of the most abundant proteins being transported in the central and peripheral bifurcations were compared. This, despite the greater sensitivity of the two-dimensional gels. In another comparative analysis, using the same methods, proteins being transported in sensory neurons were compared to those transported in motoneurons.[55] Despite these two neuronal types having very different functional and neurochemical bases, the most abundant transported species were again qualitatively similar. A recent quantitative analysis of such two-dimensional gel species has demonstrated variations in relative abundance of certain fast-transported proteins in the sensory neuron bifurcations,[56] and yet the overall constancy of the great majority of fast-transported proteins — independent of destination — provides a continuing challenge to the understanding of neuronal specificity.

An alternative approach for examining delivery to distinct membrane domains involves the comparison of fast-transported proteins destined for terminal regions of the neuron to those off-loaded along the axon. Ligating a nerve only after the front of radiolabeled fast-transported proteins has passed the point of ligature results in off-loaded species being selectively present in nerve regions immediately *distal* to the ligature. In contrast, proteins continuing toward terminal regions will be enriched in the front of labeled materials undergoing anterograde transport. Earlier studies employing such an approach revealed that a sig-

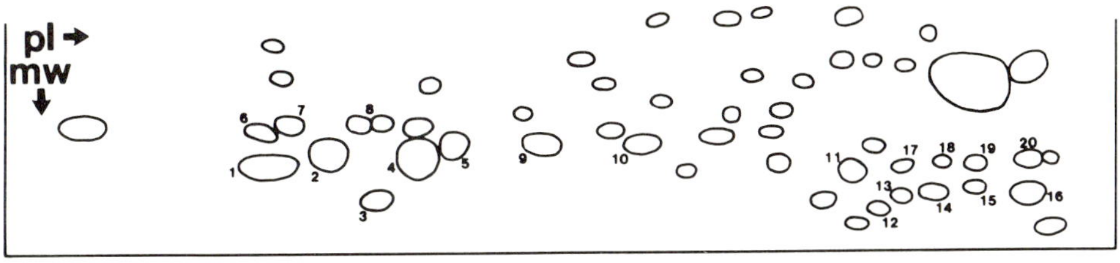

NON-SULFATED PROTEINS			SULFATED PROTEINS		
SPOT NUMBER	REL. ABUNDANCE*		SPOT NUMBER	REL. ABUNDANCE*	
	MEAN	(n)		MEAN	(n)
1	2.2	4	11	2.1	4
2	0.8	4	12	2.6	4
3	0.4	4	13	4.1	4
4	0.6	4	14	1.1	4
5	0.3	4	15	1.2	2
6	1.0	4	16	1.9	2
7	0.8	4	17	5.2	4
8	0.8	4	18	4.1	4
9	0.8	2	19	2.6	3
10	0.3	4	20	3.8	3
	0.8 ± 0.5 (S.D.)			2.9 ± 1.4 (S.D.)	

*REL. ABUNDANCE = "FRONT" + "OFF-LOADED"

FIGURE 6. Fast-transported proteins containing tyrosine sulfate are preferentially destined for terminal regions of the neuron. After labeling dorsal root ganglia with ^{35}S-met, a fast axonal transport paradigm was carried out which allowed separation of proteins off-loaded to the axolemma from those in the front of fast-transported proteins continuing toward terminals. The relative abundance of each of 10 sulfated and 10 nonsulfated species was determined in gels containing off-loaded and front fast-transported proteins. Overall means ± SD reveal that sulfated proteins are present predominantly in the front of fast-transported proteins.[34]

nificant proportion of fast-transported proteins are deposited along the axon.[57] When this protein fraction was compared on single dimension gels to those species found in the moving front, lower molecular weight species tended to be deposited while higher molecular weight species were preferentially transported toward nerve terminals.[58,59] Incorporating two-dimensional gel electrophoresis into this paradigm has offered a more highly resolved picture of differential delivery of fast-transported proteins.[34] The first intriguing finding revealed by this approach involves the subset of fast-transported proteins that contain sulfated tyrosine residues (see Section IV.). This subset of proteins appears preferentially in the front of fast-transported proteins, suggesting that they are destined for terminal regions of the neuron (Figure 6). Such a correlation suggests the possibility that the presence of tyrosine sulfate plays a role in determining ultimate intracellular destination. This proposal is consistent with results from a study of rat pheochromocytoma cells in tissue culture; the major tyrosine-$^{35}SO_4$-containing proteins were secreted in response to K^+ stimulation.[60]

A natural extension of this analysis was to examine if fast-transported glycoproteins are enriched in either the off-loaded fraction or the front. Preliminary observations indicate that glycoproteins are evenly distributed between materials delivered to the axon and to terminal regions of the neuron.[34] In view of this finding, factors other than glycomoieties may determine the destination of these fast-transported proteins once they enter the axon.

ACKNOWLEDGMENTS

It is a pleasure to thank those whose creative efforts in the frog lab have led to the findings presented in this review: Charles Bakhit, Judy Bobinski, Franci Bolen, Arlene Chiu, Anant Dravid, Mark Ellisman, Frank Longo, Pierre Lavoie, and James Lindsey. Our studies have received ongoing support from the National Multiple Sclerosis Society (RG 1296-B2), the National Science Foundation (BNS 81-12129), the National Institute of Neurological and Communicative Disorders and Stroke (NS 19636), and the Quentin L. Kopp Research Fund administered through the Beckman Research Institute of the City of Hope.

REFERENCES

1. **Hammerschlag, R. and Stone, G. C.,** Membrane delivery by fast axonal transport, *Trends Neurosci.*, 5, 12, 1982.
2. **Palade, G. E.,** Intracellular aspects of protein secretion, *Science*, 189, 347, 1975.
3. **Rotundo, R. L. and Fambrough, D. M.,** Secretion of acetylcholinesterase: relation to acetylcholine receptor metabolism, *Cell*, 22, 595, 1980.
4. **Tartakoff, A. and Vassalli, P.,** Plasma cell immunoglobulin M molecules, their biosynthesis assembly and intracellular transport, *J. Cell Biol.*, 83, 284, 1979.
5. **Dravid, A. R. and Hammerschlag, R.,** Axoplasmic transport of proteins *in vitro* in primary afferent neurons of frog spinal cord: effect of Ca^{2+}-free incubation conditions, *J. Neurochem.*, 24, 711, 1975.
6. **Stone, G. C., Wilson, D. L., and Hall, M. E.,** Two-dimensional electrophoresis of proteins in rapid axoplasmic transport, *Brain Res.*, 144, 287, 1978.
7. **Abe, T., Haga, T., and Kurokawa, M.,** Rapid transport of phosphatidylcholine occurring simultaneously with protein transport in the frog sciatic nerve, *Biochem. J.*, 136, 731, 1973.
8. **Currie, J. R., Grafstein, B., Whitnall, M. H., and Alpert, R.,** Axonal transport of lipid in goldfish optic axons, *Neurochem. Res.*, 3, 479, 1978.
9. **Rostas, J. A. P., Austin, L., and Jeffrey, P. L.,** Selective labelling of two phases of axonal transport of cholesterol in the chick optic system, *J. Neurochem.*, 32, 1461, 1979.
10. **Forman, D. S. and Ledeen, R. W.,** Axonal transport of gangliosides in the goldfish optic nerve, *Science*, 177, 630, 1972.
11. **Grafstein, F., Miller, J. A., Ledeen, R. W., Haley, J., and Specht, S. C.,** Axonal transport of phospholipid in goldfish optic system, *Exp. Neurol.*, 46, 261, 1975.
12. **Sherbany, A. A., Ambron, R. T., and Schwartz, J. H.,** Membrane glycolipids: regional synthesis and axonal transport in a single identified neuron of *Aplysia californica*, *Science*, 203, 78, 1979.
13. **Longo, F. M. and Hammerschlag, R.,** Relation of somal lipid synthesis to the fast axonal transport of protein and lipid, *Brain Res.*, 193, 471, 1980.
14. **Stone, G. C. and Hammerschlag, R.,** Evidence for sorting of proteins during the initiation of rapid axonal transport, *Soc. Neurosci. Abstr.*, 6, 82, 1980.
15. **McEwen, B. S. and Grafstein, B.,** Fast and slow components in axonal transport of protein, *J. Cell Biol.*, 38, 494, 1968.
16. **Hammerschlag, R., Dravid, A. R., and Chiu, A. Y.,** Mechanism of fast transport: a proposed role for calcium ions, *Science*, 188, 273, 1975.
17. **Hammerschlag, R., Bakhit, C., Chiu, A. Y., and Dravid, A. R.,** Role of calcium in the initiation of fast axonal transport: effects of divalent cations, *J. Neurobiol.*, 8, 439, 1977.
18. **Stone, G. C. and Hammerschlag, R.,** Differential effects of cobalt on the initiation of fast axonal transport, *Cell Mol. Neurobiol.*, 1, 3, 1981.
19. **Hammerschlag, R.,** Multiple roles of calcium in the initiation of fast axonal transport, in *Axoplasmic Transport*, Weiss, D. G., Ed., Springer-Verlag, Basel, 1982, 279.
20. **Lindsey, J. D., Hammerschlag, R., and Ellisman, M. H.,** An increase in smooth endoplasmic reticulum and a decrease in Golgi apparatus occur with ionic conditions that block initiation of fast axonal transport, *Brain Res.*, 205, 175, 1981.
21. **Lindsey, J. D. and Ellisman, M. H.,** Cobalt-induced distruption of the Golgi apparatus blocks assembly of forming face element, *Soc. Neurosci. Abstr.*, 7, 485, 1981.
22. **Stone, G. C. and Hammerschlag, R.,** Glycosylation as a criterion for defining subpopulations of fast-transported proteins, *J. Neurochem.*, 40, 1124, 1983.
23. **Stone, G. C., Hammerschlag, R., and Bobinski, J. A.,** Fast-transported glycoproteins and nonglycosylated proteins contain sulfate, *J. Neurochem.*, 41, 1085, 1983.
24. **Stone, G. C.,** Applications of two-dimensional gel electrophoresis to current problems in fast axonal transport, in *Axoplasmic Transport*, Weiss, D. G., Ed., Springer-Verlag, Basel, 1982, 454.
25. **Morré, D. M., Kartenbeck, J., and Franke, W. W.,** Membrane flow and interconversions among endomembranes, *Biochim. Biophys. Acta*, 559, 71, 1979.

26. **Droz, B. and Koenig, H. L.,** Localization of protein metabolism in neurons, in *Protein Metabolism of the Nervous System,* Lajtha, A., Ed., Plenum Press, New York, 1970, 93.
27. **Tartakoff, A. M.,** The Golgi complex: crossroads for vesicular traffic, *Int. Rev. Exp. Pathol.,* 22, 227, 1980.
28. **Tartakoff, A. M.,** Perturbation of vesicular traffic with the carboxylic iononophore monensin, *Cell,* 32, 1026, 1983.
29. **Hammerschlag, R., Stone, G. C., Bolen, F. A., Lindsey, J. D., and Ellisman, M. H.,** Evidence that all newly-synthesized proteins destined for fast axonal transport pass through the Golgi apparatus, *J. Cell Biol.,* 93, 568, 1982.
30. **Olden, K., Parent, J. B., and White, S. L.,** Carbohydrate moieties of glycoproteins: a re-evaluation of their function, *Biochim. Biophys. Acta,* 650, 209, 1982.
31. **Hasilik, A. and Neufeld, E. F.,** Biosynthesis of lysosomal enzymes in fibroblasts: phosphorylation of mannose residues, *J. Biol. Chem.,* 255, 4946, 1980.
32. **Waechter, C. J., Schmidt, J. W., and Catterall, W. A.,** Glycosylation is required for maintenance of functional sodium channels in neuroblastoma cells, *J. Biol. Chem.,* 258, 5117, 1983.
33. **Elam, J. S.,** Axonal transport of complex carbohydrates, in *Complex Carbohydrates of Nervous Tissue,* Margolis, R. U. and Margolis, R. K., Eds., Plenum Press, New York, 1979, 235.
34. **Stone, G. C., Hammerschlag, R., and Bobinski, J. A.,** Differential delivery of sulfated proteins by fast axonal transport, *Soc. Neurosci. Abstr.,* 9, 149, 1983.
35. **Margolis, R. K. and Margolis, R. U.,** Sulfated glycopeptides from rat brain glycoproteins, *Biochemistry,* 9, 4389, 1970.
36. **Huttner, W. B.,** Sulphation of tyrosine residues: a widespread modifications of proteins, *Nature (London),* 299, 273, 1982.
37. **Bettleheim, F. R.,** Tyrosine-O-sulfate in a peptide from fibrinogen, *J. Am. Chem. Soc.,* 76, 2838, 1954.
38. **Mutt, V. and Jorpes, E.,** Hormonal polypeptides of the upper intestine, *Biochem. J.,* 125, 57p, 1971.
39. **Gardner, J. D., Walker, M. D., Martinez, J., Priestley, G. P., Natarajan, S., and Bodanszky, M.,** The importance of the amino acid in position 27 of cholecystokinin in determining its biological activity on pancreatic acini, *Biochem. Biophys. Acta,* 630, 323, 1980.
40. **Morré, D. J.,** Intracellular vesicular transport: vesicles, guide elements, and mechanisms, in *Axoplasmic Transport,* Weiss, D. G., Ed., Springer-Verlag, Basel, 1982, 2.
41. **Pearse, B. M. F. and Bretscher, M. S.,** Membrane recycling by coated vesicles, *Ann. Rev. Biochem.,* 50, 85, 1981.
42. **Stone, G. C., Hammerschlag, R., and Bobinski, J. A.,** Involvement of coated vesicles in the initiation of fast axonal transport., *Brain Res.,* 291, 219, 1984.
43. **Haley, J. E., Tirri, L. J., and Ledeen, R. W.,** Axonal transport of lipids in the rabbit optic system, *J. Neurochem.,* 32, 727, 1979.
44. **Toews, A. D., Goodrum, J. F., and Morell, P.,** Axonal transport of phospholipids in rat visual system, *J. Neurochem.,* 32, 1165, 1979.
45. **Berry, R. W.,** Evidence for multiple somatic pools of individual axonally transported proteins, *J. Cell Biol.,* 87, 379, 1980.
46. **Goodrum, J. F. and Morell, P.,** Axonal transport, deposition, and metabolic turnover of glycoproteins in the rat optic pathway, *J. Neurochem.,* 38, 696, 1982.
47. **Larivière, L. and Lavoie, P. -A.,** Calcium requirement for fast axonal transport in frog motoneurons, *J. Neurochem.,* 39, 882, 1982.
48. **Hammerschlag, R., Bolen, F. A., and Carlsen, R. C.,** Accumulation of fast-transported AChE and adenylate cyclase at a nerve ligature is unaffected by conditions that inhibit accumulation of the ^{3}H-protein, *Soc. Neurosci. Abstr.,* 9, 149, 1983.
49. **Kelly, P. T. and Montgomery, P. R.,** Subcellular localization of the 52,000 molecular weight, major postsynaptic density protein, *Brain Res.,* 233, 265, 1981.
50. **Huttner, W. B., Schiebler, W., Greengard, P., and de Camilli, P.,** Synapsin I (Protein I), a nerve terminal-specific phosphoprotein. III. Its association with synaptic vesicles studied in a highly purified synaptic vesicle preparation, *J. Cell Biol.,* 96, 1374, 1983.
51. **Reichardt, L. F. and Matthew, W. D.,** Monoclonal antibodies; applications to studies on the chemical synapse, *Trends Neurosci.,* 5, 24, 1982.
52. **Hammerschlag, R.,** How do neuronal proteins know where they are going? Speculations on the role of molecular address markers, *Dev. Neurosci.,* 6, 2, 1984.
53. **Anderson, L. E. and McClure, W. O.,** Differential transport of protein in axons: comparison between the sciatic nerve and corsal columns of cats, *Proc. Natl. Acad. Sci. U.S.A.,* 70, 1521, 1973.
54. **Wilson, D. L. and Stone, G. C.,** Axoplasmic transport of proteins, *Ann. Rev. Biophys. Bioeng.,* 8, 27, 1979.
55. **Stone, G. C. and Wilson, D. L.,** Qualitative analysis of proteins rapidly transported in ventral horn motoneurons and bidirectionally from dorsal root ganglia, *J. Neurobiol.,* 10, 1, 1979.

56. **Rulli, R. D. and Wilson, D. L.,** Selective routing of fast-transported proteins in branches of sensory axons, *Soc. Neurosci. Abstr.*, 9, 148, 1983.
57. **Muñoz-Martinez, E. J., Nuñez, R., and Sanderson, A.,** Axonal transport: a quantitative study of retained and transported protein fractions in the cat, *J. Neurobiol.*, 12, 15, 1981.
58. **Cancalon, P. and Beidler, L. M.,** Differences in the composition of the polypeptides deposited in the axon and the nerve terminals by fast axonal transport in the garfish olfactory nerve, *Brain Res.*, 121, 215, 1977.
59. **Weiss, D. G., Krygier-Brevart, V., Gross, G. W., and Kreutzberg, G. W.,** Rapid axoplasmic transport in the olfactory nerve of the pike. II. Analysis of transported proteins by SDS gel electrophoresis, *Brain Res.*, 139, 77, 1978.
60. **Huttner, W. B. and Lee, R. W. H.,** Protein sulfation on tyrosine residues, *J. Cell Biol.*, 95, 389a, 1982.

Chapter 4

IONIC REQUIREMENTS FOR FAST AXONAL TRANSPORT

Martin Kanje

TABLE OF CONTENTS

I. INTRODUCTION

The effects of a variety of ions and ionic conditions on fast axonal transport (AXT) have been studied in order to elucidate the mechanism behind the AXT-machinery and the possible connection between AXT and neuronal activity.

It is known that Ca^{2+} deprivation or, conversely conditions which increase the intra-axonal Ca^{2+} level, inhibit AXT in cat and frog sciatic nerves.[1-5] AXT is also sensitive to increased axonal Na^+ levels whereas Na^+-deprivation seem to inhibit AXT in some but not all nerves.[6-14]

The significance of these effects to the AXT-machinery is difficult to evaluate. It is not possible to exclude that the various ions or ionic conditions affect AXT indirectly, i.e., by interference with energy metabolism.

From the current models of AXT all of which include a transport substrate (microtubles (MT), actin, smooth endoplasmic reticulum or the microtrabecular system), and an energy transforming system (ATPase activity), specific ionic requirements can be inferred or postulated.[15-27] Thus, models which involve MT depend on a careful control of the axonal Ca^{2+} level, to which MT are sensitive. The ATPase activity is likely to require Mg^{2+} and or Ca^{2+}. Ochs has postulated a role of Ca^{2+} in his transport filament model of AXT.[5,21-23,28] In this model Ca^{2+} is used for the calmodulin modulated activation of a Ca^{2+}-Mg^{2+}-ATPase activity associated with MT.

AXT also involves an initiation phase, which includes the passage of materials destinated for AXT through the Golgi apparatus.[2,29,30] This step is sensitive to Ca^{2+}-deprivation, Co^{2+}-ions and the Na^+-ionophore monensine.[2,29-31,32]

The effects of various ions and ionic conditions on AXT will here be reviewed.

II. Na^+

A. Na^+-Deprivation

At the axonal level Na^+-deprivation does not affect AXT in sheathed frog and cat sciatic nerves.[1,6] However, Kirpekar et al.[10] and Garcia et al.[9] found that incubation of the entire cat hypogastric nerve in Na^+-deprived medium inhibited transport. An explanation to the discrepant results were given by Lavoie who studied the effects of Na^+-deprivation in frog sciatic nerves.[11,12,14] He found that replacement of NaCl by sucrose effectively inhibited AXT at the ganglionic — but not at the axonal — level unless the nerve was desheathed. The differential effects of Na^+ could therefore be ascribed to the perineurium, which acts as an effective permeability barrier to ions and which is less developed at the ganglionic level. Further studies by Lavoie indicated that mainly Cl^- was responsible for the inhibitor action of NaCl deprived medium at the axonal level (see below).[11,12,14] The observation that Choline-Cl could substitute for NaCl to maintain AXT at the axonal but not at the ganglionic level indicated that the initiation phase of AXT and the transport process could have different requirements for Na^+.

In desheathed cat sciatic nerve the replacement of NaCl by sucrose failed to affect AXT at the axonal level.[1] The cause of the discrepant effects by Na^+ on the frog and cat sciatic nerve is not known, but was ascribed to the shorter period of exposure of the cat nerve to Na^+ deficient medium.[11]

B. Increased Axonal Na^+ Content

Increased levels of axonal Na^+ can be obtained by various means including electrical stimulation, the exposure of nerves to batrachotoxin, ouabain, or Na^+-ionophores.

1. Batrachotoxin

BTX, which increase the Na^+ permeability of the nerve, is the most potent AXT-inhibitor reported so far and inhibits transport in the nanomolar range.[8,28,33-35] The increased Na^+ permeability appear to be responsible for its effect since tetrodotoxin, (TTX) antagonizes both the BTX induced inhibition of AXT and the increased Na^+ permeability. Consequently BTX fails to affect AXT in Na^+-deprived or low Na^+ Ringer solutions.[8] Although the result appeared clear, an intracellular effect of BTX could not be excluded. In Aplysia neurons BTX has been reported to inhibit AXT without interfering with the resting membrane potential which suggests a site of action different from that of the Na^+-permeability.[36]

2. Ouabain

Ouabain increases intra-axonal Na^+ content by inhibition of the Na^+-K^+-ATPase. Ouabain is an effective inhibitor of AXT in the sheathed and desheathed frog and cat sciatic nerves and also in the cat hypogastric nerve.[6,7,9,23] The effects of ouabain is diminished in Na^+-free medium which suggests that the increased axonal Na^+-level is responsible for the inhibited AXT.[23] Other causes of the inhibition of AXT by ouabain can not be excluded but neither a decrease of high energy phosphate compounds nor depolarization are likely to explain the inhibited AXT.

3. Monensine

Hammerschlag et al. have investigated the effects of a Na^+-ionophore monensine on the initiation of AXT.[37] Monensine at concentrations which effectively suppressed the export of label, led materials into the axon and had no effect on AXT at the axonal level. Monensine was found to cause drastic ultrastructural changes in the Golgi apparatus and the smooth endoplasmic reticulum which suggested that all materials destined for fast AXT must pass through these structures before entering the axon. It is likely that monensine affects intracellular Na^+ levels. The relation of these changes to the derangement of the smooth endoplasmic reticulum and the Golgi apparatus and the inhibited AXT remain unknown.

4. Electrical Activity

Changes of ionic fluxes during electrical activity result in increased axonal levels of Na^+ and Ca^{2+}. Electrical activity does not affect the rate or the amount of labeled, axonally transported materials until very high frequencies are used, which could affect the levels of high energy phosphates and/or the ionic composition of the axoplasm.[38] On the other hand electrical activity and concomitant ionic fluxes may be involved in the guidance or regulation of the amount of transmitter related substances.[39,40] Thus, the amount of transported norepinephrine related enzymes change with activity and AXT of adenosine decrease in TTX inactivated optic nerves of goldfish.[40,41]

C. Mechanism of Na^+ Action

The mechanism by which Na^+-deprivation or increased axonal Na^+-level exerts it inhibitory action on AXT is not known. Although metabolic effects can not be excluded, the ATP content does not drop sufficiently in nerves incubated in Na^+-deprived medium or medium which increases the axonal Na^+ content (BTX).[8,28] Both Na^+-deprivation and increased axonal Na^+ content decrease the Na^+ gradient and thereby increase the influx of Ca^{2+}.[42] Since axonal fluxes of Ca^{2+} are in part regulated by Na^+ it is usually assumed that Ca^{2+} is responsible for the inhibited AXT. The increased Ca^{2+} level is know to affect MT and activate intra-axonal proteases which in turn could be the cause of the inhibited AXT.[43,44] If an increase in Ca^{2+} is responsible for the Na^+ effects it could be expected that increased extracellular Ca^{2+} levels should potentiate the effects of Na^+-deprivation or conditions which increase Na^+ influx. However, increased extra-cellular levels of Ca^{2+} does not potentiate the effect of NaCl-free Ringer in frog sciatic nerve at the axonal level and in the cat sciatic nerve Ca^{2+} acts antagonistic to BTX.[8,12]

III. K^+

K^+ deprivation inhibits AXT in the cat hypogastric nerve but fails to affect transport in the sheathed sciatic nerve of the frog and the cat.[1,6,9,10] In analogy with Na^+ a differential effect of K^+ at the ganglionic and axonal level can therefore not be excluded. In the desheathed cat sciatic nerve AXT can be maintained in the absence of K^+. However, Chan et al. found that the addition of a small amount of K^+ (5 m*M*) reduced the demand of extracellular Ca^{2+} for the maintenance of AXT.[1] Furthermore with moderate concentrations of K^+ (40 to 100 m*M*) extracellular Ca^{2+} was no longer necessary for AXT. Thus, K^+ acts in cooperation with Ca^{2+}. Very high concentrations of K^+ impairs AXT in the frog and cat sciatic nerve but AXT is not affected by K^+ concentrations sufficient to depolarize these nerves.[1,6,45]

IV. Cl^-

The effects of anions on AXT has not been studied in detail. Lavoie found that choline chloride could substitute for NaCl at the axonal level in the frog sciatic nerve whereas neither sucrose nor Na^+ isothionate could replace NaCl.[11,14]

At the axonal level Cl^- therefore seemed to be more important than Na^+. In contrast choline chloride could not replace NaCl at the somal level.

The mechanism of the Cl^--action is obscure Lavoie assumed that Cl^- acted via an effect on the sequestration of Ca^{2+}.[14] The observation that anions like Cl^- affect the macroviscosity of axoplasm, a phenomenon which is likely to affect the translocation of substances in the axon is in this context interesting.[46]

V. Ca^{2+}

A. Ca^{2+}-Deprivation

Ca^{2+} is transported axonally.[47] Ca^{2+} has been suggested to play an important role in several models of AXT and for the integrity and dynamics of various components of the AXT-machinery (MT and the microtrabecular system). A role of Ca^{2+} in the initiation of AXT which is distinct from that of the Ca^{2+} requirement in the axon has been suggested by Hammerschlag et al.[29-31] Ca^{2+}-deprivation fails to affect AXT in the sheathed frog and cat sciatic nerve unless the ganglia are deprived of Ca^{2+}.[1-4] In contrast Ca^{2+}-free medium blocks AXT in the sheathed cat hypogastric nerve (which in general appear to be sensitive to ionic changes) and the desheathed frog and cat sciatic nerve.[1,4,48,49]

B. Ca^{2+} Substitutes

We have tested the ability of various divalent cations (Mg^{2+}, Sr^{2+}, Ba^{2+}, Co^{2+}, and Mn^{2+}) to restore AXT in Ca^{2+}-deprived, chemically desheathed frog sciatic nerves.[4] None of the ions were able to restore AXT. The inhibition of AXT by Ca^{2+} deprivation in the desheathed cat sciatic nerve is delayed in the presence of Mg^{2+}. However neither Mg^{2+} nor Sr^{2+} could replace Ca^{2+} in this preparation.[1] In contrast, Lavoie et al. found that both Mg^{2+} and Sr^{2+} could replace Ca^{2+} in the partially desheathed frog sciatic nerve.[49] At the somal level of frog sciatic nerve Sr^{2+} but neither Mg^{2+} nor Co^{2+} or Mn^{2+} could substitute for Ca^{2+}. The somal Ca^{2+} requirement therefore appear distinct from that of the axonal site in this preparation.[50]

C. Ca^{2+} Antagonists

In frog sciatic nerve both organic (verapamil) and inorganic Ca^{2+} antagonists (Mn^{2+}, Co^{2+}, La^{3+}, ruthenium red) block AXT.[4,31,51-53,67] Since some of the inhibitors only interferes with the passage of Ca^{2+} through the cell membrane the results suggest an extracellular need of Ca^{2+} for AXT analogous to that of muscle contraction.

Co^{2+} ions, which enter the nerve cell, have been used to study the Ca^{2+} sensitive step in the soma of frog sciatic nerve.[29,30,32,54] The somal Co^{2+} sensitivity was found to be different from that of the axonal one. Furthermore, morphological studies of the Golgi apparatus and the smooth endoplasmic reticulum following Co^{2+} treatment and a differential effect of Co^{2+} on AXT of glycosylated proteins suggested that Ca^{2+} was needed at several sites in the soma.

D. Calmodulin

Calmodulin is axonally transported and has been shown to activate a Ca^{2+}-Mg^{2+}-ATPase in mammalian nerve at relevant axonal Ca^{2+} levels.[18,55-57] Agents which bind to and inactivate calmodulin like phenothiazines also inhibits AXT.[52,58,59] This inhibition occurs without appreciable decrease of the neural content of high energy phosphate compounds and therefore imply that calmodulin and hence also Ca^{2+} play a role in AXT.[59]

E. Increased Axonal Ca^{2+} Content

High levels of Ca^{2+} inhibits transport in most preparation.[1,3,9] In sheathed preparations Ca^{2+} concentrations in the range of 0.1 *M* is necessary, in desheathed preparations or at the ganglionic level much lower concentrations may be used (10 to 20 m*M*). AXT is also inhibited by other conditions which increase intra-axonal Ca^{2+} levels, i.e., the ionophores X537 A and A23187 which are very potent AXT-inhibitors.[48,60]

F. Mechanism of Ca^{2+} Action

A decreased axonal Ca^{2+} content has been suggested to account for the inhibited AXT after incubations of nerves in Ca^{2+}-deprived medium. Ca^{2+} is then no longer available for calmodulin, contractile proteins, or other postulated Ca^{2+} requiring steps and AXT is inhibited. However, recent experiments indicate that organelle movement in electrically punctured axons and in the extruded squid giant axoplasm occur in the apparent absence of Ca^{2+}.[61] If so AXT is in this respect distinct from other types of intracellular movements and muscle contraction. These experiments do not rule out the possibility that Ca^{2+} has a regulatory role in AXT.

In the soma, the experimental evidence indicates that Ca^{2+} is needed at several sites. By analogy of the somal events of AXT and secretion, two of the Ca^{2+} sensitive steps were suggested to be the fusion of newly synthesized proteins with the Golgi apparatus and then delivery of vesicles to the AXT-system respectively.[50]

The cause of inhibited AXT at elevated axonal Ca^{2+} levels is not known. A Ca^{2+} induced degradation of MT is one possibility for the inhibited AXT. Other suggestions include the activation of proteases which degrades the AXT-machinery.[52] Extensive proteolysis can be observed in the nerves exposed to Ca^{2+} concentrations sufficient to block AXT.[44] Ca^{2+} induced activation of intra-axonal proteases can be blocked by protease inhibitors like TLCK. If activation of proteases is responsible for the AXT-inhibition one could assume that TLCK would counteract the effects of (for instance) Ca^{2+}-ionophores. We have recently tested this possibility and have found that the inhibitory action of the Ca^{2+}-ionophore X-537 A is potentiated by TLCK.[62] Therefore a Ca^{2+} mediated effects on MT and/or energy metabolism seem more likely mechanisms for the inhibitory effects of high Ca^{2+}.

VI. Mg^{2+}

It is reasonable to assume that Mg^{2+} acts as a substrate together with ATP for the ATPase activity involved in AXT. Still, AXT is unaffected in Mg^{2+}-free medium, even after extensive attempts to remove divalent cations.[3] However, more than 50% of the neuronal Mg^{2+} remain after a 17 hr incubation of frog sciatic nerve in Mg^{2+}-free Ringer.[4] It is therefore possible that the applied treatments has not reduced Mg^{2+} to levels incompatible with AXT.

In the desheathed cat sciatic nerve Mg^{2+} has been shown to act synergistically with Ca^{2+}. Although Mg^{2+} cannot replace Ca^{2+} in this preparation, the block of AXT seen after Ca^{2+}-deprivation is delayed and AXT can be maintained at lower extracellular Ca^{2+} concentrations.[1]

VII. Zn^{2+}

Zn^{2+} is present in relatively large amounts in MT containing structures and has been shown to activate a number of ATP:ases. The relation of Zn^{2+}-MT and AXT was studied by Edström and Mattsson.[63] They found that Zn^{2+} at low concentrations ($10^{-6}M$) stimulated AXT, whereas it was inhibited by higher concentrations. This biphasic response could also be observed in the presence of other sulfydryl reagents at low concentrations.[64] The Zn^{2+} induced stimulation of AXT was paralled by a stimulation of a Ca^{2+}-Mg^{2+} ATPase recovered from the microsomal fraction of the same nerve.[65] Zn^{2+} has also been shown to stimulate the ATPase activity which copurifies with MT. These results indicated that sulfydryl groups could be important for the regulation of AXT and that factors which affect such groups, including Zn^{2+}, may play a role in AXT.

VIII. CONCLUSIONS AND PROSPECTS

Studies of the ionic requirements of AXT, at least when it concerns the direct ionic needs of the AXT-machinery are hampered by several experimental difficulties. In sheathed preparations the perineurium constitutes an effective permeability barrier to ions and even if this barrier can be destroyed, the axolemma still persists as a barrier to- and regulator of ionic fluxes. Pharmacological tools, including various drugs which affect ionic fluxes may be more or less specific. A considerable period of time is usually required to resolve movement of transported materials. The complexity and dependence of AXT on other cellular processes including energy metabolism makes it difficult to rule out indirect effects of drugs and ions. Furthermore the actual level of ions within the axon has so far not been monitored during the treatments used to block AXT. Thus, conclusions regarding the ionic needs of AXT must be drawn with caution. Extracellular Ca^{2+} and possibly Na^{+} and Cl^{-} is necessary for both the initiation and maintenance of AXT. These steps shows a differential sensitivity to Ca^{2+} and Na^{+}. Other ions including Mg^{2+} and K^{+} are not necessary per se but act synergistically or in cooperation with Ca^{2+}.

At the intra-cellular level a careful regulation of Na^{+} and Ca^{2+} seem necessary for the maintenance of AXT. Although experimental evidence are scant, Mg^{2+} is probably involved in the supply of ATP derived energy to the AXT-machinery. Ca^{2+} ions also appear to play multiple roles for the initiation of AXT.

Several ions including Zn^{2+} could be involved in the regulation of the amount and/or the selection of transported materials.

Recently the movement of minute particles within the squid giant axon was demonstrated by a refined optical technique.[66,67] The particles moved preferentially in the orthograde direction with a velocity compatible with rapid AXT. Furthermore the same movements could be observed in extruded axoplasm. Thus, rapid AXT can be studied in a system devoid of permeability barriers. The developed system has already given important results with respect to the role of actin in AXT. It is also conceivable that this system will bring to light the ionic needs of the AXT-machinery.

ACKNOWLEDGMENT

The present work was supported by grants from Naturvetenskapliga forskningsrådet.

REFERENCES

1. **Chan, S. Y., Ochs, S., and Worth, R. M.,** The requirement for calcium ions and the effects of other ions on axoplasmic transport in mammalian nerve, *J. Physiol.*, 301, 477, 1980.
2. **Dravid, A. R. and Hammerschlag, R.,** Axoplasmic transport of proteins in vitro in primary afferent neurons of frog spinal cord:effect of Ca^{2+}-free incubation conditions, *J. Neurochem.*, 24, 711, 1975.
3. **Edström, A.,** Effects of Ca^{2+} and Mg^{2+} on rapid axonal transport of proteins in vitro in frog sciatic nerves, *J. Cell. Biol.*, 61, 812, 1974.
4. **Kanje, M., Edström, E., and Ekström, P.,** Divalent cations and fast axonal transport in chemically desheathed (triton X-treated) frog sciatic nerve, *Brain Res.*, 241, 67, 1982.
5. **Ochs, S., Worth, R. M., and Chan, S-Y.,** Calcium requirement for axoplasmic transport in mammalian nerve, *Nature (London)*, 270, 748, 1977.
6. **Edström, A.,** Ionic requirements for rapid axonal transport in vitro in frog sciatic nerves, *Acta Physiol. Scand.*, 93, 104, 1975.
7. **Wolburg, H.,** Evidence for the involvement of the Na^+-K^+-ATP:ase in the mechanism of axonal protein and nucleoside transport, *Z. Naturforsh.*, 31, 683, 1976.
8. **Worth, R. M. and Ochs, S.,** Dependence of batrachotoxin block of axoplasmic transport on sodium, *J. Neurobiol.*, 13, 537, 1982.
9. **Garcia, A. G., Kirpekar, S. M., Prat, J. C., and Wakade, A. R.,** Metabolic and ionic requirements for the axoplasmic transport of dopamine beta-hydroxylase, *J. Physiol.*, 241, 809, 1974.
10. **Kirpekar S. M., Prat, J. C., and Wakade, A. R.,** Metabolic and ionic requirements for the intra-axonal transport of noradrenalin in the cat hypogastric nerve, *J. Physiol.*, 228, 173, 1973.
11. **Lavoie, P. A.,** Importance of monovalent ions for the fast axonal transport of proteins, *Can. J. Physiol. Pharmacol.*, 59, 31, 1981.
12. **Lavoie, P. A.,** On the mechanism of the inhibition of fast axonal transport caused by exposure of frog spinal ganglia to a sodium-deficient medium, *Neurosci. Lett.*, 27, 131, 1981.
13. **Lavoie, P. A.,** Ionic requirements for in vitro retrograde axonal transport of acetylcholinesterase, *Neurosci. Lett.*, 33, 213, 1982.
14. **Lavoie, P. A.,** Effects of sodium- or chloride-deficient medium on fast axonal transport in frog spinal nerves, *Can. J. Physiol. Pharmacol.*, 60, 95, 1982.
15. **Bisby, M. A.,** Axonal transport, *Gen. Pharmacol.*, 7, 387, 1976.
16. **Droz, B. and Rambourgh, A.,** Axonal smooth endoplasmic reticulum and fast orthograde transport of membrane constituents, in *Axoplasmic Transport*, Weiss, D. G., Ed., Springer-Verlag, Basel, 1982.
17. **Grafstein, B. and Forman, D. S.,** Intracellular transport in neurons, *Physiol. Rev.*, 60, 1167, 1980.
18. **Iqbal, Z., Garg, B. P., and Ochs, S.,** Calmodulin activation of brain and nerve Ca^{2+}-Mg^{2+}-ATP:ase, *Soc. Neurosci. Abstr.*, 5, 60, 1979.
19. **Kerkut, G. A.,** Axoplasmic transport, *Comp. Biochem. Physiol.*, 51, 701, 1975.
20. **Lasek, R. J.,** Axonal transport: a dynamic view of neronal structure, *TINS*, 3, 87, 1980.
21. **Ochs, S.,** Characteristics and a model for fast axoplasmic transport in nerve, *J. Neurobiol.*, 2, 331, 1971.
22. **Ochs, S.,** Fast transport of materials in mammalian nerve fibers, *Science*, 176, 252, 1972.
23. **Ochs, S.,** On the mechanism of axoplasmic transport, in *Axoplasmic Transport*, Weiss, D. G., Ed., Springer-Verlag, Basel, 1982.
24. **Rambourg, A. and Droz, B.,** Smooth endoplasmic reticulum and axonal transport, *J. Neurochem.*, 35, 16, 1980.
25. **Schwartz, J. H.,** Axonal transport: components, mechanism, and specificity, *Ann. Rev. Neurosci.*, 2, 467, 1979.
26. **Weiss, D. G. and Gross, G. W.,** Intracellular transport in nerve cell processes: the chromatographic dynamics of axoplasmic transport, in, *Biological Structures and Coupled Flows*, Oplata, A. and Balaban, M., Eds., Academic Press, New York, 1980.
27. **Wilson, D. L. and Stone, G. C.,** Axoplasmic transport of proteins, *Ann. Rev. Biophys. Bioeng.*, 8, 27, 1979.
28. **Ochs, S. and Worth, R.,** Batrachotoxin block of fast axoplasmic transport in mammalian nerve fibers, *Science*, 21, 1087, 1975.

29. **Hammerschlag, R., Bakhit, C., Chiu, A. Y., and Dravid, A. R.,** Role of calcium in the initiation of fast axonal transport of proteins: effects of divalent cations, *J. Neurobiol.*, 8, 439, 1977.
30. **Hammerschlag, R. and Lavoie, P. -A.,** Initiation of fast axonal transport: involvement of calcium during transfer of proteins from golgi apparatus to the transport system, *Neuroscience*, 4, 1195, 1979.
31. **Hammerschlag, R.,** The role of calcium in the initiation of fast axonal transport, *Fed. Proc. Fed. Am. Soc. Exp. Biol.*, 39, 2809, 1980.
32. **Lindsey, J. D., Hammerschlag, R., and Ellisman, M. H.,** An increase in smooth endoplasmic reticulum and a decrease in Golgi apparatus occur with ionic conditions that block initiation of fast axonal transport, *Brain Res.*, 205, 275, 1981.
33. **Albuquerque, E. X., Daly, J. W., and Witkop, B.,** Batrachotoxin: chemistry and pharmacology, *Science*, 172, 995, 1971.
34. **Boegman, R. J. and Albuquerque, E. X.,** Axonal transport in rats rendered paraplegic following a single subararachnoid injection of either batrachotoxin or 6-aminonicotinamide into the spinal cord, *J. Neurobiol.*, 11, 283, 1980.
35. **Boegman, R. J., Deshpande, S. S., and Albuquerque, A. X.,** Consequences of axonal transport blockade induced by batrachotoxin on mammalian neuromuscular junction. I. Early pre- and postsynaptic changes, *Brain Res.*, 187, 183, 1980.
36. **Kumahara-Siri, M. H.,** Batrachotoxin inhibits axonal transport without affecting membrane potential in single neurons of Aplysia Californica, *J. Neurobiol.*, 10, 509, 1979.
37. **Hammerschlag, R., Stone, G. C., Bolen, F. A., Lindsey, J. D., and Ellisman, M. H.,** Evidence that all newly synthesized proteins destined for fast axonal transport pass through the golgi apparatus, *J. Cell. Biol.*, 93, 568, 1982.
38. **Ignelzi, R. J. and Nyquist, J. K.,** Observations on fast axoplasmic transport in peripheral nerve following repetitive electrical stimulation, *Pain*, 7, 313, 1979.
39. **Alvarez., J., Arredondo, F., Espejo, F., and Williams, V.,** Regulation of axonal microtubules: effect of sympathetic hyperactivity elicited by reserpine, *Neuroscience*, 7, 2551, 1982.
40. **Grafstein, B., Edwards, D. L., and Alpert, R. M.,** Axonal transport and neuronal activity, in *International Cell Biology 1980-1981*, Schweiger, H. G., Ed., Springer-Verlag, Basel, 1982.
41. **Reis, D. J., Jon, T. H., and Rosse, R. A.,** Effects of reserpine on activities and amounts of tyrosine hydroxylase and dopamine-β-hydroxylase in catecholamine neuronal systems in rat brain, *J. Pharmacol. Exp. Ther.*, 193, 775, 1975.
42. **Stout, M. A. and Diecke, F. P. J.,** Monovalent cations and the sodium-sensitive calcium transport in frog sciatic nerve, *Mol. Physiol.*, 2, 213, 1982.
43. **Pant, C. H., Terakawa, S., and Gainer, H.,** A calcium activated protease in squid axoplasm, *J. Neurochem.*, 32, 99, 1979.
44. **Schlaepfer, W. W.,** Calcium induced degradation of axoplasm in isolated segments of rat peripheral nerve, *Brain Res.*, 69, 203, 1974.
45. **Partlow, L. M., Rosse, C. D., Motwani, R., and McDougal, D. B.,** Transport of axonal enzymes in surviving segments of frog sciatic nerve, *J. Gen. Physiol.*, 60, 388, 1972.
46. **Rubinsson, K. A. and Baker, P. F.,** The flow properties of axoplasm in a defined chemical environment: influence of anions and calcium., *Proc. R. Soc. London*, 205, 323, 1979.
47. **Neale, J. H. and Barker, J. L.,** Bidirectional axonal transport of $45Ca^{2+}$: studies in isolated frog sensory, motor and sympathetic neurons, Aplysia cerebral ganglion and the goldfish visual system, *Brain Res.*, 129, 45, 1977.
48. **Esqurro, E., Garcia, A. G., and Sanchez-Garcia, P.,** The effects of the calcium ionophore, A 23187, on the axoplasmic transport of dopamine beta-hydroxylase, *Br. J. Pharmacol.*, 70, 375, 1980.
49. **Lavoie, P-A., Bolen, F., and Hammerschlag, R.,** Divalent cation specificity of the calcium requirement for fast transport of proteins in axons of desheathed nerves, *J. Neurochem.*, 32, 1745, 1979.
50. **Hammerschlag, R.,** Multiple roles of calcium in the initiation of fast axonal transport, in *Axoplasmic Transport*, Weiss, D. G., Ed., Springer-Verlag, Basel, 1982.
51. **Deinum, J., Wallin, M., Kanje, M., and Lagercrantz, C.,** The effects of ruthenium red on the assembly and disassembly of microtubules and on rapid axonal transport, *Biochim. Biophys. Acta*, 675, 209, 1981.
52. **Kanje, M., Edström, A., and Ekström, P.,** The role of Ca^{2+} in rapid axonal trasnport, in *Axoplasmic Transport*, Weiss, D. G., Ed., Springer-Verlag, Basel, 1982.
53. **Lavoie, P. A., Hammerschlag, R., and Tjan, A.,** Cobalt ions inhibit fast axonal transport of 3H glycoproteins but not glycosylation, *Brain Res.*, 149, 535, 1978.
54. **Stone, G. C. and Hammerschlag, R.,** Differential effects of cobalt on the initiation of fast axonal transport, *Cell. Mol. Neurobiol.*, 1, 3, 1981.
55. **Erickson, P. F., Seamon, K. B., Moore, B. W., Lasher, R. S., and Minier, L. N.,** Axonal transport of the Ca^{2+}-dependent protein modulator of 3:5-cyclic AMP phosphodiesterase in the rabbit visual system, *J. Neurochem.*, 35, 242, 1980.

56. **Iqbal, Z. and Ochs, S.,** Fast axoplasmic transport of a calcium-binding protein in mammalian nerve, *J. Neurochem.*, 31, 409, 1978.
57. **Iqbal, Z. and Ochs, S.,** Characterization of mammalian nerve calmodulin and its possible relationship to axoplasmic transport, in, *Axplasmic Transport,* Weiss, D. G., Ed., Springer-Verlag, Basel, 1982.
58. **Edström, A., Hansson, H-A., and Norström, A.,** Inhibition of axonal transport in vitro in frog sciatic nerves by chlorpromazine and lidocaine, *Z. Zellforsch.*, 143, 53, 1973.
59. **Ekström, P., Kanje, M., and Edström, A.,** Effects of phenothiazines and dibenzazepines on axonal transport and microtubule assembly in vitro, *Acta Physiol. Scand.*, 116, 121, 1982.
60. **Kanje, M., Edström, E., and Hanson, M.,** Inhibition of rapid axonal transport in vitro by the ionophores X-537 A and A 23187, *Brain Res.*, 204, 43, 1981.
61. **Adams, R. J.,** Organelle movement in axons depends on ATP, *Nature (London)*, 297, 327, 1982.
62. **Edström, A., Ekström, P., and Kanje, M.,** Is protease activity involved in fast axonal transport, *Acta Physiol. Scand.*, in press.
63. **Edström, A. and Mattsson, H.,** Small amounts of zinc stimulate rapid axonal transport in vitro, *Brain Res.*, 86, 162, 1975.
64. **Edström, A. and Mattsson, H.,** Inhibition and stimulation of rapid axonal trasnport in vitro by sulfhydryl blockers, *Brain Res.*, 108, 381, 1976.
64a. **Edström, A., Hanson, M., Prus, K., and Wallin, M.,** Ca^{2+}- or Mg^{2+}-dependent enzymic ATP hydrolysis associated with the microsomal fraction of frog sciatic nerves, *J. Neurochem.*, 35, 297, 1980.
65. **Allen, R. D., Metuzals, J., Tasaki, I., Brady, S. T., and Gilbert, S. P.,** Fast axonal transport in squid giant axon, *Science*, 218, 1127, 1982.
66. **Brady, S. T., Lasek, R. J., and Allen, R. D.,** Fast axonal transport in extruded axoplasm from squid giant axon, *Science*, 218, 1129, 1982.
67. **Kanje, M.,** unpublished observations.

Chapter 5

CALMODULIN AND ITS ROLE IN AXOPLASMIC TRANSPORT

Zafar Iqbal

TABLE OF CONTENTS

I. INTRODUCTION

Calcium has been shown to play a significant role in the regulation of a number of key cellular physiological processes. It was recently found to be required for the maintenance of fast axoplasmic transport.[1-4] A description for the role of Ca^{2+} and other ions in axoplasmic transport is given in Chapters 4 and 6. The biochemical mechanisms mediating the Ca^{2+} effects in physiological responses including the axoplasmic transport, are not clearly understood. Calcium-binding protein(s) have been suggested to act as intracellular Ca^{2+}-acceptors in the mediation of Ca^{2+} effects.[5] As the number of enzymatic systems which are known to be regulated by Ca^{2+} is large, one has to speculate that the number of such calcium-binding proteins involved with one or more specific functions is large. Alternatively, there could be one or more protein(s), each with multiple functions. A calcium-binding protein of 17,000 daltons was recently found to be able to modulate a number of different biochemical reactions. As this protein which binds Ca^{2+} with high affinity is ubiquitous, highly conserved, and is capable of regulating a large number of different enzymatic reactions and thus, manifesting varying physiological responses, has been suggested to be the intracellular-calcium-receptor.[6] Originally described as an activator protein causing an increase in the activity of brain cyclic nucleotide phosphodiesterase (PDE)[7,8] the protein has been referred to as modulator protein, protein activator factor, activator protein, and calcium-dependent regulator protein (CDR) before acquiring its current name calmodulin (CaM). A number of excellent reviews dealing with the discovery and its involvement in cellular processes has appeared recently.[6,9-20] This chapter describes, briefly, the principal characteristics of the protein and its possible role in the fast-transport of intracellular components in peripheral nerve.

II. CHARACTERISTICS OF CALMODULIN

A. Physico-Chemical

The CaM is a low molecular weight (17,000 M_r) protein which, in the presence of Ca^{2+}, is resistant to denaturation by heating (99°C), urea (8*M*), detergents, high salt concentration, and acidic pH. These properties have been very beneficial in the isolation and purification of the protein from a large number of different sources. It is a highly conserved protein; amino acid compositions and sequences of the protein isolated from the lowest- to the highest-order plants and animals have been found to be very similar, if not identical. In addition to the conservation of the primary amino acid sequence, calmodulins from different sources appear to be identical immunologically.

The CaM has 148 amino acid residues and is characterized by the presence of one trimethyl-lysine,[21] with a few exceptions.[22,23] It also lacks in tryptophan and cystein which allows the use of tyrosine fluorescence to study conformational changes induced by Ca^{2+}. Aspartic and glutamic acids constitutes about 25% of the total amino acids thus giving its low isoelectric point (3.9).

The CaM has four Ca^{2+} binding sites and depending on the procedures and experimental conditions employed, different authors have found different dissociation constants which are in general, in the range of $10^{-6}M$. Binding of Ca^{2+} induces a more helical conformation in the protein molecule,[24] possibly by a change in the microenvironment of the two tyrosine residues. In the absence of free Ca^{2+}, both tyrosine residues 99 and 138 are accessible to acetylation, whereas in the presence of Ca^{2+}, only residue 99 can be acetylated. This change in the accessibility of tyrosine to acetylation could be an important factor in determining the biological activity of the protein.

B. Biological

Calmodulin, the intracellular Ca^{2+} receptor protein plays an important role in the mod-

Table 1
CELLULAR PROCESSES MEDIATED BY CALMODULIN

- I. Enzyme modulation
  - A. Cyclic nucleotide metabolism
    - 1. 3′,5′ cyclic AMP phosphodiesterase
    - 2. Adenylate cyclase
    - 3. Guanylate cyclase
  - B. Protein kinases
    - 1. Phosphorylase kinase
    - 2. Glylogen synthase kinase
    - 3. Protein I kinases
    - 4. Myosin light chain kinase
    - 5. Tyrosine monooxygenase kinase
    - 6. Fodrin kinase
    - 7. Tubulin kinase
    - 8. Microtubule associated protein 2 kinase
  - C. Other enzymes
    - 1. Phsophofructo kinase
    - 2. NAD kinase
    - 3. Ca^{2+}-Mg^{2+}-ATPase
      - a. Microsomal
      - b. Microtubular
    - 4. Protein phosphatase
    - 5. Glucose 1, 6-bisphosphatase
    - 6. Phospholipase A_2
    - 7. Glucocerebrosidase
    - 8. Quinate: NAD^+ 3-oxido reductase
    - 9. Tryptophan hydroxylase
- II. Calcium transport
- III. Exocytosis
  - A. Insulin release
  - B. Neurotransmitter release
  - C. Histamine release
  - D. Somatostatin release
  - E. Renin release
  - F. Amylase release
  - G. Paramecium secretion
- IV. Cell division
- V. Microtubule assembly/disassembly
- VI. DNA synthesis
- VII. Axoplasmic transport?

ulation of a number of key biochemical reactions, which in turn result in physiological responses. A list of some of the cellular processes modulated by the CaM are given in the Table 1. It is beyond the scope of this article to discuss these processes here. They have been reviewed by a number of authors.[6,9-20]

III. REGULATION BY THE CALMODULIN

The exact regulatory process involving CaM has been and continues to be an area of intensive investigation. The protein itself is not active; its active form is the calcium-calmodulin complex. therefore, the presence of calcium is an absolute requirement for its action. A simplified general scheme for calcium-calmodulin function can be summarized as follows:

$$Ca + CaM \rightarrow Ca.CaM^*$$

$$Ca.CaM^* + P \rightarrow Ca.CaM^*.P$$

$$CaCaM^{*}.P \rightarrow \text{Biochemical reactions} \rightarrow \text{Physiological response}$$

Calmodulin binds calcium to form an active Ca.CaM* complex which then binds to the receptor or effector protein (P) resulting in the formation of an active complex capable of triggering biochemical reactions, and subsequently, physiological response. The amount of calcium required for the formation of Ca.CaM* complex is in the micromolar range, slightly higher than the cytoplasmic levels of calcium present under physiological conditions. Any stimulus or signal that causes an elevation in cytoplasmic levels of calcium, either by increased uptake or release of calcium from calcium-storing components such as mitochondria or endoplasmic reticulum, is likely to activate calmodulin to exhibit its modulatory action.

A. Regulation of Calmodulin

An important possibility that seems to be dominant in the cell for the regulation of CaM levels and thereby the regulation of calmodulin-mediated processes is the level of calcium in the cell. Binding of calcium to calcium-binding sites results in conformational change which, as exemplified in the scheme above, is required for CaM function. Recent observations with various calmodulin-mediated processes suggest that some reactions can be activated by CaM with only one or two calcium bound to it (for example, cyclic nucleotide phosphodiesterase) whereas other processes would require all four calcium-binding sites to be occupied (e.g., microtubule disassembly). Thus, it is possible that the fractional occupancy of the calcium-binding sites may dictate the systems regulated by CaM and help to explain how this protein can be involved in such diverse biological processes. Another equally important possibility for the regulation of activity is through the involvment of calmodulin-binding proteins (CaMBP) as described below.

Calmodulin, as described earlier, has been observed to be present in all cells from the lowest- to the highest-order plants and all animal phyla tested. It constitutes between 0.1 and 1% of the total protein in cells, the highest being in brain and testes. However, only a fraction of the above amount is required for the CaM-dependent enzyme activities and most of the CaM in cell may be assigned to CaMBP. A number of such CaMBP have been identified in a variety of cells,[25-37] mostly by employing radioactive-labeled calmodulin[27,28] or by calmodulin-affinity chromatography.[30,31] Except for those enzymes which require binding of CaM for their activities (Table 1), most of the abundant CaMBP, in addition to their undetermined functions are perhaps involved in the regulation of available CaM in the cell.

IV. CALMODULIN IN MAMMALIAN NERVE

A prominent feature of CaM is its lack specificity to any tissue or species. Our results have indicated that the CaM, isolated from cat brain and nerve as well as dog nerve, exhibits identical properties.[38] Therefore, we isolated the protein from siatic nerves of dog. The tissue was homogenized in a Tris buffer containing Ca^{2+}-chelator, ethyleneglycol-*bis*-(β aminoethyl-ether)-*N*,*N*,*N'*,*N'*-tetraacetic acid (EGTA) at pH 7.5 and the soluble proteins obtained after high speed centrifugation were precipitated with $(NH_4)_2SO_4$ to 50% saturation and the contents heated at 90°C for 5 min before centrifugation at 20,000 rpm for 30 min. The supernatant so obtained was fractionated on a Sephadex G-100 column. The position of the calmodulin eluates was determined by the activation of CaM-dependent cyclic nucleotide phosphodiesterase (PDE) as described earlier.[38] The fractions containing CaM were further purified by affinity column chromatography.

A. Affinity Chromatography of Calmodulin

Calmodulin binds effectively with phenothiazines in the presence of Ca^{2+}.[39] Taking ad-

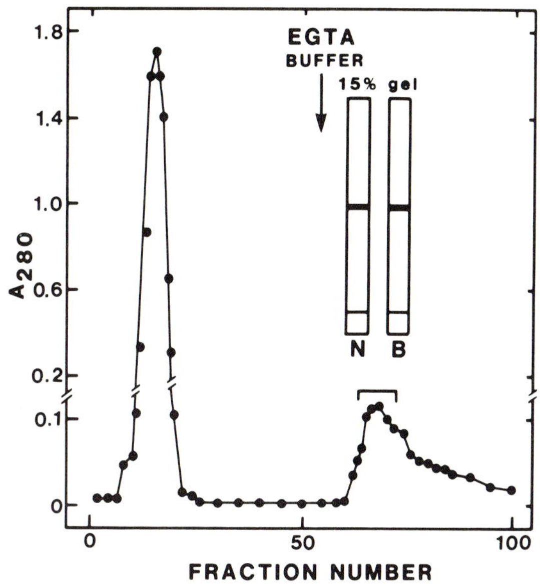

FIGURE 1. Affinity column chromatography of calmodulin. A partially purified sample of CaM was chromatographed on CAP-Sepharose column as described in the text. In the presence of Ca^{2+}, CaM binds to the column. The protein was eluted from the column by replacing Ca^{2+} with EGTA. The activity of CaM in the eluate was monitored by the activation of CaM-PDE 38. Inset shows polyacrylamide gel electrophoresis of nerve (N) and brain (B) CaM preparations on 15% gels.

vantage of this property, a CN-Br activated Sepharose 4B column was tagged with 2-chloro-10 (3-aminopropyl) phenothiazine hydrochloride (CAP) as described by Jamieson and Vananman.[40] The CAP-sepharose column was washed with a 0.20 *M* MES [(2(*N*-morpholino)-ethane sulfonic acid] buffer at pH 7.0 containing 1 m*M* dithiothreitol (DDT), 300 m*M* NaCl and 1 m*M* $CaCl_2$ (buffer A). Partially purified CaM obtained after Sephadex G-100 filtration was loaded onto the column and the column eluted with buffer A. In the presence of Ca^{2+}, CaM binds to the CAP-sepharose column, while most of the other proteins are eluted (Figure 1). Calmodulin was then eluted from the column by replacing Ca^{2+} with 10 m*M* EGTA in buffer A. The detection of CaM in the eluates was performed by the activation of PDE and the purity of calmodulin checked after its electrophoresis on polyacrylamide gels (inset in Figure 1).

B. Characterization of Mammalian Nerve Calmodulin

Purified calmodulin showed a single band on polyacrylamide gels of 10, 12.5, and 15% concentrations, and its moledular weight under denaturation conditions on polyacrylamide gels (12.5%) was found to be similar to that which we had earlier reported for cat nerve CaBP and partially purified calmodulin from cat brain.[41,42] The protein was highly acidic in nature with pI of 3.9 as determined by isoelectrofocusing on polyacrylamide gels. An analysis of the amino acid composition revealed that about 25% of total amino acids were comprised of aspartic and glutamic acids. The protein which we have isolated from nerve does not show any similarity to the low molecular weight component of the modulator-binding protein, also known as calcineurin.[34]

As indicated earlier, CaM binds up to four Ca^{2+} molecules per molecule of protein and the configuration of the protein changes when Ca^{2+} is bound to it. When analyzed electrophoretically on analytical polyacrylamide gels in the presence of denaturing agents, the mobility of CaM has been observed to be different in the presence and in the absence of Ca^{2+}. Mammalian nerve CaM showed a similar mobility characteristics on polyacrylamide gels; in the presence of Ca^{2+} chelator EGTA, CaM showed an apparent molecular weight of 21 to 22 K whereas in the presence of Ca^{2+}, the mobility of nerve CaM was identical to that of other CaM.[1,42]

V. CALMODULIN, MICROTUBULES, AND Ca^{2+}-Mg^{2+}-ATPase

The concentration of tubulin, the structural unit of microtubules is very high in neural tissue. A number of neuronal processes, including the axoplasmic transport have been demonstrated to require a proper maintenance of microtubular integrity. Exposure of nerve fibers, in vitro, to high Ca^{2+} (50 to 100 m*M*) causes a block of axoplasmic transport and disassembly of microtubules.[2,43] In the light of recent finding it seems likely that a Ca^{2+} concentration in the low micromolar range is sufficient for disassembly of microtubules. In the presence of CaM, microtubules have been shown to disassemble at 10 μM Ca^{2+}. While this interaction in vitro occurs at calmodulin:tubulin in ratios much higher than the ratios of calmodulin:enzyme necessary to activate CaM-dependent enzymes, the ratios correlate well with the distribution pattern of CaM in the cell. CaM has been found to be concentrated more in those areas where depolymerization might be expected to occur.[45] A more direct evidence pertaining to the interaction of CaM and microtubules in the living cell has now been obtained by injecting micro quantities of CaM into living fibroblastic cells at levels roughly stoichiometric with cellular tubulins.[46] In these experiments microtubules were depolymerized only when Ca^{2+}-CaM complex was injected and the depolymerization was localized to the area of injection, injections of Ca^{2+} or CaM alone were ineffective. These observations strongly suggest that at physiological concentrations of Ca^{2+}, CaM participates actively in the maintenance of assembly/disassembly equilibrium in the tissue. One possible way of CaM participation in microtubule polymerization-depolymerization is likely through the convalent modulation of microtubular proteins, tubulin, and microtubule-associated proteins (MAPs). The enzymes responsible for phosphorylation of these proteins have been shown to be CaM-dependent.[47,48] Additionally, MAP-2 and tau proteins also possess a capability to bind CaM in the presence of Ca^{2+}.[31] Whether this binding contributes to the regulation of polymerization-depolymerization is yet to be explored.

Vinblastine (VBL), an antimicrotubular agent which was earlier found to bind tubulin, after being taken up by the desheathed nerve fibers[49] and subsequently block the transport of labeled proteins,[50] has now been shown to bind CaM. In the presence of Ca^{2+}, VBL was found to inhibit specifically CaM-dependent fraction of Ca^{2+}-Mg^{2+}-ATPase. Basel Ca^{2+}-Mg^{2+}-ATPase and Na^{+}-K^{+}-ATPase were not affected.[51] Furthermore the inhibition of Ca^{2+}-Mg^{2+}-ATPase was parallel to the inhibition in Ca^{2+} transport. A block in Ca^{2+} pump would result in the elevation of intracellular Ca^{2+} which indirectly could effect the microtubular polymerization-depolymerization equilibrium.

Recently a number of laboratories have demonstrated the presence of a CaM-Ca^{2+}-Mg^{2+}-ATPase in the neuronal tissue, mainly in the synaptsomal fractions.[53-57] A CaM-dependent Ca^{2+}-Mg^{2+}-ATPase was isolated from mammalian nerve microsomal fraction after treatment of the microsomes with EGTA and Triton X-100 and fractionation of the solubulized enzyme by gel filtration on Sephedex G-100.[55] The concentration of Ca^{2+} required for optimum activity of the enzyme was in μM range and the activity was inhibited by calmodulin antagonists trifluoperazine (TFP).[1] Calmodulin-independent ATPase required Ca^{2+} in the range of 3 to 5 m*M* and was not affected by the addition of CaM-antagonists.

A CaM-dependent Ca^{2+}-Mg^{2+}-ATPase activity was also demonstrated in microtubules prepared by the warming-cooling procedure.[58] In the presence of CaM, the enzyme was optimally active at Ca^{2+} concentrations range of 10^{-8} to $10^{-5}M$, a level at which free Ca^{2+} is likely to be present in the axon. At this low Ca^{2+} concentration, the enzyme activity was blocked by TFP, while the basal enzyme activity observed at 3 mM Ca^{2+} was uneffected by TFP. The enzyme, however, does not appear to be an integral part of microtubular preparation; the activity decreases with successive polymerization-depolymerization steps. Similar results have also been obtained by other investigators.[59,60] By histochemical methods, most of the enzyme activity associated with microtubules has been found to be membrane-associated localized primarily in vesicles and membrane fragments present in microtubules as well as MAPs preparation.

VI. CaM-ANTAGONISTS AND AXOPLASMIC TRANSPORT

The concentration of CaM in nervous tissue is very high and the protein has been found to modulate a large number of key biochemical enzymes and physiological responses in the nervous system (Table 1). A simple method to varify the involvement of CaM in any such given system has been to block the activity of CaM with CaM-antagonists. If the system is dependent on CaM, addition of excess CaM or removal of the inhibitor should reverse the block. Employing this criterion phenothiazine antipsychotic agents, especially TFP was found to be a potent and specific inhbitor of CaM-dependent reactions.[39]

Using a desheathed preparation of cat sciatic nerve,[2] incorporation of TFP at a concentration of about 3×10^{-3} M in the incubation medium was found to block axoplasmic transport.[61] Uptake studies with ^{3}H-labeled TFP have, however, indicated that less than 1% of TFP is taken up by the desheathed nerve preparation.[62] Therefore, the effective concentration of TFP within the nerve fibers is at the micromolar level. As this concentration, TFP is not likely to cause significant damage in the cell. Similar effects of TFP on fast transport have recently been obtained in nerves permeabilized with low concentrations of dete gents.[63] Although phenothiazine derivative, TFP is the most commonly used anti-calmodu.in compound, it is not the most potent inhibitor of the protein. Nephthalene sulfonamide W7 (*N*-(6 amino hexyl)-5-chloro-1-naphthalenesulfonamide),[64] and compound R 24571 (calmidazolium),[65] and compound 48/80[66] have been shown to be more specific and potent calmodulin-antagonists. None of these inhibitors have been tried for their effect on axoplasmic transport in nerve fibers. However, the activity of CaM-dependent Ca^{2+}-Mg^{2+}-ATPase isolated from cat sciatic nerve was observed to be inhibited at 0.1 μM concentration of calmidazolium.[68] Recently the retrograde transport of acetylcholinesterase has been shown to be blocked by two other calmodulin-antagonists dibenzazepine and amitriptyline.[63,67]

VII. AXOPLASMIC TRANSPORT OF CALMODULIN

Employing a radiolabeled tracer technique, we had found that Ca^{2+} was transported in mammalian nerve at a fast rate in association with a calcium-binding protein (CaBP) of 15,000 M_r.[69] An analysis of the fast-transported component containing CaBP later suggested the presence of CaM in the fraction. The electrophoretic mobility of CaBP was similar to that observed for CaM and additionally CaBP could activate CaM-dependent PDE.[70] Our later studies using a combination of tracer technique coupled with ligation revealed that the amount of CaM transport in the fast-transported components was small but significant.[62,68] In these studies ^{3}H-lencine or ^{35}S-methionine was injected into the dorsal root ganglion of cat and 2 hr after the injection a postganglionic ligation was placed to stop the flow of slowly transported components. A second ligation was placed in the sciatic nerve 7 to 8 cm distal to the ganglion and the accumulation of fast-transported labeled proteins allowed for further 3 to 5 hr. The portion of the nerve containing crest of the fast-transported components was subjected to heat treatment in the presence of EGTA and high salt concentration. The extracted proteins were then fractionated on CAP-affinity columns, as described earlier. In the presence of Ca^{2+}, most of the radioactivity was eluted out, whereas the radioactivity associated with CaM was eluted with EGTA buffer. The presence of CaM in the eluate was confirmed by the activation of CaM-PDE and by its electrophoresis on polyacrylamide gels.[42] The amount of CaM transported at fast rate in mammalian nerve was found to be very small. Fast-transport of a small fraction of CaM was also suggested by the results obtained by Erickson et al. in rabbit visual system.[71] Our results however, are at variance with the observations made in the optic nerve of guinea pig where in the investigators have shown that CaM was transported solely as part of the SCb complex of proteins which are transported at a slow rate of about 2 mm/day.[72] Recently with the help of radioimmunoassay, CaM was found to be transported at an average rate of 5.5 mm/day. The authors have reported that the rate of CaM accumulation was not linear and in the first 24 hr the rate of orthograde transport was in the range of 10 to 18 mm/day,[73] a rate which is 5 to 10 times greater than that observed in guinea pigs. As there were no observations made for short time periods, (<24 hr) in the previous study,[73] the rate of fast wavefront of transported CaM cannot be determined.

VIII. ROLE OF CALMODULIN IN AXOPLASMIC TRANSPORT

Calmodulin, by virtue of its multifunctional properties mediates a number of key biochemical reactions (Table 1) some of which are of significance to the process of axoplasmic transport. A block in axoplasmic transport both in orthograde and retrograde, observed in the nerves exposed to CaM-antagonists[61-63] is most likely through the inhibition of CaM-dependent processes. The effective intracellular concentration of the antagonist (TFP) in exposed nerves is in the micromolar range.[62] At this concentration TFP binding to CaM is highly specific.

There are two major CaM associated reactions which are of prime importance to the axoplasmic transport. First is the regulation of Ca^{2+}-Mg^{2+}-ATPase — an enzyme considered to be involved in the liberation of energy needed for the continuation of the transport process.[43] The enzyme has been shown to be associated, at least functionally, with microtubules.[1,58] A small amount of CaM observed to be fast-transported could serve as the regulator of the enzyme activity.[62] Secondly, the maintenance of microtubular integrity is also very important for the transport process.[43] CaM has been demonstrated to play an important role in the assembly/disassembly of microtubules at the intracellular concentrations (μM) of Ca^{2+}.[44-46] Calmodulin, which is a calcium-binding protein and present in high amounts in neuronal tissue, could also be considered to act as a regulator of Ca^{2+} concentrations in the cell in collaboration with Ca^{2+} storage organelles like the mitochondria and ER. On the other hand, it is equally important to consider that it is the Ca^{2+} concentration in the cell which is more likely to control the activity of CaM and the observed effects of Ca^{2+} on the axoplasmic transport are mediated through CaM. An additional control over CaM activity in the cell is exercised through the interaction of CaM with its binding proteins.[25-37] The amount of some of these proteins is fairly high in neuronal tissue[37] and a few of them are known to inhibit CaM-dependent reactions.[33,34] Whether these CaMBP are involved in axoplasmic transport through the regulation of CaM activity, is not known. With the help of better techniques and more targeted efforts, it is hoped that we shall be able to gain more precise evidence concerning the role of CaM in the axoplasmic transport.

ACKNOWLEDGMENTS

The work reported here was conducted at the Physiology Department of the Indiana University School of Medicine, Indianapolis and the author is indebted to Dr. S. Ochs for his help and guidance. I also thank Miss Cle-Aster M. Combest, and Ms. Delilah Ideozu for typing the manuscript. Supported by grants from the NIH PHS ROI NS 8706-13 and in part by the American Diabetes Association, Indiana Affiliate.

REFERENCES

1. **Ochs, S. and Iqbal, Z.,** The role of calcium in axoplasmic transport, in *Calcium and Cell Function,* Cheung, W. Y., Ed., Academic Press, New York, 1982, 325.
2. **Chan, S. Y., Ochs, S., and Worth, R. M.,** The requirement for calcium ions and the effect of other ions on axoplasmic transport in mammalian nerve, *J. Physiol.,* 301, 477, 1980.
3. **Hammerschlag, R., David, A. R., and Chiu, A. Y.,** Mechanism of axonal transport. A proposed role for calcium ions, *Science,* 188, 273, 1975.
4. **Kanje, M., Edstrom, A., and Ekstrom, P.,** The role of Ca^{2+} in rapid axonal transport, in *Axoplasmic Transport,* Weiss, D. G., Ed., Springer-Verlag, Basel, 1982, 294.
5. **Kretsinger, R. H.,** Calcium binding proteins, *Annu. Rev. Biochem.,* 45, 239, 1976.
6. **Means, A. R. and Dedman, J. H.,** Calmodulin — an intracellular calcium receptor, *Nature (London),* 285, 73, 1980.
7. **Cheung, W. Y.,** Cyclic nucleotide phosphodiesterase-demonstration of an activator, *Biochem. Biophys. Res. Commun.,* 38, 533, 1970.
8. **Kakiuchi, S. and Yamazaki, R.,** Calcium dependent phosphodiesterase activity and its activating factor (PAF) from brain: studies on cyclic 3′,5′-nucleotide phosphodiesterase, *Biochem. Biophys. Res. Commun.,* 41, 1104, 1970.
9. **Wang, J. H. and Waisman, D. M.,** Calmodulin and its role in the second messenger system, *Curr. Top. Cell Regul.,* 15, 47, 1979.
10. **Woff, D. J. and Brostrom, C. O.,** Properties and functions of the calcium-dependent regulator protein, *Adv. Cyclic Nucleotide Res.,* 11, 27, 1979.
11. **Means, A. R. and Dedman, J. R.,** Calmodulin in endocrine cells and its multiple roles in hormone action, *Mol. Cell Endocrinol.,* 19, 215, 1980.
12. **Cheung, W. Y.,** Calmodulin plays a pivotal role in cellular regulation, *Science,* 207, 19, 1980.
13. **Klee, C. B., Crouch, T. H., and Richman, P. G., Calmodulin,** *Annu. Rev. Biochem.,* 49, 489, 1980.
14. **Iqbal, Z.,** Calmodulin and axoplasmic transport and their possible involvement in peripheral neuropathy, *Prog. Clin. Biol. Res.,* 39, 369, 1980.
15. **Iqbal, Z.,** Calmodulin and fast axoplasmic transport, *J. Sci. Indust. Res.,* 39, 269, 1980.
16. **Stoclet, J. C.,** Calmodulin. An ubiquitous protein which regulates calcium-dependent cellular functions and calcium movements, *Biochem. Pharmacol.,* 30, 1723, 1981.
17. **Brostrom, C. O. and Wolff, D. J.,** Properties and functions of calmodulin, *Biochem. Pharmacol.,* 30, 1395, 1981.
18. **Watterson, D. M. and Vincenzi, F. F.,** Calmodulin and cell function, *Ann. N.Y. Acad. Sci.,* 1, 1980.
19. **Cheung, W. Y. Ed.,** *Calcium and Cell Function,* Vol. 1, Academic Press, New York, 1980, 1.
20. **Dieter, P.,** Calmodulin and calmodulin-mediated processes in plants, *Plant Cell Environ.,* 7, 371, 1984.
21. **Watterson, D. M., Sharief, F., and Vanaman, T. C.,** The complete amino acid sequence of the Ca^{2+}-dependent modulator protein (calmodulin) of bovine brain, *J. Biol. Chem.,* 255, 962, 1980.
22. **Hergenhahn, H. G., Kegel, G., and Sedlmeier, D.,** Ca^{2+} binding proteins in crayfish abdominal muscle. Evidence for a calmodulin lacking trimethyllysine, *Biochem. Biophys. Acta,* 787, 196, 1984.
23. **Bodnaryk, R. P.,** Isolation and characterization of a calmodulin lacking trimethyllysine from the fat body of the silkworm, *Bombyx Mori., Insect Biochem.,* 14, 11, 1984.
24. **Burger, D., Cox, J. A., Comte, M., and Stein, E. A.,** Sequential conformational changes in calmodulin upon binding of calcium, *Biochemistry,* 23, 1966, 1984.
25. **Wang, J. H., Sharma, R. K., and Tam, S. W.,** *Calmodulin-Binding Proteins in Calcium and Cell function,* Vol. 1, Cheung, W. Y., Ed., Academic Press, New York, 1980, 305.
26. **Wang, J. H. and Desai, R.,** Modular binding protein: bovine brain protein exhibiting the Ca^{2+}-dependent association with the protein modulator of cyclic nucleotide phsophodiesterase, *J. Biol. Chem.,* 252, 4175, 1977.
27. **LaPorte, D. C. and Storm, D. R.,** Detection of CDR protein binding components using ^{125}I label CDR protein, *J. Biol. Chem.,* 253, 3374, 1978.
28. **Flanagan, S. D. and Yost, B.,** Calmodulin-binding proteins: visualization by ^{125}I-calmodulin overlay on blots quenched with Tween 20 or bovine serum albumin and poly-(ethylene oxide), *Anal. Biochem.,* 140, 510, 1984.

29. **Sobue, K., Muramoto, Y., Yamazaki, R., and Kakiuchi, S.,** Distribution in rat tissue of calmodulin-binding protein of particulate nature. Studies with ^{3}H-modulator protein, *FEBS Lett.*, 105, 105, 1979.
30. **Gazzott, P., Gloor, M., and Carafoli, E.,** Calmodulin-binding proteins in rat liver mitochondria, *Biochem. Biophys. Res. Commun.*, 119, 343, 1984.
31. **Wolff, J. and Lee, Y. C.,** Calmodulin binds to both microtubule-associated protein-2 and r-protein, *J. Biol. Chem.*, 259, 1226, 1984.
32. **Richman, P. G. and Klee, C. B.,** Interaction of ^{125}I-labeled Ca-dependent regulator protein with phosphodiesterase and its inhibitory protein, *J. Biol. Chem.*, 253, 6323, 1978.
33. **Sharma, R. K., Desai, R., Waisman, D. M., and Wang, J. H.,** Purification and subunit structure of bovine brain modulator binding protein, *J. Biol. Chem.*, 254, 4276, 1979.
34. **Klee, C. B., Crouch, T. H., and Krinks, M. H.,** Calcineurin: a calcium- and calmodulin binding protein of the nervous system, *Proc. Natl. Acad. Sci. U.S.A.*, 76, 6270, 1979.
35. **Iwasa, F. and Mohri, H.,** Calmodulin-binding proteins in the cytosol of sea urchin eggs, *J. Biochem.*, 94, 575, 1983.
36. **Mayr, G. W. and Heilmeyer, L. M. G., Jr.,** Phosphofructokinase is a calmodulin binding protein, *FEBS Lett.*, 159, 51, 1983.
37. **Kakiuchi, S.,** Calmodulin-binding proteins in brain, *Neurochem. Intern.*, 5, 159, 1983.
38. **Iqbal, Z. and Ochs, S.,** Calmodulin in mammalian nerve, *J. Neurobiol.*, 11, 311, 1980.
39. **Levin, R. M. and Weiss, B.,** Specificity of the binding of trifluoperazine to the calcium dependent activator of PDE and to a series of other, CaBP's, *Biochim. Biophys. Acta*, 540, 197, 1978.
40. **Jamieson, G. A., Jr. and Vanaman, T. C.,** Calcium-dependent affinity chromatography of calmodulin on an immobilized phenothiazine, *Biochem. Biophys. Res. Commun.*, 90, 1048, 1979.
41. **Iqbal, Z. and Ochs, S.,** Characterization of a calcium regulator protein in mammalian nerve, *Proc. Intern. Soc. Neurochem. Symp. Calcium (Jerusalem)*, 7, 391, 1979.
42. **Iqbal, Z. and Ochs, S.,** Characterization of mammalian nerve calmodulin and its possible relationship to axoplasmic transport, in, *Axoplasmic Transport*, Weiss, D. G., Ed., Springer-Verlag, Basel, 287, 1982.
43. **Ochs, S.,** *Axoplasmic Transport and Its Relation to Other Nerve Functions, John Wiley & Sons*, New York, 1982.
44. **Marcum, J. M., Dedman, J. R., Brinkley, B. R., and Means, A. R.,** Control of microtubule assembly by calcium-dependent regulator protein, *Proc. Natl. Acad. Sci. U.S.A.*, 75, 3771, 1978.
45. **Means, A. R., Tash, J. S., and Chafouleus, J. G.,** Physiological implications of the presence, distribution and regulation of calmodulin in endocrine cells, *Physiol. Rev.*, 62, 1, 1982.
46. **Keith, C., DiPaola, M., Maxfield, F. R., and Shelanski, M. L.,** Microinjection of Ca-calmodulin causes a localized depolymerication of microtubules, *J. Cell Biol.*, 97, 1918, 1983.
47. **Burke, B. E. and DeLorenzo, R. J.,** Ca- and calmodulin-stimulated endogenous phosphorylation of neurotubulin, *Proc. Natl. Acad. Sci. U.S.A.*, 78, 991, 1981.
48. **Schulman, H.,** Phosphorylation of microtubule-associated proteins by a Ca-, calmodulin-dependent kinase, *J. Cell Biol.*, 99, 11, 1984.
49. **Iqbal, Z. and Ochs, S.,** Uptake of vinca alkaloid into mammalian nerve and its subcellular components, *J. Neurochem.*, 34, 59, 1980.
50. **Chan, S. Y., Worth, R., and Ochs, S.,** Block of axoplasmic transport *in vitro* by vinca alkaloids, *J. Neurobiol.*, 11, 251, 1980.
51. **Gietzen, K., Wuthrich, A., and Bader, H.,** Effects of microtubular inhibitors on plasma membrane calmodulin-dependent Ca-transport ATPase, *Mol. Pharmacol.*, 22, 413, 1982.
52. **Watanabe, K., Williams, E. F., Law, J. S., and West, W. L.,** Specific inhibition of a calcium dependent activation of brain cyclic AMP phosphodiesterase activity by vinblastine, *Experientia (Basel)*, 35, 1487, 1979.
53. **Ross, D. H. and Cardenas, H. L.,** Calmodulin stimulation of Ca^{2+} transport in synaptic membrane, *J. Neurochem.*, 41, 161, 1983.
54. **Sobue, K., Ichida, S., Yoshida, H., Yamazaki, R., and Kakiuchi, S.,** Occurrence of a Ca^{2+}- and modulator protein-activable ATPase in the synaptic plasma membranes of brain, *FEBS Lett.*, 99, 199, 1979.
55. **Iqbal, Z., Garg, B. P., and Ochs, S.,** Calmodulin activation of brain and nerve Ca^{2+}-Mg^{2+}-ATPase, *Soc. Neurosci.*, 5,(Abstr.), 60, 1979.
56. **Sorenson, R. G. and Mahler, H. R.,** Calcium-stimulated adenosine triphosphatase in synaptic membranes, *J. Neurochem.*, 37, 1407, 1981.
57. **Goldin, S. M., Chan, S. Y., Papazian, D. M., Hess, E. J., and Rahamimoff, H.,** Purification and characterization of ATP-dependent calcium pumps from synaptosomes, *Cold Spring Harbor Symp. Quantitative Biology, Vol.* 48, Cold Spring Harbor Laboratory, Cold Spring Harbor, New York, 287, 1983.
58. **Ochs, S. and Iqbal, Z.,** Calmodulin and calcium activation of tubulin associated Ca^{2+}-Mg^{2+}-ATPase, *Soc. Neurosci.*, 6,(Abstr.), 501, 1980.
59. **Prus, K. and Wallin, M.,** The ATPase activity in brain microtubule preparation is membrane-associated, *Histochemistry*, 78, 181, 1983.

60. **Tominaga, S. and Kaziro, Y.,** Adenosine triphosphatases associated with bovine brain microtubules. II. Properties of ATPase I and II, *J. Biochem. (Japan)*, 93, 1093, 1983.
61. **Iqbal, Z. and Ochs, S.,** CDR protein in mammalian nerve, *Fed. Proc. Fed. Am. Soc. Exp. Biol.*, 38, 849, 1979.
62. **Iqbal, Z. and Ochs, S.,** The role of calcium-binding protein in axoplasmic transport, *J. Neurochem.*, 41, S549, 1983.
63. **Ekstrom, P., Kanje, M., and Edstrom, A.,** Effects of phenothiazines and dibenzazepomes on axonal transport and microtubule assembly *in vitro*, *Acta Physiol. Scand.*, 116, 121, 1982.
64. **Hidaka, H.,** Biopharmacological assessment of calmodulin function. Utility of calmodulin antagonist naphthalene sulfonamide, *Proc. West. Soc.*, 24, 203, 1981.
65. **Gietzen, K., Wuthrich, A., and Bader, H.,** R24571: a new powerful inhibitor of red blood cell Ca^{2+} transport ATPase and of calmodulin-regulated functions, *Biochem. Biophys. Res. Commun.*, 101, 418, 1981.
66. **Gietzen, K., Adamczyk-Engelmann, P., Wuthrich, A., Konstantinova, A., and Bader, H.,** Compound 48/80 is a selective and powerful inhibitor of calmodulin-regulated functions, *Biochim. Biophys. Acta*, 736, 109, 1983.
67. **Tiberi, M. and Lavoie, P. -A.,** Inhibition of retrograde axonal transport of acetylcholinesterase by the anti-calmodulin agents amitriptyline and desipramine, *J. Neurobiol.*, 16, 245, 1985.
68. **Iqbal, Z. and Ochs, S.,** unpublished results.
69. **Iqbal, Z. and Ochs, S.,** Fast axoplasmic transport of a calcium-binding protein in mammalian nerve, *J. Neurochem.*, 31, 409, 1978.
70. **Iqbal, Z. and Ochs, S.,** Similarities of fast transport calcium binding protein in nerve to calmodulin, in, *Calcium-Binding Protein and Calcium Function, Siegel, F. L., Carafoli, E., Kretsinger, R. H., McLennan, D. H., Eds.*, Elsevier/North-Holland, Amsterdam, 1980, 217.
71. **Erickson, P. F., Seamon, K. N., Moore, B. W., Lasher, R. S., and Minier, L. N.,** Axonal transport of the Ca^{2+}-dependent protein modulator of 3′:5′-cyclic-AMP-phosphodiesterase in the rabbit visual system, *J. Neurochem.*, 35, 242, 1980.
72. **Brady, S. T., Tytell, M., Heriot, K., and Lasek, R. J.,** Axonal transport of calmodulin: a physiologic approach to identification of long term association between proteins, *J. Cell Biol.*, 89, 607, 1981.
73. **Cena, V., Garcia, A. G., Gonzalez-Garcia, C., and Kirpekar, S. M.,** Orthograde and retrograde axonal transport of calmodulin in cat noradrenergic neurone, *Br. J. Pharmacol.*, 82, 143, 1984.

Chapter 6

METABOLIC AND IONIC PROPERTIES OF AXOPLASMIC TRANSPORT IN RELATION TO ITS MECHANISM

S. Ochs, J. C. J. Gaziri, and R. A., Jersild, Jr.

TABLE OF CONTENTS

I. INTRODUCTION

The essential characteristics of axoplasmic transport have been described in recent reviews and a monograph,[1-5] and touched on in various chapters in this book. In brief, the fastest rate found in mammalian nerve and in the nerves of a number of nonmammalian species is close to 410 mm/day, with the rate independent of the function of the nerve fiber, its diameter, or myelination.[5] The rate of transport is also independent of nerve activity within physiological rates of excitation or of a block of excitation or a reduction in membrane potential.[5] However, changing the permeability of the membrane to various ions, with a resulting change in their concentration in the axon, can have an effect on axoplasmic transport, as will be discussed below.

The aim of this review is to focus on several interrelated cardinal processes underlying the transport mechanism; the metabolic supply of ATP needed to maintain the transport, the role of Ca^{2+} and other ions in transport, the mechanisms regulating the level of Ca^{2+}, and the relation of Ca^{2+} to microtubules with respect to the role of microtubules as an integral part of the transport mechanism.[5] In the transport filament model proposed to account for the fast-transport mechanism,[6,7] the fast-transport of a wide range of molecular species of different types and sizes at the same fast rate, the need for ATP and Ca^{2+} to maintain transport, and the evidence that microtubules are likely to be necessary elements were taken into account. In analogy with the sliding filament theory of muscle, the microtubules were proposed to be the stationary rails along which carriers, the transport filaments move. Vesicles, particulates, soluble proteins, polypeptides, and lipids are bound to these carriers and thus are carried down the nerve fibers at the same fast rate. The energy required for the movement of the carriers and their bound components is provided by ATP which is utulized by a calmodulin activated Ca-Mg-ATPase associated with the microtubules.[8]

We consider here the metabolites responsible for a supply of the ATP needed for transport, the role of Ca^{2+} and its regulation in the fibers to keep it at an optimal level for activating the transport mechanism, and some recent findings pertaining to the relation of microtubules to transport.

II. METABOLISM — SUPPLY OF ATP

A. The Dependence of Transport on ATP

The same rate of transport shown in vivo as an outflow in sciatic nerve fibers of ^{3}H-leucine labeled proteins and polypeptides of 410 mm/day[9] is also seen when the sciatic nerves are removed and placed in a suitable medium oxygenated with 95% O_2 and 5% CO_2.[10] Transport was shown using the in vitro preparation to be dependent on oxidative phosphorylation,[11] with a rapid block of transport appearing within 15 to 30 min when anoxia was initiated by N_2, or by addition of CN, dinitrophenol, or azide to block oxidative phosphorylation.[12-14] when the tricarboxylic acid cycle was blocked with fluoroacetate a block was seen to occur later, after 1 hr, and when iodoacetic acdi (IAA) was used to block glycolysis, after 1.5 to 1.75 hr. In each case a relation of ATP to axoplasmic transport was shown by the fall in the level of ~P [combined ATP and creatine phosphate (CP)] from control levels of 1.2 to 1.4 μmol/g to a critical level of approximately half that amount, 0.5 to 0.6 μmol/g, at the time when axoplasmic transport was blocked.[13,14]

In nerves placed in chambers where maximum action potential responses could also be elicited, the initiation of N_2 anoxia results in a failure of both transport and electrical responses at approximately 20 min when the ~P level was reduced from a control of 1.4 μmol/g to the critical level of 0.5 to 0.6 μmol/g.[15] The Na-pump in the membrane maintains the ionic asymmetry on which the resting membrane potential and excitation of action potentials depends and the correlation of a block of excitability and of transport suggests that the ATP exists in a pool in the fibers which is common to both the sodium pump and transport mechanisms.

Recent studies made in giant fibers, neuroblasts, and other cells, have verified the dependence of particle movement, a form of axoplasmic transport, on an adequate level of ATP.[16,17]

B. Glucose as Metabolite

In in vitro experiments carried out in our laboratory for a number of years, transport was seen to be maintained for at least 4 to 5 hr without addition of glucose to the medium. To determine just how long in vitro transport can be maintained without added substrate, transport was first initiated as usual by injecting the L7 dorsal root ganglia of cats with ^{3}H-leucine and after 2 hr the sciatic nerves removed, the peroneal nerve desheathed (the tibial nerve remaining sheathed) and the nerve placed in a balanced ionic medium without glucose present. After allowing various periods of time for downflow, small 2 cm portions were taken for assay of ATP and CP using the sensitive luciferin-luciferase method and transport assessed in the remainder of the sciatic nerve branches. Those studies showed that in vitro transport continued without the presence of exogenous glucose for 5 to 6 hr before a complete block of transport occurs.[18] At the time of block the level of ~P dropped from a control level of 1.4 μmol/g to the critical level of 0.5 to 0.6 μmol/g.

After the survival time of transport without exogenous glucose of 5 to 6 hr had been established, the effect of an addition of glucose to the in vitro medium was examined. With either 5 or 15 m*M* glucose added, close to control levels of ATP and CP were measured and transport was maintained for considerably longer times, for at least 13 hr.[18] Several experiments made using the longer sciatic nerves taken from large dogs have shown an even longer lasting downflow in nerves incubated with glucose present.

C. Nonglycolytic Metabolic Supply of ATP

While glucose is the normal metabolite maintaining a supply of ATP in nerve in vivo, nerves incubated in vitro without exogenous glucose present have also indicated that some other endogenous source can also provide ATP. In the nerves incubated in a glucose free medium from which 15 to 20 mm portions of nerve were taken for analysis of ATP and CP, an assay of glucose, glycogen, and lactate levels was also carried out using a NAD-NADH sensitive fluorometric method. Glucose and glycogen were seen to fall to very low levels within 1.5 to 2 hr, while ATP and CP levels remained above the critical level for 5 to 6 hr while downflow was maintained.[18] These results suggest that some endogenous metabolites other than glucose are supplying ATP via a path other than one through glycolysis for times of incubation in the glucose-free medium longer than 1.5 to 2 hr and up to 3.5 to 5 hr.

The alternate source of metabolites could perhaps be fatty acids supplied by lipids present in the nerve fibers. Very little, however, is known about lipid as a metabolic source of energy in nerve or even in brain where suggestive finding have indicated that a supply of lipid metabolites may occur.[19] It is known that during fasting ketones are utilized by the brain. There is a considerable uptake of β-hydroxybutyrate and acetoacetate by the brain, with over 50% of the total oxygen consumption accounted for by ketone utilization.[20] Enzymes involved in the utilization of these metabolites, β-hydroxybutyric dehydrogenase, 3-oxoacid CoA transferase, and acetyl-CoA thiolase are present in amounts sufficient for conversion of ketones to acetyl Co-A, which then enters the Krebs cycle to supply ~P.[21,22]

In nerve in vitro, β-hydroxybutyrate was seen to prevent decreases in oxygen uptake and a fall in ~P.[23] In our studies we found that the addition of 5 m*M* β-hydroxybutyrate to the incubation medium was able to maintain ~P levels and transport in vitro for times considerably longer than the 5 to 6 hr limit seen without exogenous glucose, for at least twice as long.[18] Thus, as in brain, β-hydroxybutyrate can be utilized by nerve.

There were, however, variations seen in the effectiveness with which β-hydroxybutyrate was able to maintain transport in vitro. Such variations are perhaps similar to the effectiveness with which brain can utilize β-hydroxybutyrate.[21,22] As had been noted for brain, β-hydroxybutyrate is incapable of maintaining or restoring normal function in the complete absence of glucose in that the first product of β-hydroxybutyrate oxidation, acetoacetate, is metabolized to acetoacetyl CoA only when some glucose is available to provide succinyl CoA so as to permit the synthesis of acetoacetyl CoA. The utilization of β-hydroxybutyrate was therefore studied when glucose in a concentration of 1 m*M* was added to the in vitro medium, an amount which by itself was too low to extend transport beyond the 5 to 6 hr limit seen without an exogenous supply of metabolite. Such additions appeared to give more regularity to the maintenance by β-hydroxybutyrate of transport for longer times.[24]

The possibility that fatty acids from lipid stores in nerve may serve as an endogenous metabolic source was further suggested by some earlier studies made in our laboratory that transport in vitro is blocked by D(+) dodecanoyl carnitine.[25] Metabolic studies in liver and other cells had shown that fatty acids are first esterified by CoA to fatty acyl-CoA, the acyl groups of the activated fatty acids then reacting with carnitine as a carrier for their transport across the inner membrane of the mitochondria as acylcarnitine.[26-29] The acyl groups released from the carnitine in the mitochondria then undergo subsequent β-oxidation, the resulting acetyl CoA entering the Krebs cycle. D(+) dodecanoyl carnitine by competitively blocking the transport of acyl carnitine into the mitochondrial matrix interferes with this metabolic source of ~P and can thus, give rise to a block of transport when glucose is no longer available.

III. A NOTE ON THE PERINEURAL SHEATH

An unexpected finding made using glucose and β-hydroxybutyrate as exogenous metabolites was that the perineural sheath did not appear to act as a permeability barrier to these two metabolites. Transport was maintained in the sheathed nerve branches beyond the 5 to 6 hr limit almost as well as in the desheathed branches.[18]

We also saw in our studies of sheathed nerves in vitro that there was an efflux of glucose from the nerves into the medium.[30] This may follow because of a relatively high glucose level originally present in the axons due to a normally inhibited state of hexokinase.[31] The possibility of a carrier for glucose in the perineural sheath was examined in nerves exposed to phloretin, an agent which blocks the symmetrical Na-independent carrier for glucose transport across the blood-brain barrier.[32,33] We found that the movement of glucose out of and into the nerves was blocked by phloretin and that the flux was also insensitive to Na.[24] These observations thus support the presence of a glucose carrier in the perineural nerve sheath of peripheral nerves similar to that at the blood-brain barrier in brain.

IV. DEPENDENCE OF TRANSPORT ON CALCIUM

The dependence of axoplasmic transport on Ca^{2+} has been shown in in vitro studies when using the desheathed nerve preparation,[34,35] the perineural sheath being relatively impermeable to Ca^{2+}. When desheathed nerve preparations were placed in a Ca-free medium a decrement of axoplasmic transport was seen with a complete block at 2.6 hr, while with physiological levels of Ca^{2+} present, a normal transport was maintained.[34,35] Potassium has a facilitatory effect on the ability of Ca^{2+} to support transport. Transport is maintained without K^+, but then Ca^{2+} must be present in the somewhat higher concentrations of 3 to 5 m*M* to maintain a normal outflow pattern.[35] With the addition of 4 m*M* K^+ to the medium a physiological concentration of 1.5 m*M* Ca^{2+} suffices. The requirement for Ca^{2+} to maintain axonal transport has been verified by other investigators.[36,37]

The dependence of transport on Ca^{2+} is also indicated by the block of transport seen with the use of the calcium channel blocking agents Co^{2+}, Ni^{2+}, La^{3+}, and verapamil.[38] We consider that the reduction in the concentration of Ca^{2+} in the axoplasm by these agents, or by the deletion of Ca^{2+} from the incubating medium, is brought about by a continued operation of a Ca-Na exchange mechanism or an ATP supported Ca-pump in the membrane. When free Ca^{2+} in the axoplasm is below the critical level needed for an activation of Ca-Mg-ATPase, there is a failure of utilization of ATP and a block of transport ensues.

The Ca-ATPase found in frog nerve[39] and the Ca-ATPase present in squid nerve[40] were shown to be similar in their properties to the Ca-ATPase we had earlier found in cat nerve.[41] In all these cases the Ca-ATPase was found to be maximally activated by Ca^{2+} when it was present in a concentration range of 1 to 3 m*M*. However, we found Ca-Mg-ATPase to be also activated by Ca^{2+} in the micromolar range,[8] the range of free Ca^{2+} normally present in the giant axon and in other large cells.[42-44] Calmodulin, a protein found in a wide range of cells,[45-47] was also found present in mammalian nerve.[48,49] This protein has a special property in that when free Ca^{2+} is present in the optimal (micromolar) range, the Ca^{2+} binds to it, enabling it to activate a number of enzymes,[45-47] including Ca-Mg-ATPase from nerve.[50] As evidence of an intermediation of calmodulin, the Ca-Mg-ATPase enzymatic activity of nerve microsomal preparations was seen to be blocked by addition of trifluoroperazine (TFP), an agent binding to and disabling calmodulin.[8,50] Further evidence of a calmodulin intermediation was indicated by the block of axoplasmic transport in nerves in vitro seen when TFP was added to the incubating medium.[5,7,8,37] While relatively high amounts of TFP of 1 to 3 m*M* were required in the incubation medium, we found using ^{3}H-TFP[50] that only micromolar amounts of the TFP enters the axons, an amount adequate to block calmodulin at the concentrations of calmodulin and Ca^{2+} calculated to be present in the fibers.[8,50] The TFP block of transport does not appear to be an indirect one on metabolism, as indicated by the maintained ~P levels seen when transport was blocked by TFP.[8,37]

V. THE ROLE OF OTHER IONS IN TRANSPORT

A. Magnesium

While Ca^{2+} subserves transport in mammalian nerve fibers, other neurons and motile cells may have other ionic requirements. Magnesium appears to subserve particle movement in the giant axon[16,17] and in neuroblastoma cells.[51] The generality of this finding remains in question in that there may be differences in the underlying mechanisms responsible for motility in the different cell types. Magnesium was reported able to completely replace Ca^{2+} in maintaining transport in the frog nerve,[36] though this has not been confirmed.[52] Our studies made on mammalian nerve have shown that Mg^{2+} can only partially prolong transport, for at most 1 hr beyond that seen with a Ca-free medium before transport block occurs.[35] On the other hand, our studies showed Mg^{2+} to have a marked facilitating action with respect to Ca^{2+} in that transport is maintained with levels of Ca^{2+} below the minimal concentration usually needed when Mg^{2+} was present. We would expect that Mg^{2+} is required along with Ca^{2+} for Ca-Mg ATPase activity. So far, however, prolonged periods of incubation in media without Mg^{2+} present have not led to block, suggesting that Mg^{2+} is not readily depleted from the axons by the procedures we have used.

B. Strontium

Strontium (Sr^{2+}), which is similar in its chemical properties to Ca^{2+}, is able to replace Ca^{2+} in supporting transport *in vitro*.[5,53] A characteristic outflow pattern at the usual transport rate of 410 mm/day was seen in the desheathed cat peroneal nerve preparation with Sr^{2+} present at the same concentrations as found needed for Ca^{2+}.[5,36,53]

Strontium by binding to calmodulin as does Ca^{2+} would allow calmodulin to activate Ca-

Mg-ATPase and thus support transport. Strontium in micromolar amounts binds to calmodulin as shown by a subsequent activation of phosphodiesterase[54] and as indicated in our studies, of Ca-Mg-ATPase as well.[8,50]

C. Sodium

Batrachotoxin (BTX) opens Na^+ channels in the membrane[55] and it also causes a rapid block of transport.[56] The effect is due to the entry of Na^+ into the axons. When desheathed nerves are placed in a BTX-containing media no block was seen when the Na^+ in the medium was deleted and replaced with sucrose.[57] A similar effect was seen with BTX and Na^+ on the motility of particles in neuroblastoma cells.[58] We have also found veratridine, which causes an increase in the Na^+ permeability of the nerve membrane, to effect a similar transport block.[59] Ouabain which blocks the Na-pump, also blocks transport in desheathed nerves, a block which was partially ameliorated when Na^+ was deleted from the medium.[60] The increase in the concentration of Na^+ in the axoplasm resulting from the action of these agents might effect the block of transport by causing a release of Ca^{2+} from the mitochondria,[61,62] and perhaps other Ca^{2+} sequestration sites. Higher concentrations of Ca^{2+} in the axoplasm can cause a block of transport, as indicated in the following section. The role of sodium and other monovalent ions is also discussed.

VI. REGULATION OF Ca^{2+} IN NERVE FIBERS

A. Effect of High Concentrations of Ca^{2+} and Sr^{2+}

Although the level of free Ca^{2+} normally present in the axoplasm of mammalian nerve fibers has not as yet been measured, it is likely to be close to 10^{-7} to $10^{-6}M$, the concentration measured in the cephalopod giant axon[42-44,63] and other cells. The total amount of Ca^{2+} present in the giant axon is much higher, with the bulk of Ca^{2+} sequestered in the mitochondria,[42,61-63] smooth endoplasmic reticulum (SER),[64-66] and bound in the axoplasm,[42] most likely by calcium binding proteins (CaBP).[8,67] The low permeability of the neuron axolemma to Ca^{2+} along with the organelles sequestering Ca^{2+} or binding Ca^{2+} at sites in the axoplasm all act to keep the concentration of free Ca^{2+} in the axoplasm regulated at some optimal near micromolar level. When excess Ca^{2+} is present in the axon, it is not only sequestered by these Ca^{2+} regulatory organelles but it is eventually removed from the axon by a Na-Ca exchange mechanism[42-44,63] or by a Ca-pump requiring ATP[68,69] in the membrane.

The regulatory mechanisms are seen to come into play when desheathed nerves in which transport has been initiated are exposed to high concentrations of $Ca^{2+} > 30$ m*M*. Transport is maintained for a time until Ca^{2+} regulation is overcome and a complete block of axoplasmic transport occurs. With 50 to 60 m*M* Ca^{2+} this takes place in approximately 2.5 hr.[34,35] The block of transport is at first reversible, as shown by returning those nerves to a normal Ca^{2+} medium and finding a resumption of transport.[35] Later, after 6 hr, when the Ca regulatory mechanisms have completely failed, there is marked destruction of axonal contents, including the microtubules and neurofilaments, and the block is irreversible.[71]

High concentrations of Sr^{2+} are similar to Ca^{2+} in causing a block of transport but it takes longer for block to occur than with Ca^{2+} and there is not the disruption of microtubules and loss of other axonal organelles seen after 6 hr as was the case with high Ca^{2+}.

B. Localization of Ca^{2+} and Sr^{2+} in the Axon

The presence of Ca^{2+} and of Sr^{2+} within the organelles related to Ca^{2+} regulation (the mitochondria, ER, and the Ca-binding sites scattered within the axoplasm) was studied using pyroantimonate precipitation and energy dispersive X-ray microanalysis.[53,70,72] Pyroantimonate added to glutaraldehyde in the process of fixation of the nerves for electron microscopy enters the fibers to form insoluble Ca^{2+} precipitates. Thin sections of the nerves so prepared show the presence of electron-dense particles in those sites in which Ca^{2+} is expected to be retained.[70,72] Such localization was similar to that seen by others using this method.[73,74] While there have been differences of opinion regarding the use of pyroantimonate to localize Ca^{2+} in cells,[75] recent studies have shown that Ca^{2+} is well retained in various cellular organelles including the mitochondria when the technique is properly used.[76]

The identification of the presence of Ca^{2+} within the electron-dense particles was made using X-ray microanalysis.[70] Because of the overlap between the Ca-K and the Sb-L peaks, the latter arising from the pyroantimonate, a deconvolution (peak separation) procedure[77] was required to separate these peaks, a procedure which we have carried out with a new EDAX software program. The presence of Ca^{2+} in the electron dense particles shown by this procedure was further demonstrated by treatment of thin sections with EGTA resulting in a removal of Ca^{2+} as subsequently shown by their reexamination with X-ray microanalysis.[72]

A similar pattern of electron-dense particles in the same Ca^{2+} sequestering and binding sites, i.e., in the axoplasm and within mitochondria was seen after exposure of desheathed nerves to high concentrations of Sr^{2+}.[53] This indicates that Sr^{2+} follows the same path into those organelles sequestrating and binding Ca^{2+} for its regulation. Strontium has its spectral lines far removed from that of Sb^{2+} and other possible competing elements,[78] and thus is readily shown present in the electron dense particles with X-ray microanalysis without the need for deconvolution. The presence of both Ca^{2+} and Sr^{2+} in the same particles in nerves loaded with Sr^{2+} was observed,[53] the Sr^{2+} apparently adding to the endogenous Ca^{2+} already present in those Ca-regulatory sites.

The lability of Ca^{2+} in its regulatory sites was indicated by the increased number of particles in those organelles and binding sites as a function of the time of exposure of desheathed nerves to high levels of Ca^{2+} and their decrease on washout after Ca^{2+} loading.[70,72] A 4 hr exposure of desheathed nerves to 60 m*M* Ca^{2+} resulted in a large number of particles, the number increasing over this period of time. Nerves so loaded with Ca^{2+} were then placed back into a low Ca^{2+} medium for depletion of the ion. The particles fell off in number after a lag time of 1 hr or so and then, over the next few hours the number of electron dense particles returned to control levels in the various sequestration sites considered to participate in Ca^{2+} regulation.

Of particular interest was the greater number of electron dense particles seen in the nodal region as compared to internodal portions of the fiber.[72] In nerves exposed to high concentrations of Ca^{2+}, a greater number of large sized electron dense particles were seen just outside and an increased number of the smaller diameter particles usually seen in axons within the nodal region. The larger number and differing sizes of electron-dense particles in this region might be due to the different natures of the materials binding Ca^{2+}. Some material present within the nodes near the axolemma which is stained by ferric ion-ferrocyanide might have some relation to the Ca^{2+} binding present in the nodal region[79] and the binding materials for other ions indicated as being present in this region of the fiber[80] might well have Ca-binding properties. On comparing the distribution of electron dense particles in the nodal region of the axon as compared to the internodal region, the impression gained was that following an uptake of Ca^{2+} into the nodal region, there was a subsequent spread of Ca^{2+} into the internodes which, however, did not display a gradient as might be expected of a diffusion-like process. The cation might possibly be transported from the nodes into the internodal region.

VII. THE ROLE OF MICROTUBULES

While microtubules generally are considered to be implicated in transport,[81,85] some reports have suggested that the axolemma,[86] the ER interior,[87] or the ER membrane,[88] might serve as a route of anterogradely fast-transported materials in the fibers rather than one associated with the microtubules. As evidence against a role for microtubules, transport was reported to continue after the microtubules had been disassembled by colchicine[86] or by exposure of nerve to a high concentration of Ca^{2+}.[89] However, the position of grains in autoradiographs at the fast transported front of advancing radioactivity due to labeled proteins indicated that the transport mechanism was distributed throughout the whole cross-sectional area of the axons, as would be the case of a transport mechanism based on microtubules,[9,90] rather than a transport associated with the axolemma or the relatively sparse ER. The correlation of autoradiographic grains of fast-transported radioactivity to the central aggregation of microtubules seen in the axons of nerves treated with IDPN is also in accord with a close relation of microtubules to transport.[91,92] Evidence recently gained using the Allen video enhancement technique has shown particle movement in keratocytes to occur along linear structures which were later identified by immune labeling to be microtubules, again supporting the role of microtubules in transport.[93] Particle transport was observed using the Allen video enhancement technique in invertebrate giant nerve fibers and along the microtubules in the axoplasm extruded from the giant axon.[94,95] This preparation and technique allows important insights into the underlying mechanism to be made. Movement in both directions on the same microtubule was seen[96] and the dependence of transport on ATP verified with, in addition, a soluble ATPase isolated from the preparation which supports the translocation of particles.[97] It remains possible that different mechanisms may subserve particle movement in the giant axons or in nonvertebrate nerve fibers and nonneural cells, as compared to vertebrate nerve fibers.

A report which appeared to negate the necessary role of microtubules in transport was that following exposure of nerves to high Ca^{2+} transport still remained when the microtubules were disassembled.[89] For such a study it is important to determine the time when a block of transport occurs and the loss of the microtubules. The relation of microtubules to transport block was studied in desheathed cat peroneal nerves by determining the density of microtubules in the nerve fibers at times before and after axoplasmic transport had been blocked by exposure to high (> 50 m*M*) concentrations of Ca^{2+}. After 2.5 hr of exposure of nerve to 60 m*M* Ca^{2+}, when a transport block appears,[9] microtubules showed only a small reduction in density in the fibers.[71] The microtubules present could support a resumption of transport as shown by returning the nerves to a normal Ca^{2+} medium.[9] Only after a 6 hr exposure to high concentrations of Ca^{2+} was a marked destruction of microtubules and other organelles seen,[71] a change likely indicative of a Ca^{2+} induced proteolysis.[98,99] In no case did transport occur in the absence of microtubules.

In studies of the effect of maytansine and vinca alkaloids in the desheathed nerve preparation, a small reduction of microtubule density was seen to appear along with or just preceeding a block of transport.[100] Other studies have shown a similar reduction of microtubules when transport was slowed in nerves exposed to lower temperatures and more so when the critical blocking temperature of 11 to 13°C was reached.[101,102] These reductions in microtubule density do not appear to be sufficient in themselves to cause a block of transport, although they may well signal changes in the microtubules related to the block of transport.

The reduction of microtubular densities seen after cooling or exposure to mitotic blocking agents may be explained by recent findings showing how the microtubules are disposed in the nerve fiber, and the possibility of a turnover of tubulin in them. The microtubules are present in the axon as short segment several microns to several hundred microns in length staggered in the axial direction.[103-105] The segments are oriented in the axon with one end of the microtubule segment differentiated from the other with respect to their rates of assembly and disassembly as shown in in vitro studies.[81-85] This could be the case as well in vivo with an ongoing assembly and disassembly in the segments allowing for a turnover of tubulin in the microtubules.[5] Such a process in vivo is likely to be related to the level of intracellular Ca^{2+} present[46] and with an increase in Ca^{2+} concentration could lead to an alteration of the microtubules. A reduction in the temperature or the action of mitotic blocking agents could be responsible for a block of assembly and/or an augmented disassembly and the reduction in the microtubule density seen. More evidence bearing on the turnover of tubulin in the microtubules in nerves in vivo should help explain the effects of these agents and the block of transport they produce. Such changes may be related to how the Ca-Mg-ATPase, which is considered associated with the microtubule side-arms, is inhibited in its utilization of ATP to maintain transport on the basis of the transport filament model.[5]

REFERENCES

1. **Wilson, D. L. and Stone, G. C.,** Axoplasmic transport of proteins, *Annu. Rev. Biophys. Bioeng.*, 8, 27, 1979.
2. **Schwartz, J. H.,** Axonal transport: components mechanisms and specificity, *Annu. Rev. Neurosci.*, 2, 467, 1979.
3. **Grafstein, B. and Forman, D. S.,** Intracellular transport in neurons, *Physiol. Rev.*, 60, 1167, 1980.
4. **Thoenen, H. and Kreutzberg, G. W., Eds.,** The role of fast transport in the nervous system, *Neurosci. Res. Prog. Bull.*, 20, 1, 1981.
5. **Ochs, S.,** *Axoplasmic Transport and Its Relation to Other Nerve Functions,* John Wiley & Sons, New York, 1982.
6. **Ochs, S.,** Fast transport of materials in mammalian nerve fibers, *Science,* 176, 252, 1972.
7. **Ochs, S.,** Calcium requirement for axoplasmic transport and the role of the perineurial sheath, in *Nerve Repair and Regeneration: Its Clinical and Experimental Basis,* Jewett, D. L. and McCarroll, H. R., Eds., C. V. Mosby, St. Louis, 1980, 77.
8. **Ochs, S. and Iqbal, Z.,** The role of calcium in axoplasmic transport in nerve fibers, in *Calcium and Cell Function,* Vol. 3, Cheung, W. Y., Ed., Academic Press, New York, 1983, 325.
9. **Ochs, S.,** Rate of fast axoplasmic transport in mammalian nerve fibres, *J. Physiol. (London),* 227, 627, 1972.
10. **Ochs, S. and Smith, C. B.,** Fast axoplasmic transport in mammalian nerve in vitro after block of glycolysis with iodoacetic acid, *J. Neurochem.*, 18, 833, 1971.
11. **Ochs, S. and Hollingsworth, D.,** Dependence of fast axoplasmic transport in nerve on oxidative metabolism, *J. Neurochem.*, 18, 107, 1971.
12. **Sabri, M. I. and Ochs, S.,** Relation of ATP and creatine phosphate to fast axoplasmic transport in mammalian nerve, *J. Neurochem.*, 19, 2821, 1972.
13. **Ochs, S.,** Energy metabolism and supply of ~P to the fast axoplasmic transport mechanism in nerve, *Fed. Proc. Fed. Am. Soc. Exp. Biol.*, 33, 1049, 1974.
14. **Ochs, S.,** Axoplasmic transport, in *Basic Neurochemistry,* 3rd. ed., Siegel, G. J., Albers, R. W., Agranoff, B. W., and Katzman, R., Eds., Little Brown and Company, Boston, 1981, chap. 22.
15. **Leone, J. and Ochs, S.,** Anoxic block and recovery of axoplaic transport and electrical excitability of nerve, *J. Neurobiol.*, 9, 229, 1978.
16. **Adams, R. J.,** Organelle movement in axons depends on ATP, *Nature (London),* 297, 327, 1982.
17. **Brady, S. T., Lasek, R. J., and Allen, R. D.,** Fast axonal transport in extruded axoplasm from squid giant axon, *Science,* 218, 1129, 1982.
18. **Gaziri, L. C. J. and Ochs, S.,** Increased duration of axoplasmic transport in vitro with added glucose or β-hydroxybutyrate, *Soc. Neurosci. Abstr.*, 9, 150, 1983.
19. **Agardh, C. -D., Chapman, A. G., Nilsson, B., and Siesjö, B. K.,** Endogenous substrates utilized by rat brain in severe insulin-induced hypoglycemia, *J. Neurochem.*, 36, 490, 1981.
20. **Owen, O. E., Morgan, A. P., Kemp, H. G., Sullivan, J. M., Herrera, M. G., and Cahill, G. F., Jr.,** Brain metabolism during fasting, *J. Clin. Invest.*, 46, 1589, 1967.
21. **Krebs, H. A., Williamson, D. H., Bates, M. W., Page, M. A., and Hawkins, R. A.,** The role of ketone bodies in caloric homeostasis, *Adv. Enzyme Reg.*, 9, 387, 1971.
22. **Sokoloff, L., Fitzgerald, G. G., and Kaufman, E. E.,** Cerebral nutrition and energy metabolism, in *Nutrition and the Brain,* Vol. 1, Wurtman, R. J. and Wurtman, J. J., Eds., Raven Press, New York, 1977, 87.

23. **Greene, D. A. and Winegrad, A. I.,** In vitro studies of the substrates for energy production and the effects of insulin on glucose utilization in the neural components of peripheral nerve, *Diabetes*, 28, 878, 1979.
24. **Gaziri, L. C. J. and Ochs, S.,** unpublished observations.
25. **Ochs, S.,** unpublished observations.
26. **Bremer, J.,** Carnitine in intermediary metabolism. The metabolism of fatty acid esters of carnitine by mitochondria, *J. Biol. Chem.*, 237, 3628, 1962.
27. **Fritz, J. B. and Yue, K. T. N.,** Long-chain carnitine acyltransferase and the role of acylcarnitine derivatives in the catalytic increase of fatty acid oxidation induced by carnitine, *J. Lipid Res.*, 4, 279, 1963.
28. **Ramsay, R. R. and Tubbs, P. K.,** The mechanism of fatty acid uptake by heart mitochondria: an acylcarnitine-carnitine exchange, *FEBS Lett.*, 54, 21, 1975.
29. **Tubbs, P. K., Ramsay, R. R., and Edwards, M. R.,** Inhibitors of carnitine transport and metabolism, in *Carnitine Biosynthesis, Metabolism, and Functions*, Frenkel, R. A. and McGarry, J. D., Eds., Academic Press, New York, 1980, 207.
30. **Gaziri, L. C. J. and Ochs, S.,** unpublished observations.
31. **Fromm, H. J.,** Mechanism and mode of regulation of brain hexokinase, in *The Regulation of Carbohydrate Formation and Utilization in Mammals*, Veneziale, C. M., Ed., University Park Press, Baltimore, 1981, 45.
32. **Pardridge, W. M.,** Transport of nutrients and hormones through the blood-brain barrier, *Fed. Proc. Fed. Am. Soc. Exp. Biol.*, 43, 201, 1984.
33. **Pardridge, W. M. and Oldendorf, W. H.,** Kinetics of blood-brain barrier transport of hexoses, *Biochim. Biophys. Acta*, 382, 377, 1975.
34. **Ochs, S., Worth, R. M., and Chan, S. Y.,** Calcium requirement for axoplasmic transport in mammalian nerve, *Nature (London)*, 270, 748, 1977.
35. **Chan, S. Y., Ochs, S., and Worth, R. M.,** The requirement for calcium ions and the effect of other ions on axoplasmic transport in mammalian nerve, *J. Physiol. (London)*, 301, 477, 1980.
36. **Lavoie, P. -A., Bolen, F., and Hammerschlag, R.,** Divalent cation specificity of the calcium requirement for fast transport of proteins in axons of desheathed nerves, *J. Neurochem.*, 32, 1745, 1979.
37. **Kanje, M., Edström, A., and Ekström, P.,** The role of Ca^{2+} in rapid axonal transport, in *Axoplasmic Transport*, Weiss, D., Ed., Springer-Verlag, Basel, 1982, 294.
38. **Ochs, S.,** Block of axoplasmic transport by agents interfering with calcium flux: cobalt, nickel, lanthanum, verapamil; and the maintenance of transport in calcium-free media by strontium, *Soc. Neurosci. Abstr.*, 8, 826, 1982.
39. **Hammerschlag, R. and Bobinski, J. A.,** Ca^{2+}- or Mg^{2+}-stimulated ATPase activity in bullfrog spinal nerve: relation to Ca^{2+} requirements for fast axonal transport, *J. Neurochem.*, 36, 1114, 1981.
40. **Shecket, G. and Lasek, R. J.,** Mg^{2+}- or Ca^{2+}-activated ATPase in squid giant fiber axoplasm, *J. Neurochem.*, 38, 827, 1982.
41. **Khan, M. A. and Ochs, S.,** Magnesium or calcium activated ATPase in mammalian nerve, *Brain Res.*, 81, 413, 1974.
42. **Baker, P. F.,** Transport and metabolism of calcium ions in nerve, *Prog. Biophys. Mol. Biol.*, 24, 179, 1972.
43. **Brinley, F. J., Jr.,** Calcium and magnesium metabolism in cephalopod axons, *Fed. Proc. Fed. Am. Soc. Exp. Biol.*, 35, 2572, 1976.
44. **Blaustein, M. P.,** The interrelationship between sodium and calcium fluxes across cell membranes, *Rev. Physiol. Biochem. Pharmacol.*, 70, 33, 1974.
45. **Yagi, K., Yazawa, M., Kakauchi, S., Ohshima, M., and Uenishi, K.,** Identification of an activator protein for myosin light chain kinase as the Ca^{2+}-dependent modulator protein, *J. Biol. Chem.*, 253, 1338, 1978.
46. **Means, A. R. and Dedman, J. R.,** Calmodulin — an intracellular calcium receptor, *Nature (London)*, 285, 73, 1980.
47. **Cheung, W. Y.,** Calmodulin plays a pivotal role in cellular regulation, *Science*, 207, 19, 1980.
48. **Iqbal, Z. and Ochs, S.,** CDR protein in mammalian nerve, *Abstr. Fed. Proc.*, 38, 849, 1979.
49. **Iqbal, Z. and Ochs, S.,** Calmodulin in mammalian nerve, *J. Neurobiol.*, 11, 311, 1980.
50. **Iqbal, Z. and Ochs, S.,** Characterization of mammalian nerve calmodulin and its possible relationship to axoplasmic transport, in *Axoplasmic Transport*, Weiss, D., Ed., Springer-Verlag, Basel, 1982, 287.
51. **Forman, D. S.,** Saltatory organelle movement and the mechanism of fast axonal transport, in *Axoplasmic Transport*, Weiss, D., Ed., Springer-Verlag, Basel, 1982, 234.
52. **Kanje, M., Edström, A., and Ekström, P.,** Divalent cations and fast axonal transport in chemically desheathed (Triton X-treated) frog sciatic nerve, *Brain Res.*, 241, 67, 1982.
53. **Ochs, S., Jersild, R. A., Jr., Breen, T., and McKitrick, L.,** The maintenance of axoplasmic transport by strontium and its localization in nerve fibers, *J. Neurobiol.*, 17, 55, 1986.
54. **Cox, J. A., Malnoe, A., and Stein, E. A.,** Regulation of brain cyclic nucleotide phosphodiesterase by calmodulin, *J. Biol. Chem.*, 256, 3218, 1981.

55. **Albuquerque, E. X., Daly, J. W., and Witkop, B.,** Batrachotoxin: chemistry and pharmacology, *Science*, 172, 995, 1971.
56. **Ochs, S. and Worth, R. M.,** Batrachotoxin block of fast axoplasmic transport in mammalian nerve fibers, *Science*, 187, 1087, 1975.
57. **Worth, R. M. and Ochs, S.,** Dependence of batrachotoxin block of axoplasmic transport on sodium, *J. Neurobiol.*, 13, 537, 1982.
58. **Forman, D. S. and Shain, W. G., Jr.,** Batrachotoxin blocks saltatory organelle movement in electrically excitable neuroblastoma cells, *Brain Res.*, 211, 242, 1981.
59. **Ochs, S. and Gaziri, L. C. J.,** unpublished observations.
60. **Ochs, S.,** unpublished observations.
61. **Carafoli, E. and Crompton, M.,** The regulation of intracellular calcium by mitochondria, *Ann. N.Y. Acad. Sci.*, 307, 269, 1978.
62. **Carafoli, E. and Crompton, M.,** The regulation of intracellular calcium, in *Current Topics in Membranes and Transport. Membrane Properties: Mechanical Aspects, Receptors, Energetics and Calcium Dependence on Transport*, Vol. 10, Bronner, F. and Kleinzeller, A., Eds., Academic Press, New York, 1978, 151.
63. **Baker, P. F.,** The regulation of intracellular calcium, *Symp. Soc. Exp. Biol.*, 30, 67, 1976.
64. **Henkart, M. P., Reese, T. S., and Brinley, F. J., Jr.,** Endoplasmic reticulum sequesters calcium in the squid giant axon, *Science*, 202, 1300, 1978.
65. **Henkart, M.,** Identification and function of intracellular calcium stores in axons and cell bodies of neurons, *Fed. Proc. Fed. Am. Soc. Exp. Biol.*, 39, 2783, 1980.
66. **Duce, I. R. and Keen, P.,** Can neuronal smooth endoplasmic reticulum function as a calcium reservoir?, *Neuroscience*, 3, 837, 1978.
67. **Iqbal, Z. and Ochs, S.,** Fast axoplasmic transport of a calcium-binding protein in mammalian nerve, *J. Neurochem.*, 31, 409, 1978.
68. **Baker, P. F. and Glitsch, H. G.,** Does metabolic energy participate directly in the Na^+-dependent extrusion of Ca^{2+} ions from squid giant axons?, *J. Physiol. (London)*, 233, 44P, 1973.
69. **DiPolo, R.,** Ca pump driven by ATP in squid axons, *Nature (London)*, 274, 390, 1978.
70. **Chan, S. Y., Ochs, S., and Jersild, R., Jr.,** Calcium localization in mammalian nerve fibers in relation to its regulation and axoplasmic transport, *Soc. Neurosci. Abstr.*, 5, 59, 1979.
71. **Ochs, S., Jersild, R. A., Jr., and McAdoo, V.,** unpublished observations.
72. **Chan, S. Y., Ochs, S., and Jersild, R. A., Jr.,** Localization of calcium in nerve fibers, *J. Neurobiol.*, in press.
73. **Wick, S. M. and Hepler, P. K.,** Localization of Ca^{++}-containing antimonate precipitates during mitosis, *J. Cell Biol.*, 68, 500, 1980.
74. **Stoeckel, M. E., Hindelang-Gertner, C., Dellmann, H. -D., Porte, A., and Stutinsky, F.,** Subcellular localization of calcium in the mouse hypophysis. I. Calcium distribution in the adeno- and neurohypophysis under normal conditions, *Cell Tissue Res.*, 157, 307, 1975.
75. **Slocum, R. D. and Roux, S. J.,** An improved method for the subcellular localization of calcium using a modification of the antimonate precipitation technique, *J. Histochem. Cytochem.*, 30, 617, 1982.
76. **Wick, S. M. and Hepler, P. K.,** Selective localization of intracellular Ca^{2+} with potassium antimonate, *J. Histochem. Cytochem.*, 30, 1190, 1982.
77. **Chandler, J. A.,** X-ray microanalysis in the electron microscope, in *Practical Methods in Electron Microscopy*, Vol. 5, Glauert, A. M., Ed., North-Holland, Amsterdam, 1977, 315.
78. **Somlyo, A. P., Somlyo, A. V., Devine, C. E., Peters, P. D., and Hall, T. A.,** Electron microscopy and electron probe analysis of mitochondrial cation accumulation in smooth muscle, *J. Cell Biol.*, 61, 723, 1974.
79. **Waxman, S. G. and Foster, R. E.,** Ionic channel distribution and heterogeneity of the axon membrane in myelinated fibers, *Brain Res. Rev.*, 2, 205, 1980.
80. **Landon, D. N. and Langley, O. K.,** The local chemical environment of nodes of Ranvier: a study of cation binding, *J. Anat.*, 108, 419, 1971.
81. **Shelanski, M. L. and Feit, H.,** Filaments and tubules in the nervous system, in *The Structure and Function of Nervous Tissue*, Vol. 6, Bourne, G. H., Ed., Acadmic Press, New York, 1972, 47.
82. **Stephens, R. E. and Edds, K. T.,** Microtubules: structure, chemistry, and function, *Physiol. Rev.*, 56, 709, 1976.
83. **Soifer, D., Ed.,** The biology of cytoplasmic microtubules, *Ann. N.Y. Acad. Sci.*, 253, 1, 1975.
84. **Goldman, R., Pollard, T., and Rosenbaum, J., Eds.,** *Cell Motility*, Book C, Cold Spring Harbor Laboratory, Cold Spring Harbor, New York, 1976.
85. **Dustin, P.,** *Microtubules*, Springer-Verlag, Basel, 1978.
86. **Byers, M. R.,** Structural correlates of rapid axonal transport: evidence that microtubules may not be directly involved, *Brain Res.*, 75, 97, 1974.
87. **Droz, B., Rambourg, A., and Koenig, H. L.,** The smooth endoplasmic reticulum: structure and role in the renewal of axonal membrane and synaptic vesicles by fast axonal transport, *Brain Res.*, 93, 1, 1975.

88. **Hammerschlag, R. and Stone, G. C.,** Membrane delivery by fast axonal transport, *Trends Neurosci.*, 5, 12, 1982.
89. **Brady, S. T., Chrothers, S. D., Nosal, C., and McClure, W. O.,** Fast axonal transport in the presence of high Ca^{2+}: evidence that microtubules are not required, *Proc. Natl. Acad. Sci. U.S.A.*, 77, 5909, 1980.
90. **Ochs, S. and Jersild, R. A., Jr.,** Fast axoplasmic transport in nonmyelinated mammalian nerve fibers shown by electron microscopic radioautography, *J. Neurobiol.*, 5, 373, 1974.
91. **Griffin, J. W., Price, D. L., Hoffman, P. N., and Cork, L. C.,** The axonal cytoskeleton: alterations of organization and axonal transport in models of neurofibrillary pathology, *J. Neuropath. Exp. Neurol.*, 40, 316, 1981.
92. **Papsozomenos, S. C., Autilio-Gambetti, L., and Gambetti, P.,** Reorganization of axoplasmic organelles following β, β-iminodipropionitrile administration, *J. Cell Biol.*, 91, 866, 1981.
93. **Hayden, J. H., Allen, R. D., and Goldman, R. D.,** Cytoplasmic transport in keratocytes: direct visualization of particle translocation along microtubules, *Cell Motility*, 3, 1, 1983.
94. **Vale, R. D., Schnapp, B. J., Reese, T. S., and Sheetz, M. P.,** Movement of organelles along filaments dissociated from the axoplasm of the squid giant axon, *Cell*, 40, 449, 1985.
95. **Brady, S. T., Lasek, R. J., and Allen, R. D.,** Video microscopy of fast axonal transport in extruded exoplasm: a new model for study of molecular mechanisms, *Cell Motility*, 5, 81, 1985.
96. **Schnapp, B. J., Vale, R. D., Sheetz, M. P., and Reese, T. S.,** Single microtubles from squid axoplasm support bidirectional movement of organelles, *Cell*, 40, 455, 1985.
97. **Vale, R. D., Schnapp, B. J., Reese, T. S., and Sheetz, M. P.,** Organelle, bead, and microtubule translocations promoted by soluble factors from the squid giant axon, *Cell*, 40, 559, 1985.
98. **Schlaepfer, W. W.,** Experimental alterations of neurofilaments and neurotubules by calcium and other ions, *Exp. Cell Res.*, 67, 73, 1971.
99. **Schlaepfer, W. W.,** Structural alterations of peripheral nerve induced by the calcium ionophore A23187, *Brain Res.*, 136, 1, 1977.
100. **Ghetti, B. and Ochs, S.,** On the relation between microtubule density and axoplasmic transport in nerves treated with maytansine in vitro, in *Peripheral Neuropathies*, Vol. 1, Canal, N. and Pozza, G., Eds., Elsevier, Amsterdam, 1978, 177.
101. **Ochs, S. and Smith, C.,** Low temperature slowing and cold-block of fast axoplasmic transport in mammalian nerves in vitro, *J. Neurobiol.*, 6, 85, 1975.
102. **Brimijoin, S., Olsen, J., and Rosenson, R.,** Comparison of the temperature-dependence of rapid axonal transport and microtubules in nerves of the rabbit and bullfrog, *J. Physiol. (London)*, 287, 303, 1979.
103. **Chalfie, M. and Thomson, J. N.,** Organization of neuronal microtubules in the nematode *Caenorhabditis elegans*, *J. Cell Biol.*, 82, 278, 1979.
104. **Bray, D. and Bunge, M. B.,** Serial analysis of microtubules in cultured rat sensory axons, *J. Neurocytol.*, 10, 589, 1981.
105. **Tsukita, S. and Ishikawa, H.,** The cytoskeleton in myelinated axons: serial section study, *Biomed. Res.*, 2, 424, 1981.

Chapter 7

INTRA-AXONAL TRANSPORT OF CYTOSKELETAL PROTEINS

Masanori Kurokawa

TABLE OF CONTENTS

I. INTRODUCTION

Proximo-distal convection of a bulk of axoplasm at a velocity of 1 to 2 mm/day was first described by Weiss and Hiscoe[1] in 1948 on the basis of their experiments using nerve constriction. Weiss put forward the concept of *perpetual neuronal growth* by stating that: neuroplasm is subject to continuous degradation, which, unless offset by commensurate replacement, leads to progressive atrophy. Replacements are constantly synthesized within the confines of the perikaryon at a rate corresponding to peripheral demands and then are somehow conveyed down the length of the axon at a rate adjusted to the normal width of the channel, i.e., the diameter of the nerve fiber. The constriction of the channel at a bottleneck reduces the rate of passage of supplies and thereby keeps all farther peripheral levels under-supplied, whereas surplus accumulates proximally.[2]

II. CYTOSKELETAL PROTEINS AS MAJOR COMPONENTS OF SLOW-TRANSPORT

Extensive introduction of radioisotopes and electron microscopy in the 1960s greatly facilitated the subsequent study of axoplasmic transport. Quantitative radioautography by Droz[3] has shown that a great majority of the [^{3}H]-labeled proteins newly synthesized in rat spinal ganglion cells and migrating down the axon, is associated with axoplasm rich in neurofilaments and microtubules. Additional experiments with leeches confirmed the association of radioactivity with these cytoskeletal components. Radioautography further showed that the bulk of labeled proteins distributed in the axoplasm migrates at a slow rate.[4] In contrast to slow-transport, fast-transport appeared to be associated with membraneous profiles,[4] in agreement with the results of biochemical studies showing that the radioactivity conveyed by fast-transport is bound essentially to membranes of a microsomal nature.[5,6]

Despite the radioautographic findings that slow-transport is associated with the axonal cytoskeleton, the bulk of slowly migrating materials was found in the *soluble* compartment when examined by the conventional cell fractionation procedure.[7] There were, however, several indications that tubulin, or the colchicine-binding component, slowly migrates down the axon in various experimental systems (for review, see Reference 8).

More recent studies using SDS-polyacrylamide gel electrophoresis and fluorography firmly established the relation of slow-transport with the axonal cytoskeleton. Using SDS-PAGE, Hoffman and Lasek[9] have demonstrated that more than 75% of the total radioactivity in slow-transport in the rat ventral motor neurone and cat spinal ganglion sensory neurone is accounted for by five polypeptides. Of these polypeptides, two have been tentatively identified as tubulin on the basis of their molecular weights (57 and 53 kDa). The three remaining polypeptides (210, 160, and 68 kDa) have been collectively referred to as the slow component triplet, and were tentatively associated with neurofilaments because of their prominence and identical migration in slow-transport. The nature of the polypeptide that constitutes mammalian neurofilaments had been a matter of controversy for some years, mainly because of the contamination of glial fibrillary acidic (GFA) protein in the neurofilament fraction prepared by conventional cell fractionation procedures. The view that the triplet represents neurofilaments was corroborated by later demonstrations that (1) neurofilaments purified from the rat sciatic nerve are composed exclusively of the triplet polypeptides with molecular weights of 200, 160, and 68 kDa,[10,11] and (2) the triplet is selectively lost from the transected nerve during the same time interval in which neurofilaments are undergoing granular disintegration.[12,13] Also, degeneration of the mouse optic nerve eliminates simultaneously the triplet and neurofilaments, leaving the astroglial filaments intact.[14]

Lasek and Hoffman[15] have defined two subcomponents in the slow-transport, designated slow component a (SCa) and slow component b (SCb). In the rat ventral motor neurones,

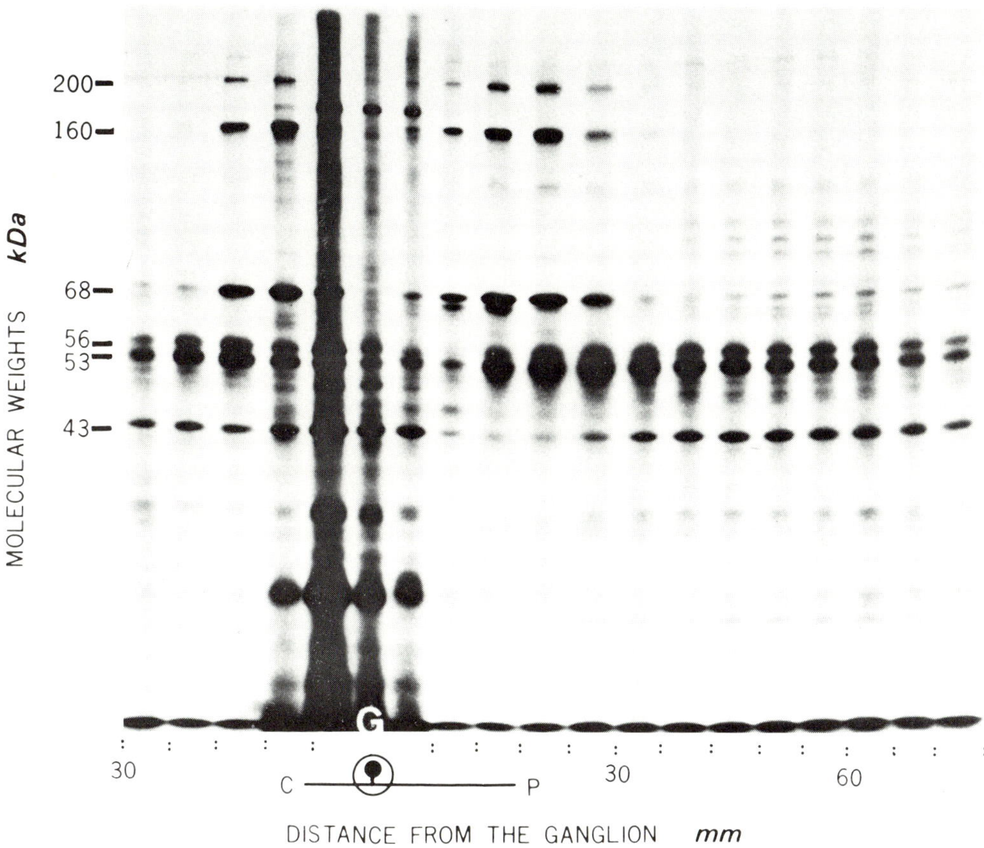

FIGURE 1. Fluorographic visualization of polypeptides migrating down the axons.[23] Polypeptides in consecutive nerve segments were analyzed by SDS-gel electrophoresis 14 days after labeling the dorsal root ganglion (L_5) of the adult rat with L-[^{35}S]methionine. G, ganglion. P, peripheral branch of axons (sciatic nerve). C, central branch of axons (dorsal root).

SCa is defined as a 1 to 2 mm/day peak containing tubulin and the triplet, and SCb as a 3 to 4 mm/day peak containing tubulin and a number of other polypeptides including a 43 kDa polypeptide, which was later identified as actin.[16,17] In axons of guinea pig retinal ganglion cells and hypoglossal and phrenic neurones, tubulin is not contained in SCb, and it was proposed that SCa corresponds to microtubule-neurofilament network, and SCb represents the transport of microfilament network with proteins complexed with microfilaments.[18,19]

Though actin is a conspicuous member in the slowly migrating axonal polypeptides (Figure 1), there have been few demonstrations that it exists in the form of microfilaments in mature axons.[20] Using nitrobenzooxadiazol phallacidin as a specific probe, F-actin has recently been located in the subaxolemmal space in quick-frozen and deep-etched frog axons.[21]

In some neurones, a portion of tubulin migrates faster than the neurofilament triplet. Using SDS-PAGE and fluorography, we have observed that in the peripheral branch of rat dorsal root ganglion cell axons, fronts of the triplet, tubulin, and actin migrate at 2 to 3 mm/day, 9 to 13 mm/day, and ~19 mm/day, respectively, and the flow profiles of the triplet and tubulin, and of tubulin and actin overlap[22,23] (Figure 1). The interpretation of this overlapping is complicated by the observation that half of the axons in this sensory nerve are reported to be unmyelinated.[24] However, the observation that SCa and SCb are not clearly separated in the dorsal root sensory neurones (Figure 1) does not seem to be attributable solely to the

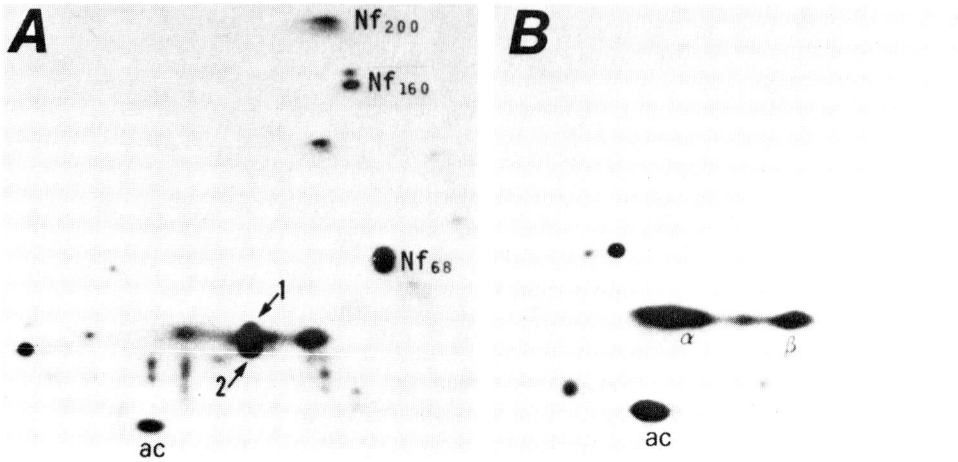

FIGURE 2. Two populations of tubulin transport.[27] The L_5 ganglion was labeled as in Figure 1, and polypeptides in stretches of the sciatic nerve 15 to 30 mm (A), and 39 to 60 mm (B) away from the ganglion were analyzed by two-dimensional electrophoresis, and then visualized by fluorography. Nf_{200}, Nf_{160}, and Nf_{68} represent the 200, 160 and 68 kDa components of the neurofilament triplet. ac, Actin. α, α-Tubulin. β β-Tubulin.

difference in axon population, since in the mammalian ventral neurones where a large majority of motor axons is believed to be myelinated,[25] a portion of tubulin migrates faster than the triplet.[9,26] It has recently been observed that there are two distinct populations of tubulin transport in the peripheral branch of the rat dorsal root ganglion cell axons (Figure 2): one migrating with the triplet, and the other migrating faster independently of the triplet.[27] In two-dimensional electrophoresis, the latter is composed solely of α- and β-tubulin, whereas the former contains at least two other components distinct from α- and β-tubulin (spots 1 and 2, Figure 2A). Though the nature of these components remains to be elucidated, the observation indicates that a considerable portion of axonal tubulin migrates separately from the neurofilament triplet. In electron microscopy, neurofilament domains can be distinguished from microtubule domains, though these two types of domains are contiguous.[28]

Lasek and his colleagues have proposed the hypothesis that a group of polypeptides in the axoplasm are transported as a supramolecular structural assembly (SCb), rather than simply moving towards the terminal independently. Proteins so far assigned to SCb include neurone-specific enolase, creatine phosphokinase,[29] calmodulin,[30] and clathrin.[31] Ellisman and Porter speculate that SCb components are likely candidates for some of the major constituents of the microtrabecular system.[32] Though actin and clathrin are assigned to the SCb complex,[31] these two polypeptides were found to migrate at different velocities.[33] An interpretation for this apparent contradiction has recently been given, where the SCb complex is divided into three subclasses on the basis of profile analysis of the SCb radioactivity; actin is placed in class I, and clathrin in class III.[34] Classification is based on the assumption that the individual polypeptides have differing affinity for an SCb group.

The quick-freeze, deep-etching method revealed the presence of cross-links which interconnect microtubules, neurofilaments, membraneous organelles, and plasmalemma.[21,28,35] In addition, many granular structures are seen to be attached to cytoskeletal elements. On the basis of experiments where the nerve was extracted with Triton X-100[35] or with water,[21] suggestion has been made that these granular structures represent the slowly migrating soluble proteins which are fixed *in situ* by quick freezing. It remains to be seen whether the granular structures, together with other elements, constitute a unitary complex relevant to SCb as hypothesized from the transport study.

The behavior of *fodrin* (heterodimer of 250 and 240 kDa polypeptides) in the axonal transport is peculiar in that it is not confined to any one of the five velocity groups of Willard, but is found in four groups, excluding the most rapidly migrating group (group I).[36-38] Fodrin proved to be a protein ubiquitously present in the cortical cytoplasm of not only neurones but also of many nonneuronal cells.[39] As a part of their studies on calmodulin-binding proteins, Kakiuchi and his colleagues have purified from the brain a protein that comprises two subunits (240 and 235 kDa), and termed it *calspectin*.[40] There are several lines of evidence to indicate that calspectin is identical to fodrin (for reviews, see References 41 and 42), and this protein now proves to be a member of the spectrin-like proteins widely distributed in various types of cells. Calspectin is not only a prominent calmodulin-binding protein in the brain, but also an actin-binding protein. Although the dimer and tetramer of calspectin are both bound to actin filaments, only the tetramer can cross-link to form a viscous gel.[43,44] Analogy between the fodrin movement in the lymphocyte capping process and in axonal transport (group IV) has recently been pointed out.[45] Both processes involve the redistribution of actin, myosin, calmodulin, and fodrin; the presence of myosin-like protein in group IV transport has previously been shown.[46] If the cortical cytoplasm contains all these proteins and constitutes a mobile lining, this might provide the motive force of other axonal organelles when they become linked to the lining.[45]

III. COMPOSITIONAL VARIATIONS IN THE AXONAL CYTOSKELETON

Axonal cytoskeleton as seen by slow transport exhibits compositional variation to a considerable extent. In guinea pig vagal nerve, the neurofilament triplet is virtually undetectable by fluorography even after a prolonged exposure,[26,47-49] which appears to be in accord with the morphological observation that vagal axons are mostly unmyelinated and small in diameter.[50] Electron microscopic examination indicates that as axons become smaller, the neurofilaments become fewer and less prominent, until in very thin unmyelinated axons neurofilaments are rare and generally only microtubules are present.[51]

There is an obvious asymmetry of the axonal transport in two branches of bifurcating axons from the dorsal root ganglion cells. Whereas the rate of fast-transport is equivalent in the two branches, the rate of slow-transport is lower in the central branch (dorsal root) than in the peripheral branch (sciatic nerve).[52] Also, the net amount of transported materials is less in the central branch.[53,54] Fluorographic and radiometric analyses have shown that the triplet, tubulin, and actin migrate more rapidly (Figure 1), and also in greater amount in the peripheral axons.[22,23] Also, peripheral axons regenerate two times faster than the central axons.[55] Another striking aspect of central-peripheral asymmetry relates to the tubulin/triplet (or actin + tubulin/triplet) ratio that is significantly higher in the peripheral branch.

The wealth of total cytoskeletal proteins in the peripheral branch relative to the central branch may relate to the total volume of axoplasm, which is most likely greater in the peripheral axons than in the central axons. However, the peripheral/central ratio is not uniform in the triplet, tubulin, and actin, and as a consequence the tubulin/triplet (or actin + tubulin/triplet) ratio is significantly higher in the peripheral branch. There is a considerable body of evidence that actin and tubulin are related to axon growth more directly than neurofilaments. Tubulin and actin are specifically increased in cultured sympathetic neurones when outgrowth is stimulated by nerve growth factor.[56] In regenerating toad[57] and goldfish[58] retinal ganglion cells, significant increase in the synthesis of actin and tubulin was seen, in addition to increase of a certain species of glycoproteins that are assignable to the rapid phase of transport. It is possible that a greater amount of actin and tubulin relative to the neurofilament triplet is required to maintain a steady state of mature axons that undergo perpetual growth in the sense of Weiss[2] and this requirement is greater in the peripheral axons because of their larger volume. The greater demand for cytoskeletal proteins, particularly for tubulin and actin, of the peripheral branch[23] seems to provide at least one explanation for the well established morphological observation that dorsal root ganglion cells show profuse chromatolytic changes after section of their peripheral axons, but little if any reaction after section of their central axons (for review, see Reference 59).

IV. FACTORS THAT AFFECT SLOW-TRANSPORT

Changes in the rate of axoplasmic transport during development have been examined by several workers (for review, see Reference 60). In one of the carefully controlled studies, Hendrickson and Cowan[61] have shown in the rabbit optic system that slow-transport declines in rate from approximately 5 mm/day in 6-day-old animals to about 2 mm/day in animals over 4 weeks of age, whereas rate of fast-transport is increased from 120 mm/day in 6-day-old rabbits to 150 mm/day by the end of the third week postnatally, and attains the adult level of about 240 mm/day at the end of the fourth week. These workers considered the fourth postnatal week to be a critical period in the development of the rabbit visual system in that it may coincide with the establishment of definitive synaptic connections by the retinal ganglion cell axons. Decrease in velocity of slow-transport by a factor of two between immature and adult animals has been observed also in rat sciatic nerve,[62] and in cat dorsal root ganglion cells.[63]

To interpret changes in velocity of the axoplasmic transport in growing axons, many variables should be taken into account, particularly changes in the number and dimensions (length and diameter) of axons, and changes in relative proportions of materials migrating at different velocities. Levine et al.[64] have recently shown that decrease in rate of slow transport in the retinal ganglion cells of growing rabbits can be explained by the following observations. First, the velocity of group IV polypeptides, including tubulin and actin,[36] decreases by a factor of about two during 18 to 25 days postnatally. Second, the 200 kDa component in the neurofilament triplet is not detected until the twelfth postnatal day and in the absence of this component, the remaining two components of the triplet (145 and 73 kDa) are transported at about 8 mm/day, coordinately with group IV polypeptides; about 3 weeks after birth their velocity decreases eightfold to their adult (group V) velocity. The induction of transport of the 200 kDa component at about 3 weeks after birth appears to cross-link the neurofilaments (Reference 65) and the drag of the cross-linked neurofilaments on the axoplasmic matrix (group IV) slows its velocity.

Since the initial demonstration by Dahlström[66] and Kreutzberg,[67] blockade of rapid transport, including retrograde transport,[68,69] by antimitotic agents has been documented.[70,71] However, whether slow-transport is also interrupted by these agents has been controversial. In some experiments colchicine blocked slow-transport only partially,[72,73] whereas in others slow-transport was more profoundly affected by colchicine than fast-transport.[74,75] However, in all of these studies, the components in slow-transport affected by colchicine were not adequately characterized. In our experiment using SDS-PAGE followed by radiometry and fluorography, colchicine injected into the dorsal root ganglion of the adult rat blocked the tubulin transport, with little if any blockade of the neurofilament triplet and actin.[76] By contrast, vinblastine sulfate similarly applied blocked the transport of all cytoskeletal proteins. Under the experimental conditions used, blockade of tubulin transport by colchicine was not complete, which may be accounted for, at least in part, by the relatively short duration of colchicine action. Since the inhibitory action of colchicine on slow transport lasted for 2 days after its application, but no more than 4 days, and axons once damaged by colchicine can resume tubulin transport,[77] the newly synthesized protein may have been detected by fluorography prepared from the nerve obtained 7 days after the colchicine application.[76] The labeled protein seen in the tubulin region on SDS-PAGE of the colchicine-treated nerve has recently been shown to represent a neurofilament-associated protein (NAP; spot 2 in Figure 2A) distinct from tubulin.[102,103]

β,β′-Iminodipropionitrile (IDPN) has been known to induce argentophilic ballooning of the proximal portions of axons, especially in the anterior horn of the spinal cord and brainstem, and primary *axostasis* is assumed to be its mechanism of action.[78,79] The action of IDPN is peculiar in that it selectively blocks transport of the neurofilament triplet, without any detectable effect on transport of tubulin and actin,[26,80] or on fast anterograde and retrograde transport,[80] Within 24 hr after a single intraperitoneal administration of IDPN, axonal distention with an enormous accumulation of neurofilaments inside is seen at the boundary between the intramedullary axons and the ventral root. The axonal distention progresses towards the cell body with the result that the intramedullary axons are thickened along their entire length by day 6. The ultrastructure of axons in the sciatic nerve remains unaltered during these 6 days.[26] However, in a rat that received IDPN by intraperitoneal injection four times every 3 days, and was kept alive for a further 3 weeks, a characteristic reorganization of axoplasmic organelles was observed in the sciatic nerve, i.e., displacement of microtubules, agranular reticulum, and mitochondria towards the center, and of neurofilaments towards the periphery of the axons.[81,82] In such an axon, radioautographic silver grains representing fast-transport of radioactive proteins were located mostly in the central region, whereas in the control axon silver grains showed no preferential localization.[82] Association of radioautographic grains of fast-transport with the microtubule-rich peripheral region of the axons, and their absence in the neurofilament-rich central region have previously been shown in regenerating neurones.[83] These observations are in favor of the notion that microtubules are related to fast-transport more directly than, and independently of, neurofilaments.

The rate of slow-transport is decreased by lowering the ambient temperature; a Q_{10} of 3.3 to 3.7 was reported for the garfish olfactory nerve,[84] which is significantly higher than the value of 2.2 reported for fast-transport.[85]

V. DEGRADATION OF AXONAL CYTOSKELETON

Radioautographic analyses indicate that the great majority of slowly moving materials is retained in the axons, whereas fast-transport is mainly confined to the synaptic terminal.[4,86] Fast-transport is assigned to the renewal of various membraneous components such as synaptic vesicles, presynaptic and axolemmal plasma membranes, mitochondria, and endoplasmic reticulum.[4] The function of cytoskeletal proteins in slow-transport is to maintain the axon per se, ensuring the structure involved in fast-transport and possibly also generating the driving force of the transport. Neurofilaments, microtubules, and actin filaments are not normally observed in axon terminals. However, some minor components assigned to slow-transport, e.g., clathrin,[31,33] calmodulin,[30,48,87] nourone-specific enolase,[29,88] choline acetyltransferase, and lactate dehydrogenase[60] must enter the terminals to exert their physiological action.

To explain the selective degradation of the axonal cytoskeleton in terminals, Lasek and Hoffman[15] have proposed a hypothesis that neurofilaments advancing into the terminal are subject to breakdown by a Ca^{2+}-activated proteolytic process. Since the protease itself would be transported down the axon without exerting its action on neurofilaments during migration, the hypothesis supposes that Ca^{2+} concentration must be greater in the terminal than in the axon proper. Microtubules may also be degraded by Ca^{2+}-protease, or directly depolymerized by Ca^{2+}, since microtubule disassembly in vitro can be brought about by concentrations of Ca^{2+} as low as 1 to 30 μM,[89,90] in the presence of calmodulin.[91,92]

There is a growing body of evidence that Ca^{2+}-activated protease or proteases acting on neurofilaments exist in both invertebrate[93,94] and mammalian[95,96] neurones or the nervous system. In most of these reports, protease activity requires a rather high concentration of Ca^{2+} and its subcellular distribution is not known. Pant et al.[97] have recently observed that, whereas the stellate ganglion cells and retinal fibers from squid contain significant Ca^{2+}-

protease activity, synaptosomes from the optic lobe are completely devoid of the activity. This observation led them to postulate that the axonal neurofilaments are degraded by Ca^{2+}-protease at the axon-nerve ending junction, followed by an autoinactivation of the Ca^{2+}-protease. However, as the authors themselves point out, it must be proven in further studies that nerve terminals derived from retinal fibers are significantly present in their synaptosomal preparations.

Malik et al.[98] have recently purified two forms of a Ca^{2+}-activated neutral protease (CANP) to near homogeneity from calf brain. One (CANP I) is maximally active at 700 μM Ca^{2+}, and the other (CANP II) at 2 μ*M* Ca^{2+}. Tubulin, microtubule-associated proteins, neurofilament triplet, and desmin were extensively hydrolyzed by both of the proteases. The requirement for low Ca^{2+} concentration of CANP II makes it likely that this enzyme has physiological significance, though its cellular and subcellular distribution remains to be elucidated.

Apart from these apparently soluble Ca^{2+}-proteases, we have observed that Ca^{2+}-protease activity is bound to the neurofilament-rich cytoskeleton prepared from the rat spinal cord. The enzyme is maximally active with 200 μ*M* Ca^{2+}, and preferentially degrades the 160 kDa component of the neurofilament triplet, in contrast to the soluble protease activity, which degrades a variety of cytoskeletal proteins, including the neurofilament triplet, GFA protein, actin, tubulin, and a microtubule-associated protein. The protease is released from the cytoskeleton only at a high ionic strength, and can be reassociated with the protease-free cytoskeleton when the ionic strength is lowered to a physiological level.[99,100] An inhibitory factor for this protease was demonstrated in the soluble fraction from the rat spinal cord.[100] The enzyme may exert a subtle influence on the filament-filament or filament-membrane interaction, since the 160 kDa component, together with the 200 kDa component, is thought to be located on the outer surface of the neurofilaments.[65] Modification of the 140 kDa component, apparently proteolytic in nature, has been reported to occur during the migration of the neurofilament triplet down the axons.[101]

REFERENCES

1. **Weiss, P. and Hiscoe, H. B.,** Experiments on the mechanism of nerve growth, *J. Exp. Zool.*, 107, 315, 1948.
2. **Weiss, P.,** The concept of perpetual neuronal growth and proximo-distal substance convection, in *Regional Neurochemistry*, Kety, S. S. and Elkes, J., Eds., Pergamon Press, New York, 1961, 220.
3. **Droz, B.,** Synthèse et transfert des protéines cellulaires dans les neurones ganglionnaries. Étude radioautographique quantative en microscopie électronique, *J. Micros.*, 6, 201, 1967.
4. **Droz, B., Koenig, H. L., and Di Giamberardino, L.,** Axonal migration of protein and glycoprotein to nerve endings. I. Radioautographic analysis of the renewal of protein in nerve endings of chicken ciliary ganglion after intracerebral injection of [^{3}H]lysine, *Brain Res.*, 60, 93, 1973.
5. **Abe, T., Haga, T., and Kurokawa, M.,** Rapid transport of phosphatidylcholine occurring simultaneously with protein transport in the frog sciatic nerve, *Biochem. J.*, 136, 731, 1973.
6. **Grafstein, B., Miller, J. A., Ledeen, R. W., Haley, J., and Specht, S. C.,** Axonal transport of phospholipid in goldfish optic system, *Exp. Neurol.*, 46, 261, 1975.
7. **Karlsson, J. -O. and Sjöstrand, J.,** Synthesis, migration and turnover of protein in retinal ganglion cells, *J. Neurochem.*, 18, 749, 1971.
8. **Heslop, J. P.,** Axonal flow and fast transport in nerves, *Adv. Comp. Physiol. Biochem.*, 6, 75, 1975.
9. **Hoffman, P. N. and Lasek, R. J.,** The slow component of axonal transport. Identification of major structural polypeptides of the axon and their generality among mammalian neurons, *J. Cell Biol.*, 66, 351, 1975.
10. **Mori, H. and Kurokawa, M.,** Purification of neurofilaments and their interaction with vinblastine sulphate, *Cell Struc. Funct.*, 4, 163, 1979.

11. **Mori, H. and Kurokawa, M.,** Morphological and biochemical characterisation of neurofilaments isolated from the rat peripheral nerve, *Biomed. Res.*, 1, 24, 1980.
12. **Schlaepfer, W. W. and Micko, S.,** Chemical and structural changes of neurofilaments in transected rat sciatic nerve, *J. Cell Biol.*, 78, 369, 1978.
13. **Schlaepfer, W. W. and Micko, S.,** Calcium-dependent alterations of neurofilament proteins of rat peripheral nerve, *J. Neurochem.*, 32, 211, 1979.
14. **Tsukita, S., Ishikawa, H., and Kurokawa, M.,** Isolation of 10-nm filaments from astrocytes in the mouse optic nerve, *J. Cell Biol.*, 88, 245, 1981.
15. **Lasek, R. J. and Hoffman, P. N.,** The neuronal cytoskeleton, axonal transport and axonal growth, in *Cell Motility,* Goldman, R., Pollard, T., and Rosenbaum, J., Eds., Cold Spring Harbor Laboratory, Cold Spring Harbor, New York, 1021, 1976.
16. **Black, M. M. and Lasek, R. J.,** Axonal transport of actin: slow component *b* is the principal source of actin for the axon, *Brain Res.*, 171, 401, 1979.
17. **Willard, M., Wiseman, M., Levine, J., and Skene, P.,** Axonal transport of actin in rabbit retinal ganglion cells, *J. Cell Biol.*, 81, 581, 1979.
18. **Black, M. M. and Lasek, R. J.,** Slow components of axonal transport: two cytoskeletal networks, *J. Cell biol.*, 86, 616, 1980.
19. **Tytell, M., Black, M. M., Garner, J. A., and Lasek, R. J.,** Axonal transport: each major rate component reflects the movement of distinct macromolecular complexes, *Science,* 214, 179, 1981.
20. **LeBeux, Y. J. and Willemot, J.,** An ultrastructural study of the microfilaments in rat brain by means of heavy meromyosin labeling. I. The perikaryon, the dendrites and the axon, *Cell Tissue Res.*, 160, 1, 1975.
21. **Hirokawa, N.,** Cross-linker system between neurofilaments, microtubules, and membranous organelles in frog axons revealed by the quick-freeze, deep-etching method, *J. Cell Biol.*, 94, 129, 1982.
22. **Mori, H., Komiya, Y., and Kurokawa, M.,** Slow axoplasmic transport. Its asymmetry in two branches of bifurcating axons, *Proc. Jpn. Acad. Ser B,* 53, 252, 1977.
23. **Mori, H., Komiya, Y., and Kurokawa, M.,** Slowly migrating axonal polypeptides. Inequalities in their rate and amount of transport between two branches of bifurcating axons, *J. Cell Biol.*, 82, 174, 1979.
24. **Andres, K. H.,** Untersuchungen über den Feinbau von Spinalganglien, *Z. Zellforsch.*, 55, 1, 1961.
25. **Coggeshall, R. E., Coulter, J. D., and Willis, W. D.,** Unmyelinated axons in the ventral roots of the cat lumbosacral enlargement, *J. Comp. Neurol.*, 153, 39, 1974.
26. **Yokoyama, K., Tsukita, S., Ishikawa, H., and Kurokawa, M.,** Early changes in the neuronal cytoskeleton caused by β,β′-iminodipropionitrile: selective impairment of neurofilament polypeptides, *Biomed. Res.*, 1, 537, 1980.
27. **Tashiro, T. and Komiya, Y.,** Two distinct components of tubulin transport in sensory axons of the rat recognized by dimethyl sulfoxide treatment, *Biomed. Res.*, 4, 443, 1983.
28. **Schnapp, B. J. and Reese, T. S.,** Cytoplasmic structure in rapid-frozen axons, *J. Cell Biol.*, 94, 667, 1982.
29. **Brady, S. T. and Lasek, R. J.,** Nerve-specific enolase and creatine phosphokinase in axonal trasnport: soluble proteins and the axoplasmic matrix, *Cell,* 23, 515, 1981.
30. **Brady, S. T., Tytell, M., Heriot, K., and Lasek, R. J.,** Axonal transport of calmodulin: a physiologic approach to identification of long-term associations between proteins, *J. Cell Biol.*, 89, 607, 1981.
31. **Garner, J. A. and Lasek, R. J.,** Clathrin is axonally transported as part of slow component b: the microfilament complex, *J. Cell Biol.*, 88, 172, 1981.
32. **Ellisman, M. H. and Porter, K. R.,** Microtrabecular structure of the axoplasmic matrix: visualization of cross-linking structures and their distribution, *J. Cell Biol.*, 87, 464, 1980.
33. **Mori, H. and Kurokawa, M.,** Intra-axonal transport of clathrin and actin, *Biomed. Res.*, 2, 677, 1981.
34. **Garner, J. A. and Lasek, R. J.,** Cohesive axonal transport of the slow component b complex of polypeptides, *J. Neurosci.*, 2, 1824, 1982.
35. **Tsukita, S., Usukura, J., Tsukita, S., and Ishikawa, H.,** The cytoskeleton in myelinated axons: a freeze-etch replica study, *Neuroscience,* 7, 2135, 1982.
36. **Lorenz, T. and Willard, M.,** Subcellular fractionation of intra-axonally transported polypeptides in the rabbit visual system, *Proc. Natl. Acad. Sci. U.S.A.*, 75, 505, 1978.
37. **Willard, M. and Hulebak, K. L.,** The intra-axonal transport of polypeptide H: evidence for a fifth (very slow) group of transported proteins in the retinal ganglion cells of the rabbit, *Brain Res.*, 136, 289, 1977.
38. **Willard, M., Cowan, W. M., and Vagelos, P. R.,** The polypeptide composition of intra-axonally transported proteins: evidence for four transport velocities, *Proc. Natl. Acad. Sci. U.S.A.*, 71, 2183, 1974.
39. **Levine, J. and Willard, M.,** Fodrin: axonally transported polypeptides associated with the internal periphery of many cells, *J. Cell Biol.*, 90, 631, 1981.
40. **Kakiuchi, S., Sobue, K., Morimoto, K., and Kanda, K.,** A spectrin-like calmodulin-binding protein (calspectin) of brain, *Biochem. Int.*, 5, 755, 1982.
41. **Kakiuchi, S.,** Calmodulin-binding proteins in brain, *Neurochem. Int.*, 5, 159, 1983.

42. **Kakiuchi, S. and Sobue, K.,** Control of the cytoskeleton by calmodulin-binding proteins, *Trends Biochem. Sci.*, 8, 59, 1983.
43. **Kakiuchi, S., Sobue, K., Kanda, K., Morimoto, K., Tsukita, S., Tsukita, S., Ishikawa, H., and Kurokawa, M.,** Correlative biochemical and morphological studies of brain calspectin: a spectrin-like calmodulin-binding protein, *Biomed. Res.*, 3, 400, 1982.
44. **Tsukita, S., Tsukita, S., Ishikawa, H., Kurokawa, M., Morimoto, K., Sobue, K., and Kakiuchi, S.,** Binding sites of calmodulin and actin on the brain spectrin, calspectin, *J. Cell Biol.*, 97, 574, 1983.
45. **Levine, J. and Willard, M.,** Redistribution of fodrin (a component of the cortical cytoplasm) accompanying capping of cell surface molecules, *Proc. Natl. Acad. Sci. U.S.A.*, 80, 191, 1983.
46. **Willard, M.,** The identification of two intra-axonally transported polypeptides resembling myosin in some respects in the rabbit visual system, *J. Cell Biol.*, 75, 1, 1977.
47. **Tashiro, T., Kasai, H., and Kurokawa, M.,** A calmodulin-related polypeptide rapidlly migrates within the mammalian nerve, *Biomed. Res.*, 1, 292, 1980.
48. **Tashiro, T. and Kurokawa, M.,** Rapid transport of a calmodulin-related polypeptide in the vagal nerve, in *Axoplasmic Transport,* Weiss, D. G., Ed., Springer-Verlag, Basel, 1982, 153.
49. **McLean, W. G., McKay, A. L., and Sjöstrand, J.,** Electrophoretic analysis of axonally transported proteins in rabbit vagus nerve, *J. Neurobiol.*, 14, 227, 1983.
50. **Agostoni, E., Chinnock, J. E., Daly, M. de Burgh, and Murray, J. G.,** Functional and histological studies of the vagus nerve and its branches to the heart, lungs and abdominal viscera in the cat, *J. Physiol.*, 135, 182, 1957.
51. **Peters, A., Palay, S. L., and Webster, H. De F.,** The fine structure of the nervous system, in *The Cells and Their Processes,* Hoeber, New York, 1970, chap. 4.
52. **Komiya, Y. and Kurokawa, M.,** Asymmetry of protein transport in two branches of bifurcating axons, *Brain Res.*, 139, 354, 1978.
53. **Lasek, R. J.,** Axoplasmic transport in cat dorsal root ganglion cells as studied with [^{3}H]-L-leucine, *Brain Res.*, 7, 360, 1968.
54. **Ochs, S.,** *Axoplasmic Transport and Its Relation to Other Nerve Functions,* John Wiley & Sons, New York, 1982, chap. 11.
55. **Wujek, J. R. and Lasek, R. J.,** Correlation of axonal regeneration and slow component *b* in two branches of a single axon, *J. Neurosci.*, 3, 243, 1983.
56. **Fine, R. E. and Bray, D.,** Actin in growing nerve cells, *Nature New Biol.*, 234, 115, 1971.
57. **Skene, J. H. P. and Willard, M.,** Changes in axonally transported proteins during axon regeneration in toad retinal ganglion cells, *J. Cell Biol.*, 89, 86, 1981.
58. **Giulian, D., Ruisseaux, H. D., and Cowburn, D.,** Biosynthesis and intraaxonal transport of proteins during neuronal regeneration, *J. Biol. Chem.*, 255, 6494, 1980.
59. **Cragg, B. G.,** What is signal for chromatolysis?, *Brain Res.*, 23, 1, 1970.
60. **Grafstein, B. and Forman, D.,** Intracellular transport in neurons, *Physiol. Rev.*, 60, 1167, 1980.
61. **Hendrickson, A. E. and Cowan, W. M.,** Changes in the role of axoplasmic transport during postnatal development of the rabbit's optic nerve and tract, *Exp. Neurol.*, 30, 403, 1971.
62. **Droz, B. and Leblond, C. P.,** Axonal migration of proteins in the central nervous system and peripheral nerves as shown by radioautography, *J. Comp. Neurol.*, 121, 325, 1963.
63. **Lasek, R. J.,** Axonal transport of proteins in dorsal root ganglion cells of the growing cat: a comparison of growing and mature neurons, *Brain Res.*, 20, 121, 1970.
64. **Levine, J., Simon, C., and Willard, M.,** Mechanistic implications of the behavior of axonally transported proteins, in *Axoplasmic Transport,* Weiss, D. G., Ed., Springer-Verlag, Basel, 1982, 275.
65. **Willard, M. and Simon, C.,** Antibody decoration of neurofilaments, *J. Cell Biol.*, 89, 198, 1981.
66. **Dahlström, A.,** Effect of colchicine on transport amine granules in sympathetic nerves of rat, *Eur. J. Pharmacol.*, 5, 111, 1968.
67. **Kreutzberg, G. W.,** Neuronal dynamics and axonal flow. IV. Blockage of intra-axonal enzyme transport by colchicine, *Proc. Nat. Acad. Sci. U.S.A.*, 62, 722, 1969.
68. **Abe, T., Haga, T., and Kurokawa, M.,** Retrograde axoplasmic transport: its continuation as anterograde transport, *FEBS Lett.*, 47, 272, 1974.
69. **Kristensson, K. and Sjöstrand, J.,** Retrograde transport of protein tracer in the rabbit hypoglossal nerve during regeneration, *Brain Res.*, 45, 175, 1972.
70. **Hanson, M. and Edström, A.,** Mitosis inhibitors and axonal transport, *Int. Rev. Cytol. Suppl.*, 7, 373, 1978.
71. **Brimijoin, S.,** Microtubules and the capacity of the system for rapid axonal transport, *Fed. Proc. Fed. Am. Soc. Exp. Biol.*, 41, 2312, 1982.
72. **Karlsson, J. -O., Hansson, H. -A., and Sjöstrand, J.,** Effect of colchicine on axonal transport and morphology of retinal ganglion cells, *Z. Zellforsch.*, 115, 265, 1971.
73. **Sjöstrand, J. and Hansson, H. -A.,** Effect of colchicine on the transport of axonal protein in the retinal ganglion cells of the rat, *Exp. Eye Res.*, 12, 261, 1971.

74. **James, K. A. C., Bray, A. C., Morgan, I. G., and Austin, L.,** The effect of colchicine on the transport of axonal protein in the chicken, *Biochem. J.*, 117, 767, 1970.
75. **McGregor, A. M., Komiya, Y., Kidman, A. D. and Austin, L.,** The blockage of axoplasmic flow of proteins by colchicine and cytochalasins A and B, *J. Neurochem.*, 21, 1059, 1973.
76. **Komiya, Y. and Kurokawa, M.,** Preferential blockade of the tubulin transport by colchicine, *Brain Res.*, 190, 505, 1980.
77. **Komiya, Y.,** Recovery of slow axonal transport after colchicine treatment in rat sciatic nerve, *J. Neurobiol.*, 14, 87, 1983.
78. **Chou, S. -M. and Hartmann, H. A.,** Axonal lesions and waltzing syndrome after IDPN administration in rats. With a concept "axostasis", *Acta Neuropathol.*, 3, 428, 1964.
79. **Chou, S. -M. and Hartmann, H. A.,** Electron microscopy of focal neuroaxonal lesions produced by β,β′-iminodipropionitrile (IDPN) in rats. I. The advanced lesions, *Acta Neuropathol.*, 4, 590, 1965.
80. **Griffin, J. W., Hoffman, P. N., Clark, A. W., Carrol, P. T., and Price, D. L.,** Slow axonal transport of neurofilament proteins; impairment by β,β′-iminodipropionitrile administration, *Science*, 202, 633, 1978.
81. **Papasozomenos, S. Ch., Autilio-Gambetti, L., and Gambetti, P.,** Reorganization of axoplasmic organelles following β,β′-iminodipropionitrile administration, *J. Cell Biol.*, 91, 866, 1981.
82. **Papasozomenos, S. Ch., Yoon, M., Crane, R., Autilio-Gambetti, L., and Gambetti, P.,** Redistribution of proteins of fast axonal transport following administration of β,β′-iminodipropionitrile: a quantitative autoradiographic study, *J. Cell Biol.*, 95, 672, 1982.
83. **Lentz, T. L.,** Distribution of leucine-^{3}H during axoplasmic transport within regenerating neurons as determined by electron-microscope radioautography, *J. Cell Biol.*, 52, 719, 1972.
84. **Cancalon, P.,** Influence of temperature on the velocity and on the isotope profile of slowly transported labeled proteins, *J. Neurochem.*, 32, 997, 1979.
85. **Gross, G. W. and Beidler, L. M.,** A quantitative analysis of isotope concentration profiles and rapid transport velocities in the C-fibers of the garfish olfactory nerve, *J. Neurobiol.*, 6, 213, 1975.
86. **Hendrickson, A. E.,** Electron microscopic distribution of axoplasmic transport, *J. Comp. Neurol.*, 144, 381, 1972.
87. **Erickson, P. F., Seamon, K. B., Moore, B. W., Lasher, R. S., and Minier, L. N.,** Axonal transport of the Ca^{2+}-dependent protein modulator of 3′:5′-cyclic AMP phosphodiesterase in the rabbit visual system, *J. Neurochem.*, 35, 242, 1980.
88. **Erickson, P. F. and Moore, B. W.,** Investigation of the axonal transport of three acidic, soluble proteins (14-3-2, 14-3-3, and S-100) in the rabbit visual system, *J. Neurochem.*, 35, 232, 1980.
89. **Weisenberg, R. C.,** Microtubule formation *in vitro* in solutions containing low calcium concentrations, *Science*, 177, 1104, 1972.
90. **Haga, T., Abe, T., and Kurokawa, M.,** Polymerization and depolymerization of microtubules *in vitro* as studied by flow birefringence, *FEBS Lett.*, 39, 291, 1974.
91. **Nishida, E., Kumagai, H., Ohtsuki, I., and Sakai, H.,** The interaction of calcium-dependent regulator protein of cyclic nucleotide phosphodiesterase and microtubule proteins. I. Effect of calcium-dependent regulator protein on the calcium sensitivity of the microtubule assembly, *J. Biochem.*, 85, 1257, 1979.
92. **Marcum, J. M., Dedman, J. R., Brinkley, B. R., and Means, R. R.,** Control of microtubule assembly-disassembly by calcium-dependent regulator protein, *Proc. Natl. Acad. Sci. U.S.A.*, 75, 3771, 1978.
93. **Gilbert, D. S., Newby, B. J., and Anderton, B. H.,** Neurofilament disguise, destruction and discipline, *Nature (London)*, 256, 586, 1975.
94. **Pant, H. C. and Gainer, H.,** Properties of a calcium-activated protease in squid axoplasm which selectively degrades neurofilament proteins, *J. Neurobiol.*, 11, 1, 1980.
95. **Kamakura, K., Ishiura, S., Sugita, H., and Toyokura, Y.,** Identification of Ca^{2+}-activated neutral protease in the peripheral nerve and its effects on neurofilament degeneration, *J. Neurochem.*, 40, 908, 1983.
96. **Zimmerman, U. -J. P. and Schlaepfer, W. W.,** Characterization of a brain calcium-activated protease that degrades neurofilament proteins, *Biochemistry*, 21, 3877, 1982.
97. **Pant, H. C., Gallant, P. E., Gould, R., and Gainer, H.,** Distribution of calcium-activated protease activity and endogenous substrates in the squid nervous sytem, *J. Neurosci.*, 2, 1578, 1982.
98. **Malik, M. N., Fenko, M. D., Iqbal, K., and Wisniewski, H. M.,** Purification and characterization of two forms of Ca^{2+}-activated neutral protease from calf brain, *J. Biol. Chem.*, 258, 8955, 1983.
99. **Ishizaki, Y., Tashiro, T., and Kurokawa, M.,** A calcium-activated protease which preferentially degrades the 160-kDa component of the neurofilament triplet, *Eur. J. Biochem.*, 131, 41, 1983.
100. **Tashiro, T. and Ishizaki, Y.,** A calcium-dependent protease selectively degrading the 160 000 M_r component of neurofilaments is associated with the cytoskeletal preparation of the spinal cord and has an endogenous inhibitory factor, *FEBS Lett.*, 141, 41, 1982.
101. **Nixon, R. A., Brown, B. A., and Marotta, C. A.,** Posttranslational modification of a neurofilament protein during axoplasmic transport: implications for regional specialization of CNS axons, *J. Cell Biol.*, 94, 150, 1982.

102. **Tashiro, T., Komiya, Y., and Kurokawa, M.,** Colchicine blocks exclusively and completely the intra-axonal transport of tubulin, *Biomed. Res.*, 5, 381, 1984.
103. **Tashiro, T., Kurokawa, M., and Komiya, T.,** Two populations of axonally transported tubulin differentiated by their interaction with neurofilaments, *J. Neurochem.*, 43, 1220, 1984.

Chapter 8

NUCLEIC ACID TRANSPORT

N. A. Ingoglia, G. Chakraborty, and N. F. Zanakis

TABLE OF CONTENTS

I. INTRODUCTION

One of the most widely used techniques for studying axonal transport of macromolecules is to inject a radioactively labeled precursor into the vicinity of a neuronal cell body and then examine the distribution of radioactive product along the length of the nerve. When this technique is used for the study of proteins, the findings are in many ways quite uncomplicated. According to most investigators, following injection of radioactive amino acids, labeled proteins are either in the axon or perhaps the post-synaptic cell, but not in Schwann or glial cells surrounding the axon,[1] and while labeled protein moves down the nerve according to fixed time courses, their amino acid precursors are generally considered not to be transported[1] (but see Weiss[2]).

The same type of experiments performed using nucleic acid or lipid precursors, leads to significantly different results. In the case of lipids there is transport of unbound precursors and depending on the precursor used, the Schwann or glial cells may be labeled.[3,4] When labeled nucleosides are used to examine the axonal transport of nucleic acids, there is a movement along the axons of unbound trichloroacetic acid (TCA) soluble material and a variety of studies have reported that all cells in the vicinity of the axon (Schwann, or glia, neurons, and ependymal cells) are capable of utilizing those precursors for nucleic acid synthesis.[5] This has been true for experiments using uridine, adenosine, guanosine, and cytidine, but not for experiments using thymidine to specifically investigate the axonal transport of DNA.

II. AXONAL TRANSPORT OF DNA

The most definitive study on the axonal transport of DNA has been published only as an abstract.[6] These experiments were performed in the goldfish visual system and sought to answer two questions: first, is there a transport of DNA? And second, does the axon transport and supply thymidine to glia surrounding the axon to be utilized for DNA synthesis during reactive gliosis (as has been described for RNA precursors)? Injections of ^{3}H-thymidine into the right eye of fish resulted in significant differences in left vs. right tectum radioactivity (suggesting axonal transport) for both free ^{3}H-thymidine and ^{3}H-DNA. Analysis of tissue radioactivity by subcellular fractionation indicated that the difference in the amount of ^{3}H DNA in left vs. right tecta could be accounted for by a concentration of ^{3}H-DNA in mitochondria.[6] In a subsequent analysis of the distribution of ^{3}H-DNA in the tissue by EM autoradiography, virtually all of the radioactivity was found localized to the axon.[55] Thus, these studies indicate that there is an axonal transport of DNA as part of axonal mitochondria. But, unlike other nucleosides related to glial RNA synthesis, the axon does not support the axonal transport of free thymidine to be transferred to surrounding glia for DNA synthesis. This is an important point in discussion of the possible trophic regulatory roles of transcellularly transferred nucleosides to be used for nucleic acid synthesis by glia.[5]

III. AXONAL TRANSPORT OF RNA

The axonal transport of RNA has proved to be significantly more complex and intriguing than DNA transport. The remainder of this chapter is devoted to a review of the literature related to the characteristics and function of RNA in axons.

A. Presence of tRNA in Squid Axoplasm

While there had been several reports suggesting that RNA was present in axons, the first definitive evidence was supplied by Lasek and colleagues in 1973.[7] The experiments were simple and conclusive due in large part to the system used, the giant axon obtained from the squid, *Loligo pealei*. The experiments entailed collecting axoplasm extruded from the giant axon, extracting presumptive RNA by phenol extraction and ethanol precipitation, and then fractionating the sample by SDS polyacrylamide gel electrophoresis. When axoplasm was isolated free of glial contamination it was found that RNA was, in fact, present in the sample, but that the predominant if not the only species of RNA present was 4S RNA. These results were exciting because they fit with the morphological evidence that axons lack ribosomes, and by inference should also lack ribosomal RNA.[8] This original finding by Lasek and collaborators has since been replicated by his own group[9] and by two independent laboratories.[10,11] RNA of the 4S species is generally considered to be transfer RNA and this proved to be the case in axoplasm of the squid. Using somewhat different approaches Black and Lasek[9] and Giuditta et al.,[10] each showed that 4S RNA in the giant axon of the squid did have the properties of tRNA. Several years later we repeated these experiments using a variety of labeled amino acids and were able to show that not only did squid axoplasm contain a variety of species of tRNA, but that their cognate aminoacyl tRNA synthetases also were integral components of axoplasm.[12] Thus, axoplasm of the giant axon of the squid contains the elements necessary for the aminoacylation of tRNA. This is not mitochondrial tRNA,[7] but is tRNA contained in the soluble compartment of axoplasm. The question can be raised, of course, of how applicable this finding is to vertebrate axons.

B. Axonal Transport of tRNA in Vertebrate Axons

Prior to Lasek's finding of 4S RNA in the giant axon, many reports had been published on the presence of RNA in vertebrate axons. The difficulty with these experiments was either that the biochemical measurements were indirect and inferential and did not preclude the possibility that glial RNA as well as axonal RNA was being measured, or in morphological studies that the silver grains (^{3}H-RNA) could not unequivocally be shown to be intra-axonal.[13] Gambetti et al.,[14] using EM autoradiography first reported the presence of small amounts of RNA in rabbit optic axons. Following intraocular injections of ^{3}H-uridine less than 20% of the total ^{3}H-RNA in the optic nerve was inside the axon (the remainder being in surrounding glia). Since these experiments were done at the EM level, however, it was certain that the silver grains (^{3}H-RNA) were in fact originating from axoplasm and not from glia.

Several years later, in a collaborative effort between our two laboratories, we were able to show the presence of ^{3}H-RNA within regenerating optic axons of goldfish.[15] The results were conclusive and exciting because (1) they unequivocally demonstrated the presence of RNA in vertebrate axons, (2) the amount of RNA in regenerating axons was greatly increased over nonregenerating controls, suggesting a special function for RNA during axonal growth, and (3) much of the RNA was found in the most advanced tip of the growing axons, the axonal growth cones. The finding of RNA in axons of a variety of vertebrate nerves both regenerating and intact has since been confirmed by a number of laboratories and has been reviewed elsewhere.[13]

It would seem logical that the RNA present in vertebrate axons would, like invertebrates, be composed primarily of 4S RNA. This has been the conclusion reached by our laboratory and by Austin's laboratory in Australia.[16] Experiments in goldfish, rat, and chicken optic nerves, and in rat and chicken sciatic nerves, indicate that the RNA present within the axons of each of those systems is likely to be 4S RNA.[13] Over the past several years both of our laboratories have attempted to demonstrate that the 4S RNA in vertebrate axons, like invertebrates is transfer RNA. Using similar approaches but different analytical techniques we have reached the same conclusion, i.e., that the 4S RNA present in axons of chick optic nerves[16] and regenerating optic axons of goldfish[17] is transfer RNA. Based on the available data, primarily from Austin's, Gambetti's, and our laboratories, we have proposed that tRNA is likely to be present in all axons both central and peripheral and from invertebrates to the higher vertebrates.[13]

The origin of axonal RNA in vertebrate nerves is also fairly well established. The possible sites of synthesis are (1) intra-axonal (mitochondrial), (2) glial or Schwann cell synthesis followed by transcellular transfer, or (3) neuronal somal synthesis followed by axonal transport. Mitochondrial synthesis is unlikely since biochemical analysis of squid axoplasmic tRNA indicates that it is not associated with axonal mitochondria[7] and in autoradiographic experiments involving vertebrate axons, ^{3}H-RNA is rarely found associated with mitochondria.[15] Local synthesis by glia followed by transcellular transfer is also unlikely. In several experiments we attempted to label RNA in axons by injecting ^{3}H-uridine into the fiber pathways (intracerebral injections in fish), but found no ^{3}H-RNA within axons.[15] Gould has recently injected ^{3}H-uridine directly into the rat sciatic nerve and has also been unable to observe ^{3}H-RNA in axons of that nerve.[18] There is, however, good evidence for the axonal transport of RNA. In all of the experiments by Austin's and our laboratories, as well as those of earlier investigators axonal RNA can be labeled by injecting ^{3}H-uridine into the vicinity of the neuronal cell body and monitoring the appearance of ^{3}H-4S RNA along distal segments of the nerve or in nerve terminals. Whether the analysis is biochemical or morphological the conclusion has been the same, i.e., that 4S RNA must arrive in axons and axon terminals as a result of axonal transport.

C. Axonal Transport of tRNA in Regenerating Nerves

The facet of this work which is most exciting and likely to be most relevant to clinical neurology, is the effect of nerve growth on the axonal transport of RNA. In goldfish, crushing the optic nerve results in complete regeneration and restoration of function.[19-21] During regrowth of the nerve, the amount of 4S RNA which is transported increases dramatically. There may be as much as a 15-fold increase in RNA reaching optic terminals as fibers are reconnecting[22,23] and a severalfold increase in the nerves well before axons have reconnected.[24] This change in RNA transport can also be seen in other species where nerves are growing, i.e., in optic nerves of neonatal rats[25] and in regenerating motor fibers of the rat sciatic nerve.[26,27] In the latter two cases the difference between intact and regenerating nerves was not as pronounced as in goldfish, but there were clear indications of an increase in transport of RNA in growing axons compared with nongrowing controls.

An important question raised by these studies relates to nerves which are not capable of regeneration: i.e., does a central neuron of mammals, for example, increase its export of 4S RNA in response to damage to the axon? We do not have a definitive answer to this question, but preliminary results of experiments performed in our laboratory, indicate that although mature optic axons of rats do transport RNA, they do not increase their export of RNA when the nerve is damaged. If this proves to be true, it may turn out to be an important link in our attempts to understand the reasons for the lack of regeneration of injured central axons in humans.

D. Transfer RNA in Axons as an Amino Acid Donor to Acceptor Proteins: Post-Translational Protein Modification

The central question to the findings summarized above is the identification of the function of tRNA in axons.

Before Lasek and collaborators first reported the presence of 4S RNA in axons its function had essentially been predicted. In 1971 Droz and Koenig reported the results of experiments in which they found a low-level incorporation of ^{3}H-amino acids into proteins in synaptosomes.[28] In their concluding speculations they suggested that this incorporation was unlikely to be protein synthesis, but that it could represent the post-translational addition of amino acids to proteins as described in other systems 8 years earlier by Kaji and associates.[29] This reaction requires ATP, K^+, Mg^{+2}, and tRNA but not ribosomal RNA and is able to transfer arginine in eukaryotes (phenylalanine, leucine, and tryptophan in prokaryotes), from a donor tRNA to an acceptor protein.[30] Thus, the findings of Droz and Koenig predict the presence of tRNA in axons 2 years prior to its demonstration by Lasek and colleagues in 1973. Most of the papers of Austin's, Lasek's, and our groups also proposed this as the most likely role of tRNA in axons, but until late in 1982 its proof remained elusive.

In the summers of 1981 and 1982 we collaborated with Antonio Giuditta at the Marine Biological Laboratories in Woods Hole, Mass., to examine tRNA in squid axoplasm, with the ultimate goal of identifying its function in axons. Initially we found that tRNAs were present in axoplasm for a variety of amino acids (and not just arginine) and that their cognate aminoacyl tRNA synthetase enzymes were also natural constituents of axoplasm.[12] This seemed to argue against a role for tRNA as a participant in post-translational protein modification since, as was mentioned above, only arginine has been reported to play this role in eukaryotes. However, several related experiments performed in the brains of rats, persuaded us that that the arginine reaction might not be the only amino acid involved in post-translational modification.[31]

Next, we performed experiments in which we attempted to demonstrate the transfer of amino acids from tRNA to proteins under in vitro conditions and under conditions in which ^{3}H-aminoacylated tRNA was injected directly into the axon. Under neither condition could we find evidence for the incorporation of amino acids into proteins,[11] confirming results reported previously.[32] We then attempted to elicit the reaction (the incorporation of ^{3}H-amino acids into proteins) in a 150,000 $\times$ g supernatant fraction of axoplasm, a fraction which theoretically would contain all the elements needed for the reaction (tRNA and associated soluble proteins) but would lack axonal organelles and precipitable proteins. Again, we were not able to demonstrate the incorporation of amino acids into proteins in this high speed supernatant fraction of squid axoplasm.[11]

As part of the same experiment, we passed the supernatant through a Sephacryl S200 column, and removed all molecules of less than 25,000 daltons. Fractions were collected containing molecules of (1) greater than 120,000 daltons (2) between 60,000 and 90,000 daltons, (3) between 25,000 and 40,000 daltons. Each of these fractions was incubated with a variety of ^{3}H-amino acids and a reaction mixture as described by Barra et al.[33]

For each amino acid, the removal of molecules under 20,000 mol wt resulted in the incorporation of ^{3}H-amino acids into proteins in amounts significantly greater than heat inactivated controls (40 times for arg and lys to 2.5 times for asp). The largest incorporation of amino acids into protein occurred in the highest molecular weight fraction, but the other fractions also demonstrated activity. Thus, we were able to show that the incorporation of amino acids into proteins does occur in axoplasm, but can be demonstrated, only after small molecules are removed. This reaction is not protein synthesis since it occurs in the high speed supernatant of axoplasm and also in the absence of endogenous amino acids (removed by the S200 column). Thus, these experiments demonstrate the post-translational addition of a variety of amino acids to acceptor proteins in axoplasm of the squid giant axon and the presence of molecules in axoplasm capable of inhibiting this reaction.[11]

In other experiments we have found that (1) the reaction also occurs in rat sciatic nerves and that the reaction ingredients are transported axonally,[34] (2) that the reaction also occurs in the goldfish optic nerve,[24] and (3) that the level of amino acids incorporated into proteins in both increases dramatically during nerve regeneration. Other experiments indicate that activity is lost if tRNA is removed from the reaction suggesting that the reaction is dependent on tRNA.[35]

The data we have collected to this point lead us to conclude that axoplasm contains the elements necessary for the post-translational addition of a variety of amino acids to acceptor proteins, that these reactions are likely to be dependent on the presence of tRNA as the amino acid donor, that the reaction is greatly affected by the functional state of the axon, i.e., nerve regeneration; and that in vivo the reactions are inhibited by a substance (or

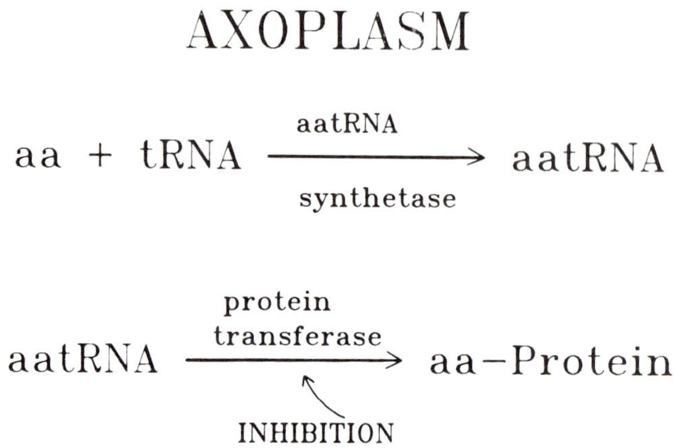

FIGURE 1. The reaction shown above has been demonstrated to occur in axoplasm of squid giant stellate axons,[11] rat sciatic nerve axons,[35] and goldfish optic axons.[24] The aminoacylation of tRNA in axons has been described previously,[9,10,12,16,17] but the transfer of a variety (and perhaps all) amino acids from tRNA to proteins has been described only in the past year. The reactions appear to be under the control of an inhibitory substance or substances which must be removed before the reaction can be observed.[11,35,24]

substances) whose molecular weight is less than 20,000 daltons. We do not know yet what the inhibitor is but it is not Ca^{++}; it is heat stable, and its activity is not lost by incubation with proteinase K, thus it is likely not to be a protein.[35] The reaction is summarized in Figure 1.[35]

IV. PHYSICAL PROPERTIES OF AXONAL tRNA

Several questions arise when one realizes that tRNA in any axon, if synthesized in the neuronal cell body as the evidence suggests, must be transported long distances over extended periods of time and thus, must be resistant to degradation during transport. Understanding how this is accomplished by an axon is best approached by first understanding some of the physical properties of axonal tRNA.

The rate of transport of tRNA has been difficult to estimate using standard tracer methodology because of the local synthesis of RNA from transported ^3H precursors in cells surrounding the axon. The best approximation is that it migrates in the faster component of the slow wave, slow component b.[36] This is also the component that contains the bulk of the soluble proteins transported by the axon;[1] this brings us to one of the properties of axonal tRNA. In most of our experiments in squid, rat sciatic nerve, and in optic nerves of goldfish, the greatest activity (ability to incorporate amino acids into proteins) was found in the highest molecular weight fraction of the S200 column.[11,24,35]

The fraction which we have called FI has a molecular weight of greater than 120,000. Since transfer RNA has a molecular weight of 22,000, one would predict that it would not be found in FI. However, following the procedures of Bandyopadhyay and Deutscher,[37] i.e., homogenizing the tissue very gently and making sure that the samples were not frozen, we found that like rat liver homogenates, tRNA in rat sciatic nerves, is part of FI. We have reached this conclusion based on experiments in which FI was treated with phenol: chloroform, isoamyl alcohol, and then nucleic acids were precipitated with ethanol and finally dissolved in buffer. When this sample was layered on a 10% SDS polyacrylamide gel, an optical density peak was found which coincided with an *E. coli* tRNA marker, indicating the presence of 4S(t)RNA.[35] Thus, FI, although it excludes molecules under 120,000 mol wt, does contain tRNA. FI also is capable of aminoacylation of tRNA and of transfering amino acids to proteins[11] and therefore must contain aminoacyl tRNA synthetases, and aminoacyl (protein) transferases, the enzymes responsible for mediating these reactions.[29] These proteins and tRNA appears to be part of a high molecular weight protein, nucleic acid complex in axons, a complex which is transported axonally as part of slow component b. This complex may form a structure similar to the granules identified morphologically by Schnapp and Reese in turtle optic nerves.[38] Within this complex it is likely that tRNA is acylated, deacylated, and then reacylated, according to environmental signals. The nature of these signals is, at present, unknown. The amino acids which are removed from tRNAs are then transferred to acceptor proteins; we are currently attempting to characterize these acceptor proteins.

There are likely to be other proteins and smaller molecules associated with this high molecular weight complex. We have demonstrated the association of the polyamine spermine with tRNA transported in regenerating optic axons of goldfish.[39] This was not unexpected since spermine is known to be physically bound to biologically active tRNAs.[40] The binding of spermine to the anticodon stem and the variable loop region of tRNAs are known to be critical for aminoacylation of tRNA.[40,41] But spermine and perhaps other polyamines, may have additional functions as part of this complex. One of the most intriguing is as a potential inhibitor of axonal ribonucleases which might destroy tRNAs in the axon. We have recently reported that the tRNA present in FI appears to be resistant to the action of bovine pancreatic ribonuclease (50 or 100 μg/mℓ), but that ribonuclease is effective if FI is broken down so that the high molecular weight complex is destroyed.[11] We speculated that the protein complex of which tRNA is a part might be sufficient to confer protection to tRNA molecules from ribonuclease. While this may be true, the polyamines may also play a role in this function. Thus, it has been shown that yeast tRNA is resistant to bovine pancreatic ribonuclease in the presence of small amounts of spermine.[42] It is thought that the polycationic nature of spermine allows it to form electrostatic or hydrogen bonds with tRNA causing the tRNA to be more compact and more tightly folded and hence less susceptible to ribonucleases. Other experiments have also implicated the polyamines in maintaining the stability of RNA.[43] Thus, polyamines may function in association with axonal tRNA not only as requisite cations for efficient biological activity, (aminoacylation and amino acid transfer reactions), but also as agents capable of protecting tRNA from enzymatic degradation during its axonal sojourn.

IV. SPECULATION ON THE ROLE OF tRNA MEDIATED POST-TRANSLATIONAL PROTEIN MODIFICATION IN AXONAL METABOLISM

It seems likely that tRNA mediated protein modification by amino acid addition is a metabolic reaction which occurs to some extent in all axons and involves several acceptor proteins. The finding that a variety of amino acids and not just arginine are involved in this reaction is somewhat surprising, considering the previous literature (reviewed by Soffer[30]). But this being the case, the way seems clear for the investigation of the role of a range of amino acids in the modification of a host of acceptor proteins, under a variety of experimental conditions. The area of research most intriguing to us is protein modification during nerve regeneration. Our original observation, that regenerating optic axons of goldfish transport axonally large amounts of tRNA,[15,44] suggests that protein modification is likely to be a particularly important function during axonal growth. The precise role(s) for modification reactions in axonal elongation are not known, and can, at this point, only be speculated upon.

1. During axonal elongation new memebrane is inserted into the growing axon at the base of the advancing growth cone[45] as well as along the length of the axon itself.[46,47]

The insertion of new membrane fragments requires a local biochemical event in the axon, the nature of which is poorly understood. We speculate that the modification, by amino acid addition, of axonally transported membrane precursors may be a prerequisite for their insertion into the membrane. The presence of large amounts of tRNA in axonal growth cones[15] and the report of post-translational modification of membrane proteins by arginylation[48] in erythrocyte ghost membranes are consistent with this speculation.

2. During axonal elongation the course of growing axons may be influenced by a variety of local factors including NGF, Ca^{++}, dibutyryl cyclic AMP, or cyclic GMP,[49] by the presence or absence of target tissue,[50] and in general by a variety of chemical trophic substances released either by target neurons or glia in the pathway of the growing axon.[51] The response of the growing axon to any of these factors is undoubtedly a local reaction of the growing tip of the axon to environmental cues and not of the cell body which in many cases is a significant distance from the growth cones. The growth cone contains large amounts of tRNA, presumably actively involved in protein modification in the advancing axon tip. We speculate that one of the ways in which a growing axon may respond to external guidance cues is to modify axonal proteins which may in some as yet unexplained way, (perhaps by selective membrane insertion) be able to influence directionality of the growing neuron.
3. Another way in which post-translational protein modification might be involved in nerve regeneration is in neuronal recognition. It is well established that growing neurons will choose preferred pathways and avoid inappropriate ones, and that neurons capable of making the best regenerative responses, goldfish and frog optic fibers for example, do so with remarkable fidelity to their original projections.[52] This ability of a neuron to recognize an appropriate pathway for growth, as well as an appropriate post-synaptic neuron for synapse formation are processes which are likely also to be mediated at the level of the growth cone, rather than neuronal soma, and thus again are candidates for a function of tRNA mediated protein modification.
4. The ultimate goal of the growing nerve is to form functional synapses with appropriate target neurons. Synaptogenesis is likely to involve a complex series of biochemical steps, none of which have been described. Recently it was reported that during synaptogenesis, 4 of 300 identified axonally transported proteins changed their relative abundance in presynaptic neurons.[53] Other experiments did not identify changes in the amount of axonally transported proteins during synaptogenesis in the goldfish visual system but did report the loss of a protein spot (on 2D gels) and the appearance of a new one, which they conclude must have arisen from post-translational protein modification of an axonally transported protein in growth cones of regenerating axons.[54] The type of modification observed in these experiments was not apparent. However, tRNA mediated amino acid addition must be considered as a possible explanation of their data. Certainly, one could construct a model in which the contact of a growing axon with an appropriate postsynaptic neuron would result in the transfer of a signal from the target neuron, to the growing axon, a signal capable of eliciting the modification of a protein or family of proteins which would then lock-in that synaptic connection. It is likely that the tRNA dependent transfer of amino acids to acceptor proteins will turn out to occur under a variety of circumstances in both normal and pathological states. We believe that the further description and characterization of these reactions will mark the opening of a new field of investigation in the molecular biology of the axon.

REFERENCES

1. **Grafstein, B. and Forman, D. S.,** Intracellular transport in neurons, *Physiol. Rev.,* 60, 1167, 1980.
2. **Weiss, D.,** this volume, chap. 20.
3. **Droz, B. DiGiamberardino, L., Koenig, H. L., Boyenwal, J., and Hassig, R.,** Axon-myelin transfer of phospholipid components in the course of their axonal transport as visualized by radioautography, *Brain Res.,* 155, 347, 1978.
4. **Gould, R. M., Spivak, W. D., Sinatra, R. S., Lindquist, T. D., and Ingoglia, N. A.,** Axonal transport of choline lipids in normal and regenerating rat sciatic nerve, *J. Neurochem.,* 39, 1569, 1982.
5. **Ingoglia, N. A., Sharma, S. C., Pilchman, J., Baranowski, K., and Sturman, J. A.,** Axonal transport and transcellular transfer of nucleosides and polyamines in intact and regenerating optic nerves of goldfish: speculation of axonal regulation of periaxonal cell metabolism, *J. Neurosci.,* 2(10), 1412, 1982.
6. **Papasozomenous, S., Autilio-Gambetti, L., and Gambetti, P.,** Axonal transport of DNA during regeneration, *J. Neuropathol. Exp. Neurol.,* 39, 380, 1981.
7. **Lasek, R. J., Dabrowski, C., and Nordlander, R.,** Analysis of axoplasmic RNA from invertibrate giant axons, *Nature (London),* 244, 1188, 1973.
8. **Peters, A., Palay, S. L., and Webster, H. de,** *The Fine Structure of the Nervous System,* W. B. Saunders, Philadelphia, 1976.
9. **Black, M. M. and Lasek, R. J.,** The presence of transfer RNA in the axoplasm of the squid giant axon, *J. Neurobiol.,* 8, 229, 1977.
10. **Giuditta, A., Metatora, S., Felsami, A., and Del Rio, A.,** Factors of protein synthesis in the axoplasm of squid giant axons, *J. Neurochem.,* 28, 1393, 1977.
11. **Ingoglia, N. A., Giuditta, A., Zanakis, M. F., Babigian, A., Tasaki, I., Chakraborty, G., and Sturman, J.,** Incorporation of ^{3}H-amino acids into proteins in a partially purified fraction of axoplasm: evidence for transfer RNA mediated, post-translational protein modification in squid giant axons, *J. Neurosci.,* 3(12), 2463, 1983.
12. **Ingoglia, N. A. and Giuditta, A.,** Multiple species of transfer RNA are present in squid axoplasm, *Trans. Am. Soc. Neurochem.,* 13(1), 235, 1982.
13. **Ingoglia, N. A. and Zanakis, M. F.,** Axonal transport of 4S RNA, in *Axoplasmic Transport,* Weiss, D. G., Ed., Springer-Verlag, Basel, 1982, 161.
14. **Gambetti, P. C., Autilio-Gambetti, L., Shafer, B., and Pfaff, L.,** Quantitative autoradiographic study of labeled RNA in rabbit optic nerve after intraocular injection of ^{3}H-uridine, *J. Cell Biol.,* 59, 677, 1973.
15. **Gambetti, P., Ingoglia, N. A., Autilio-Gambetti, L., and Weis, P.,** Distribution of ^{3}H-RNA in goldfish optic tectum following intraocular or intracranial injection of ^{3}H-uridine: evidence of axonal migration of RNA in regenerating optic fibers, *Brain Res.,* 154, 284, 1978.
16. **Scheffer, J. W., Howe, N., Gunning, P. W., and Austin, L.,** The axoplasmic transport of transfer RNA in the chick optic system, *J. Neurochem.,* 43, 698, 1984.
17. **Zanakis, M. F., Eskin, B., and Ingoglia, N. A.,** Evidence that multiple species of aminoacylated transfer RNA are present in regenerating optic axons of goldfish, *Neurochem. Res.,* 9, 249, 1984.
18. **Gould, R. M.,** personal communication.
19. **Sperry, R. W.,** Optic nerve regeneration with return of vision in anurans, *J. Neurophysiol.,* 7, 57, 1944.
20. **Murray, M.,** Regeneration of retinal axons into the goldfish optic tectum, *J. Comp. Neurol.,* 168, 175, 1976.
21. **Murray, M.,** A quantitative study of regenerative sprouting by optic axons in goldfish, *J. Comp. Neurol.,* 209, 352, 1982.
22. **Ingoglia, N. A., Weis, P., and Mycek, J.,** Axonal transport of RNA during regeneration of the optic nerve of goldfish, *J. Neurochem.,* 6, 549, 1975.
23. **Ingoglia, N. A.,** 4S RNA in regenerating optic axons of goldfish, *J. Neurosci.,* 2, 331, 1982.
24. **Chakraborty, G., Leach, T., Zanakis, M. F., and Ingoglia, N. A.,** Posttranslational protein modification by tRNA dependent amino acid addition in regenerating optic nerves of goldfish, *J. Neurochem.,* in press, 1986.
25. **Politis, M. and Ingoglia, N. A.,** Axonal transport of nucleosides, nucleotides and 4S RNA in the neonatal rat visual system, *Brain Res.,* 169, 343, 1979.
26. **Lindquist, T. D. and Ingoglia, N. A.,** Evidence that 4S RNA is axonally transported in normal and regenerating rat sciatic nerves, *Brain Res.,* 166, 95, 1979.
27. **Lindquist, T. D., Ingoglia, N. A., and Gould, R.,** 4S RNA is transported axonally in normal and regenerating axons of the sciatic nerve of rats, *Brain Res.,* 230, 181, 1981.
28. **Droz, B. and Koenig, H. L.,** Dynamic condition of protein in axons and axon terminals, *Acta Neuropathol.,* 5, 109, 1971.
29. **Kaji, H., Novelli, G. D., and Kaji, A.,** A soluble amino acid incorporating system from rat liver, *Biochim. Biophys. Acta,* 76, 477, 1963.

30. **Soffer, R. L.,** Biochemistry and biology of aminoacylated tRNA protein transferase, in *Transfer RNA: Biological Aspects,* Soll, D., Abelson, J. N., and Shimmel, P. R., Eds., Cold Spring Harbor Symposium No. 913, 493, 1980.
31. **Chakraborty, G., and Ingoglia, N. A.,** Posttranslational incorporation of arginine and sine into rat brain soluble proteins in vitro, *Soc. Neurosci.,* 11(Abstr.) 1097, 1985.
32. **Lasek, R. J., Gainer, H., and Barker, J. I.,** Cell to cell transfer of glial proteins to the squid giant axon, *J. Cell Biol.,* 74, 501, 1977.
33. **Barra, H. S., Rodriguez, J. A., Arce, C. A., and Caputto, R.,** A soluble preparation from rat brain that incorporate into its own proteins ^{14}C arginine by a ribonuclease-sensitive system and ^{14}C tyrosine by a ribonuclease-insensitive system, *J. Neurochem.,* 20, 97, 1973.
34. **Zanakis, M. F., Chakraborty, G., and Ingoglia, N. A.,** Axonal transport in rat sciatic nerves of a high molecular weight complex containing elements capable of transfer RNA dependent posttranslational protein modification, *Soc. Neurosci. Abstr.,* 9, 145, 1983.
35. **Zanakis, M. F., Chakraborty, G., Sturman, J. A., and Ingoglia, N. A.,** Postranslational protein modification by amino acid addition in intact and regenerating axons of the rat sciatic nerve, *J. Neurochem.,* 43, 1286, 1984.
36. **Lindquist, T. D.,** Studies of the Axonal Transport of RNA and Taurine in the Normal and Regenerating Rat Sciatic Nerve, Ph.D. thesis, Graduate School of Biomedical Sciences, U.M.D.N.J., Newark, N.J.
37. **Bandyopadhyay, A. K. and Deutscher, M. P.,** Complex of aminoacyl-transfer RNA synthetase, *J. Mol. Biol.,* 60, 113, 1971.
38. **Schnapp, B. J. and Reese, T. S.,** Cytoplasmic structure of rapid frozen axons, *J. Cell Biol.,* 94, 667, 1982.
39. **Ingoglia, N. A., Sturman, J. A., Jaggard, P., and Perez, C.,** Association of spermine and 4S RNA during axonal transport in regenerating optic nerve of goldfish, *Brain Res.,* 238, 341, 1982.
40. **Quigley, G. J., Teeter, M. M., and Rich, A.,** Structural analysis of spermine and magnesium ion binding to yeast phenylalanine transfer RNA, *Proc. Natl. Acad. Sci. U.S.A.,* 75, 64, 1978.
41. **Takeda, Y.,** Role of polyamines in aminoacyl transfer-RNA formation, in, *Advances in Polyamine Research,* Campbell, R. A., Morris, D. R., Bartos, D., Daves, G. D., and Bartos, F., Eds., Raven Press, New York, 1978, 255.
42. **Kedracki, R. and Szer, W.,** A note on the effect of spermine on degradation of pyrimidine polynucleotides by pancreatic ribonuclease, *Acta Biochem. Pol.,* 14, 163, 1967.
43. **Hunter, A. R., Farrell, P. J., Jackson, R. J., and Hunt, T.,** The role of polyamines in cell free protein synthesis in the wheat-germ system, *Eur. J. Biochem.,* 75, 149, 1977.
44. **Ingoglia, N. A. and Tuliszewski, R.,** Transfer RNA may be axonally transported during regeneration of goldfish optic nerves, *Brain Res.,* 112, 371, 1976.
45. **Pfenninger, K. H.,** Axonal transport in the sprouting neuron transfer of newly synthesized membrane components to the cell surface, in *Axoplasmic Transport in Physiology and Pathology,* Weiss, P. G. and Gorio, A., Eds., Springer-Verlag, Basel, 1982, 52.
46. **Tessler, A., Autilio-Gambett, L., and Gambetti, P.,** Axonal growth during regeneration: a quantitative autoradiographic study, *J. Cell Biol.,* 87, 197, 1980.
47. **Griffin, J. W., Price, D. L., Drachman, D. B., and Morris, J.,** Incorporation of axonally transported glycoproteins into axolemma during nerve regeneration, *J. Cell Biol.,* 88, 205, 1981.
48. **Kaji, H. and Prema, R.,** Membrane modification by arginyl tRNA, *FEBS Lett.,* 66, 194, 1976.
49. **Gundersen, R. W. and Barrett, J. N.,** Characterization of the turning response of dorsal root neurites towards nerve growth factor, *J. Cell Biol.,* 87, 546, 1980.
50. **Gaze, R. M. and Sharma, S. C.,** Axial differences in the reimnervation of the goldfish optic tectum by regenerating optic nerve fibres, *Exp. Brain Res.,* 10, 171, 1970.
51. **Barde, Y. A., Edger, D., and Thurman, H.,** Neurotropic factors, *Burne, R. M. and Hoffman, J. F., Eds., Ann. Rev. Physiol.,* 45, 601, 1983.
52. **Meyer, R. L.,** Ordering of retinotectal connections: a multivariable operational analysis, in *Current Topics in Developmental Biology,* Mosdora, A. A. and Murray, A., Eds., Academic Press, 1982, 101.
53. **Sonderegger, P., Fishman, M. C., Bokoum, M., Bauer, H. C., and Nelson, P. G.,** Axonal proteins of presynaptic neurons during synaptogenesis, *Science,* 221, 1294, 1983.
54. **Skene, P., Willard, M., and Freeman, J. A.,** Modification of axonally transported protein in toad retinotectal terminals, *Soc. Neurosci. Abstr.,* 8, 864, 1982.
55. **Gambettii, P.,** personal communication.

Chapter 9

AXONAL TRANSPORT OF GLYCOPROTEINS AND PROTEOGLYCANS

John S. Elam

TABLE OF CONTENTS

I. INTRODUCTION

Proteins containing covalently attached carbohydrate chains have increasingly become a center of experimental investigation (for review see References 1 thru 3). These molecules, by definition, include the glycoproteins and the proteoglycans. The glycoproteins possess relatively short oligosaccharide chains composed of a variable combination of *N*-acetylglucosamine, *N*-acetylgalactosamine, mannose, galactose, fucose, *N*-acetylneuraminic acid (sialic acid), and occasionally glucose.[4,5] Attachment to the protein core is generally either through binding of *N*-acetylglucosamine to asparagine or binding of *N*-acetylgalactosamine to serine or threonine. The asparagine linked species are further subdivided into low molecular weight polymannosyl chains (composed of ASN-*N*-acetylglycosamine-*N*-acetylglucosamine-(n)mannose and higher molecular weight complex chains with additional sugars branching off the mannose core. The presence of terminal sialic acid or inorganic sulfate frequently renders these complex chains anionic.

The proteoglycans contain covalently attached glycosaminoglycans (GAGs) in addition to variable numbers of lower molecular weight oligosaccharides.[6,7] The GAGs are high molecular weight linear polymers consisting of alternating hexosamine and uronic acid residues and in most instances containing high levels of inorganic sulfate. The proteoglycans show considerable diversity in structure, ranging from the very large highly aggregated species characteristic of cartilage to lower molecular weight lipophylic species found embedded in plasma membranes.[7] Proteoglycans are affiliated with the extracellular matrix of a large variety of tissues[8] and in some instances exist in freely soluble cytoplasmic form.[9]

Because of their pervasive distribution in tissues, considerable effort has been expended on determining the various functions of glycoproteins and proteoglycans. The glycoproteins have for many years been implicated in cell to cell and cell to substrate adhesion processes.[10-12] These interactions in some instances appear to specifically depend upon the presence of the carbohydrate portion of the molecules.[11-13] Glycosylation of proteins has also been associated with processes of intracellular routing[14,15] and protection against proteolytic breakdown.[15,16] There are further reports of glycoproteins serving as cell surface receptors[17,18] and mediating transport through the plasma membrane.[19]

The proteoglycans have long been known to be important structural components of connective tissue.[6-8,20] In other tissues, proteoglycans associated with the extracellular matrix may, along with the glycoproteins, mediate attachments of the cell surface to the substrate.[8] The proteoglycans have also been implicated in the binding of physiologically regulated cations.[20]

II. NERVOUS SYSTEM GLYCOPROTEINS AND PROTEOGLYCANS

The cytological observations of a carbohydrate rich coat on the surface of neurons and other nervous tissue cells[21,22] has not escaped the attention of glycoconjugate biochemists. Initial characterization of nervous tissue centered primarily on preparative studies conducted in mammalian brain.[23] Carbohydrate chains attached to brain glycoproteins were found to be predominately *N*-glycosidically linked to asparagine and include both polymannosyl and complex forms.[5,23] Brain specific compositional attributes of the latter include relatively high levels of fucose, frequent di-sialic acid sequences and numerous terminal galactose and *N*-acetylglucosamine groups.[5,24]

The localization of specific glycoproteins to particular cells of the nervous system remains largely unexplored. Analysis of bulk isolated brain cells have shown a higher glycoprotein content in glia than in neurons, but with similar gross carbohydrate content in both cell types.[25] Recent use of immunocytochemical techniques has produced evidence for glycoproteins specific to purkinje cells[26] and astrocytes,[27] respectively. Use of SDS gel analysis has indicated differences also exist in cell surface glycoproteins between adrenergic and cholinergic sympathetic neurons[28] and between cerebellar interneurons and astrocytes.[29]

At the subcellular level, the glycoproteins of brain show a predominately membranous localization. Particular attention has focused on characterization of molecules associated with synaptic plasma membranes, synaptic vesicles, and myelin. The synaptic membranes have been the most extensively studied and have shown a large diversity in glycoproteins, having molecular weights ranging from 3,000 to greater than 200,000 daltons and binding to a variety of plant lectins.[30-33] A smaller number of molecules are specifically localized at the outer synaptic membrane surface[34,35] and in the immediate synaptic junctional region.[36,37] The carbohydrate chains associated with synaptic plasma membrane are enriched in mannose[38] and synaptic junction glycoprotein binding Concanavalin A (Con A) have been demonstrated to contain polymannosyl chains.[39] Recent evidence also suggests that several of the Con A binding molecules may share the same protein core.[40]

Synaptic vesicles were originally reported to contain a more limited number of glycoproteins than synaptic plasma membrane, with some molecules apparently in common with the plasma membrane.[41,42] More recent studies utilizing lectins indicate the presence of a much larger variety of glycoproteins in whole brain synaptic vesicles than originally thought.[43] Vesicles specific to Aplysia single neurons[44] and chromaffin cells[45] appear to have a more limited glycoprotein composition. As in the case of synaptic plasma membranes, synaptic vesicle glycoproteins are specifically located on either the internal or external membrane surface.[46,47]

Myelin from the mammalin CNS is found to contain a variety of glycoproteins.[48,49] These include the myelin associated glycoprotein (MAG) that has been purified and immunohistochemically localized to the periaxonal junction region.[50] Peripheral myelin is known to contain the Po glycoprotein as a major constituent[51] as well as a variety of less concentrated species.[52]

A continuing issue has been whether some of the glycoproteins of CNS or PNS myelin fractions may be contributed by adherent axonal or other plasma membrane.[53,54] Axolemma enriched fractions isolated from CNS myelinated axons contain at least ten different molecular weight classes of glycoproteins ranging from 28,000 to 250,000 kDa.[55] High molecular weight glycoproteins have also been detected in isolated axolemmal fractions from nonmyelinated crab sensory nerve[56] and garfish olfactory nerves.[57] In no case has a rigorous comparison been made between axolemmal and synaptic glycoproteins in the same preparation.

Soluble glycoproteins of brain have also been the subject of a limited number of studies. These represent a small fraction of the total glycoprotein pool and have a relatively rapid turnover.[58-60] Some success in isolating individual soluble glycoproteins by combined chromatographic techniques has been reported.[9,61]

Studies on nervous system proteoglycans initially focused on the more easily isolated GAGs.[23] Those found to be present in nervous tissue include hyaluronic acid, chondroitin 4 and 6 sulfate, heparan sulfate, and dermatan sulfate. The proteoglycans containing the GAG chains are present in both PNS and CNS,[62] and in both neurons and glia of CNS.[25] Unlike the glycoproteins, the majority of the GAG containing molecules of mammalian brain are found in soluble fractions or fractions easily dissociated from the membrane. An exception is heparan sulfate proteoglycan, which appears to be tightly membrane bound. More recent studies have begun the isolation and characterization of intact proteoglycans. A soluble chondroitin sulfate proteoglycan has been isolated from rat brain[9] and from cultured glial cells.[64] Preliminary characterization of membrane bound proteoglycans of nervous tissue has been undertaken but has not yet resulted in extensive purification.[65,66] Localization of particular proteoglycans to neurons is also at a preliminary state. A heparan sulfate proteoglycan released in cell culture appears to be localized on the neuronal cell surface.[67] Proteoglycans are common constituents of secretory vesicles[68-70] and their presence in synaptic vesicles has been reported.[71,72]

III. AXONAL TRANSPORT OF GLYCOPROTEINS AND PROTEOGLYCANS

Despite a growing literature on nervous tissue glycoconjugates, much remains to be learned about their specific distribution among different types of neurons and their differential association with axonal and synaptic substructures of a given neuron. An experimental approach that has allowed a relatively direct determination of glycoproteins and proteoglycans within specific neurons and their subcompartments is the evaluation of axonally transported molecules. A particular advantage of this method is that the molecules are unequivically neuronal and questions of intrinsic nonneuronal contamination of isolated fractions (axolemma, synaptic plasma membrane, synaptic vesicles, etc.) are to a large degree circumvented. Study of axonally transported molecules can also contribute to our basic understanding of the transport process and the mechanisms of intracellular routing.

A. Axonal Transport of Glycoproteins

1. Methodology

By far the most common approach to identifying axonally transported glycoproteins is to present appropriate radioactive precursors to the cell bodies and measure the outflow of labeled molecules into associated nerves or nerve terminals (for reviews see References 73 and 74). Glycoprotein transport in specific nerve tracts can also be inferred by orthograde or retrograde movement of exogenously applied lectins.[75-77] However, beyond providing inferences about the nature of the lectin binding sites, these studies have not allowed an analysis of the composition of the glycosylated molecules to which the lectins are attached.

a. Choice of Precursor

Axonal transport of glycoproteins has been documented after use of a variety of carbohydrate precursors including ^{3}H-fucose, ^{3}H-glucosamine, ^{3}H-galactosamine, and ^{3}H-galactose. A subpopulation of sulfated glycoprotein can also be labeled with $^{35}SO_4$.[78,79] Fucose has been the most commonly employed precursor, probably because of its localization almost exclusively in glycoproteins and its tendency not to be metabolized into other labeled forms.[80,81] A potential disadvantage in using fucose is that molecules containing only simple oligosaccharides might be unlabeled or poorly labeled.

b. Experimental Preparations

Most studies on transported glycoproteins have been conducted in the vertebrate optic nerve[82-87] and in vertebrate peripheral nerve.[88-93] The identified neurons of Aplysia have also been utilized very effectively to profile glycoprotein transport in single axons.[94,95] Increasingly, characterization of transport in specific pathways of vertebrate brain has been achieved by use of highly localized precursor injection.[96-99] Some studies have assessed putative transported glycoproteins by evaluation of label in isolated synaptosomes or synaptosomal subfractures following generalized labeling of the brain.[100-102] Unfortunately, this approach suffers the limitations of a mixed cellular source of neuronal molecules and the potential for contribution of label by nonneuronal contaminants of the fractions.

2. Rates and Direction of Transport

To a first approximation, the rates of anterograde transport of glycoprotein have been found to be similar to the most rapid migration of amino acid labeled protein in the same system.[88,92,103,104] However, two discrete rates of rapid transport of glycoprotein have been observed in the locus coeruleus to hypothalamus pathway in brain[96] and in single axons of Aplysia.[105] It is uncertain to what extent distinct transport populations migrating at different rapid rates may be averaged out in other studies analyzing large populations of neurons. A prolonged and biphasic cell body release of glycoproteins that migrate at the same rapid rate has been observed in mammalian visual systems.[87,106] Later released molecules show some compositional differences from earlier released molecules, however these may be the result of differential turnover (see Section III.A.5b).

Retrograde transport of labeled glycoproteins has been detected in both peripheral nerve and optic nerve preparations.[107,108] In the goldfish optic nerve, turnaround of transported ^{3}H-fucose labeled glycoproteins occurs within 2 hours of arrival in the tectum and accounts for at least 40% of the rapid loss of nerve terminal label.[108] Retrograde transport of lectins, presumably attached to glycoproteins, has also been widely documented.[76,77,109]

3. Subcellular Distribution

a. During Transport

The question of localization of glycoproteins during the act of transport has obvious implications for rapid transport mechanisms. In microscopic studies on single axons of Aplysia, autoradiographic grains derived from ^{3}H-fucose labeled glycoproteins undergoing transport are highly concentrated over vesicular profiles,[110] suggesting that vesicles are the principle vehicle of movement. In contrast, Markov et al.[111] concluded from autoradiographic analysis that ^{3}H-fucose radioactivity undergoing transport in axons of the ciliary ganglion is largely associated with the endoplasmic reticulum. Glycoprotein molecules undergoing rapid axonal transport in the garfish olfactory nerve were found, upon subcellular fractionation, to be predominately affiliated with membrane having density and enzyme marker characteristics of axolemma.[112] This observation was interpreted as evidence of transport of membrane precursor vesicles. Taken together, these studies support the possibility that glycoproteins undergoing transport are in a state of dynamic equilibrium between the smooth endoplasmic reticulum and vesicular elements. Smaller amounts of glycoprotein radioactivity in transit are recovered in post-microsomal supernatant[112] and may represent soluble molecules that are sequestered within membranous elements in the intact tissue.

b. After Deposition

Results from a number of preparations have indicated that transported glycoproteins are distributed to axolemma,[54,112,113] synaptic plasma membrane,[85,90,114] synaptic vesicles,[90,110] and synaptic cytoplasmic fractions.[78,82,84,87,114] The proportion of molecules remaining along the axon is greater (up to 85%) in relatively long nerves[115] and the axonally deposited label appear to be concentrated in axolemma.[112,116,117] Some of this label may serve as a marker for axolemmal contamination in biochemical evaluation of isolated myelin fractions.[53,54,118] Although synaptic plasma membrane and synaptic vesicles are found to contain the highest concentration of label delivered to synapses,[85,90,114,119] there are significant amounts in soluble constituents that may play a role in the secretory activities of the nerve ending.[78,82,84,87,119,120]

4. Composition of Axonally Transported Glycoproteins

a. General Composition

A large number of studies have evaluated the composition of axonally transported glycoproteins on SDS acrylamide gels.[73,84,87,93,95,98,99,112,121-125] Although several studies report a peak of radioactivity between 15,000 and 25,000 daltons,[84,93,121] the great majority of transported molecules range in molecular weight between 30,000 and >200,000 daltons. Higher resolution two-dimensional gel analysis has indicated a minimum of 50 to 100 separable glycoproteins transported in a single nerve.[123,125] The single cell preparation of Aplysia shows a less complex profile, with most transported fucose label localized in five high molecular weight peaks.[95]

Transported glycoproteins have also been assessed according to their ability to bind to a variety of lectins. A series of studies by Karlsson and associates have shown that significant subpopulations of different sized glycoproteins transported in the rabbit optic nerve have affinity for Con A,[126] lentil-lectin,[127] wheat germ agglutinin,[126] or Aleuria aurantia lectin.[128] Binding to these particular lectins infers the presence to differing degrees of exposed mannose, *N*-acetylglucosamine and fucose. Interestingly, binding of lectins specific to galactose or *N*-acetylgalactsamine was minimal,[126] despite the presence of terminal galactose in whole brain glycoproteins.[5] Binding of transported glycoproteins to wheat germ agglutinum has also been observed in guinea pig auditory and optic nerves[123] while CON A binding has also been shown in guinea pig auditory, guinea pig optic, and garfish olfactory nerves.[123,129]

Further information on the nature of the carbohydrate chains attached to transported glycoproteins has been gained by subjecting labeled molecules to proteolytic digestion and evaluating the composition of the remaining carbohydrate enriched glycopeptides.[73] In the goldfish optic nerve, glycopeptides from transported glycoproteins labeled with $^{35}SO_4$ eluted as a broad retarded peak on Sephadex G-50 columns, suggestive of molecular weights ranging from approximately 3,000 to 10,000 daltons.[78] Analysis of ^{3}H-fucose labeled glycopeptides in the same system showed the additional presence of lower molecular weight glycopeptides that are dialyzable.[120] Somewhat more detailed analysis has been made of ^{3}H-glucosamine labeled glycopeptides that were isolated from glycoproteins undergoing axonal transport in the garfish olfactory nerve.[112,129-131] These molecules have a very broad size range, extending from unusually high molecular weight chains that are excluded from Sephadex G-50 columns to low molecular weight dialyzable chains. Compositional analysis combined with Con A affinity chromatography and ion exchange chromatography indicates that the large glycopeptides include typical acidic and nonacidic asparagine linked complex chains and that the low molecular weight glycopeptides include a subpopulation of simple polymannosyl chains. The very large (Sephadex G-50 excluded) chains are mainly nonacidic and have a composition similar to the recently described polyglycosyl or lactosaminoglycan chains which contain alternating galactose-*N*-acetylglucosamine sequences.[132] These high molecular weight glycopeptides have not been observed in other vertebrate neural preparations.

The overall conclusion from these studies is that the variety in transported glycoprotein size and lectin binding affinity is a reflection of similar variety in the attached carbohydrate chains.

b. Cell-Specific Composition

A direct acrylamide gel comparison of glycoproteins transported in different tracts of the same animal has been made in the rat and guinea pig. Padilla and Morell[98] noted many similarities and a few clear-cut differences in the unidimensional SDS gel labeling pattern for glycoproteins transported in rat optic nerve and hippocampal-commisural pathway. The level of resolution did not permit a determination of whether the similarities reflect quantitative or qualitative differences in transported molecules in the two tracts. Interestingly, more differences were seen in ^{3}H-fucose labeled glycoproteins than in ^{35}S-methionine labeled total protein. This was interpreted as a possible reflection of a special role of glycoproteins in maintaining the tract-specific synaptic connections. Subsequently, Goodrum and Morell[99] utilized higher resolution fluorography of SDS gels to show some quantitative but no qualitative differences in transported molecules of the rat optic, geniculocortical and nigrostriatal tracts. Wenthold and McGarvey[123] compared proteins and glycoproteins transported in guinea pig optic and auditory nerves and concluded that there are numerous tract specific differences, again more common among the glycoproteins than total proteins. The use of two-dimensional gels permitted the conclusion that some glycoproteins of very similar identical size and PI are common to both tracts, while others are qualitatively different. The general conclusion from these studies is that many transported glycoproteins are common to different cells of the CNS but that cell specific differences in type and amount of particular glycoproteins do exist.

c. Composition in Axonal and Synaptic Subfractions

Studies on single axons of Aplysia,[113] sciatic,[124] and optic[99] nerves of rat indicate that certain axonally transported glycoproteins are preferentially deposited in the axon, possibly as functional constituents of the axolemma. A 49,000 mol wt glycoprotein is found deposited in both optic and sciatic nerve and has been hypothesized to be a subunit of Na^+, K^+ ATPase.[99,124] Analysis of glycoproteins deposited in axons of the garfish olfactory nerve indicates that axolemma enriched fractions contain a prevalence of molecules between 100,000 and 150,000 mol wt, while soluble fractions contain a greater fraction of molecules between 50,000 and 100,000 mol wt.[112] This difference in composition suggest differential roles of axonal glycoproteins in membranous and cytoplasmic functions.

Relatively few studies have directly evaluated the composition of axonally transported glycoproteins in isolated synaptic plasma membrane, synaptic vesicles, or synaptic soluble fractions. Specht[133] reported that synaptic plasma membrane of hamster optic nerve terminal contains six major SDS gel peaks and a variety of minor peaks, ranging in molecular weight from 30,000 to near 300,000 daltons. Heaviest labeling was in a molecule of 50,000 daltons. Elam and Ripellino[119] utilized SDS gels to compare ^{3}H-fucose and $^{35}SO_4$-labeled molecules transported to purified synaptic plasma membrane and synaptic vesicles of goldfish optic nerve terminals. ^{3}H-fucose labeled molecules in both fractions showed a similar broad molecular weight distribution between 30,000 and 200,000 daltons with heaviest labeling between 50,000 and 100,000 daltons. In contrast, use of $^{35}SO_4$ showed extensive labeling of synaptic plasma membrane molecules greater than 100,000 daltons that was not observed in synaptic vesicles. This compositional difference may be related to a unique functional role of sulfated molecules in each fraction.

Synaptic plasma membrane labeling patterns have also been examined after generalized intracraneal labeling of the brain. It should be remembered that conclusions from such studies are tempered by the mixed cellular source of the membrane and some degree of contamination by postsynaptic and nonneuronal elements. Studies on rat brain[134,135] have shown that after intracraneal injection of ^{3}H-fucose, isolated synaptic plasma membrane contained labeled molecules ranging from 40,000 to $>$150,000 daltons, with major peaks of label at 74,000 and 50,000 mol wt. These compare in size to major glycoprotein constituents of the membrane that have been determined by other methods.[33] Higher molecular weight ^{3}H-fucose labeled molecules were found to be concentrated in synaptic junctional complexes, but it is unknown whether their synthesis was in presynaptic or postsynaptic cells.

Taken together, the studies on synaptic membrane suggest transport of a large size range of molecules but with some concentration of molecules between 50,000 to 100,00 mol wt. The possible difference in glycoproteins delivered to plasma membrane and synaptic vesicles remains largely unexplored but may be significant for sulfated molecules.

5. Fate of Axonally Transported Glycoproteins

A number of investigators have measured the timecourse of removal of radioactivity from axonally transported glycoproteins.[53,73,78,79,84,90,98,102,103,106,114,124,136,137] Loss of the ^{3}H-fucose radioactivity is frequently found to be biphasic[85,87,138] with estimated half-life ranging from several hours to several weeks. Rates of turnover of label also vary considerably between individual glycoproteins of a particular transport population, with a very rapid half-life of radioactivity reported for certain glycoproteins of the auditory[122] and optic nerves.[87,99] A more rapid loss of transported ^{3}H-fucose radioactivity in soluble as opposed to particular glycoproteins has been observed in goldfish optic terminals.[120]

The interpretation of this body of data is made very difficult by the potentially multiple fates of the transported molecules. These include retrograde transport, partial or complete metabolic breakdown, further addition of carbohydrate residues, secretion, and transfer of intact glycoproteins or metabolized label to perineuronal cells.

a. Retrograde Transport

As previously discussed (Section III., A.2), the retrograde transport process can account for a large fraction of the rapid loss of transported radioactivity from nerve terminals.[108] Since the rules governing the turnaround of axonally transported glycoproteins are not known, it cannot be said if retrograde transport could cause a rapid loss of label in particular individual glycoproteins. Analysis of retrogradely transported amino acid labeled protein indicates that most anterogradely transported molecules undergo retrograde flow,[139] suggesting that this process alone would not account for unusually rapid turnover of a particular molecule.

b. Metabolic Turnover

Evidence that rapid loss of transported radioactivity is at least partially due to rapid metabolic breakdown is provided by recent experiments of Morell and Goodrum.[140] They found that different ^{3}H-fucose labeled carbohydrate chains obtained from the same parent glycoprotein turn over at either rapid or slower rates corresponding to the gross biphasic turnover rates of the total population of transported glycoproteins. The authors also point out that the apparent biphasic cell body release of rapidly transported proteins in some systems may in fact be the result of combined prolonged continuous release and biphasic turnover of individual carbohydrate chains. Nerve terminal enzymes acting on endogenous sialoglycoproteins of rat synaptic membrane preparations have been described.[141,142] Differential turnover of specific carbohydrate chains can also be inferred from the observation that ^{35}S-labeled axonally transported glycoproteins show relatively rapid turnover.[78,79,87,138]

c. Further Glycosylation of Transported Glycoproteins

The addition of carbohydrate residues to previously transported glycoproteins could affect their mobility in gels and lead to an erronous conclusion that there was loss of particular molecules due to degradation. This type of modification would also raise the interesting question of the role of such additional glycosylation in the ultimate function of the molecules.

Several lines of evidence support the concept of glycosylation of proteins in nerve endings. Glycosyltransferase has been found in isolated synaptic plasma membrane fractions[143,144] and synaptic junctional complexes.[145,146] In addition, a variety of synaptosomal preparations have been shown to incorporate radioactive sugars into glycoproteins.[147-150] Incorporation into glycoproteins of both synaptic soluble fraction and synaptic plasma membrane has also been shown following in vitro incubation in $^{35}SO_4$.[151,152] An ultimate problem in interpreting these studies is the possibility that low levels of postsynaptic or contaminating enzymes account for the activity.[33] A major advance in this important area of inquiry would be the immunocytochemical demonstration of presynaptic glycoprotein-glycosyltransferases.

Evidence for glycosylation within axons has come from studies on the R2 cell of Aplysia. Ambron et al.[95] noted the appearance in the axon of a labeled glycoprotein not seen in the cell body, suggesting possible axonal modification. Localized injection of ^{3}H-*N*-acetyl galactosamine into the same axon resulted in incorporation into a variety of axonal glycoproteins that was not dependent on new protein synthesis.[153]

d. Secretion and Transfer to Perineuronal Cells

Secretion of axonally transported glycoproteins could theoretically be required for the maintenance of the extracellular matrix or occur as a byproduct of secretion of neurotransmitter. However, as yet direct experimental evidence for such events has not been produced. Glycoproteins have been shown to be released from cultured neuroblastoma cells[154,155] and amino acid labeled axonally transported proteins have recently been shown to be released from regenerating peripheral nerve.[156]

There is evidence that radioactivity from transported glycoproteins can be transferred to periaxonial and postsynaptic cells. Toews et al.[124] observed a time dependent buildup of axonally transported ^{3}H-fucose radioactivity in the Po glycoprotein of rat sciatic nerve myelin.

This was interpreted as resulting from the breakdown of the initially transported molecules and reutilization of released fucose by adjacent glia. Such extraneuronal reincorporation processes may cause overestimates of transported ^{3}H-fucose turnover times and makes more difficult the detection of possible redistribution of fucose residues among molecules within the axonal or nerve terminal compartments.

There is also evidence for the transfer of transported ^{3}H-fucose radioactivity to postsynaptic cells.[86,157,158] This phenomenon, documented primarily in the mammalian visual system, can result in the transsynaptic transfer and transport in secondary neurons of up to 15% of the original transported radioactivity.[106] The critical question of whether this reflects the spread of free fucose from metabolic breakdown or the transfer of intact glycoproteins has not been resolved.

B. Axonal Transport of Proteoglycans

Initial attempts to document proteoglycan transport centered on the demonstration of labeling of GAGs isolated from the transported molecules. Studies conducted on $^{35}SO_4$-labeled molecules transported in both goldfish optic system[78,159] and garfish olfactory nerve[115] showed that pronase released label was approximately equally divided between sufated glycopeptides and sulfated glycosaminoglycans. Further characterization of labeled GAGs by differential hydrolytic enzyme treatment combined with cellulose acetate electrophoresis revealed that heparan sulfate and chondroitin sulfate were associated with transported proteoglycans in both nerves, with heparan sulfate being the more predominant constituent. Labeling of comparable preparations of olfactory nerve with ^{3}H-glucosamine showed that GAGs contain only 8% of total protein bound hexosamine label.[112] This indicates that their concentration in the transport pool is relatively low in comparison to glycoprotein oligosaccharides. Subcellular fractionation of transported GAGS of both optic tectum and olfactory nerve showed that between 85 to 90% of the labeled molecules are affiliated with particulate subfractions.[78,112] In nerve, very little additional label was found to be solubilized after repeated washes with buffered sucrose or 0.5 *M* NaCl,[160] indicating a tight association with membranes and a considerably lower degree of solubility than is observed for GAGs of whole brain.[63] Some concentration of transported GAGs is observed in the axolemma enriched subfraction,[112] suggesting a possible role on the neuronal cell surface. Independent studies providing evidence for axonal transport of GAG containing molecules has been obtained from mammalian optic[79,87] and peripheral motor[161] nerves. Preliminary compositional analysis indicates the likely presence of transported chondroitin sulfate in both preparations.

More recent studies have centered on attempts to isolate intact axonally transported proteoglycans. Stadler and Dowe[71] have preliminarily characterized a heparan sulfate containing proteoglycan fraction that is axonally transported to cholinergic nerve endings of Torpedo electric organ. Immunological studies indicate that the proteoglycan(s) is localized in the interior of the vesicle and is exposed upon electrical stimulation.[162] Elam and Ripellino[119,160] have partially purified a soluble proteoglycan fraction that is transport labeled in the goldfish optic terminals. Apparent molecular weight on gel filtration ranges from 150,000 to 400,000 and both heparan sulfate and chondroitin sulfate appear to be present. The large apparent size and constant cooling during isolation make it unlikely that this fraction is derived from in vitro proteolysis of the more prevalent membrane bound species. Characterization of transported membrane affiliated proteoglycans in the goldfish system has not been performed. It remains possible that some of the aforementioned high molecular weight $^{35}SO_4$-label associated with synaptic plasma membrane fraction (Section III., A.4c) resides in proteoglycans.

Clearly the characterization of axonally transported proteoglycans remains at a very preliminary state. However, their presence as a minority constituent of the rapid transport phase in a variety of systems appears to be well established.

IV. POSSIBLE FUNCTIONS OF AXONALLY TRANSPORTED GLYCOPROTEINS AND PROTEOGLYCANS

It is not yet possible to assign specific functional roles to the majority of individual glycoproteins or proteoglycans that undergo axonal transport. Speculation about possible functions is based on our combined knowledge of glycoconjugate function in other systems and the properties exhibited by transported molecules.

A. Subcellular Routing

A possible role of glycosylation in sorting out molecules destined for axonal transport has been postulated by Stone and Hammerschlag.[125] These authors have observed from double labeled 2D gel analysis that approximately 90% of axonally transported proteins of greater than 35,000 daltons are glycosylated. This suggests that addition of carbohydrate may be a general requirement for shunting proteins of this size into the axon. Beyond axonal entry there are presumably further routing signals required to deliver particular molecules to specific axonal and synaptic subcompartments.[14,15] To the limited extent that compositional differences between such fractions have been identified, (Section III., A.4c) they may also reflect a role of glycosylation in such routing.

B. Synaptic Physiology

The relatively high axonal transport labeling of glycoproteins in synaptic plasma membrane and synaptic vesicles imply a role in some aspects of normal synaptic physiology. The observation that removal or blockage of sialic acid residues in nerve terminal glycoproteins inhibits synaptic transmission[163] suggests a role in neurotransmitter storage or release. The discovery of axonally transported proteoglycans in synaptic vesicles[71] and their exposure upon nerve stimulation[162] suggests a similar function for the transported proteoglycans. Nerve terminal glycoproteins have also been implicated in the uptake of extracellular amines.[164]

Correlation between amounts of axonal transport of glycoproteins and electrophysiological activity has been inferred. Levin[165] reported that reserpine induced depletion of norepinephrine stores leads to an initial suppression followed by greater than normal axonal transport of ^{3}H-fucose labeled glycoproteins. The excess transport possibly reflects the replacement of the affected synaptic vesicles. Other studies have shown that an increase in ^{3}H-carbohydrate labeling of optic nerve terminals follows exposure to light.[166,167]

A more clearly established function of synaptic glycoproteins is in the form of neurotransmitter receptors (for review see Reference 168). Some of these are presynaptic and are presumably provided by axonal transport. Con A has been found to separate Ach receptors from lobster plasma membranes having different pharmacological properties,[169] while tunicamycin mediated under-glycosylation of cultured muscle cell Ach receptors alters quantitative aspects of ligand interaction.[170] These observations suggest a direct or indirect effect of sugar residues on the conformation of receptor binding sites. Interestingly, tunicamycin treatment in embryonic muscle cells leads to greatly accelerated turnover of Ach receptor molecules.[170] This phenomenon may reflect a general role of glycosylation in synaptic as well as other systems in decreasing susceptability to proteolysis.[15,16]

C. Axonal Growth and Synaptogenesis

The widespread implication of glycoconjugates in processes of cell-cell and cell-substrate interaction raises a question as to whether axonally transported glycoproteins and proteoglycans play such a role in axonal ontogeny and regeneration. In support of this possibility, amounts of transported glycoprotein are found to increase in a variety of growing nerves[91,116,117,129,171] and tunicamycin has been shown to inhibit further outgrowth.[172] Retrograde transport of glycoprotein has also been found to increase in regeneration.[107,108] Although this may be an inevitable consequence of the temporary shortening of the nerve, the returning molecules may serve as retrograde signals to the cell body, monitoring the site of axonal lesion and state of axonal regrowth.[173]

A variety of glycoproteins are inserted into the plasmalemma of growing neurites,[116,117,174-177] with lectin binding differences detected between growth cones and maturing axon shafts.[176] Transient increases in Con A binding have been noted in several developing axonal preparations[178-180] and direct compositional analysis of glycoproteins transported into regenerating olfactory nerve axons has shown a major increase in the presence of Con A binding polymannosyl chains.[129] These molecules appear to be particularly concentrated in a high density fraction which may be enriched in glycoconjugates from the outer cell coat. Glycoproteins and proteoglycans from cultured neurons are found in the extracellular matrix[181] and appear to contribute to the production of the basal lamina.[155] Some have been specifically implicated in axonal adhesion processes.[67,182] Taken in total, these observations are compatible with a role for axonally transported glycoproteins and proteoglycans in the cell-cell and cell-substrate adhesion regulating the directed outgrowth of neurites.

An additional body of evidence has implicated glycoprotein in cell-cell interactions associated with synaptogenesis. Studies on embryonic chick visual system have shown major increases in amounts of transported glycoproteins at the time of optic nerve connection to the optic tectum.[83,183-185] Changes in the relative transport labeling of particular glycoproteins have been noted at the time of synaptogenesis in developing hamster optic nerve[133] and in regenerating goldfish[171] and toad[186] optic nerves. This may reflect the need for specific glycosylated molecules in the cell-cell recognition process. It can be speculated that similar requirements for particular molecules to maintain specific synaptic connections in mature nerve account for some of the observed cell specific differences in transported glycoprotein composition (Section III., A. 4b).

D. Learning and Memory

If glycoconjugates do in fact play regulatory roles in synaptic physiology and axonal connectively, it is not a great conceptual leap to suggest that they may also function in alterations of these processes associated with learning. A number of studies have noted increases in ^{3}H-sugar incorporation into brain in association with learning, but have generally not adequately dealt with the complications of potentially altered blood flow and precursor availability (for discussion see Reference 187). Somewhat more encouraging are recent reports of learning related increase in glycoprotein synthesis in partially purified synaptic membrane that are restricted to particular brain regions and that, in some instances, are restricted to particular SDS gel separated bands.[187-189] To the extent that these results represent alterations in intrinsic presynaptic constituents, it implies a minimal role of axonal transport in delivery of learning related molecules to the terminals and possibly a more specific role in altering rates of delivery or subcellular localization.

REFERENCES

1. **Spriro, R. G.,** Glycoproteins, in *Advances in Protein Chemistry,* Vol. 27, Anfinien, C. B., Edsall, J. T., and Richards, F. M., Eds., Academic Press, New York, 1973, 349.
2. **Lennarz, W.,** *The Biochemistry of Glycoproteins and Proteoglycans,* Plenum Press, New York, 1980.
3. **Sharon, N. and Lis, H.,** Glycoproteins: research booming on long ignored ubiquitous compounds, *Mol. Cell Biochem.,* 42, 167, 1982.
4. **Kornfeld, R. and Kornfeld, S.,** Comparative aspects of glycoprotein structure, *Ann. Rev. Biochem.,* 45, 217, 1976.

5. **Krusius, T. and Finne, J.,** Structural features of tissue glycoproteins. Fractionation and methylation analysis of glycopeptides derived from rat brain, kidney, and liver, *Eur. J. Biochem.*, 78, 369, 1977.
6. **Lindahl, V. and Hook, M.,** Glycosaminoglycans and their binding to biological macromolecules, *Ann. Rev. Biochem.*, 47, 385, 1978.
7. **Hascall, V. C. and Kimura, J. H.,** Proteoglycans: isolation and characterization, *Methods Enzymol.*, 82, 769, 1982.
8. **Aplin, J. and Hughes, R.,** Complex carbohydrates of the extracellular matrix: structures, interactions and biological roles, *Biochim. Biophys. Acta*, 694, 375, 1982.
9. **Kiang, W.-L., Margolis, R. U., and Margolis, R. K.,** Fractionation and properties of a chondroitin sulfate proteoglycan and the soluble glycoproteins of brain, *J. Biol. Chem.*, 256, 10529, 1981.
10. **Turner, R. S.,** Sponge cell adhesions, in *Specificity of Embryological Interactions*, Garrod, D. R., Ed., Chapman and Hall, London, 1978, 199.
11. **Kinsey, W. H. and Lennarz, W. J.,** Isolation of a glycopeptide fraction from the surface of the sea urchin egg that inhibits sperm-egg binding and fertilization, *J. Cell Biol.*, 91, 325, 1981.
12. **Damsky, C. H., Knudsen, K. A., and Buck, C. A.,** Integral membrane glycoproteins related to cell-substratum adhesion in mammalian cells, *J. Cell. Biochem.*, 18, 1, 1982.
13. **Pippia, P., Ivaldi, G.,and Cogoli, A.,** Identification of carbohydrate and functional groups involved in the adhesion of neoplastic cells, *FEBS Lett.*, 116, 281, 1980.
14. **Fitting, T. and Kabat, D.,** Evidence for a glycoprotein "signal" involved in transport between subcellular organelles, *J. Biol. Chem.*, 257, 14011, 1982.
15. **Olden, K., Parent, J. B., and White, S. L.,** Carbohydrate moieties of glycoproteins. A reevaluation of their function, *Biochim. Biophys. Acta*, 650, 209, 1982.
16. **Milenkovic, A. G. and Johnson, T. C.,** The relationship between glycosylation and glycoprotein metabolism of mouse neuroblastoma N18 cells, *Biochem. J.*, 191, 21, 1980.
17. **Bar-Shavit, Z., Ofek, I., Goldman, R., Mirelman, D., and Sharon, N.,** Mannose residues on phagocytes as receptors for the attachment of Escherichia coli and Salmonella typhi, *Biochem. Biophys. Res. Commun.*, 78, 455, 1977.
18. **Bahl, O. P.,** Human chorionic gonadotropin, its receptor and mechanism of action, *Fed. Proc. Fed. Am. Soc. Exp. Biol.*, 36, 2117, 1977.
19. **Olden, K., Pratt, R., Jaworski, C., and Yamada, K.,** Evidence for role of glycoprotein carbohydrates in the membrane transport: specific inhibition by tunicamycin, *Proc. Natl. Acad. Sci. U.S.A.*, 76, 791, 1979.
20. **Comper, W. and Laurent, T.,** Physiological function of connective tissue polysaccharides, *Physiol. Rev.*, 58, 255, 1978.
21. **Rambourg, A. and LeBlond, C. P.,** Electron microscope observations on the carbohydrate rich coat present at the surface of cells in the rat, *J. Cell Biol.*, 32, 27, 1967.
22. **Castejon, H. V.,** A sulfated polysaccharide in some nerve-cell coats in brains of different species of vertebrates, *J. Histochem. Cytochem.*, 18, 685, 1970.
23. **Margolis, R. K. and Margolis, R. U.,** Structure and distribution of glycoproteins and glycosaminoglycans, in *Complex Carbohydrates of Nervous Tissue*, Margolis, R. U. and Margolis, R. K., Eds., Plenum Press, New York, 1979, 45.
24. **Finne, J., Krusius, T., Rauvala, H., and Hemminki, K.,** The disialosyl group of glycoproteins: occurence in different tissues and cellular membrane, *Eur. J. Biochem.*, 77, 319, 1977.
25. **Margolis, R. U. and Margolis, R. K.,** Distribution and metabolism of mucopolysaccharides and glycoproteins in neuronal perikarya, astrocytes, and oligodendroglia, *J. Biochem.*, 13, 284, 1974.
26. **Reeber, A., Vincedon, G., and Zanetta, J. P.,** Isolation and immunohistochemical localization of a Purkinje cell specific glycoprotein subunit from rat cerebellum, *Brain Res.*, 229, 53, 1981.
27. **Ghandour, M. S., Langley, O. K., Gombos, G., Vincedon, G., and Warecka, K.,** Cellular localization of the brain specific A2-glycoprotein in rat cerebellum: an immunohistological study, *Neuroscience*, 7(1), 231, 1982.
28. **Braun, S. J., Sweadner, K. J., and Patterson, P. H.,** Neuronal cell surfaces: distinctive glycoproteins of cultured adrenergic and cholinergic sympathetic neurons, *J. Neurosci.*, 1, 1397, 1981.
29. **Webb, M.,** Cell surface sialoglycoproteins of cultured rat cerebellar interneurons, *J. Neurochem.*, 40, 769, 1983.
30. **Zanetta, J.-P., Reeber, A., Vincendon, G., and Gombos, G.,** Synaptosomal plasma membrane glycoproteins. II. Isolation of fucosyl-glycoproteins by affinity chromatography on the *Ulex Europeus* lectin specific for L-fucose, *Brain Res.*, 138, 317, 1977.
31. **Gurd, J. W.,** Synaptic plasma membrane glycoproteins: molecular identification of lectin receptors, *J. Biochem.*, 16, 369, 1977.
32. **Goodrum, J. F. and Tanaka, R.,** Protein and glycoprotein composition of subsynaptosomal fractions. Implications for exocytosis and recycling of synaptic vesicles, *Neurochem. Res.*, 3, 599, 1978.

33. **Mahler, H.,** Glycoproteins of the synapse, in *Complex Carbohydrates of Nervous Tissue,* Margolis, R. V. and Margolis, R. K., Eds., Plenum Press, New York, 1979, 45.
34. **Wang, Y.-J. and Mahler, H. R.,** Topography of the synaptosomal membrane, *J. Cell Biol.,* 71, 639, 1976.
35. **Mahadik, S. P., Hungund, B., and Rapport, M. M.,** Topographic studies of glycoproteins of intact synaptosomes from rat brain cortex, *Biochim. Biophys. Acta,* 511, 240, 1978.
36. **Rostas, J. A. P., Kelly, P. T., Pesin, R. H., and Cotman, C. W.,** Protein and glycoprotein composition of synaptic junctions prepared from discrete synaptic regions and different species, *Brain Res.,* 168, 151, 1979.
37. **Gurd, J.,** Molecular and biosynthetic heterogeneity of fucosyl glycoproteins associated with rat brain synaptic junctions, *Biochim. Biophys. Acta,* 555, 221, 1979.
38. **Brunngraber, E. G. and Javaid, J. I.,** Subcellular and anatomical distribution in rat brain of glycoproteins that contain mannose-rich heteropolysaccharide chains, *Biochim. Biophys. Acta,* 404, 67, 1975.
39. **Gurd, J. W. and Fu, S. C.,** Concanavalin A receptors associated with rat brain synaptic junctions are high mannose-type oligosaccharides, *J. Neurochem.,* 39, 719, 1982.
40. **Mena, E. E. and Cotman, C. W.,** Synaptic cleft glycoproteins contain homologous amino acid sequences, *Science,* 216, 422, 1982.
41. **Morgan, I. G., Zanetta, J.-P., Breckenridge, W. C., Vincendon, G., and Gombos, G.,** The chemical structure of synaptic membranes, *Brain Res.,* 62, 405, 1973.
42. **Richter-Landsberg, C., Neuhoff, V., and Waehneldt, T. V.,** Membrane protein and glycoprotein composition of beef brain synaptic vesicles, *J. Neurosci. Res.,* 3, 103, 1977.
43. **Zanetta, J. P., Reeber, A., and Vincendon, G.,** Glycoproteins from adult rat brain synaptic vesicles: fractionation on four immobilized lectins, *Biochim. Biophys. Acta,* 670, 393, 1981.
44. **Ambron, R. T., Sherbany, A. A., Shkolnik, L. J., and Schwartz, J. H.,** Distribution of membrane glycoproteins among the organelles of a single identified neuron of Aplysia. II. Association of a [^{3}H] glycoprotein with vesicles, *Brain Res.,* 207, 17, 1981.
45. **Huber, E., Konig, P., Schuler, G., Aberer, W., Plattner, H., and Winkler, H.,** Characterization and topography of the glycoproteins of adrenal chromaffin granules, *J. Neurochem.,* 32, 35, 1979.
46. **Wagner, J. and Kelly, R.,** Topological organization of proteins in an intracellular secretory organelle: the synaptic vesicle, *Proc. Natl. Acad. Sci. U.S.A.,* 76, 4126, 1978.
47. **Smith, A. P. and Loh, H. H.,** Organization of brain synaptic vesicle proteins, *J. Neurochem.,* 36, 1749, 1981.
48. **Zanetta, J.-P., Sarlieve, L. L., Mandel, P., Vincendon, G., and Gombos, G.,** Fractionation of glycoproteins associated to adult rat brain myelin fractions, *J. Neurochem.,* 29, 827, 1977.
49. **Poduslo, J. F., Harman, J. L., and McFarlin, D. E.,** Lectin receptors in central nervous system myelin, *J. Neurochem.,* 34, 1733, 1980.
50. **Sternberger, N. H., Quarles, R. H., Itoyama, Y., and Webster, H. D.,** Myelin-associated glycoprotein demonstrated immunocytochemically in myelin and myelin-forming cells of developing rat, *Proc. Natl. Acad. Sci. U.S.A.,* 76, 1510, 1979.
51. **Roomi, M. W., Ishaque, A., Khan, N. R., and Eylar, E. H.,** The Po protein — the major glycoprotein of peripheral nerve myelin, *Biochim. Biophys. Acta,* 536, 112, 1978.
52. **Singh, H., Silberlicht, I., and Singh, I. J.,** A comparative study of the polypeptides of mammalian peripheral nerve myelin, *Brain Res.,* 144, 303, 1978.
53. **Monticone, R. and Elam, J.,** Isolation of axonally transported glycoproteins with visual system myelin, *Brain Res.,* 100, 61, 1975.
54. **Mattieu, J.-M., Webster, H. DeF., DeVries, G. H., Corthay, S., and Koellreutter, B.,** Glial versus neuronal origin of myelin proteins and glycoproteins studied by combined intraocular and intracranial labelling, *J. Neurochem.,* 31, 93, 1978.
55. **DeVries, G. H., Payne, W., and Saul, R. G.,** Composition of axolemma enriched fractions isolated from bovine CNS myelinated axons, *Neurochem. Res.,* 6, 521, 1981.
56. **Balerna, M., Fosset, M., Chicheportiche, R., Romey, G., and Lazdunski, M.,** Constitution and properties of axonal membrane of crustacean nerves, *Biochemistry,* 14, 5500, 1975.
57. **Chacko, G. K., Villegas, G. M., Barnola, F. U., Villegas, R., and Goldman, D. E.,** The polypeptide and the phospholipid components of axon plasma membranes, *Biochem. Biophys. Acta,* 443, 19, 1976.
58. **Margolis, R. K. and Gomex, Z.,** Rapid turnover of fucose in the water-soluble glycoproteins of brain, *Biochim. Biophys. Acta,* 313, 226, 1973.
59. **Margolis, R. K., Margolis, R. U., Preti, C., and Lai, D.,** Distribution and metabolism of glycoproteins and glycosaminoglycans in subcellular fractions of brain, *Biochemistry,* 14, 4797, 1975.
60. **Krusius, T., Finne, J., Margolis, R. U., and Margolis, R. K.,** Structural features of microsomal, synaptosomal, mitochodrial, and soluble glycoproteins of brain, *Biochemistry,* 17, 3849, 1978.
61. **Langley, O. K.,** Fractionation of soluble brain glycoproteins by affinity and hydroxylapatite chromatography, *J. Neurochem.,* 33, 221, 1979.

62. **Chandrasekaran, E. V. and Bachhawat, B. K.,** Isolation and characterization of glycosaminoglycans in peripheral nerve and spinal cord of monkey, *J. Neurochem.*, 16, 1529, 1969.
63. **Kiang, W.-L., Crockett, C. P., Margolis, R. K., and Margolis, R. U.,** Glycosaminoglycans and glycoproteins associated with microsomal subfractions of brain and liver, *Biochemistry*, 17, 3841, 1978.
64. **Norling, B., Glimelius, B., Westermark, B., and Wasteson, A. A.,** Chondroitin sulphate proteoglycan from human cultured glial cells aggregates with hyaluronic acid, *Biochem. Biophys. Res. Comm.*, 84, 914, 1978.
65. **Margolis, R. K., Crockett, C. P., and Margolis, R. U.,** Dissociative extraction of brain proteoglycans, *J. Neurochem.*, 30, 1177, 1978.
66. **Branford White, C. J.,** Identification of glycosaminoglycans in nerve terminals, *J. Neurol. Sci.*, 41, 261, 1979.
67. **Lander, A. D., Fujii, D. K., Gospodarowicz, D., and Reichardt, L. F.,** Characterization of a factor that promotes neurite outgrowth: evidence linking activity to a Heparan sulfate proteoglycan, *J. Cell Biol.*, 94, 574, 1982.
68. **Geissler, D., Martinek, A., Margolis, R. U., Margolis, R. K., Skrivanek, J. A., Ledeen, R., Konig, P., and Winkler, H.,** Composition and biogenesis of complex carbohydrates of ox adrenal chromaffin granules, *Neuroscience*, 2, 685, 1977.
69. **Kronguist, K. E., Elmahdy, A., and Ronzio, R. A.,** Synthesis and subcellular distribution of heparan sulfate in the rat exocrine pancreas, *Arch. Biochem. Biophys.*, 182, 188, 1977.
70. **Zanini, A., Giannattasio, G., Nussdorfer, G., Margolis, R. K., Margolis, R. U., and Meldolesi, J.,** Molecular organization of prolactin granules. II. Characterization of glycosaminoglycans and glycoproteins of the bovine prolactin matrix, *J. Cell Biol.*, 86, 260, 1980.
71. **Stadler, H. and Dowe, G.,** Identification of a heparan sulfate-containing proteoglycan as a specific core component of cholinergic synaptic vesicles from *Torpedo marmorata*, *EMBO J.*, 1, 1381, 1982.
72. **Walker, J. H., Obrocki, J., and Zimmermann, C.,** Identification of proteoglycan antigen characteristic of cholinergic synaptic vesicles, *J. Neurochem.*, 41, 209, 1983.
73. **Elam, J. S.,** Axonal transport of complex carbohydrates, in *Complex Carbohydrates of Nervous Tissue*, Margolis, R. U. and Margolis, R. K., Eds., Plenum Press, New York, 1979, 235.
74. **Grafstein, B. and Forman, D. S.,** Intracellular transport in neurons, *Physiol. Rev.*, 60, 1167, 1980.
75. **Margolis, T. P., Marchand, C. M.-F., Kistler, H. B., Jr., and LaVail, J. H.,** Uptake and anterograde axonal transport of wheat germ agglutinin from retina to optic tectum in the chick, *J. Cell Biol.*, 89, 152, 1981.
76. **Nennesmo, I. and Kristensson, K.,** Somatopetal axonal transport of fluorescent lectins: distribution pattern and cytophotometric quantification in mouse peripheral neurons, *Neurosci. Lett.*, 27, 243, 1981.
77. **Phillipson, O. T. and Griffiths, A. C.,** Anterograde and retrograde labelling of CNS pathways with unconjugated lectins using the unlabelled antibody enzyme method, *Brain Res.*, 265, 199, 1983.
78. **Elam, J. S. and Agranoff, B. W.,** Transport of proteins and sulfated mucopolysaccharides in the goldfish visual system, *J. Neurobiol.*, 2, 379, 1971.
79. **Karlsson, J.-O. and Linde, A.,** Axonal transport of ^{35}S sulphate in retinal ganglion cells of the rabbit, *J. Neurochem.*, 28, 293, 1977.
80. **Quarles, R. H. and Brady, R. O.,** Synthesis of glycoproteins and gangliosides in developing rat brain, *J. Neurochem.*, 18, 1809, 1971.
81. **Webster, J. C. and Klingman, J. C.,** Incorporation of radiolabeled sugars into synaptic junctional macromolecules from chick brain, *Brain Res. Bull.*, 5, 31, 1980.
82. **McEwen, B. S., Forman, D. S., and Grafstein, B.,** Components of fast and slow axonal transport in the goldfish optic nerve, *J. Neurobiol.*, 2, 361, 1971.
83. **Bondy, S. C. and Madsen, C. J.,** Development of rapid axonal flow in the chick embryo, *J. Neurobiol.*, 2, 279, 1971.
84. **Karlsson, J.-O. and Sjostrand, J.,** Rapid intracellular transport of fucose-containing glycoproteins in retinal ganglion cells, *J. Neurochem.*, 18, 2209, 1971.
85. **Marko, P. and Cuenod, M.,** Contribution of the nerve cell body to renewal of axonal and synaptic glycoproteins in the pigeon visual system, *Brain Res.*, 62, 419, 1973.
86. **Specht, S. and Grafstein, B.,** Accumulation of radioactive protein in mouse cerebral cortex after injection of ^{3}H fucose into the eye, *Exp. Neurol.*, 41, 705, 1973.
87. **Goodrum, J. F., Toews, A. D., and Morell, P.,** Axonal transport and metabolism of [^{3}H] fucose- and [^{35}S]-sulfate-labeled macromolecules in the rat visual system, *Brain Res.*, 176, 255, 1979.
88. **Ochs, S.,** Fast transport of materials in mammalian nerve fibers, *Science*, 176, 252, 1972.
89. **Edstrom, A. and Mattsson, H.,** Rapid axonal transport *in vitro* in the sciatic system of the frog of fucose-, glucosamine-, and sulfate-containing material, *J. Neurochem.*, 19, 1717, 1972.

90. **Bennett, G., DiGiamberardino, L., Koenig, H. L., and Droz, B.,** Axonal migration of protein and glycoprotein to nerve endings. II. Radioautographic analysis of the renewal of glycoproteins in nerve endings of chicken ciliary ganglion after intracerebral injection of ^{3}H fucose and ^{3}H glucosamine, *Brain Res.*, 60, 129, 1973.
91. **Frizell, M. and Sjostrand, J.,** The axonal transport of ^{3}H fucose labelled glycoproteins in normal and regenerating peripheral nerves, *Brain Res.*, 78, 109, 1974.
92. **Gross, G. W. and Beidler, L. M.,** A quantitative analysis of isotope concentration and rapid transport velocities in the C-fibers of the garfish olfactory nerve, *J. Neurobiol.*, 6, 213, 1975.
93. **Barker, J. L., Hoffman, P. N., Gainer, H., and Lasek, R. J.,** Rapid transport of proteins in the sonic motor system of the toadfish, *Brain Res.*, 97, 291, 1975.
94. **Ambron, R. and Schwartz, J. H.,** Regionsl aspects of neuronal glycoprotein and glycolipid synthesis in *Complex Carbohydrates of Nervous Tissue*, Margolis, K. V. and Margolis, R. K., Eds., Plenum Press, New York, 1979, 269.
95. **Ambron, R. T., Goldman, J. E., and Schwartz, J. H.,** Axonal transport of newly synthetized glycoproteins in a single identified neuron of *Aplysia californica*, *J. Cell Biol.*, 61, 655, 1974.
96. **Levin, B. E.,** Axonal transport of ^{3}H fucosyl glycoproteins in noradrenergic neurons in the rat brain, *Brain Res.*, 130, 424, 1977.
97. **Geinisman, Y., Bondareff, W., and Telser, A.,** Transport of ^{3}H fucose labelled glycoproteins in the septo-hippocampal pathway of young adult and senescent rats, *Brain Res.*, 125, 182, 1977.
98. **Padilla, S. S. and Morell, P.,** Axonal transport of [^{3}] fucose-labeled glycoproteins in two intra-brain tracts of the rat, *J. Neurochem.*, 35, 444, 1980.
99. **Goodrum, J. F. and Morell, P.,** Axonal transport, deposition, and metabolic turnover of glycoproteins in the rat optic pathway, *J. Neurochem.*, 38, 696, 1982.
100. **Zatz, M. and Barondes, S. H.,** Rapid transport of fucosyl glycoproteins to nerve endings in mouse brain, *J. Neurochem.*, 18, 1125, 1971.
101. **Marinari, U. F., Morgan, I. G., Mack, G., and Gombos, G.,** Synthesis of synaptic glycoproteins. II. Delayed labelling of the glycoproteins of synaptic vesicles and synaptosomal plasma membranes, *Neurobiology*, 2, 176, 1972.
102. **Langley, O. K. and Kennedy, P.,** The metabolism of nerve terminal glycoproteins in the rat brain, *Brain Res.*, 130, 109, 1977.
103. **Forman, D. S., Grafstein, B., and McEwen, B. S.,** Rapid axonal transport of ^{3}H fucosyl glycoproteins in the goldfish optic system, *Brain Res.*, 48, 327, 1972.
104. **McLean, W. G., Frizell, M., and Sjostrand, J.,** Axonal transport of labelled proteins in sensory fibres of rabbit vagus nerve *in vitro*, *J. Neurochem.*, 25, 695, 1975.
105. **Goldberg, D. J. and Ambron, R. T.,** Two rates of fast axonal transport of [^{3}H] glycoprotein in an identified invertebrate neuron, *Brain Res.*, 229, 445, 1981.
106. **Specht, S. C. and Grafstein, B.,** Axonal transport and transneural transfer in mouse visual system following injection of ^{3}H fucose into the eye, *Exp. Neurol.*, 54, 352, 1977.
107. **Frizell, M., McLean, W. G., and Sjostrand, J.,** Retrograde axonal transport of rapidly migrating labelled proteins and glycoproteins in regenerating peripheal nerves, *J. Neurochem.*, 27, 191, 1976.
108. **Whitnall, M. H., Currie, J. R., and Grafstein, B.,** Bidirectional axonal transport of glycoproteins in goldfish optic nerve, *Exp. Neurol.*, 75, 191, 1982.
109. **Kistler, H. B. and Schwartz, J. H.,** Retrograde axonal transport of a wheat germ agglutinin horseradish peroxidase conjugate in *Aplysia californica*, *Brain Res.*, 244, 343, 1982.
110. **Thompson, E. B., Schwartz, J. H., and Kandel, E. R.,** A radioautographic analysis in the light and electron microscope of identified *Aplysia* neurons and their processes after intrasomatic injection of L-^{3}H fucose, *Brain Res.*, 112, 251, 1976.
111. **Markov, D., Rambourg, A., and Droz, B.,** Smooth endoplasmic reticulum and fast axonal transport of glycoproteins, an electron microscope radioautographic study of thick sections after heavy metals impregnation, *J. Microsc. Biol. Cell.*, 25, 57, 1976.
112. **Elam, J.,** Composition and subcellular distribution of glycoproteins and glycosaminoglycans undergoing axonal transport in garfish olfactory nerves, *J. Neurochem.*, 39, 1220, 1982.
113. **Ambron, R.,** Differences in the distribution of specific glycoproteins among the regions of a single identified neuron, *Brain Res.*, 239, 489, 1982.
114. **DiGiamberardino, L., Bennett, G., Koenig, H. L., and Droz, B.,** Axonal migration of protein and glycoprotein to nerve endings. III. Cell fraction analysis of chicken ciliary ganglion after intracerebral injection of precursors of proteins and glycoproteins, *Brain Res.*, 60, 147, 1973.
115. **Elam, J. S. and Peterson, N. W.,** Axonal transport of sulfated glycoproteins and mucopolysaccharides in the garfish olfactory nerve, *J. Neurochem.*, 26, 845, 1976.
116. **Tessler, A., Antilio-Gambetti, L., and Gambetti, P.,** Axonal growth during regeneration: A quantitative autoradiographic study, *J. Cell Biol.*, 87, 197, 1980.

117. **Griffin, J. W., Price, D. L., Drachman, D. B., and Morris, J.,** Incorporation of axonally transported glycoproteins into axolemma during nerve regeneration, *J. Cell Biol.*, 88, 205, 1981.
118. **Antillo-Gambetti, L., Gambetti, P., and Shafer, B.,** Glial and neuronal contribution to proteins and glycoproteins recovered in myelin fractions, *Brain Res.*, 84, 336, 1975.
119. **Elam, J. and Ripellino, J.,** Subsynaptic distribution of axonally transported sulfated glycoconjugates, *Trans. Am. Soc. Neurochem.*, 13, 224, 1982.
120. **Ripellino, J. A. and Elam, J. S.,** Differential turnover of axonally transported glycoproteins, *Neurochem. Res.*, 5, 351, 1980.
121. **Edstrom, A. and Mattsson, H.,** Electrophoretic characterization of leucine-, glucosamine- and fucose-labelled proteins rapidly transported in frog sciatic nerve, *J. Neurochem.*, 21, 1499, 1973.
122. **Tytell, M., Gulley, R. L., Wenthold, R. J., and Lasek, R. J.,** Fast axonal transport in auditory neurons of the guinea pig: a rapidly turned-over glycoprotein, *Proc. Natl. Acad. Sci. U.S.A.*, 77, 3042, 1980.
123. **Wenthold, R. J. and McGarvey, M. L.,** Different polypeptides are rapidly transported in auditory and optic neurons, *J. Neurochem.*, 39, 27, 1982.
124. **Toews, A. D., Saunders, B. F., and Morell, P.,** Axonal transport and metabolism of glycoproteins in rat sciatic nerve, *J. Neurochem.*, 39, 1348, 1982.
125. **Stone, G. and Hammerschlag, R.,** Glycosylation as a criterion for defining subpopulations of fast transported proteins, *J. Neurochem.*, 40, 124, 1983.
126. **Karlsson, J. O.,** Proteins of axonal transport: interaction of rapidly transported proteins with lectins, *J. Neurochem.*, 32, 491, 1979.
127. **Karlsson, J. O.,** Proteins of axonal transport: polypetides interacting with the lectin from Lens Culinaris, *J. Neurochem.*, 34, 11184, 1980.
128. **Gustavsson, S., Ohlson, C., and Karlsson, J. O.,** Glycoproteins of axonal transport: affinity chromatography on fucose-specific lectins, *J. Neurochem.*, 38, 852, 1982.
129. **Cole, G. and Elam, J.,** Characterization of axonally transported glycoproteins in regenerating garfish olfactory nerve, *J. Neurochem.*, 41, 691, 1983.
130. **Elam, J. S. and Peterson, N. W.,** Axonal transport of glycoproteins in the garfish olfactory nerve: isolation of high molecular weight glycopeptides labeled with ^{3}H-fucose and ^{3}H-glucosamine *J. Neurochem.*, 33, 571, 1979.
131. **Cole, G. and Elam, J.,** Axonal transport of glycoproteins in regenerating olfactory nerve: enhanced glycopeptide concanavalin A binding, *Brain Res.*, 222, 437, 1981.
132. **Krusius, T., Finne, J., and Rauvala, H.,** The poly(glycosyl) chains of glycoproteins, *Eur. J. Biochem.*, 92, 289, 1978.
133. **Specht, S. C.,** Postnatal changes in [^{3}H] fucosyl glycopeptides of hamster optic nerve synaptosomal membrane, *Dev. Br. Res.*, 4, 109, 1982.
134. **Gurd, J. W. and Mahler, H. R.,** Fractionation of synaptic plasma membrane glycoproteins by lectin affinity chromatography, *Biochemistry*, 13, 5193, 1974.
135. **Fu, S. C., Druz, T. F., and Gurd, J. W.,** Development of synaptic glycoproteins: effect of postnatal age on the synthesis and concentration of synaptic membrane and synaptic junctional fucosyl and sialyl glycoproteins, *J. Neurochem.*, 36, 1338, 1981.
136. **Forman, D. S., McEwen, B. S., and Grafstein, B.,** Rapid transport of radioactivity in goldfish optic nerve following injections of labeled glucosamine, *Brain Res.*, 28, 119, 1971.
137. **Gurd, J. W.,** Biosynthesis of synaptic membranes. Incorporation of [^{3}H] leucine into proteins and glycoproteins of rat brain synaptic plasma membranes, *Brain Res.*, 147, 201, 1978.
138. **Elam, J.,** Composition, subcellular distribution and turnover of axonally transported glycoproteins, in *Axoplasmic Transport*, Weiss, D. G., Ed., Springer-Verlag, Basel, 1982, 125.
139. **Bisby, M. A.,** Reversal of axonal transport: similarity of proteins transported in anterograde and retrograde directions, *J. Neurochem.*, 36, 741, 1981.
140. **Morell, P. and Goodrum, J. F.,** Transport kinetics of axonally transported glycoproteins; turnover of the fucose moiety, *J. Neurochem.*, 41 (Suppl), 569, 1983.
141. **Yohe, H. C. and Rosenberg, A.,** Action of intrinsic sialidase of rat brain synaptic membranes on membrane sialolipid and sialoprotein components, *in situ*, *J. Biol. Chem.*, 252, 2412, 1977.
142. **Cruz, T. and Gurd, J.,** Identification of intrinsic sialidase and sialoglycoprotein substrates in rat brain synaptic junctions, *J. Neurochem.*, 40, 1599, 1983.
143. **Bosmann, H. B.,** Synthesis of glycoproteins in brain. Identification, purification, and properties of glycosyltransferases from purified synaptosomes of guinea pig cerebral cortex, *J. Neurochem.*, 19, 763, 1972.
144. **Den, H., Kaufman, B., McGuire, E. J., and Roseman, S.,** The sialic acids. XVIII. Subcellular distribution of seven glycosytransferases in embryonic chicken brain, *J. Biol. Chem.*, 250, 739, 1975.
145. **Goodrum, J. F., Bosmann, H. B., and Tanaka, R.,** Glycoprotein galactosyltransferase activity in synaptic junctional complexes isolated from rat forebrain, *Neurochem. Res.*, 4, 331, 1979.
146. **Rostas, J. A. P., Leung, W. N., and Jeffrey, P.,** Glycosyltransferase activities in chicken brain synaptic junctions, *Neurosci. Lett.*, 24, 155, 1981.

147. **Bosmann, H. B.,** Identification of proteins and glycoproteins synthesized by isolated rat cerebral cortex synaptosomes in vitro, *Neurobiology*, 1, 114, 1971.
148. **Festoff, B. W., Appel, S. H., and Day, E.,** Incorporation of ^{14}C-glucosamine into synaptosomes in vitro, *J. Neurochem.*, 18, 1871, 1971.
149. **Dutton, G. R., Haywood, P., and Barondes, S. H.,** ^{14}C-Glucosamine incorporation into specific products in the nerve ending fraction in vivo and in vitro, *Brain Res.*, 57, 397, 1973.
150. **Branford White, C. J. and Dawson, J. M.,** Fucose incorporation and identification of fucosylglycoproteins in synaptosomes, *J. Neurochem.*, 37, 1155, 1981.
151. **Branford White, C. J.,** Sulphated glycoproteins in synaptosomes, *Neurosci. Lett.*, 16, 307, 1980.
152. **Branford White, C. J.,** The incorporation in vitro of sulphate ions into synaptic-membrane glycoproteins, *Biochem. J.*, 200, 373, 1981.
153. **Ambron, R. T. and Treistman, S. M.,** Glycoproteins are modified in the axon of R2, the giant neuron of *Aplysia californica*, after intraxonal injection of ^{3}H N-acetylgalactosamine, *Brain Res.*, 121, 287, 1977.
154. **Truding, R., Shelanski, M. L., and Morell, P.,** Glycoproteins released into the culture medium of differentiating murine neuroblastoma cells, *J. Biol. Chem.*, 250, 9348, 1975.
155. **Alitalo, K., Kurkinen, M., Vaheri, A., Virtanen, I., Rohde, H., and Timpl, R.,** Basal lamina glycoproteins are produced by neuroblastoma cells, *Nature (London)*, 287, 465, 1980.
156. **Tedeschi, B. and Wilson, D.,** Release of fast transported material from regenerating nerve, *Soc. Neursci. Abst.*, 9, 48, 1983.
157. **Casagrande, V. A. and Harting, J. K.,** Transneuronal transport of tritiated fucose and proline in the visual pathways of tree shrew Tupaia glis, *Brain Res.*, 96, 367, 1975.
158. **Reperant, J., Miceli, D., and Raffin, J.,** Transneuronal transport of tritated fucose and proline in the avian visual system, *Brain Res.*, 121, 343, 1977.
159. **Elam, J. S., Goldberg, J. M., Radin, N. S., and Agranoff, B. W.,** Rapid axonal transport of sulfated mucopolysaccharide proteins, *Science*, 170, 458, 1970.
160. **Ripellino, J.,** The Subcellular Distribution, Metabolism and Composition of Axonally Transported Glycoconjugates, Ph.D. thesis, Florida State University, Tallahassee, 1982.
161. **Held, I. and Young, I. J.,** Transport of radioactivity derived from labelled N-acetylglucosamine in mammalian motor axons, *J. Neurobiol.*, 3, 153, 1972.
162. **Jones, R. T., Walker, J. H., Stadler, H., and Whittaker, V. P.,** Immunohistochemical localization of a synaptic-vesicle antigen in a cholinergic neurone under conditions of stimulation and rest, *Cell Tissue Res.*, 223, 117, 1982a.
163. **Baux, G., Simonneau, M., and Tauc, L.,** Transmitter release: rutheneium red used to demonstrate a possible role of sialic acid containing substrates, *J. Physiol.*, 191, 161, 1979.
164. **Massarelli, R., Wong, T. Y., Harth, S., Louis, J. C., Freysz, L., and Dreyfus, H.,** Possible role of sialocompounds in the uptake of choline into synaptosomes and nerve cell cultures, *Neurochem. Res.*, 7, 301, 1982.
165. **Levin, B. E.,** Effects of reserpine on fast, intermediate and slow axonal transport of proteins on rat locus coerulus neurons, *Brain Res.*, 187, 495, 1980.
166. **Burgoyne, R. D. and Rose, S. P. R.,** Changes in glycoprotein metabolism in the cerebral cortex following first exposure of dark-reared rats to light, *J. Neurochem.*, 34, 510, 1980.
167. **Caputto, B. L., Nores, G. A., Cemborain, B. N., and Caputto, R.,** The effect of light exposure following an intraocular injection of [^{3}H]N-Acetylmannosamine on the labeling of gangliosides and glycoproteins of retina ganglion cells and optic tectum of singly caged chickens, *Brain Res.*, 245, 231, 1982.
168. **Margolis, R. U. and Margolis, R. K.,** Metabolism and function of glycoproteins and glycosaminoglycans in nervous tissue, *Int. J. Biochem.*, 8, 85, 1977.
169. **Marquis, J. K., Hilt, D. C., and Mautner, H. G.,** Concanavalin A-Sepharose affinity chromatography of axon plasma membrane proteins. Heterogeneity of cholinergic binding sites, *J. Neurochem.*, 34, 1071, 1980.
170. **Prives, J. and Bar-Sagi, D.,** Effect of tunicamycin, an inhibitor of protein glycosylation, on the biological properties of acetylcholine receptor in cultured muscle cells, *J. Biol. Chem.*, 258, 1175, 1983.
171. **Giulian, D., Ruisseaux, H. D., and Cowburn, D.,** Biosynthesis and intra-axonal transport of proteins during neuronal regeneration, *J. Biol. Chem.*, 255, 6494, 1980.
172. **Heacock, A. M.,** Glycoprotein requirements for neurite outgrowth in goldfish retina explants: effects of tunicamycin, *Brain Res.*, 241, 307, 1982.
173. **Bisby, M. A. and Bulger, V. T.,** Reversal of axonal transport at a nerve crush, *J. Neurochem.*, 29, 313, 1977.
174. **Estridge, M. and Bunge, R.,** Compositional analysis of growing axons from rat sympathetic neurons, *J. Cell Biol.*, 70, 138, 1978.
175. **Feldman, E. L., Axelrod, D., Schwartz, M., Heacock, A. M., and Agranoff, B. W.,** Studies on the localization of newly added membrane in growing neurites, *J. Neurobiol.*, 12, 591, 1981.

176. **Pfenninger, K. H. and Maylie-Pfenninger, M.-F.,** Lectin labeling of sprouting neurons. I. Regional distribution of surface glycoconjugates, *J. Cell Biol.*, 89, 536, 1981.
177. **Pfenninger, K. H. and Maylie-Pfenninger, M.-F.,** Lectin labeling of sprouting neurons. II. Relative movement and appearance of glycoconjugates during plasmalemmal expansion, *J. Cell Biol.*, 89, 547, 1981.
178. **McLaughlin, B. J. and Wood, J. G.,** The localization of concanavalin A binding sites during photoreceptor synaptogenesis in the chick retina, *Brain Res.*, 119, 57, 1977.
179. **Zanetta, J.-P., Roussel, G., Ghandour, M. S., Vincendon, G., and Gombos, G.,** Postnatal development of rat cerebellum: massive and transient accumulation of concanavalin A binding glycoproteins in parallel fiber axolemma, *Brain Res.*, 142, 301, 1978.
180. **Denis-Donini, S., Estenoz, M., and Augusti-Tocco, G.,** Cell surface modifications in neuronal maturation, *Cell. Differ.*, 1, 193, 1978.
181. **Chernoff, E., Maresh, G., and Culp, L.,** Isolation and characterization of a large, neurite associated glycoconjugate from neuroblastoma cells, *J. Cell Biol.*, 96, 661, 1983.
182. **Rutishauser, U., Gall, W. E., and Edelman, G. M.,** Adhesion among neural cells of the chick embryo. IV. Role of the cell surface molecule CAM in the formation of neurite bundles in cultures of spinal ganglia, *J. Cell Biol.*, 79, 382, 1978.
183. **Marchisio, P. C., Sjostrand, J., Aglietta, M., and Karlsson, J.-O.,** The development of axonal transport of proteins and glycoproteins in the optic pathway of chick embryos, *Brain Res.*, 63, 273, 1973.
184. **Sjostrand, J. O., Karlsson, P. C., and Marchisio, P. C.,** Axonal transport in growing and mature retinal ganglion cells, *Brain Res.*, 62, 395, 1973.
185. **Gremo, F. and Marchisio, P. C.,** Dynamic properties of axonal transport of proteins and glycoproteins: a study based on the effects of metaphase blocking drugs in the developing optic pathway of chick embryos, *Cell Tissue Res.*, 161, 303, 1975.
186. **Skene, J. H. P. and Willard, M.,** Characteristics of growth-associated polypeptides in regenerating toad retinal ganglion cell axons, *J. Neurosci.*, 1, 419, 1981.
187. **Morgan, D. G. and Routtenberg, A.,** The incorporation of intrastriatally injected [^{3}H] fucose into electrophoretically separated synaptosomal glycoproteins. II. The influence of passive avoidance training, *Brain Res.*, 179, 343, 1979.
188. **Ramirez, G., Karlsson, B., and Perrin, C. L.,** Effects of environmental complexity on glucosamine incorporation into rat cerebral cortex hippocampus, *in vivo, Brain Res.*, 79, 296, 1974.
189. **Burgoyne, R. D. and Rose, S. P. R.,** Subcellular localization of increased incorporation of [^{3}H]fucose following passive avoidance learning in the chick, *Neurosci. Lett.*, 19, 343, 1980.

Chapter 10

PARTICLE MOVEMENTS IN AXOPLASMIC TRANSPORT

Toshifumi Takenaka

TABLE OF CONTENTS

I. INTRODUCTION

The movement of optically detectable particles in tissue-cultured neurites were observed by Masumoto[1] and since that time many works have been carried out qualitatively[2-6] and quantitatively.[7-11] Isolated single axons from nerve have also been studied by use of the light microscopy.[12-27] Darkfield, phase contrast, and Nomarski optics microscopy have been used to visualize particles movement by time-lapse cinematography or video-tape with computer. More recently, the development of video-enhanced microscopy with a computer frame memory has brought a more precise study of particle movements in axon.[27-30]

In the early study of neurite, Nakai observed the bidirectional movements of particles by using phase-contrast microscopy.[2] Smith, Kirkpatrick, Cooper and Smith have shown spherical particles to have a movement which may be saltatory, the particles stopping and starting at irregular intervals.[12-16] Rod-shaped particles which are mitochondria were also seen and they move very slowly.

This technique of microscopic observation has limitations in that the morphological and biochemcial identification of particles may not be easy because of microscopic resolution limit. However it is useful for studying the temporal and spacial analyses in the living physiological condition. A video-enhanced method described here is useful for this study.

II. PHARMACOLOGICAL EFFECTS ON THE TRANSPORTED PARTICLES IN NEURITES

Movements of transported particles in tissue cultured nerve cells were observed under Nomarski optics and analyzed with video system. Neurites in these nerve cells had no connective tissues in the early stage, so that drugs could directly act on them. These samples, therefore, are suitable for examining the role of cytoskeletal structures in axoplasmic transport with chemical drugs.

Neurites from the dorsal root ganglion explants dissected from six day-old chick embryos were used. After 3 to 4 days in culture, directional movement of particles in the neurite could be observed. Particles moved in both directions, anterogradely and retrogradely. Density of retrogradely transported particles were about twice those of anterogradely transported ones. Sizes of the particles are about 0.1 to 0.2 μm.

Neurites near the growth cones with a diameter of 1.3 to 2.5 μm were selected and the number of particles passing through a given region in the neurites were counted. During 10 min, 9 to 15 particles were seen to pass the region. Table 1 shows the time course of the particle numbers after the application of colchicine. The transported particles gradually decreased by the treatment with colchicine in concentrations higher than 1×10^{-7} *M*. At 1×10^{-7} *M*, about 50% of transported particles persisted in 60 min, but at concentrations higher than 1×10^{-6} *M*, no transport could be observed after the same period of time. The blocking time was decreased at a high concentration of colchicine. At 1×10^{-3} *M* colchicine, all particles stopped 30 min after the application. At 1×10^{-8} *M* of colchicine, no significant effects on the density of transported particles could be observed. Effect of vinblastine on the number of transported particles were similar to those of colchicine at the same concentration. These two agents which disrupt microtubules were effective in decreasing the number of transported particles at concentrations higher than 1×10^{-7} *M*. Cytochalasin D, on the other hand appeared to have no effect. At 2×10^{-5} *M* of cytochalasin D, which was high enough to retract growth cones, showed no effect on the number of transported particles.

The average velocity of particles was measured when they moved smoothly over 20 μm and was calculated to be 1.2 to 1.8 μm/sec. There were no significant differences in the average velocities of anterograde and retrograde transport.

After application of colchicine or vinblastine, the velocity of transported particles were

Table 1
EFFECT OF COLCHICINE ON THE NUMBER OF TRANSPORTED PARTICLES

Agent (*M*) colchicine	Time (min) Control	0—10	10—20	20—30	30—40	40—50	50—60
1×10^{-3}	13	10	5	3	0	0	0
1×10^{-4}	9	11	5	4	2	2	0
1×10^{-5}	9	7	9	6	3	1	0
1×10^{-6}	13	10	6	7	3	1	0
1×10^{-7}	15	—	13	10	10	5	8
1×10^{-8}	13	—	14	15	15	11	11

Note: The numbers indicate those of the transported particles passing through one region of a neurite during 10 min.

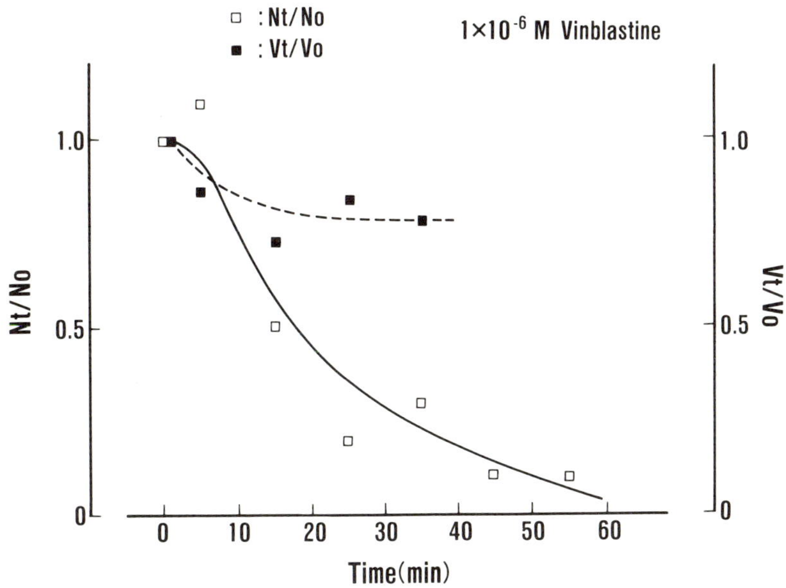

FIGURE 1. Effects of 1×10^{-6} *M* vinblastine on relative number and relative velocity of transported particles. (N_t/N_o) shows relative number of transported particles. (V_t/V_o) shows relative velocity of transported particles. This figure shows that after exposure to the agent the number of transported particles decreased but the velocity of the transported particles remained almost constant.[32]

maintained at almost the same level as in the control, although their density gradually decreased as stated above. Figure 1 shows the effects of 1×10^{-6} *M* of vinblastine on relative number and relative velocity of transported particles. It is evident that the drug did not affect the velocity of transported particles.

III. MORPHOLOGICAL CHANGE OF NEURITES BY INHIBITORS AND ITS CORRELATION TO THE PARTICLE MOVEMENT

Morphological changes of the neurites after exposure to colchicine or vinblastine were examined by Nomarski optics. Before the application of the drug, the neurites appeared

Table 2
THE DISTANCE OF SWOLLEN REGIONS IN A NEURITE

n	$\overline{X}$	SD
23	13.6 μm	3.6 μm

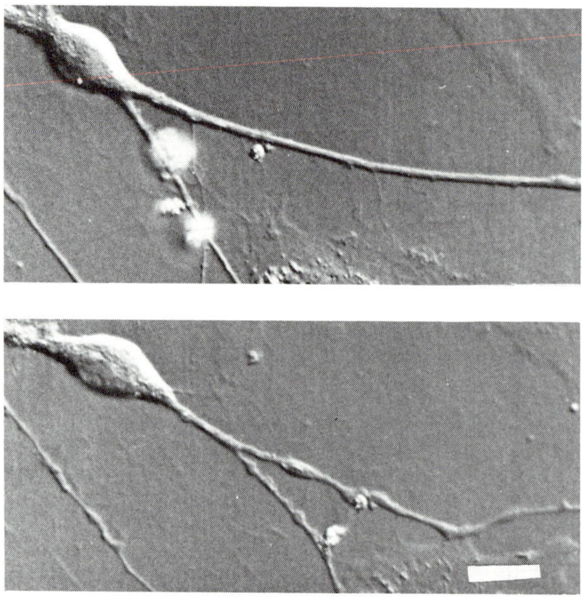

FIGURE 2. Nomarski micrographs of isolated chick dorsal root ganglion cell 3 days in culture illustrating the effect of 1×10^{-6} *M* vinblastine on the morphology of the cell body and the neurite. Upper: before the exposure to the drug, appearance of cell body and neurite are smooth. Lower: 10 min after the exposure to the drug, swelling occured in the nerve. Bar: 20 μm.

smooth and had a constant diameter. The application, however, caused rugged and swollen regions at regular intervals (10 to 24 μm). These intervals are correlated to the intervals of saltatory movements in normal neurites (Table 2). These intervals did not correlate with the diameter of the neurites, but the time course of the morphological changes appeared to correlate with the diameter. Thinner neurites changed their appearance earlier than thick ones. Figure 2 shows such morphological changes of neurites 10 min after the application of 1×10^{-6} *M* vinblastine. The neurites became thinner overall except in swollen regions. Cell bodies were not affected. The change in the movement of transported particles was correlated to the morphological changes of neurites. In the early stage of the swelling, only fractions of the particles was impeded, hesitated, or stopped at swollen regions. At the later stage, most of the particles stopped and accumulated there, and finally no particles could be observed being transported through the neurites (Figure 3). We thought that such an accumulation of particles caused the local swelling. it should be noted, however, that the present morphological changes did not affect the velocity of smoothly transported particles.

The morphological changes were also analyzed by electron microscopy. In the swollen regions, many lysosomes and mitochondria gathered whereas they were scarcely found in

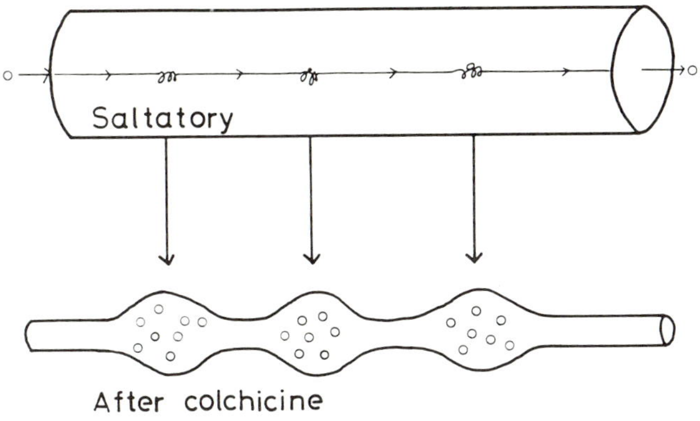

FIGURE 3. Saltatory movement and the swollen places after colchicine treatment. The places where wondering movement occures in saltatory movement correspond to the swollen places.

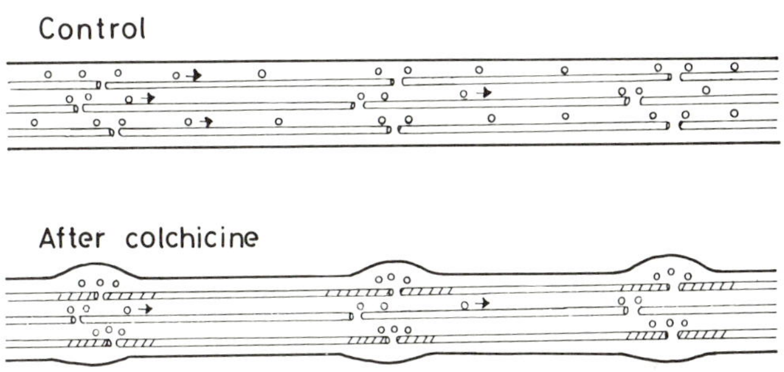

FIGURE 4. Schematic diagram of track junction theory. Microtubule might have junctions in every about 15 μm which are functional or structural. The microtubules act as a sort of track on which the particles move. The particles have to change from one track to a succeeding one in order to continue their travel.[42]

other regions of neurites. Neurofilament were well preserved althrough the neurites, but microtubules were not found in the swollen regions.

IV. SALTATORY AND TRACK JUNCTION THEORY

The places where wondering movement occurs in saltatory correspond to the swollen places. These places might be the junctions of tracks where the particles have to change from one tracks to a succeding one in order to continue their travel (track junction theory). The microtubules act as a sort of track on which the transported particles move.[33,34] When the number of microtubules in neurites is reduced by the drugs, the number of transported particles decreases. The microtubules are gradually disrupted from their ends by the drug, so that the microtubules cannot convey the particles as usual. The velocity of transported particles does not depend on the number of microtubules, but on the motile activity of the transport machinery. The velocity of the particles remains constant after the particles get on the track of the microtubule (Figure 4).

V. PARTICLES TRANSPORT IN PERMEABILIZED AXON

In order to understand the mechanism of fast-transport, one approach is to study the drugs

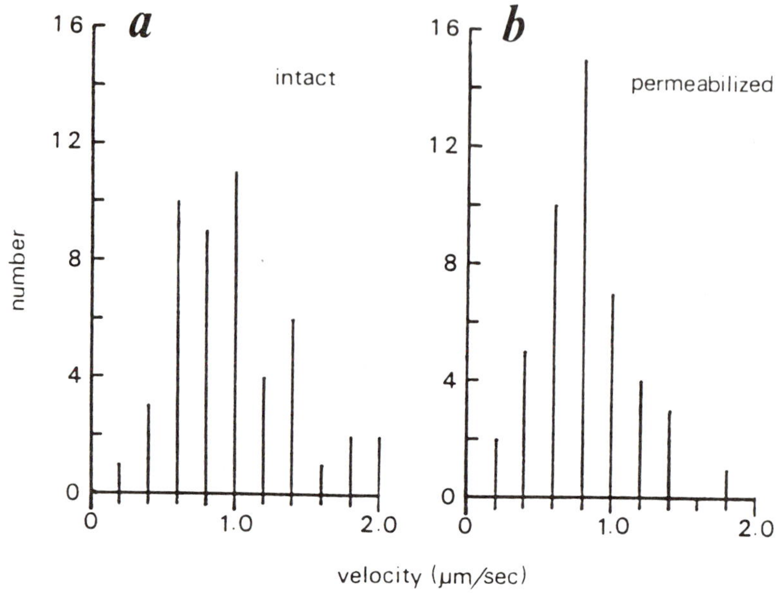

FIGURE 5. A histogram of the velocities of particles. (a) Intact axons. (b) Permeabilized axons reactivated in 5 m*M* ATP.[42]

effect on the transport. Many drugs do not cross the axonal plasma membrane. Understanding of such intracellular mechanism requires a cell model where the plasma membrane barrier is eliminated. Three kinds of cytoplasmic models for studying the axoplasmic transport have been reported. Adams punctured holes in the plasma membrane of giant axons from crab legs by a high voltage discharge and succeeded in reactivation of the particle movement.[35] In such a procedure small holes of 4 nm diameter remained open for 2 hr but his preparation still have the membrane barrier for proteins such as enzymes or antibodies. The second preparation is the extruded axoplasm from a squid giant axon which also shows the fast axonal transport.[36] However, the extruded model is applicable only for the giant axon. On the other hand the third preparation of permeabilized axon with a detergent is applicable for axons in any category and allows us to apply macromolecular agents. Saponin, one of the detergent, has a specific high affinity to cholesterol of the plasma membrane.

It was reported that saponin specifically permeabilized axolemma of squid axons leaving cytoplasmic structure intact.[37] Forman et al. have recently shown the saltatory movements of particles and also the effects of drugs on axonal transport in axons permeabilized with saponin.[38-41] Gotoh and Takenaka also reported the reactivation of particle movements and effects of drugs on axonal transport in saponin-treated axons.[42]

VI. VELOCITIES OF TRANSPORTED PARTICLES IN PERMEABILIZED AXON

A fin nerve of squid was permeabilized with saponin. Artificial intracellular fluid (AICF) was composed of 400 m*M* K glutamate, 150 m*M* glycine, 10 m*M* $MgCl_2$, 5 m*M* ATP, 1 m*M* EDTA, and 20 m*M* HEPES (pH 7.2). Axons were treated with 0.01% saponin in an ATP-free AICF for 2 min. Then they were washed with an AICF lacking both saponin and ATP. Electron-microscopical observation of the permeabilized axon revealed holes of approximatley 100 nm in diameter punctured in its plasma membrane. It should be noted that this size is large enough to introduce macromolecular nonpermeable agents in the axoplasm.

The histogram of the particle velocities in intact and permeabilized axons are shown in Figure 5. An average velocity in the controls was 1.09 ± 0.59 μm/sec at 20°C. After the treatment with saponin, the particle movement reappeared with a complete AICF and its velocity was 0.87 ± 0.39 μm/sec. Thus, the velocity of transported particles in permeabilized axons is similar to that of unlysed ones. The density of transported particles, however, sometimes decreased by permeabilization. The particle movement in the permeabilized axon continued at least for 1 hr in the presence of ATP. Inactivation of the movement in ATP-free AICF and the reactivation in the complete AICF could be reversibly repeated.

VII. DIRECTIONAL AND OSCILLATORY MOVEMENTS OF PARTICLES IN PERMEABILIZED AXON

The effects of pharmacological agents on the particle movement in permeabilized axons are shown in Figure 6. The particle movement was blocked by 1 m*M* colchicine. Cytochalasin B (10 μ/mℓ) also blocked the particle movement. Vanadate, (1 m*M*) which is known to be an inhibitor of ATPase, effectively blocked the particle movement.

The particle movement consists of two movements, directional movement and oscillation. The two types of movements can be differentiated by drugs as shown in Figure 7. Vanadate blocked both directional and oscillatory movements. However, colchicine or vinblastine dissociated these two movements. The oscillatory movements was detectable at least for 10 min after cessation of the directional movement. Oscillation is a nondirectional local movement of particles oscillating only within 2 to 3 μm. This oscillation is not a Brownian movement, because of two reasons, (1) is that the energy of this movement is higher than the thermal energy and (2) is that oscillation is ATP dependent. It is possible to examine this possibility theoretically. The force, F, required to move a spherical body at a constant velocity in a viscous medium is given by $F = 6, \pi\eta\ rv$, where η is the coefficient of viscosity of the medium, r is the radius of the body, and v is its velocity. Taking the viscosity of cytoplasm as 10^3 P, the radius of the particle as 10^{-5} cm, and its velocity as 10^{-4} cm/sec, one obtains a force of 2×10^{-5} dyn. Therefore, the energy, E, required to move the particle over the distance of 1 μm, is given by $E = F \times 1\ \mu m = 2 \times 10^{-9}$ erg. On the other hand, thermal energy at room temperature is expressed as $kT = 4 \times 10^{-14}$ erg. Thus, the thermal energy or a Browian motion is too small to explain the present oscillatory movement. The present oscillation appears to be an energy-requiring process. In fact, the oscillatory movement was found to be ATP-dependent.

These results indicate that microtubules are tracks along which organelles travel and disruption of tracks caused a loss of directional movement, the oscillatory movement will remain even after disruption of tracks. Vanadate or ATP inhibits the functioning of ATPase machinery of particle movement, so it completely blocks both the directional and oscillatory movements. The motile activity of the transport machinery correspond to the oscillatory movement and the microtubule function in transport correspond to the directional movements.

The study of the oscillatory movement is the first step in discrimination of tracks and machinery of axoplasmic transport. It is essential to study microscopic movements such as the oscillation by the direct observation of transported organelles for an understanding of the precise mechanism of axoplasmic transport.

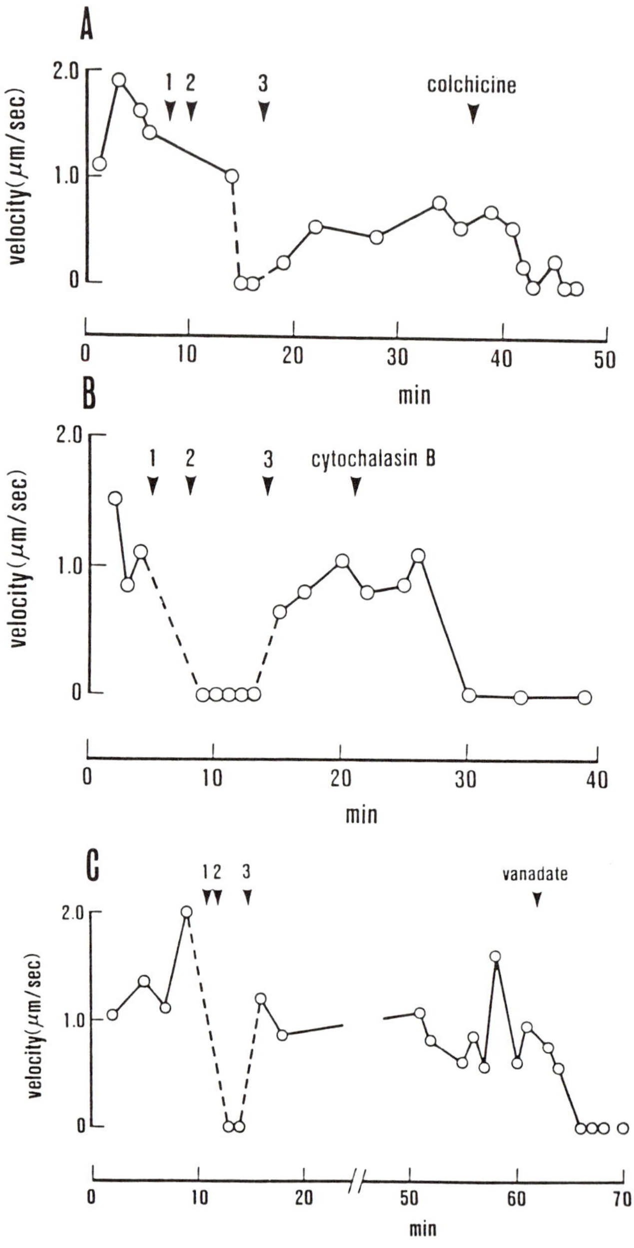

FIGURE 6. Effects of inhibitors of filamentous components and ATPase on the velocity of particles. (A) Colchicine (1 m*M*). (B) Cytochalasin B (10 μg/mℓ). (C) Vanadate (1 m*M*). Each numeral in figures indicates; (1) treatment with saponin (0.01%), (2) washing with AICF lacking ATP, and (3) substitution with complete AICF (see text).

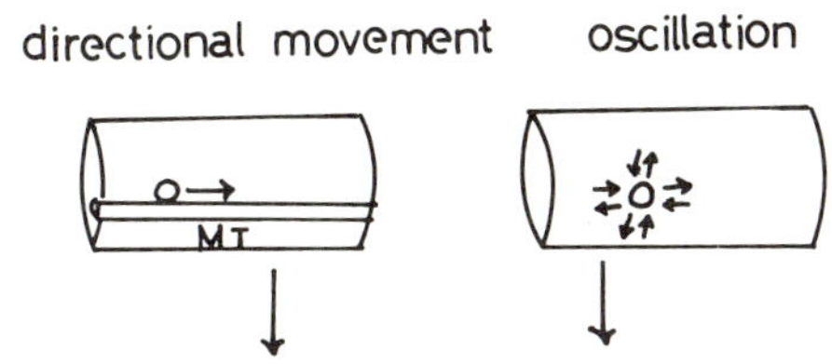

	directional movement	oscilltion
(intact)	+	+
complete AICF	+	+
ATP-free	−	−
vanadate	−	−
colchicine	−	+
vinblastine	−	+
cytochalasin B	−	+

FIGURE 7. Effects of various drugs on two modes of particle movements: directional movement and oscillation. See the text for the details. Vanadate, 1 m*M*. Colchicine, 1 m*M*. Vinblastine, 0.1 m*M*. Cytochalasin B, 10 μg/mℓ.

REFERENCES

1. **Matsumoto, T.** The granules, and mitochondria in the sympathetic nervefibers cultivated in vitro, *John Hopkins Hosp. Bull.*, 31, 91, 1920.
2. **Nakai, J.**, The movement of neurons in tissue culture, in, *Primitive Motile Systems in Cell Biology*, Allen, R. D. and Kamiya, N., Eds., Academic Press, New York, 1964, 377.
3. **Burdwood, W. O.**, Rapid bidirectional particle movements in neurons, *J. Cell Biol.*, 27, 115, 1965.
4. **Pomerat, C. M., Hendelman, W. J., Raiborn, C. W., and Massey, J. F.**, Dynamic properties of nervous tissue in vitro in, *The Neuron*, Hyden, H., Ed., Elsevier, Amsterdam, 1967, 119.
5. **Pomerat, C. M., Hendelman, W. J., Raiborn, C. W., and Massey, J. F.**, Dynamic activities of nervous tissue in vitro, in *The Neuron*, Hyden, H., Ed., Elsevier, Amsterdam, 1967, 119.
6. **McMahan, U. J. and Kuffler, S. W.**, Visual identification of synaptic boutons on living ganglion cells and of varicosities in postganglionic axons in the heart of the frog, *Proc. R. Soc. London., B*, 117, 485, 1971.
7. **Lubinska, L.**, Axoplasmic streaming in regenerating and in normal nerve fibers, in, *Mechanisms of Neural Regeneration*, Singer, M. and Schade, J. P., Eds., *Prog. Brain Res.*, 13, 1, 1964.
8. **Berlinrood, M., McGee-Russell, S. M., and Allen, R. D.**, Patterns of particle movement in nerve fibers in vitro, an analysis by photokymography and microscopy, *J. Cell Sci.*, 11, 875, 1972.
9. **Breuer, A. C., Christian, C. M., Henkart, M., and Nelson, P. G.**, Computer analyses of organelle translocation in primary neuronal cultures and continuous cell lines, *J. Cell Biol.*, 65, 562, 1975.
10. **Leestmas, J. E.**, Velocity measurements of particulate neuroplasmic flow in organized mammalian CNS tissue cultures, *J. Neurobiol.*, 7, 173, 1976.
11. **Leestmas, J. E. and Frreman, S. S.**, Computer assited analysis of particulate axoplasmic flow in organized CNS tissue cultures, *J. Neurobiol.*, 8, 453, 1977.
12. **Kirkpatrick, J. B., Bray, J. J., and Palmer, S. M.**, Visualization of axoplasmic flow in vitro by Nomarski microscopy. Comparison to rapid flow of radioactive proteins, *Brain Res.*, 43, 1, 1972.
13. **Smith, R. S.**, Centriperal movement of particles in myelinated axons, *Cytobios*, 3, 259, 1971.
14. **Smith, R. S.**, Detection of organelles in myerinated nerve fibers by darkfield microscopy, *Can. J. Physiol. Pharmacol.*, 50, 467, 1972.

15. **Smith, R. S.,** Microtubule and neurofilament densities in amphibian spinal root nerve fibers relationship to axoplasmic transport, *Can. J. Physiol. Pharmacol.*, 51, 798, 1973.
16. **Cooper, P. D. and Smith, R. S.,** The movement of optically detectable organelles in myelinated axons of Xenopus laevis, *J. Physiol. (London)*, 242, 77, 1974.
17. **Smith, D. S., Jarlfors, U., and Cameron, B. F.,** Morphological evidence for the participation of microtubules in axonal transport, *Ann. N.Y. Acad. Sci.*, 253, 472, 1975.
18. **Smith, R. S. and Koles, Z. J.,** Mean velocity of optically detected intraaxonal particles measured by a cross-correlation method, *Can. J. Physiol. Pharmacol.*, 54, 859, 1976.
19. **Hammond, G. R.,** The movement of optically detectable particles in intact myelinated axons, *J. Physiol. (London)*, 269, 74, 1977.
20. **Hammond, G. R. and Smith, R. S.,** Inhibition of the rapid movement of optically detectable axonal particles by colchicine and vinblastine, *Brain Res.*, 128, 227, 1977.
21. **Heslop, J. P. and Howes, E. A.,** Saltatory movements of organelles in intact nerves of Rhodnius prolixus Stal, *J. Exp. Biol.*, 70, 247, 1977.
22. **Forman, D. S., Padjen, A. L., and Siggins, G. R.,** Axonal transport of organelles visualized by light microscopy: cinematographic and computer analysis, *Brain Res.*, 136, 197, 1977.
23. **Forman, D. S., Padjen, A. L., and Singgins, G. R.,** Effect of temperature on the rapid retrograde transport of microscopically visible intra-axonal organelles, *Brain Res.*, 136, 215, 1977.
24. **Smith, R. S.,** Rapid particle transport in axons from decapod crustaceans, *Soc. Neurosci. Abstr.*, 3, 31, 1977.
25. **Smith, R. S.,** Axonal inclusions in the crab Hemigrasus nudus, *J. Neurocytol.*, 7, 611, 1978.
26. **Smith, R. S.,** The short term accumulation of axonally transported oranelles in the region of lacalized lesions of single myelinated axons, *J. Neurocytol.*, 9, 39, 1980.
27. **Smith, R. S. and Cooper, P. D.,** Variability and temperature dependence of the velocity of retrograde particle transport in myelinated axons, *Can. J. Physiol. Pharmacol.*, 59, 857, 1981.
28. **Allen, R. D., Allen, N. S., and Travis, K. L.,** Video-enhanced contrast, differential interfence contrast (AVEC-DIC) microscopy. A new method capable of analyzing microtubulerelated motility in the reticulopodial network of Allogromia laticollaris, *Cell Motility*, 1, 291, 1981.
29. **Allen, R. D. and Allen, N. S.,** Video-enhanced microscopy with a computer frame memory, *J. Microsc.*, 129, 3, 1983.
30. **Allen, R. D., Lasek, R. J., Girbert, S. P., Hodge, A. J., and Govind, C. K.,** Fast axonal transport in lobster axons, *Biol. Bull.*, 163, 379, 1982.
31. **Allen, R. D., Matuzals, J., Tasaki, L., Brady, S. T., and Girbert, S. P.,** Fast axonal transport in squid giant axon, *Science*, 218, 1127, 1982.
32. **Horie, H., Takenaka, T., and Kaiho, Y.,** Effect of particles and morphology in tissue cultured neurites, *Brain Res.*, 288, 85, 1983.
33. **Ochs, S.,** Characteristics and a model for fast axoplasmic transport in nerve, *J. Neurobiol.*, 2, 331, 1971.
34. **Ochs, S.,** Fast transport of materials in mammalian nerve fibers, *Science*, 176, 252, 1972.
35. **Adams, R. J.,** Organelle movement in axons: depends on ATP, *Nature (London)*, 297, 327, 1982.
36. **Brady, S. T., Lasek, R. J., and Allen, R. D.,** Fast axonal transport in extruded axoplasm from squid giant axon, *Science*, 218, 1129, 1982.
37. **Gotoh, H., Takenaka, T., and Shozushima, M.,** The fibrillar network in the cytoplasm of squid giant axon extracted with saponin, *Cell Struct. Funct.*, 8, 11, 1983.
38. **Forman, D. S.,** A permeabilized cell model of saltatory organelle movement, *J. Cell Biol.*, 91, (2), 414a, 1981.
39. **Forman, D. S.,** Vanadate inhibits saltatory organelle movement in a permeabilized cell model, *Exp. Cell Res.*, 141, 139, 1982a.
40. **Forman, D. S., Brown, K. J., and Livengood, D. R.,** Fast axonal transport in permeabilized lobster giant axons, *Soc. Neurosci. Abstr.*, 8, 827, 1982a.
41. **Forman, D. S., Brown, K. J., and Livengood, D. R.,** Fast axonal transport in permeabilized lobster giant axons is inhibited by vanadate, *J. Neurosci.*, 3, 1279, 1983.
42. **Gotoh, H. and Takenaka, T.,** Particle movement in squid axons permeabilized with saponin, *J. Neurobiol.*, 3, 239, 1984.

Chapter 11

AXONAL TRANSPORT OF NEUROTRANSMITTER ORGANELLES IN ADRENERGIC, CHOLINERGIC, AND PEPTIDERGIC NEURONS

Annica Dahlström

TABLE OF CONTENTS

I. INTRODUCTION

The neuron is a very specialized type of cell with regard to its often very elongated processes. It is a secretory cell in principle, but the secretion product is released from the cell very far from the site of production of the secretory granules or their precursors. The secretory products that have attracted most attention from investigators in pharmacology and physiology are probably the transmitters and the transmitter metabolizing enzymes, although no doubt many other secretion products (e.g., membrane coat products, receptor material, trophic substances) are formed in the cell body, transported distally along the axon towards the nerve endings and liberated there into the extraneuronal surface or space.

Like in any other cell the traffic is not only one-way. A significant amount of material is also moving in a retrograde direction, including both intraneuronal and extraneuronal substances. Worn out organelles and inactivated enzymes, as well as molecules taken up from the extra neuronal milieu move towards the cell body for repair, recycling, or neurotrophic information.

This article will focus on the axonal transport of transmitter storing organelles in pheripheral adrenergic, cholinergic, and peptidergic neurons. It may appear appropriate then to present the scientist, F. H. Scott, who in 1906[1] first introduced the idea of a proximodistal axonal transport of material involved in transmission. He studied the fatiguability of the reflex arch in the frog by electrical stimulation with varied impulse frequency of the dorsal roots. When he stimulated the central end of cut dorsal roots he found that the reflex permanently disappeared earlier when a larger number of stimuli was given than when a smaller number of impluses was given during a similar time period. However, if the same large number of impulses was given to an *intact* root, connected with its ganglion, the reflex could, if the root was left unstimulated for some time, recover.

The difference between cut and intact roots was not due to degeneration, since cut roots, if left unstimulated, were effective for far longer periods than stimulated intact roots. Other possible causes, i.e., muscle fatigue or fatigue of the motor neuron, were ruled out by Scott, who instead proposed that the cause of the permanent disability of cut stimulated roots to transmit the reflex was the result of ''using up of all the substance which passes between the root and the cells in the cord''. In his summary of this article[1] he states:

> ''It seems to me simpler to suppose that the nerve cells secrete a substance the passage of which from the nerve endings is necessary to stimulation. The recovery of effect after transient fatigue I attribute to the passage of a portion of this substance down the nerve fibre to the nerve ending.''

In these lines we meet in fact two new ideas; (1) the hypothesis on chemical transmission, first proposed by Elliott in 1904[2] but not accepted in the scientific world until much later, and (2) the hypothesis on axonal transport of transmitter organelles, which was not further studied until the 1960s, about 15 years after the demonstration of the axonal flow phenomenon by Paul Weiss in 1946.[3,4] F. H. Scott was never appreciated by his contemporary fellow scientists — his logic and correct interpretations of the results were too advanced for the rather static attitude on the nervous system of that time.

II. AXONAL TRANSPORT OF ADRENERGIC ORGANELLES

A. Historical Comments

The suggestion that NE may be transported from the perikaryon to the nerve endings was first mentioned by Schmiterlöw in 1948.[5] He measured biochemically the NE contents in sympathetic ganglia, splenic nerve, and some sympathetically innervated tissues (e.g., the spleen and the aortic wall) and found that the innervated tissues contained remarkably high concentrations of the amine when compared to the concentrations in the splenic nerve. He wrote "As it is impossible to believe that 1/10 of the aortic tissue would consist of nervous structures, the following hypothesis might be evolved. In the ganglionic cell (the perikaryon) the sympathomimetic substance is built up either as such or into the form of a precursor. This substance is then transported peripherally in the axoplasm."

This hypothesis of Schmiterlöw's was put forward long before the discovery by von Euler and Hillarp in 1956[6] that NE in splenic nerve was stored in organelles. It was later shown to be correct with the obvious exception that the material in transit is not NE itself, but the NE storing organelles (since these were discovered much later).

With the introduction of the fluorescence histochemical technique of Falck and co-workers[7] for the cellular localization of tissue CA and 5-HT (for description and methodology see also Reference 8 to 10) the study of the adrenergic system entered a new era. The method was based on two important observations in tissues fixed in formalin solution. The first was made by Erös in 1932,[11] who observed a yellow fluorescence in formalin fixed enterochromaffine cells. More than 20 years later Barter and Pearse[12,13] could show that the fluorescence was due to the presence of 5-HT in these cells. The other observation was made by Eränkö in 1952,[14] who observed that certain cells in formalin fixed adrenal medulla developed a green fluorescence, which he could later demonstrate to be due to NE.[15] Liquid formalin fixation could, however, not be used for demonstrating neuronal MA due to diffusion of MA into the fixative. When, however, formaldehyde gas was introduced onto dried tissue specimens (air-dried whole mounts or freeze-dried tissues)[16] delicate adrenergic nerves could be observed in the fluorescence microscope, with the formaldehyde induced fluorescence (FIF) method.

B. Anterograde Transport of Adrenergic Organelles

The first direct observations on the axonal transport of NE storing organelles were made using this FIF method. Dahlström and Fuxe observed in 1964[17] that in rat ligated splanchnic nerves NE fluorescence accumulated proximal to the ligature. NE and 5-HT also accumulated in the white matter of transsected spinal cords.[18] This observation was further investigated with fluorescence histochemical,[19] biochemical,[20] and pharmacological[19,21] methods, and the results of these studies demonstrated that NE was accumulating stored in granules (with a reserpine and tetrabenazine sensitive storage mechanism), which were transported at fast rates (> 200 mm/day) distally in the axons. Retrogradely accumulating NE was also observed,[19,22] but these accumulations were very small when compared to the anterogradely accumulating NE (Figure 1).

The mechanism of fast transport of NE was shown to be inherent to the axon in double crush experiments,[22] where NE in the middle segment, isolated from soma and nerve endings by the two crushes, continued to move mainly distally (Figure 2). The axonal microtubules are essential parts of the transport machinery as shown in experiments using mitotic inhibitors like colchicine[23] or vinblastine.[24] It is likely that the transport system has the same ionic and metabolic requirements as has been demonstrated for rapidly transported material in motor and sensory nerves (see Reference 26). Thus, oxidative phosphorylation, glycolysis, and possibly Ca^{++} ions are presumably important for transport of NE granules.[27]

The rate of anterograde transport of NE organelles was originally calculated to be about 120 mm/day in the rat and 240 mm/day in the cat.[20] These early calculations were based mainly on three assumptions, which were later reconsidered. The first assumption was that all NE in the nerve is rapidly transportable. This was later not found to be the case (both transportable and immobile NE fractions exist in the nerve). When the clearance of NE distal to a crush or ligation was studied in the sciatic nerve only 45% of the total NE was found to have moved into more distal segments, thus representing the mobile NE fraction.

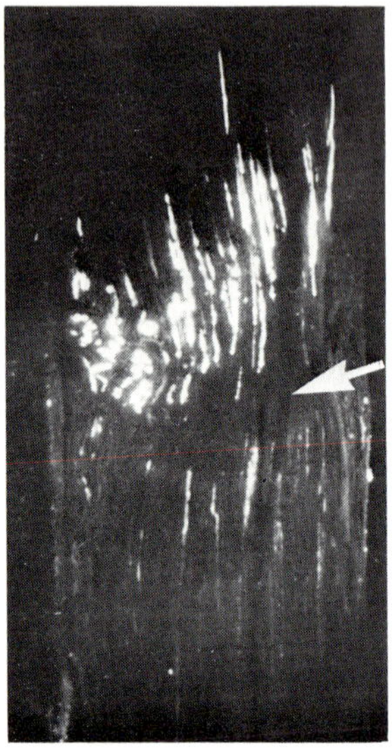

FIGURE 1. Fluorescence micrograph of a 3 hr crushed rat sciatic nerve, showing accumulations of NA-induced fluorescence proximal to the crush (→). Small but distinct retrograde accumulations are present distally. Magnification × 60.

Of the remaining stationary NE about 20% was located in nerve terminals of the *vasa nervorum* and the rest probably intra-axonally.[28] The second assumption, that retrograde transport of NE is insignificant when compared to anterograde transport, may still be valid. However, another constituent of axonal NE granules, the enzyme DBH (which converts DA into NE inside the organelles) has been demonstrated to accumulate distally as well as proximally to a crush.[29] (See below.) The third assumption was that the amine granules have a fixed number of NE molecules during their passage down the axons. This is not the case. The amine granules undergo a turnover of NE during their transport and accumulation near the crush site, as shown by superfusion of a ligated nerve with agents that block NE synthesis. When TH, the initial rate limiting enzyme in NE synthesis is blocked, the NE levels proximal to the ligation are lowered.[30] Inhibition of the DBH, located inside the granules, results in a buildup of DA in the nerve.[31] If the degrading enzyme monoamine oxidase (MAO), stationed in mitochondrial membranes, is inhibited, NE levels in ligated nerves increase. A simultaneous inhibition of TH and MAO in superfused nerves resulted in a NE level close to that found in control nerves, superfused with medium only.[30] If this continuous turnover of NE in the granules resulted in unaltered net concentrations of NE in the granules, this observation would not interfere with the calculations on the rate of transport. However, some observations show that the amine granules gradually increase their net store of NE during the proximodistal transport. First, the ratio of NE:granular protein is higher in distal segments of the axons than in more proximal segments.[32] Since the protein cannot be synthesized locally in the axons,[33,34] this must mean that the amount of NE has increased. Secondly, if the amine granules, accumulated proximal to a crush during 6 hr, are allowed to sit at the accumulation site and further transport into the area is hindered by a second more proximal crush, the accumulated granules clearly increased their NE content by a factor of about 2[35] (Figure 3).

Thus, the rate of transport of axonal NE granules, as estimated from ligation experiments, were recalculated, considering the later observations. The rate of transport in rat sciatic now

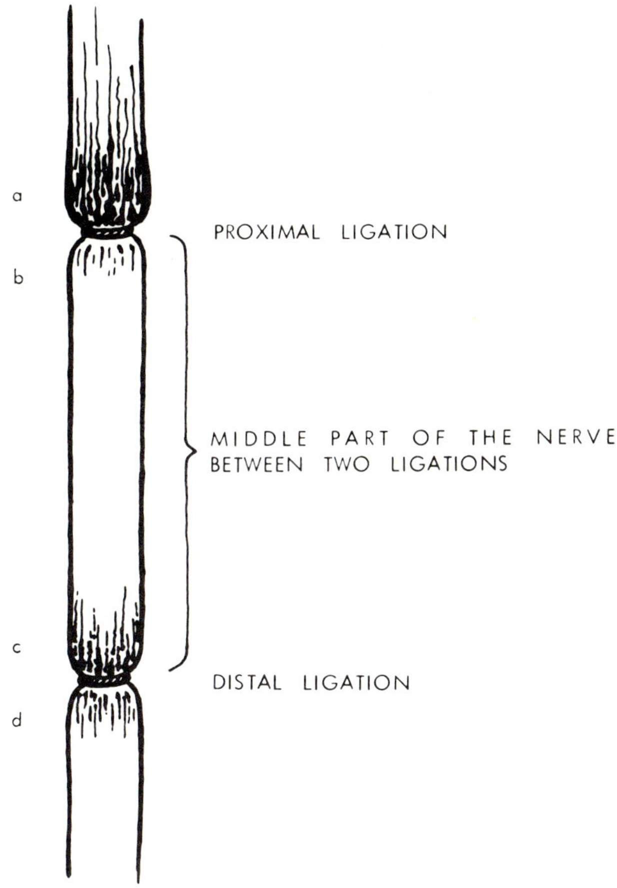

FIGURE 2. Schematic illustration on the mode of redistribution of NE relative to two ligations simultaneously performed 6 hr before killing in the rat och cat sciatic nerves. In the isolated segment most of the NE was transported in an anterograde direction and accumulated at (c), but a small amount had also migrated in retrograde direction and was present at (b). The accumulations at (a), in the nerve segment in continuity with the perikarya, was steadily increasing with time up to 18 to 24 hr, while the accumulations at (c) were fully developed at 1 to 2 hr depending upon the length of the isolated segment. The retrograde accumulations at (d) were always larger than those at (b), probably because of the continuity with distal nerve segments that could supply material for retrograde transport. The experiment demonstrated, i.e., that axonal transport is independent of nerve impulse activity, and that the mechanism for transport is inherent to the axon proper.[22]

appears to be around 200 mm/day, a figure that may be subjected to further correction. Assuming that the mobile fraction in cat sciatic nerve is similar to that in the rat, i.e., about 50%, the rate of transport in cat would be about 400 mm/day for the mobile fraction. This is in good agreement with the finding that 30 to 50% of the NE in cat peroneal nerve is mobile and that clearance of the mobile NE in a 53-mm long isolated segment requires 3.25 hr, yielding a transport rate towards the peripheral end of this segment of 392 mm/day.[31] This figure is close to the fast transport rate of 400 mm/day, proposed to be the general rate for rapidly transported material in all mammalian nerves[26] (see Chapter 1).

An untraumatic way of interrupting fast axonal transport is local cooling of a nerve. This causes depolymerization of microtubules within the cooled segment. Upon rewarming the tubules polymerize rapidly and axonal transport is resumed. This method to study axonal transport was introduced by Brimijoin,[36] who used it to study the transport of NE and DBH

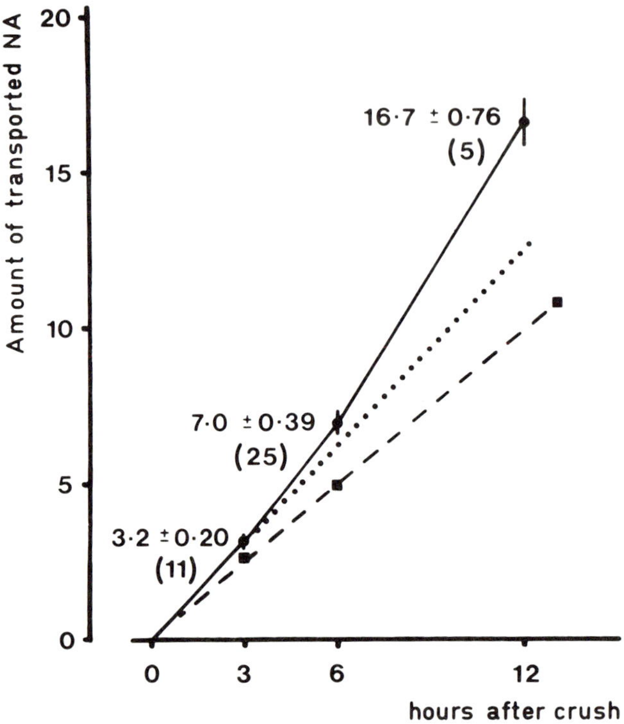

FIGURE 3. The increase in NE (μg/segment) in rat sciatic nerve proximal to a crush is indicated as a solid line. The 5 or 10 mm segments proximal to the crush were assayed for NE and compared to the NE content in an uncrushed control nerve of similar length. Figures for NE increase in the assayed segment that had occurred after crushing were obtained by subtracting the NE content in control nerve segments from the total NE content in the crushed nerve segments. Mean ± SEM are given, numbers in brackets show number of observations. The dotted line indicates the theoretical NE increase with time, had the accumulations proceeded with 3.2 μg/3 hr as during the first 3 hr period. The hatched line shows the theoretical curve for NA increase expected to occur if the transported NE granules had *not* increased their NE loading during transport and accumulation (correlation coefficient 0.98). The difference between the solid and hatched lines thus represent the extra NE loading of the accumulated NE granules.[35]

in rabbit sciatic nerve in vitro.[36-38] The nerve was incubated in a two compartment chamber and the distal compartment superfused with cold (1 to 3°C) medium while the proximal section was kept warm. NE and DBH accumulated in parallel at the interphase between the two compartments (Figures 4 and 5), and when the distal (cooled) part of the nerve was rewarmed the accumulated material started to move into the distal part of the nerve (Figures 6 and 7). The peak of material could be traced along the nerve with biochemical assays of small nerve segments at 1 to 3 hr after rewarming, and the rate of this movement could be calculated. NE and DBH moved at the same rates (Figure 7), as would be expected because of their co-localization to the amine granules, but the peak was flattened with time indicating that not all granules moved at the same rate. The maximal rate (the migration distally of the *front* of the peak) and mean rate (the migration distally of the *hump* of the peak) was calculated to 430 and 300 mm/day, respectively.[38]

C. Biochemistry and Ultrastructure of Axonal Amine Granules

The best studied axon preparation, concerning biochemistry as well as ultrastructure, is

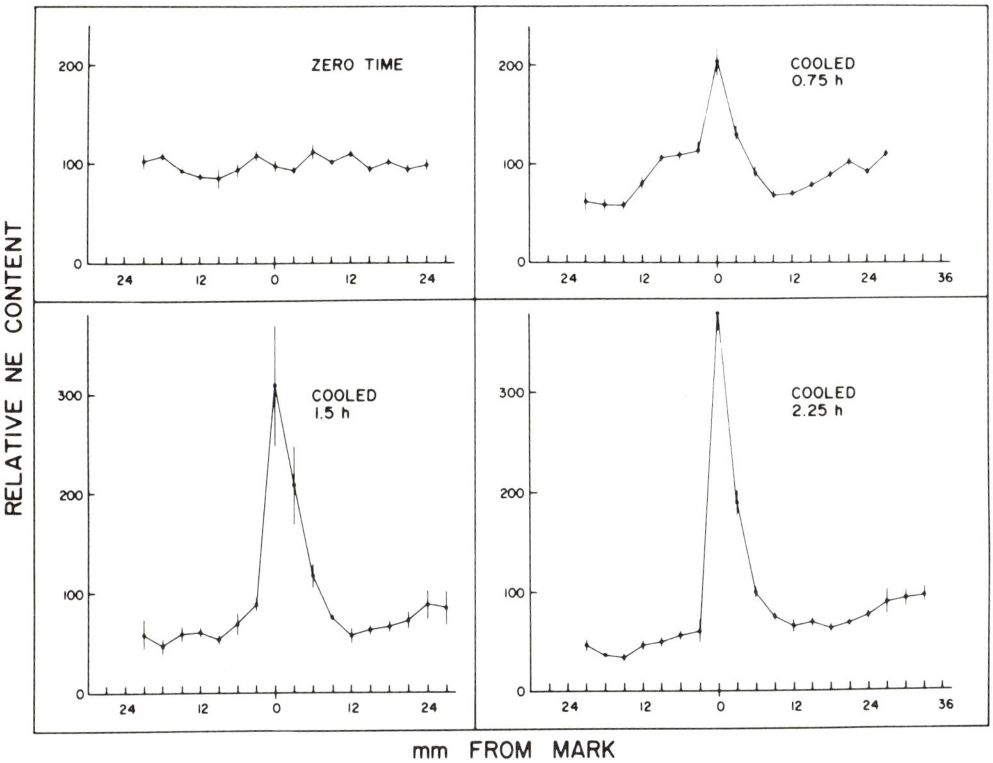

FIGURE 4. Distribution of NE along rabbit sciatic nerves locally cooled in vitro for various periods of time. Means ± SEM are given in relative units. On the abscissa proximal is to the left and distal is to the right. n = 4 − 6. NE gradually accumulated with time after onset of cooling proximal to the cold-block.[38]

the bovine splenic nerve. It was used by von Euler and Hillarp in 1956[4] when they demonstrated the granular localization of NE in adrenergic axons, and it is still the preparation of preference because it is so rich in material. Bovine splenic nerves contain NE, and ATP in a ratio NE/ATP of between 4 to 7/1, depending upon the level studied; the more distally located granules having higher NE content.[31] They further contain protein, chromogranin, DBH, phospholipids, cholesterol,[39] and an opioid like peptide[40] which cross-reacts with antisera to enkephalin.[41,42]

Electron microscopy of purified bovine splenic nerve granules shows a homogenous population of large dense core vesicles (LDV) with an average diameter of 763 Å[43] (Figure 8). In cat and guinea pig, the proximal ends of ligated adrenergic axons contain also some smaller DV (SDV) with a diameter around 400 to 500Å,[44] i.e., similar to the majority of the vesicles seen in nerve terminals. In rat adrenergic nerves both LDV and SDV are found in ligated axons,[45,46] as well as tubules, which upon $KMnO_4$ fixation contain a dense matrix suggested to be due to the presence of NE inside these structures.[47] $KMnO_4$ fixation cannot, however, be considered to unequivocally demonstrate the subcellular localization of NE in tissues, since in bovine splenic nerve preparations the density of the matrix did not vary in parallel with the NE content of the vesicles.[43] The exact ultrastructure of the NE storing organelles in adrenergic axons of the rat is therefore uncertain, but the LDV together with the SDV are good candidates.

D. Retrograde Transport of Adrenergic Organelles

The early fluorescence histochemical studies on ligated or crushed rat sciatic nerves

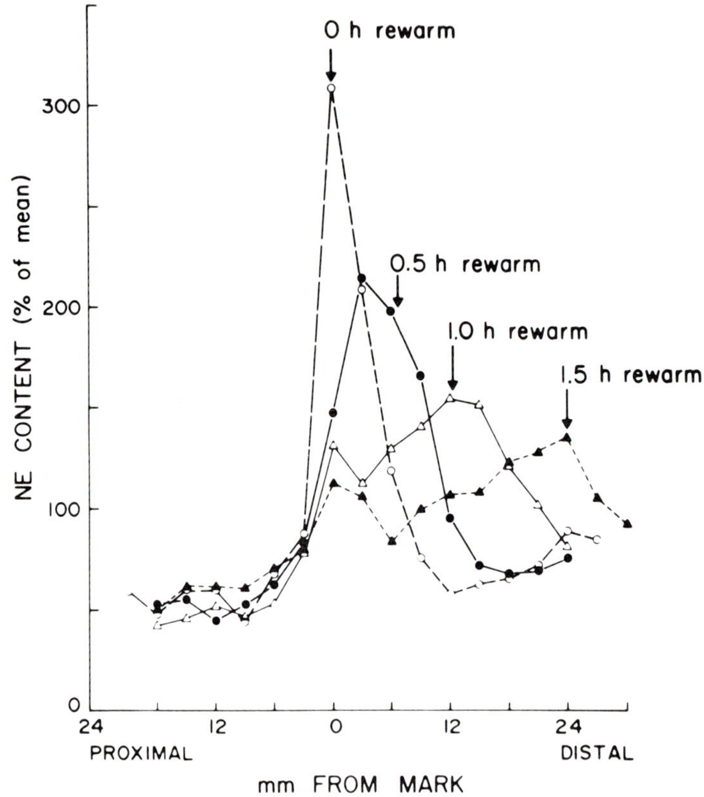

FIGURE 5. The proximo-distal transport of NE in locally cooled nerves subjected to rewarming. The NE accumulated proximal to the cold-block during 2.25 hr (0 hr rewarm peak), started to move distally after rewarming the nerve, and the peak had moved approximately 24 mm during 1.5 hr after rewarming.[38]

revealed that also distally to a crush some NE accumulated (Figure 1). The amounts were, however, small when compared with the proximal accumulations. When another constituent of the amine granules, DBH, was studied with immunocytochemical methods, significant amounts of DBH were found, not only proximal to but also distal to the crush in guinea pigs.[29] With the cold block technique and immunofluorescence methods also Brimijoin and Helland[48] found that DBH was transported in a retrograde direction. This fraction of DBH appears to have much less enzymatic activity per molecule DBH than the DBH in transit distally.[29] The DBH moving towards the soma presumably comes from the nerve terminal region where it has been active in synthesizing NE. A support for this source of retrogradely transported DBH is the observations that antibodies to DBH, when given to the living animal, are taken up by the nerve terminals[49] and transported retrogradely in adrenergic axons.[50,51] These DBH antibodies have presumably combined with the DBH molecules transiently exposed in the nerve terminal membrane during exocytosis of NE vesicles.[52]

The retrogradely moving DBH thus may be worn-out molecules (with low enzymatic activities) from the nerve endings, returning to the soma for recycling. It is in all likelihood packaged in an organelle, since the rate of retrograde transport in cold block experiments was fast (about 300 mm/day[48]). The package may also contain some NE, but it is not known if the NE storing material is localized in the same subcellular compartment as DBH. The morphology of the storage organelle is not known, typical SDV or LDV are not observed distally to a crush.[53,54]

Various models have been proposed for the mechanism underlying fast axonal transport

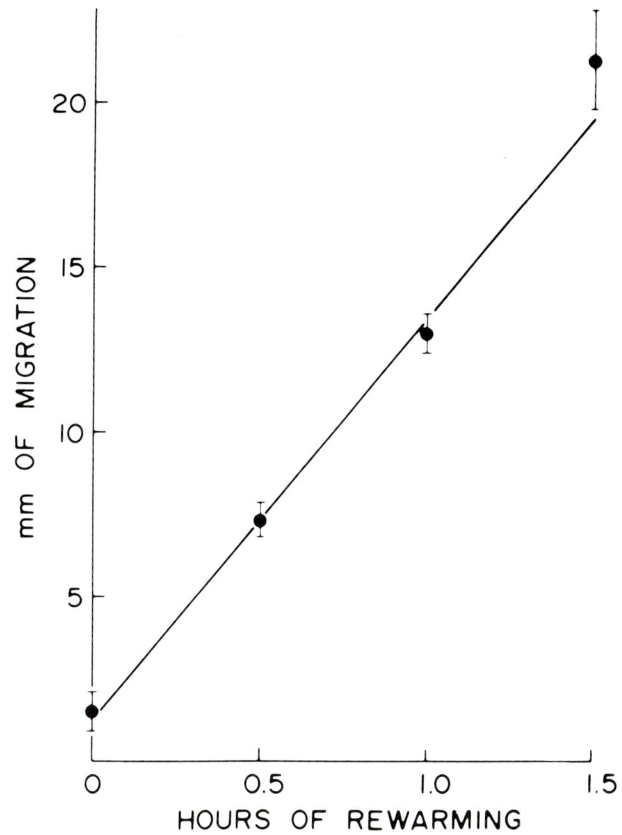

FIGURE 6. Velocity of anterograde transport of NE deducted from the experiments indicated in Figure 5. The position of the highest concentration of migrating NE was measured in each nerve that was rewarmed after 1.5 hr of local cooling. The distances between these positions and the reference mark (the area of the cold-block) are expressed as millimeter of migration. This figure indicates that most material travelled distally at a rate of 20 mm/1.5 hr, i.e., 320 mm/day (*mean* rate of transport, see text).[38]

(see discussion in Reference 26). The exact molecular events are not known, but microtubules appear to be essential. The mechanism may be pictured like a railway system: the microtubules can be considered as the stable railway track, they are slowly transported or growing in a distal direction, can be dismounted into subunits and be put together again (confer the cold block effect described above). Material packaged into organelles (wagons) can be transported rapidly as long as they are in contact with the rail. These wagons are slightly sticky (negative charge) and therefore unpackaged material in the neighborhood may attach for periods of time to the exterior of the wagons and ride piggy-back for some distance. Some wagons move steadily in contact with the rail continously and their rate of transport is maximal, around 400 mm/day. Other wagons have a slower effective rate of transport because they have left the rail track at regular or irregular intervals, their motion is then temporarily arrested. In between these arrests they move at maximal speed. The soluble material will or will not adhere to the surface of the wagons depending upon the milieu in the axoplasm (ionic strength, pH, Ca^{++} concentration), which hypothetically might differ regionally, or differ slightly from one axon to the other. Some soluble material never adhere to the wagons, these are transported at the slowest rates. This simplified way of picturing axonal transport mechanisms may explain all the various rates of transport observed in the

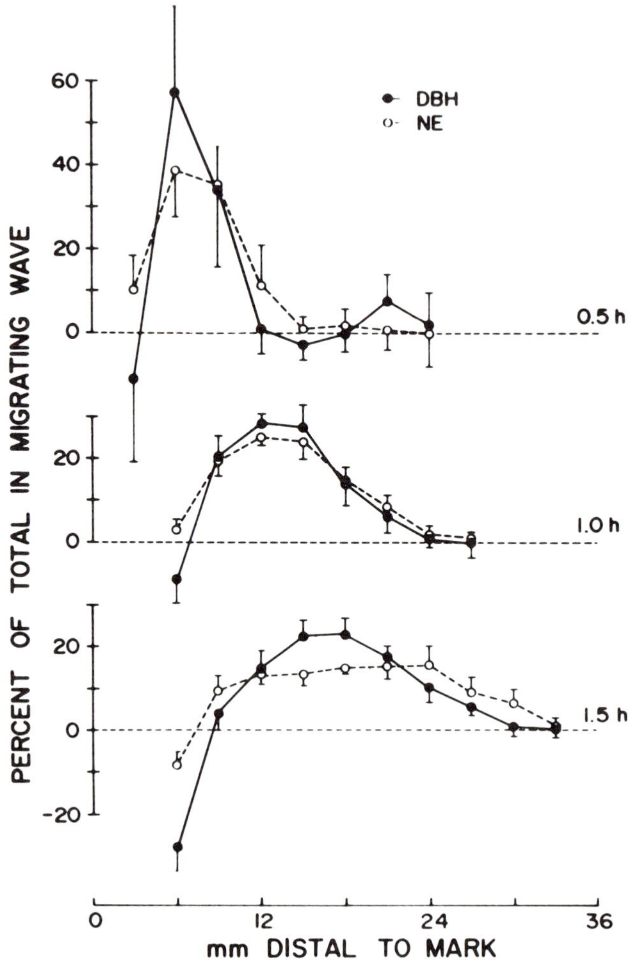

FIGURE 7. Anterograde transport of NE (A) and of NE compared with DBH (B) in rabbit sciatic nerves after rewarming of cold-block nerves. The nerves were initially cooled locally for 1.5 hr and then rewarmed for the indicated times. A wave of material was migrating distally after rewarming, and NE and DBH-activity were moving together as shown in B. The most distal point at which both NE and DBH were significantly elevated over baseline levels was 17 mm from the reference mark (the area of the cold-block) after 1.5 hr of rewarming, indicating a *maximum* rate of transport of 432 mm/day.[38]

literature, fast as well as slow. Several sophisticated models on the mechanism of axonal transport have been published, e.g., the transport filament model of Ochs[26,55] and the microstream hypothesis of Gross.[56] For a more detailed description of these models see, e.g., Ochs*.[26]

III. AXONAL TRANSPORT OF CHOLINERGIC ORGANELLES

A. Acetylcholine Studies

ACh is present in all parts of the cholinergic neuron, but concentrated to the nerve endings

* Recently a novel brain ATPase of 130 KD has been isolated. This ATPase has the predicted characteristics of the motor involved in fast axonal transport.[125,126]

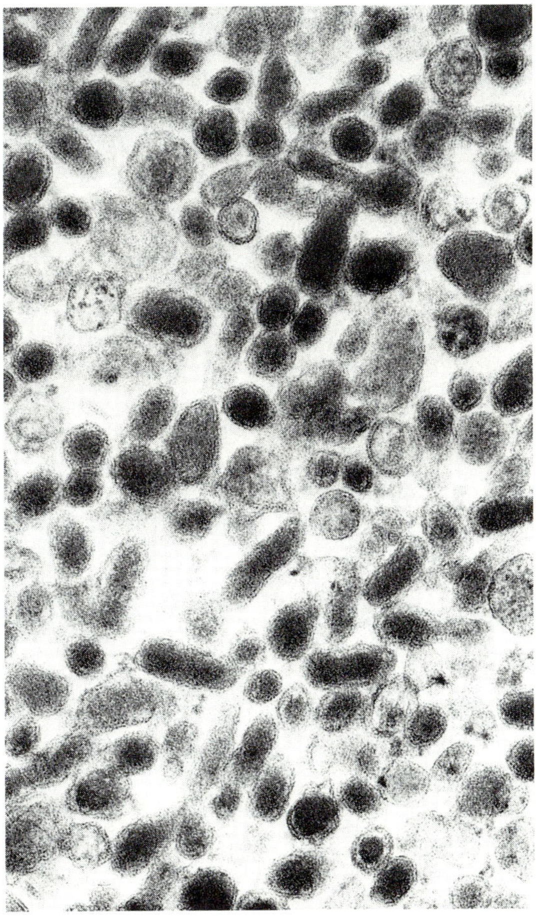

FIGURE 8. Purified fraction of large dense cored vesicles (LDV) from bovine splenic nerve. Bars indicate 2000 Å. The vesicles are spherical or somewhat elongated, with an electron dense matrix surrounded by a halo and a double-layered outer membrane. (By courtesy of Dr. Åsa Thureson-Klein.)

where it is stored in synaptic vesicles. It is synthesized from choline, which is transported into the cholinergic neuron by a high affinity carrier[57] and acetyl-Co-A, provided by the mitochondria, in a reaction catalyzed by the enzyme ChAT. ACh is degraded by the enzyme AChE, which splits the ACh into choline and acetate. (Detailed aspects on ACh-synthesis and metabolism can be found, e.g., in Reference 58).

ChAT and AChE are formed in the perikaryon and transported intra-axonally towards the nerve endings (for review see Reference 59). ChAT is thus present in all parts of the neuron, which explains why ACh can be found throughout the nerve cell. However, since AChE is also present in all parts of the neuron a compartmentation of ChAT, ACh, and AChE within the neuron may be expected to be present, especially in areas where the ACh concentration is high.

In the intact cholinergic axon the ACh content is fairly low, but no exact figure can be calculated. The study of the cholinergic neuron has been hampered in comparison with the adrenergic neuron by the fact that no sensitive fluorescence or cytochemical method has so far been worked out that could permit exact localization of ACh containing structures or measurements of intraneuronal concentrations. Biochemical analyses of homogenates from mammalian tissues has so far been used to study ACh. However, since nervous tissues contain many nonneuronal structures and cell types, which will contaminate homogenates and since mechanical or pharmacological manipulations of the tissue prior to extraction influence the ACh levels, the concentration in, e.g., axons cannot be determined.

Most of the ACh in intact peripheral cholinergic axons appears to be cytoplasmic, as in homogenized peripheral nerve subjected to ultracentrifugation most ACh is found in the supernatant.[60,61] In contrast, most ACh in the nerve endings is found in the particulate fraction.[62]

When a peripheral nerve, carrying cholinergic axons, is ligated or crushed ACh accumulates rapidly[63] in analogy with the events in adrenergic axons. In ligated sciatic nerves of rat the increase in ACh, assayed on the guinea pig ileum preparation,[64] in the 5 mm nerve segment just proximal to the crush increases with time up to 24 hr, while distally a small increase takes place during the first hours only (Figure 9). Most of this increase distally may be due to a local synthesis of ACh, caused by the crush procedure. It is known that manipulations of a nerve induces ACh synthesis locally.[65] This can be prevented if the nerve is chilled, but during the crush operation procedure the temperature is at least above room temperature. The distal increase may also be due to retrograde accumulations of ACh, but these would continue for more than the initial 3 hr, at least this is the case in other neuron systems where retrograde accumulations continue to build up distally to a crush at least for up to 8 hr (see Reference 26). The true retrograde accumulation of ACh may of course be too small to pick up with biochemical assays. This is the case for the adrenergic system, where histofluorescence clearly demonstrates retrograde accumulations of NE, but these were not detectable with biochemical assays on 5 mm segment nerves.[20]

However, since trauma induced local ACh-synthesis has indeed been demonstrated it was assumed that the increase distally to the crush was mainly due to local synthesis. A similar synthesis should have occurred also proximally to the crush, and would theoretically have contributed to the ACh increase in this segment. Therefore, the increase in the segment distally (assumed to represent local ACh synthesis) was subtracted from the ACh values assayed in the proximal segments (curve I, Figure 9). The resulting curve (curve II, Figure 9) showed a linear ACh increase with time, a curve which is probably indicating accumulation of ACh due to anterograde axonal transport.[63,66] When ACh was assayed in nerve segments further distally, a fall of 21% below control had occurred by 12 hr after crushing, which indicated that about 20% of the axonal ACh is rapidly transportable. Taking this figure into account the mean rate of anterograde ACh transport is calculated to be about 120 mm/day.[63,66]

In doubly crushed nerves ACh accumulated towards the distal ligature, demonstrating that the transport mechanism is independent of soma as well as nerve endings.[63] Axonal microtubules are involved since vinblastine as well as colchicine, injected subepineurally, could block the axonal transport.[67,68]

In the cholinergic preganglionic cervical sympathetic nerve ACh accumulated in a similar manner. Since the ACh levels in this thin nerve are low (10 to 15 pmol/5 mm)[69,70] in comparison with the sciatic (50 to 60 pmol/5 mm),[63,66,71] nerves from 5 to 15 rats had to be pooled for each experiment. After crushing the ACh amount increased in the 5 mm segment proximal to the crush in a linear fashion; at 12 hr the increase was 5 to 6 pmol over basal levels. In this nerve no increase was observed in the segment immediately distal to the crush, suggesting that local trauma-induced ACh synthesis and/or retrograde ACh transport were less pronounced than in the sciatic nerve. The clearance of ACh from a segment further distal to the crush was assayed and found to be 5% in one set of experiments[69] and somewhat higher in a later study,[70] using a similar technique. This indicates that a small fraction of ACh in rat preganglionic cervical nerve was rapidly transportable and taking this into account the rate of transport of ACh was at least 110 to 120 mm/day.

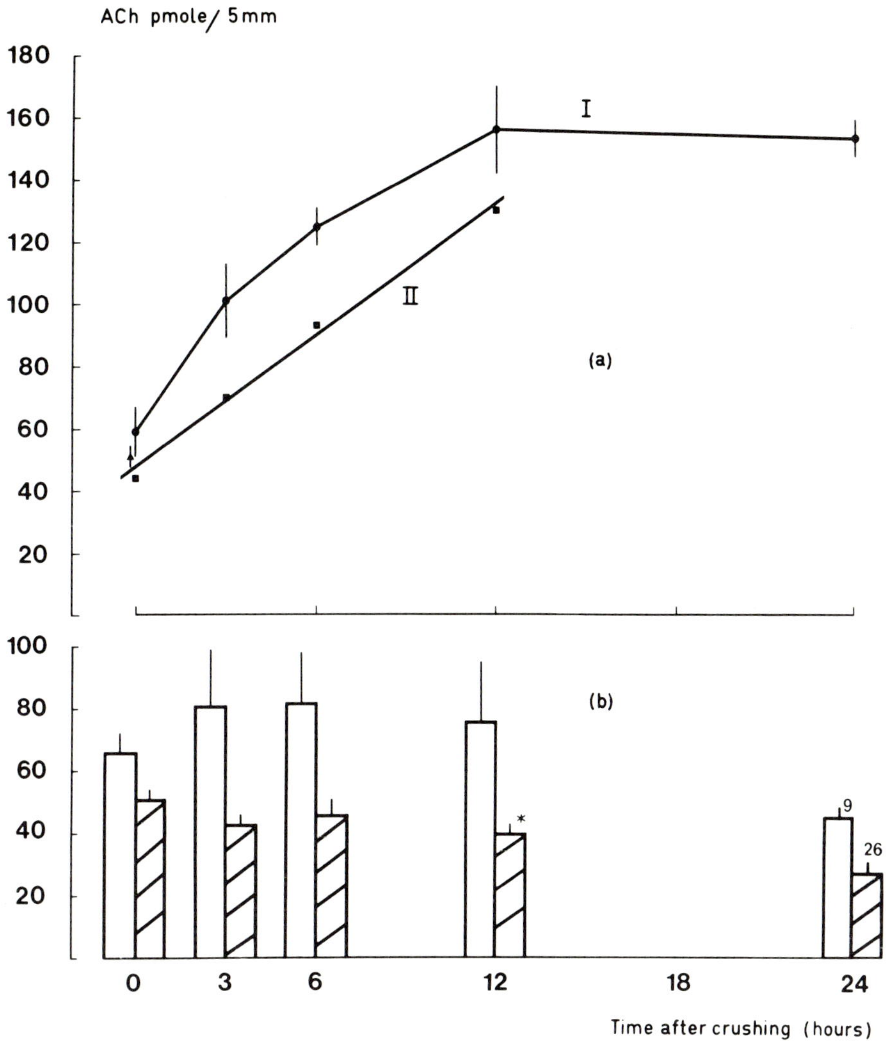

FIGURE 9. (A) Changes in ACh content in 5 mm rat sciatic nerves proximal to crush 0 to 24 hr after operation. Curve I shows the experimental ACh values assayed. The triangle at 0 hr indicates the ACh level in 5 mm unoperated nerve. Curve II was obtained by subtracting the ACh which was assumed to be locally synthesized due to the crush procedure, according to Feldberg.[65] The locally synthesized ACh fraction was assumed to be similar on both sides of the crush and the increase over normal ACh in the segment immediately distal to the crush (the open rectangles in B) was used as an estimate of local ACh synthesis. When this correction was performed the ACh increase proximal to the crush was linear (II) (n = 5). (B) The change in ACh distal to a crush 0 to 24 hr after operation. Open rectangles are the 5 mm segments immediately distal to the crush (n = 3 − 4), hatched rectangles are 5 mm segments 10 to 25 mm distal to the crush (n = 9 − 26). Vertical bars are SEM. The decrease below the 0 hr value of ACh in the hatched bars indicates the rapidly transportable fraction.[63]

Thus, measurements of ACh accumulations in crush-operated rat sciatic and cervical preganaglionic nerves indicate that a fraction of axonal ACh is rapidly transportable, that the mechanism for transport is inherent to the axon and dependent of axonal microtubules. The rate could be calculated to be at least 120 mm/day for both systems, but this may well be an underestimate. The transportable ACh fraction is probably located in organelles which are rapidly transported and which protect the ACh which accumulates from degradation by AChE, accumulating in substantial amounts in ligated or crushed cholinergic axons.

B. Axonal ACh-Storing Organelles

The presence of accumulated ACh in a sedimentable pool has been demonstrated after density gradient centrifugations of sciatic nerve segments immediately proximal to a ligature or crush.[66,72] The possible artefactual contamination of occluded cytoplasmic membrane sacs was ruled out in these experiments by the absence in the sediment of ChAT activity, which is mainly cytoplasmic in this type of nerve (see Reference 58).

The axonal ACh organelle has a rapid turnover of ACh. This is indicated by experiments where inhibition of ACh synthesis (with HC-3 which blocks the high-affinity choline carrier) and/or inhibition of AChE-activity (with eserine or soman) were investigated in an in vitro perfusion chamber where ligated or crushed sciatic nerves were incubated with these drugs.[73] Inclusion of HC-3 in the superfusion fluid reduced the accumulated ACh by about 75% during the 8 hr experiment. The AChE inhibitors soman or eserine nearly doubled the ACh proximal to the crush. When HC-3 and AChE inhibition were combined the ACh amounts were approximately normalized to control (i.e., 8 hr crush only) values. These experiments show some important issues:

1. The axonally transported and accumulating ACh is stored in an organelle which protect ACh from degradation by the large amounts of AChE present in accumulated motor axons.
2. There is a continuous leakage of ACh from the accumulated ACh organelles to the cytoplasm where it is degraded.
3. This leakage is approximately matched by a continuous incorporation of newly synthesized ACh into the organelles.
4. The organelles which contain ACh during their axoplasmic transport contain a certain net amount of ACh, which however is subjected to a rapid turnover.

Other experiments support this view. Isotope-labeled choline was injected into the ventral horn of the cat spinal cord to label the ACh in motor neurons (labeled ACh cannot be used since it is not taken up across the neuronal membrane), and the spread of labeled choline and ACh along the ventral roots was studied.[72] A small but diffuse peak of radioactive choline was moving distally in the roots, but labeled ACh was in contrast evenly distributed distally along the nerves without any distinct peak formation. After density gradient centrifugations almost all radioactive ACh was present in the supernatant. The authors suggested that this was indicative of a loss of initially incorporated ACh from organelles during their transport distally in the roots and mixing with the soluble ACh in the axoplasm.[72] Another experiment, also performed in cat, was done by Woodward and Lindström,[74] who injected ^{3}H-choline into the lumbar ventral horn nuclei and found by autoradiography that motor perikarya had a more efficient uptake of choline than the surrounding neuropil. ^{3}H-ACh was found in ventral roots and appeared to move distally at a rate of more than 25 mm/day, by a mechanism sensitive to ligation or to local application of colchicine. When ^{3}H-choline was injected into the ventral roots ^{3}H-ACh was formed, which migrated distally. The migrating ^{3}H-ACh peak, however, left a veil of residual ^{3}H-ACh along the length of the nerves, suggesting a marked turnover of initially incorporated ACh.[72]

C. Immunohistochemical Studies

As discussed in the previous chapter the presence of an axonal ACh storing organelle, which is transported distally in cholinergic axons, is strongly suggested. We know very little of this organelle when compared with the adrenergic axonal organelle, which has been identified and characterized morphologically as well as biochemically (see above). The reason for this lack of knowledge is mainly twofold. Firstly, the *synaptic* ACh vesicles have been in focus for most studies on cholinergic neurochemistry, while axonal events were less investigated. Secondly, the only marker for the axonal cholinergic vesicles is ACh and as indicated above, this is not a very stable marker.

A couple of years ago Kelly and co-workers[75,76] succeeded in producing, in rabbits, an antiserum (RASVA) against a very pure fraction of synaptic ACh vesicles from the elasmobranch Narcine Brasiliensis. This antiserum showed excellent cross-reactivity with mammalian nervous tissue.[75] Using this antiserum we have demonstrated the presence of immunoreactive organelles in motor end plates, in ligated motor axons, and in motor perikarya in the spinal cord.[79] The antiserum also reacts with antigenic determinants in other neuron systems, especially in the adrenergic neuron.[77] This is not surprising in view of the fact, (1) that RASVA is produced against an organelle with many antigenic components and (2) that adrenergic and cholinergic nerves are ontogenetically closely related (see below). Therefore the RASVA cannot be used to identify unknown cholinergic structures, but in a known system it may be very useful. In our studies we have used it to investigate the motor system in the sciatic nerve of rat, and since this nerve also carries sympathetic adrenergic and cholinergic axons all animals undergo surgical lumbar sympathectomy before they are used for experiments.

The indirect immunofluorescence method of Coon's[78] has been used. After perfusion fixation with 4% buffered paraformaldehyde and rinsing in PBS with 5% sucrose the specimens are cryostat sectioned (10 μ). The sections are placed on gelatin coated glass slides and incubated with RASVA (1:500-1:1000), SWAR-FITC, (1:30), rinsed thoroughly and finally mounted in buffered glycerine. All incubation solutions contain 0.1% Triton X-100.

In motor end plates a strong immunofluorescence was observed,[79] concentrated near the margin of the end plate processes where synaptic vesicles are enriched[80] (Figure 10a). Also motor perikarya in the spinal cord contained a weak immunofluorescence.[81]

In intact sciatic nerves no immunoreactive material can be seen, but already 1 hr after crushing the sciatic nerve axons with immunofluorescence can be seen.[81] The amounts of accumulated material increases thereafter rapidly, both proximally and distally to the crush area[82] (Figure 10b). The proximal accumulations are always larger than distally and are present in myelinated axons. The majority of these axons are probably motor in nature, for the following reasons:

1. AChE staining, with a procedure that selectively stains motor axons,[83] of sections which have first been incubated with RASVA and examined in the fluorescence microscope, reveal that nearly all axons with accumulated immunoreactive material also contain accumulated AChE, but the intra-axonal distribution is not identical,[79] and the RASVA does not cross-react with mammalian AChE.
2. Crushed ventral roots contain a majority of axons with strongly immunoreactive material (Figure 10c), while dorsal roots exhibit a lower immunoreactivity.
3. In motor perikarya a weak but significant amount of material can be seen in the cytoplasm.
4. Motor end plates, but not sensory nerve endings, are strongly immunoreactive.

In accumulated axons the material has a granular appearance (Figure 10d). This indicates either that the immunoreactive axonal organelle is comparatively large or that smaller organelles migrate together in clusters in the axon. Synaptic vesicles are rarely observed in motor axons, but large organelles with a more or less dense matrix have been observed in ligated nerves.[84] In cholinergic nerves of the feline autonomic neuron system innervating exocrine glands the coexistence of vasoactive intestinal peptide (VIP) with ACh has been demonstrated.[85] VIP has in these nerves been localized to large dense cored vesicles,[86] which in density gradient centrifugation studies have been shown also to contain ACh.[87] Thus, in this autonomic neuron system at least part of the axonally transported ACh is stored in large dense cored vesicles.

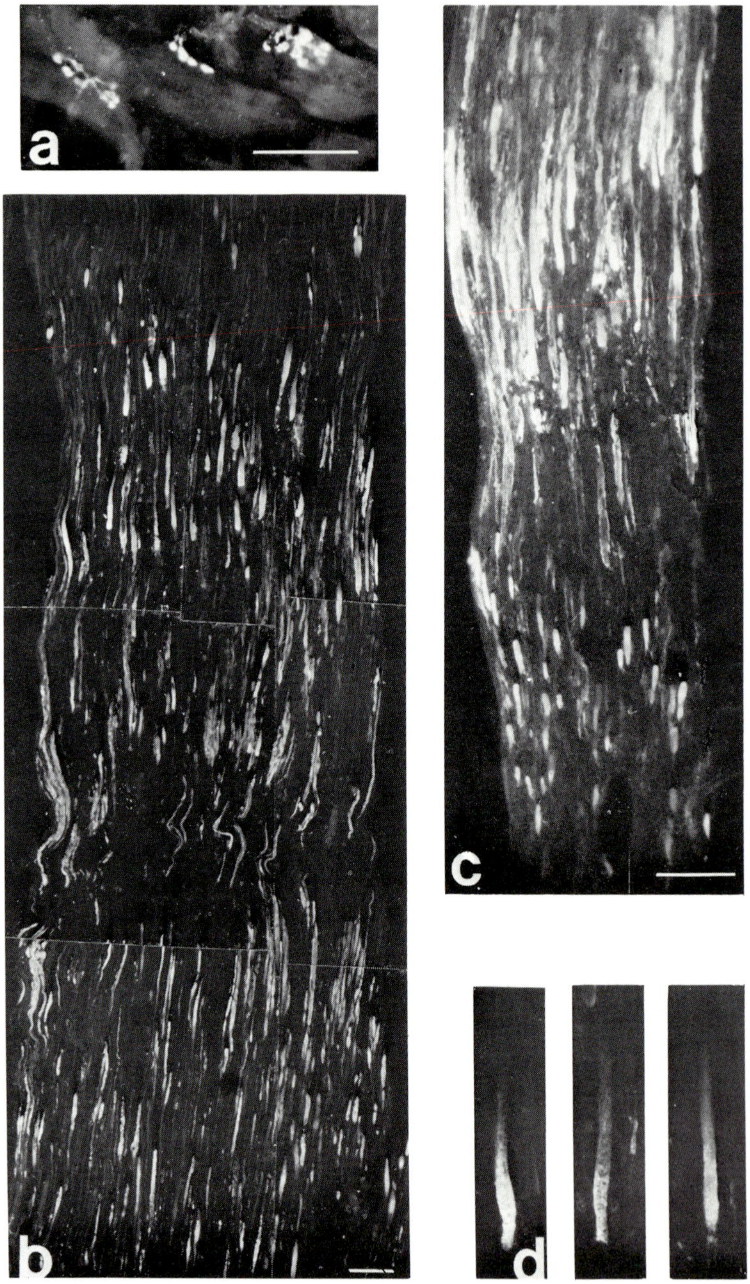

FIGURE 10. (a) Fluorescence micrograph of motor endplates in rat skeletal muscle. The perfusion fixed tissue was incubated for indirect immunofluorescence using the RASVA antiserum in the first step, and RBITC (Dakopatts, Copenhagen) in the second incubation. The end plate arborizations contain immunoreactive cholinergic vesicle material which is concentrated near the margin of the endfeet. The preterminal, incoming axons (→), contain a lower concentration of material. (b) Fluorescence micrographs from 9 hr crushed rat sciatic nerve. The section was incubated for immunofluorescence as described in 10a. Large myelinated axons contain accumulated immunoreactive cholinergic vesicle material proximal and also distal to the crush (→). Magnification × 180. (c) Fluorescence micrograph of a crushed (12 hr) ventral root of rat. The crush area is indicated (→). The secion was incubated for immunofluorescence as indicated in 10a, but with FITC-labeled SWAR in the second step. Proximal (top) to the crush, most axons contain large amounts of immunoreactive cholinergic vesicle material. Also distal accumulations are present. (d) Accumulated granular material with RASVA-immunoreactivity in large myelinated axons in rat sciatic nerve proximal to a 9 hr crush (same magnifications as in a). Bars represent 50 μ.

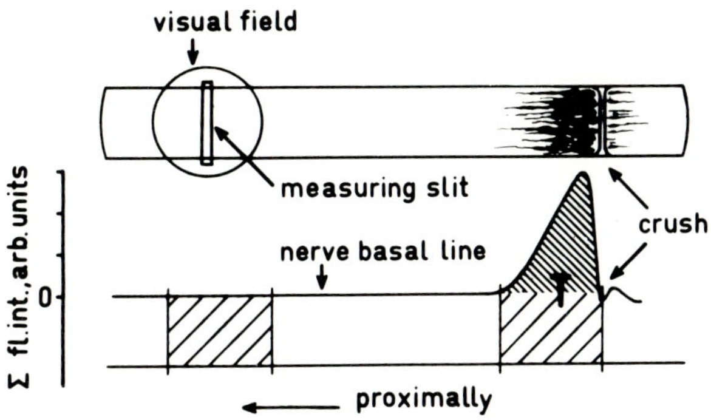

a

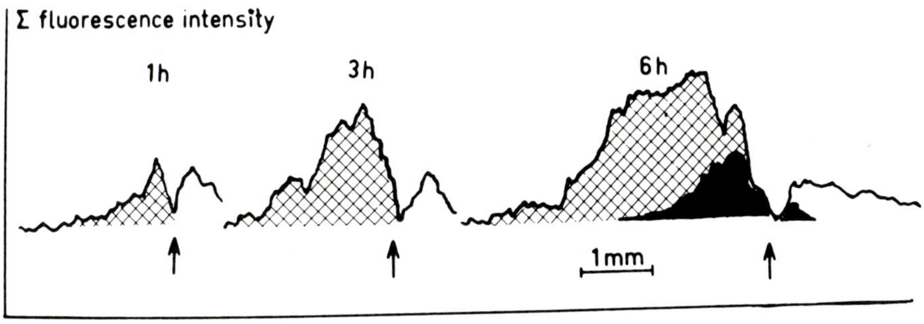

b

FIGURE 11. (a) Schematic illustration of the cytofluorimetric scanning procedure. The measuring slit allows fluorescence in the sections, passed under the slit at a constant rate, to enter a photometer which registers the fluorescence intensity on a graph (lower part). Background fluorescence (nerve basal line) and accumulated fluorescence (densely hatched area) are registered.[89] (b) Cytofluorimetric scanning profiles of sciatic nerve sections incubated for immunofluorescence with rabbit-antiserum to Narcine cholinergic vesicles. The nerves were crush-operated 1, 3, or 6 hr before dissection. The amount of immunofluorescent material which accumulated proximal (to the left in the figure) to the crush (arrows) is represented by the cross-hatched areas, which increase with time after crushing. Distal accumulations are also shown, unlabeled. The black area to the right shows the amount and longitudinal distribution of SP-like immunoreactivity in an 18 hr-crushed rat sciatic nerve (compare Figure 13).

The immunoreactive material observed in motor axons after incubation with RASVA may well represent the axonal ACh organelle which is axonally transported. Biochemical identification of ACh in isolated immunoreactive organelles is, however, needed to establish the identity of these organelles as the ACh storing axonal organelles. Some support for the view that the immunoreactive material proximal to a crush may represent ACh organelles comes from studies on the time course for accumulation of this material. The amount of accumulated immunofluorescent material in crushed nerves can be quantified using a cytofluorimetric scanning procedure[88,89] (Figure 11a). Under maximally standardized conditions an immunoincubated longitudinal nerve section is passed under a measuring slit at a constant speed. The fluorescence intensity is registered through the slit by a photometer and recorded as a curve. The integrated area under the accumulation curve (see Figures 11a and b) is proportional to the amount of immunofluorescence in the section and is given in arbitrary units. Using this procedure the rate of accumulation of RASVA immunoreactive material proximal to a crush (Figure 12) is rather similar to the time course for ACh accumulation in this nerve[90] (Figure 10a).

One major difference exists, however, between the behavior in crushed nerves of ACh and of immunoreactive organelles and that concerns the retrograde accumulation. ACh does not accumulate distally to a crush while immunoreactive material accumulates in substantial

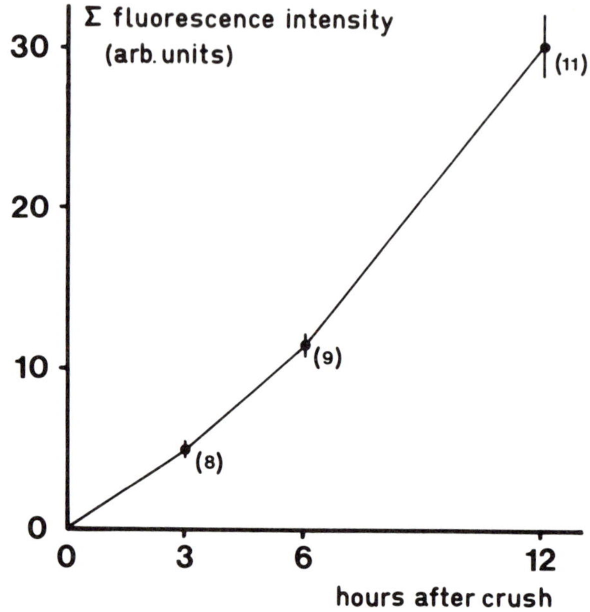

FIGURE 12. The time course of accumulation of immunoreactive cholinergic vesicle material proximal to a crush performed on rat sciatic nerves. The abscissa indicates time after crushing and the ordinate gives the intensity of fluorescence of the immunoreactive cholinergic vesicle material expressed in arbitrary units. These units represent areas under scanning profiles, as indicated in Figure 11b. Mean ± SEM.

amounts (Figure 10b). This may indicate that the organelle is subjected to a considerable degree of retrograde transport, and that the recycling organelles are different from the distally transported ones, i.e., concerning the ACh content/storage. The situation would in this case have similarities to the adrenergic system, where retrogradely migrating immunoreactive DBH has lower enzymatic activity than anterogradely moving DBH, and furthermore very little NE is affiliated with the DBH-containing organelles in retrograde transit (see above).

IV. AXONAL TRANSPORT IN PEPTIDE NEURONS

In addition to the classic transmitters, NE and ACh, several peptides have been suggested to act as neurotransmitters. Bioactive polypeptides have been isolated from the gastrointestinal tract since the beginning of this century, and the list of discovered peptides is still increasing. Of the numerous peptides present in the gut several have been localized also to neuron systems in the CNS and the PNS. The considerable increase in our knowledge about these substances is largely dependent upon the introduction of immunohistochemical techniques into general use.

In the nervous system neurons with somatostatin (SOM), vasoactive intestinal peptide (VIP), substance P (SP), enkephaline (ENK), gastrin/cholecystokinin (GA/CCK), neurotensin (NT), avian pancreatic polypeptide (APP), or bombesine have been demonstrated.[91-98] Since the possibility exists that the antisera used cross-react with other, but chemically related, substances it is considered appropriate to relate to "VIP-like immunoreactivity" etc., but in the following chapter, we will simply relate to the peptide itself although scientifically this is not quite correct.

Intra-axonal transport of peptides has been demonstrated to occur in all peptide neurons investigated. After axotomy the peptide accumulates on the proximal side. In many neurons the perikaryal levels of the peptides are too low to enable detection, probably due to the rapid export into the axon of the newly synthesized and packaged peptide. After treatment with vinblastine or colchicine, however, the intraperikaryal levels build up[99] due to the interrupted intra-axonal transport, while synthesis is unharmed.[24] Thus, production continues but export is arrested, hence the increased concentrations of the peptide which enable detection of the peptidergic somata. The phenomenon of axonal transport has thus been used for detection and mapping of peptide neuron systems.

More systematic studies on the characteristics of axonal transport in peptide neurons have included long peripheral nerves, i.e., the vagus and the sciatic, in a variety of species (e.g., cat, rabbit, rat, guinea pig). VIP, SP, SOM, and CCK have been assayed by RIA in segments of crush-operated or ligated nerves, as well as after applying double crushes to establish the transportable fraction of each peptide.

A. VIP

In motor nerves of rat[100] and cat,[101] VIP accumulated mainly proximally to the crush. In rat nerves no distal accumulation was observed,[100] while a small but significant amount of VIP was present distally to a crush in feline nerves.[101] In both species VIP was present along the intact nerve as well as in ligated nerves in one molecular species, determined by column chromatography. Furthermore no gradient was observed along the nerves. This indicates that no processing of the VIP molecule occurs during axonal transport, in contrast to the situation in the hypothalamo-pituitary peptide neuron systems, where cleavage of precursor molecule takes place in the axon.[102]

The transportable fraction of VIP in cat sciatic was determined to 28% moving distally and 15% migrating retrogradely. The stationary phase was thus about 60% of total, most of which was probably present in the perineurial sheath.[101] Both soluble and particulate VIP was present in the sciatic, but only the particulate fraction was transported. The transport was shown to be microtubule dependent, since local vinblastine injections blocked the transport.[101]

The VIP axons in cat sciatic disappear after lumbar sympathectomy,[85] thus they emanate from cell bodies in the sympathetic lumbar ganglia. These cell bodies are sympathetically cholinergic, and thus ACh and VIP coexist in these nerves (see above). VIP and ACh are probably located in large dense core vesicles, demonstrated to be VIP immunoreactive in the electron microscope.[86]

Transport of VIP has also been demonstrated in the rat cervical vagal nerve. Only proximal accumulations of VIP were observed. For technical reasons the transportable fraction could not be determined, but the average rate of the bulk flow of VIP anterogradely was in the order of 20 mm/day.[103] The true rate is probably higher, however.

B. SOM

In rat cervical vagal nerve SOM accumulated more rapidly than VIP on the proximal side of a crush. A small accumulation was found also distally. No estimate of the transportable fraction was made, but the average rate of the bulk flow was calculated to be about 65 mm/day,[103] which is probably too low a figure.

C. SP

SP-containing cell bodies are present mainly in sensory ganglia.[104-106] The central (leading to the spinal cord) as well as the peripheral processes of these bipolar sensory neurons contain and transport SP towards the endings of the processes. After crushing SP accumulates rapidly on the proximal (cell body) side of the crush in the vagus and the sciatic nerves of the guinea pig, but retrograde accumulations were scarce, and could sometimes be observed in the sciatic nerve (Figure 13) but not in the vagus nerve.[107] The transportable fraction was

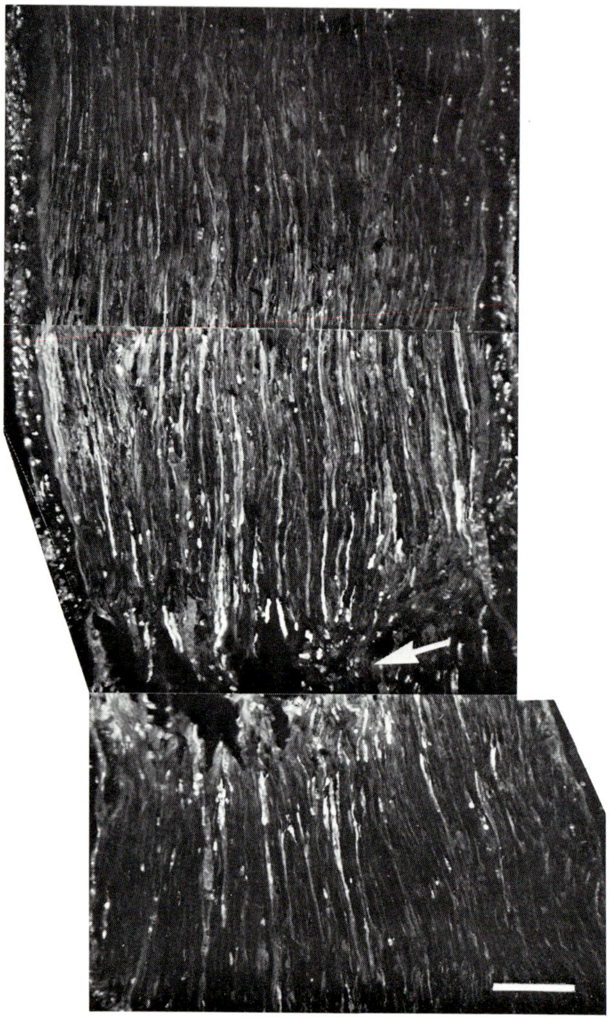

FIGURE 13. Fluorescence microphotograph of a rat sciatic nerve crushed 18 hr before sacrifice. SP-like immunoreactivity has accumulated in thin axons proximal, but also to some extent distal, to the crush (→). Bar = 50 μ.

estimated to be 26% of total SP in the vagus (moving away from the cell body) and 17% in the sciatic. The mean velocity of the transportable SP fraction was calculated to be 115 mm/day in the vagus nerve and 142 mm/day in the sciatic. Cold block experiments in the vagus nerve indicated that the transport could be even faster, as rapid as 240 mm/day.[107]

SP in both nerves was present in a particulate and a soluble fraction, but only particulate SP was subjected to axonal transport.[107]

In rat cervical vagal nerve, SP accumulated rapidly on the proximal side up to 24 hr after crushing, while distally an initial fall was seen, followed, however, by 24 hr of 75% increase over control.[103] The average rate of bulk flow anterogradely was calculated to 34 mm/day, but the transportable fraction was not estimated. In rabbit vagus nerve, however, the transportable fraction, estimated in double-crush experiments, was found to be 34% of total, migrating distally at a rate of 240 mm/day.[103] Colchicine blocked the transport, indicating that microtubules are involved.

The bipolar SP-containing cells of the dorsal ganglia send one branch towards the spinal cord, where its nerve endings are found, mainly in the outer layer of the dorsal horn. SP released from this central branch is supposed to act as a transmitter in the spinal cord. Therefore one would assume that most of the axonally transported SP would enter the central branch. However, much more material is transported distally in the peripheral branch in the sciatic nerve than in the central branch. This observation together with the virtual absence of retrograde SP transport in the peripheral branch suggests that SP is released also from the sensory nerve endings, as indeed demonstrated to occur in the dental pulp[108,109] and in the eye.[110] This suggests that SP has important physiological roles in the periphery, e.g., antidromic vasodilatation[111,112] and enhancing the exitability of neurons in the prevertebral ganglia (see Reference 113), innervated by SP-containing sensory nerve terminals.[114]

D. CCK

This peptide has been studied in feline vagal and sciatic nerves.[115] As for SP and SOM it is located in sensory axons of the two nerves. Only one form of CCK (CCK-8) was generally present in cell bodies as well as axons, and CCK-8 also accumulated in ligated nerves. About 40% of total was in the rapidly transportable fraction, and the mean velocity of anterograde transport of CCK-8 was calculated to be 192 mm/day in the sciatic nerve, and 168 mm/day in the vagus. No retrograde transport could be detected.

V. FUNCTIONAL ASPECTS ON THE AXONAL TRANSPORT OF TRANSMITTER CONTAINING ORGANELLES

In all neuron systems, adrenergic, cholinergic, as well as peptidergic, dealt with in previous sections, intra-axonal transport of the respective transmitters has been demonstrated to occur. NE accumulates in ligated nerves as a result of rapid anterograde migration of NE-storing granules, which have been well characterized biochemically and ultrastructurally. In the cholinergic neuron ACh accumulates proximally to a crush, and there is much evidence to support the hypothesis that this is a result of rapid anterograde axonal transport of organelles that store ACh. The axonal ACh containing organelles are so far unidentified, but when a more stable marker for them than ACh has been found, they can be isolated and studied. The RASVA antiserum, directed against cholinergic synaptic vesicles, or monoclonal antibodies to such organelles, may be a suitable marker (experiments in progress will give the answer).

Synthesis of NE and ACh takes place in the nerve endings from where the transmitters are released. One may therefore ask why these transmitters are also transported distally along the axons. The answer to this question is most certainly that this is an epiphenomenon to the presence in the whole neuron of enzymes which synthesize NE (TH, DDC and DBH) or ACh (ChAT). But the axon also contains degrading enzymes (MAO and AChE, respectively) and a second prerequisite for the axonal transport of NE and ACh is therefore the omnipresence of organelles, which have a capacity to take up and store the synthesized NE and ACh, thereby giving some protection against enzymatic degradation. That such a contionous synthesis and break down of transmitters occurs has been demonstrated in adrenergic as well as in cholinergic neurons with the use of inhibitors to the various enzymes (see above).

The functional importance of NE and ACh transport is thus not to supply the terminals with the transmitter. Instead, the significance of axonal transport of these organelles must be to provide the nerve endings with other important material incorporated into these organelles in the cell-body. One such type of molecule is DBH for the *adrenergic* NE organelle. This enzyme is essential for NE synthesis in the nerve endings and cannot be manufactured elsewhere than in the soma. The axonal NE granules also contain other important molecules, in certain neuron systems (e.g., the splenic nerve) enkephaline is present (see above). In addition, the axonal NE granules function as releasers of NE in the nerve terminal varicosities. A pile of evidence indicates that these newly arrived granules are more active in release, uptake, and synthesis of NE than vesicles arrived in the nerve endings earlier.[116-120] For hypothetically the large newly arrived granules are in some way transforme into smaller NE storing synaptic vesicles by mechanisms which may not clear, but they may involve a loss of material due to repeated exocytotic release.[116-120]

For the *cholinergic* neuron much less in known. The ACh synthesizing enzyme ChAT has not been demonstrated to be present in synaptic ACh vesicles, and the slow rate of accumulation in rat ligated sciatic nerve axons suggests that ChAT is not packed into the axonal ACh containing organelles in this nerve. The repeated observations that accumulated ACh (in sucrose density centrifugation experiments) appears to be stored in a heavier particle than the synaptic vesicle,[66,72] may indicate that these particles are loaded with macromolecules whose action in the nerve ending is at present unknown. Also, these axonal organelles may serve as precursors of synaptic vesicles to replenish the nerve ending with membrane material for formation of true synaptic vesicles.

Cholinergic and adrenergic neurons appear to be closely related from an ontogenetic aspect. Sympathetic neurons which express clear adrenergic characteristics at an early developmental stage change into typically cholinergic neurons at later stages. Also, in tissue culture of dissociated neurons from sympathetic ganglia of newborn rats, neurons express adrenergic or cholinergic properties depending upon the medium in which they are grown.[121-123] Thus, the choice of neurotransmitter used by these neurons is dependent on the milieu and on the type of target organ they innervate.

One would therefore expect that basal mechanisms like axonal transport of transmitter organelles or precursors may share common characteristic features. Therefore, the observation that ACh accumulates proximally to a nerve crush in essentially the same manner as NE accumulates in crushed adrenergic axons might be expected. However, there are some essential differences in the way of handling the transmitter metabolism and turnover, which are important to remember. First, ACh is synthesized in a one step enzymatic reaction, while NE synthesis requires three enzymatic steps. Secondly, the inactivation of released transmitter at the nerve endings is different. In the cholinergic ending it is achieved by splitting of released ACh by synaptic AChE and reuptake of one of the products, choline, while the adrenergic nerve ending possesses an efficient reuptake pump, whereby released NE can be brought back into the neuron for repeated use. Thirdly, about 80% of the axonal ACh appears to be cytoplasmic, while in adrenergic axons most NE is particle bound. Fourthly, the axonal NE organelle has a slow turnover of NE and a more firm storage site for NE, while the axonal ACh organelle has a rapid turnover of ACh, and continously looses ACh to the cytoplasm while replenishment from synthesis seem to balance this loss.

In peptide neurons the functional significance of an anterograde axonal transport of transmitter organelles appears more obvious. Since these macromolecules cannot be formed by the nerve endings (other than by cleavage of a larger precursor molecule as in neurosecretory neurons) they have to be supplied by transport from the factory, i.e., the cell body. All the transported peptides have been shown to be packaged in particles, a process which occurs in the Golgi system of the soma.[124] The release of peptides probably takes place by exocytosis whereby a part or all of the peptide is released to the exterior of the nerve ending.

Retrograde transport in adrenergic and cholinergic axons of material localized originally in NE granules (DBH) and probably affiliated with ACh organelles (RASVA immunoreactive substance), respectively, may return to the soma for various reasons. The lower enzymatic activity of the returning DBH suggests that the molecule is altered, may be worn out, and is moving back up to the cell body for repair or recycling. One other reason may be that the molecule, by repeated exocytotic events, has been exposed to the exterior of the nerve endings and there adsorbed molecules, which, upon delivery to the soma, have influence on the metabolic machinery (neurotrophic factors, for instance).

In peptide neurons retrograde transport was much less noticeable than in adrenergic or cholinergic nerves. This may indicate that all the material delivered to the nerve ending by anterograde transport is released from or broken down in the nerve endings. Another possibility may be that although much material is secreted from the neuron, part of the peptide is degraded by local proteases so that immunoreactivity is lost, and therefore the returning material is not detected by immunochemical procedures. So far, however, there are no indications that this might be the case.

A. Conclusion

The main functional significance of anterograde transport of transmitter containing organelles is to supply the nerve endings with material(s) which are needed there but which cannot be manufactured locally. Such materials do not include NE and ACh; these transmitters are formed mainly in the nerve endings; their axonal transport distally is an epiphenomenon to the presence in the soma and axons of synthesizing enzymes and storage sites. Peptide transmitters must, however, be delivered to the terminals by axonal transport; they cannot be manufactured locally. Retrograde transport may serve to recycle used material for repair or reuse in the soma, but may also supply to the soma exogenous substances which modulate somatal metabolism.

VI. LIST OF ABBREVIATIONS

ACh — Acetylcholine
AChE — Acetylcholine esterase
CA — Catecholamine
ChAT — Choline acetyl transferase
DA — Dopamine
DBH — Dopamine-β-hydroxylase
FIF — Formaldehyde induced fluorescence
FITC — Fluorescein isothiocyanat
5-HT — 5-Hydroxytryptamine, serotonin
MA — Monoamine
MAO — Monoamine oxidase
NE — Norepinephrine
RASVA — Rabbit antisynaptic vesicle antiserum
RBITC — Rhodamine-B-isothiocyanat
SWAR — Swine-anti-rabbit-IgG
TH — Tyrosine hydroxylase

ACKNOWLEDGMENTS

Work from the author's laboratory was supported by grants from the Swedish MRC (2207), M. Bergvall's Foundation, The Medical Faculty, University of Göteborg, W. and M. Lundgren's Foundation, Tore Nilsson's Foundation, Riksföreningen för Trafik-och Polioskadade, T. and R. Söderberg's Foundations Lars Hierta's Foundation, and Åke Wiberg's Foundation.

REFERENCES

1. **Scott, F. H.,** On the relation of nerve cells to fatigue of their nerve fibers, *J. Physiol. (London)*, 34, 145, 1906.
2. **Elliot, T. R.,** On the action of adrenaline, *J. Physiol. (London)*, 31, 20, 1904.
3. **Weiss, P. and Hiscoe, H.,** Experiments on the mechanism of nerve growth, *J. Exp. Zool.*, 107, 315, 1948.
4. **Weiss, P.** *Regional Neurochemistry*, Kety, S. S. and Elkes, J., Eds., Pergamon Press, New York, 1961, 220.
5. **Schmiterlöw, C. G.,** The nature and occurrence of pressor and depressor substances in extracts from blood vessels, *Acta. Physiol. Scand.*, 16 (Suppl.), 56, 1948.
6. **Euler, U. S. V. and Hillarp, N.-Å.,** Evidence for the presence of noradrenaline in submicroscopial structures of adrenergic axons, *Nature (London)*, 177, 44, 1956.
7. **Falck, B., Hillarp, N.-Å., Thieme, G., and Torp, A.,** Fluorescence of catecholamines and related compounds condensed with formaldehyde, *J. Histochem. Cytochem.*, 13, 147, 1965.
8. **Dahlström, A. and Fuxe, K.,** Evidence for the existence of monoamine containing neurons in the central nervous system. I. Demonstration of monoamines in the cell bodies of brain stem neurons, *Acta Physiol. Scand.*, 62 (Suppl.), 232, 1964.
9. **Falck, B. and Owman, Ch.,** A detailed methodological description of the fluorescence method for cellular demonstation of biogenic monoamines, *Acta Univ. Lund.*, II, 7, 1965.
10. **Jonsson, G.,** Flourescence methods for the histochemical demonstration of monoamines. XII. Fluorescence studies on biogenic monoamines and related compounds condensed with formaldehyde, *Histochemie*, 8, 288, 1967.
11. **Erös, G.,** Eine neue Darstellungsmethode der sogenannten "gelben" argentaffinen Zellen des Magen-darmtraktes, *Zbl. Allg. Path. Anat.*, 54, 385, 1932.
12. **Barter, R. and Pearse, A. G. E.,** Detection of 5-hydroxytryptamine in mammalian enterochromaffin cells, *Nature (London)*, 172, 810, 1953.
13. **Barter, R. and Pearse, A. G. E.,** Mammalian enterochromaffin cells as the source of serotonin (5-hydroxtryptamine), *J. Pathol. Bacteriol.*, 69, 25, 1955.
14. **Eränkö, O.,** On the histochemistry of the adrenal medulla of the rat, with special reference to acid phosphate, *Acta Anat., Suppl.*, 17, 1952.
15. **Eränkö, O.,** Histochemistry of nonadrenaline in the adrenal medulla of rats and mice, *Endocrinology*, 57, 363, 1955.
16. **Falck, B. and Torp, A.,** New evidence for the localization of noradrenaline in the adrenergic nerve terminals, *Med. Exp.*, 6, 169, 1962.
17. **Dahlstöm, A. and Fuxe, K.,** A method for the demonstration of adrenergic nerve fibers in peripheral nerves, *Z. Zellforsch.*, 62, 602, 1964.
18. **Dahlström, A. and Fuxe, K.,** Evidence for the existence of monoamine containing neurons in the central nervous system. II. Experimentally induced changes in the intraneuronal amine levels of bulbo-spinal neuron systems, *Acta. Physiol. Scand.*, 64 (Suppl.), 247, 1965.
19. **Dahlström, A.,** Observations on the accumulation of noradrenaline in the proximal and distal parts of peripheral adrenergic nerves after compression, *J. Anat.*, 99, 667, 1965.
20. **Dahlström, A., and Häggendal, J.,** Studies on the transport and life-span of amine storage granules in a peripheral adrenergic neuron system, *Acta Physiol. Scand.*, 67, 278, 1966.
21. **Dahlström, A.,** The effect of reserpine and tetrabenazine on the acccumulation of nonadrenaline in the rat sciatic nerve after ligation, *Acta. Physiol. Scand.*, 69, 167, 1967.
22. **Dahlström, A.,** The transport of nonadrenaline between two simultaneous ligations of the sciatic nerve of rat and cat, *Acta Physiol. Scand.*, 69, 158, 1967.
23. **Dahlström, A.,** Effect of colchicine on transport of amine storage granules in sympathetic nerves of rat, *Eur. J. Pharmacol.*, 5, 111, 1968,
24. **Dahlström, A.,** The effect of drugs on axonal transport of amine storage granules, *New Aspects of Storage and Release Mechanisms of Catecholamines*, Bayer Symposium II, 1970, 20.
25. **Dahlström, A.,** Effects of vinblastine and colchicine on monamine containing neurons of the rat, with special regard to the axoplasmic transport of amine granules, *Acta Neuropathol.*, V (Suppl.), 226, 1971.
26. **Ochs, S.,** *Axoplasmic Transport and its Relation to Other Nerve Functions*, John Wiley & Sons, New York, 1982.
27. **Banks, P., Mayor, D., and Mraz, P.,** Metabolic aspects of the synthesis and intra-axonal transport of nonadrenaline storage vesicles, *J. Physiol.*, 229, 383, 1973.
28. **Häggendal, J., Dahlström, A., and Larsson, P. A.,** Rapid transport of nonadrenaline in adrenergic axons of rat sciatic nerve distal to a crush, *Acta Physiol. Scand.*, 94, 386, 1975.
29. **Nagatsu, I., Kondo, Y., Kato, T., and Natgatsu, T.,** Retrograde axoplasmic transport of inactive dopamine-β-hydroxylase in sciatic nerves, *Brain Res.*, 116, 277, 1976.

30. **McLean, W. G. and Keen, P.,** Local synthesis and breakdown of noradrenaline in constricted rat sciatic nerves, *Eur. J. Pharmacol.*, 18, 74, 1972.
31. **Ben-Johnathon, N., Maxson, R. E., and Ochs, S.,** Fast axoplasmic transport of noradrenaline and dopamine in mammalian peripheral nerve, *J. Physiol. (London)*, 281, 1495, 1981.
32. **Lagercrantz, H., Kirksey, D. F., and Klein, R. L.,** On the development of sympathetic nerve vesicles during axonal transport, *J. Neurochem.*, 23, 769, 1974.
33. **Hydén, H.,** The neuron, in, *The Cell*,Vol. 4, Brachet, J. and Mirsky, A., Eds., Academic Press, New York, 1960, 215.
34. **Koenig, E.,** Synthesis mechanisms in the axon. II. RNA in myelinfree axons of the cat, *J. Neurochem.*, 12, 357, 1965.
35. **Dahlström, A., Häggendal, J., and Larsson, P. A.,** On the noradrenaline loading in axonal amine storage granules in rat crushed sciatic nerves, *Acta. Physiol. Scand.*, 94, 386, 1975.
36. **Brimijoin, S.,** Stop-flow: a new technique for measuring axonal transport and its application to the transport of dopamine-β-hydroxylase, *J. Neurobiol.*, 6, 379, 1975.
37. **Brimijoin, S.,** A histofluorescence study of events accompanying accumulation and migration of norepinephrine within locally cooled nerves, *J. Neurobiol.*, 8, 251, 1977.
38. **Brimijoin, S. and Wiermaa, M. J.,** Direct comparison of the rapid axonal transport of norepinephrine and dopamine-β-hydroxylase activity, *J. Neurobiol.*, 8, 239, 1977.
39. **DePotter W. B., Smith, A. D., and de Schaepdryver, A. F.,** Subcellular fractionation of splenic nerve: ATP, chromogranin A and dopamine-β-hydroxylase in noradrenergic vesicles, *Tissue Cell.*, 2, 529, 1970.
40. **Wilson, S. P., Klein, R. L., Chang, K.-J., Gasparis, M. S., Viveros, O. H., and Yang, W.-H.,** Are opoid peptides co-transmitters in noradrenergic vesicles of sympathetic nerves?, *Nature (London)*, 288, 707, 1980.
41. **Klein, R. L., Wilson, S. P., Dzielak, D. J., Yang, W.-H., and Viveros, O. H.,** Opioid peptides and noradrenaline co-exist in large dence-cored vesicles from sympathetic nerve, *Neuroscience*, 7, 2255, 1982.
42. **Klein, R. L.,** Chemical composition of the large noradrenergic vesicles, in, *Neuotransmitter Vesicles*, Klein, R. L., Lagercrantz, H., and Zimmerman, H., Eds., Academic Press, New York, 1982, 133.
43. **Thureson-Klein, Å., Klein, R. L., and Yen, S.-S,** Ultrastructure of highly purified sympathetic nerve vesicles: correlation between matrix and norepinephrine content, *J. Ultrastruct. Res.*, 43, 18, 1973.
44. **Tomlinson, D. R.,** Two populations of granular vesicles in constricted postganglionic sympathetic nerves, *J. Physiol.*, 245, 727, 1975.
45. **Hökfelt, T. and Dahlström, A.,** Effect of two mitosis inhibitors (colchicine and vinblastine) on the distribution and axonal transport of noradrenaline storage particles, studied by fluorescence and electron microscopy, *Z. Zellforsch.*, 119, 460, 1971.
46. **Taxi, J. and Sotelo, C.,** Cytological aspects of the axonal migration of catecholamines and of their storage material, *Brain Res.*, 62, 431, 1973.
47. **Tranzer, J. P.,** New aspects of the localization of catecholamines in adrenergic neurons, in, *Frontiers in Catecholamine Research*, Usdin, E. and Snyder, S., Eds., Pergamon Press, Oxford, 1973, 453.
48. **Brimijoin, S. and Helland, L.,** Rapid retrograde transport of dopamine-β-hydroxylase as examined by the stop-flow technique, *Brain Res.*, 102, 217, 1976.
49. **Jacobowitz, D. M., Ziegler, M. G., and Thomas, J. A.,** In vivo uptake of antibody to dopamine-β-hydroxylase into sympathetic elements, *Brain Res.*, 91, 165, 1975.
50. **Fillenz, M., Gagnon, C., Stoeckel, K., and Thoenen, H.,** Selective uptake and retrograde axonal transport of dopamine-β-hydroxylase antibodies in perpheral adrenergic neurons, *Brain Res.*, 114, 293, 1976.
51. **Ziegler, M. G., Thomas, J. A., and Jacobowitz, D. M.,** Retrograde axonal transport of antibody to dopamine-β-hydroxylase, *Brain Res.*, 104, 390, 1976.
52. **Geffen, L. B., Rush, R. A., Costa, M., Furness, J. B., and Blessing, W.,** Use of dopamine-β-hydroxylase antibodies to present transmitter specific immune lesions of noradrenergic nerves, in, *Neurotoxins: Fundamental and Clinical Advances*, Chubb, I. W. and Geffen, L. B., Eds., Flinders University of South Australia, Flinders, 1980, 227.
53. **Kapeller, K. and Mayor, D.,** Accumulation of organelles distal to the site of constriction of postganglionic sympathetic nerves, *J. Physiol.*, 194, 95P, 1967.
54. **Geffen, L. B. and Ostberg, A.,** Distribution of granular vesicles in normal and constricted sympathetic neurones, *J. Physiol.*, 204, 583, 1969.
55. **Ochs, S.,** Characterization of fast orthograde transport, *Neurosci. Res. Program Bull.*, 20, 19, 1981.
56. **Gross, G. W.,** The microstream concept of axoplasmic and dendritic transport, *Adv. Neurol.*, 12, 283, 1975.
57. **Jope, R. S.,** High-affinity choline transport and acetyl CoA production in brain and their roles in the regulation of acetylcholine synthesis, *Brain Res. Rev.*, 180, 313, 1979.
58. **Tuček, S.,** *Acetylcholine Synthesis in Neurons*, Chapman & Hall, London, 1978.
59. **Dahlström, A.,** The presence, metabolism and axonal transport of transmitters in peripheral mammalian axons, in, *Handbook of Neurochemistry* Vol. 5, Lajtha, A., Ed., Plenum Press, New York, 1983, 405.

60. **Prajmovsky, M. and Welsh, J. H.,** Total and free acetylcholine in rat perpheral nerve, *J. Neurophysiol.*, 11, 1, 1948.
61. **Carlini, E. A. and Green, J. P.,** Acetylcholine activity in the sciatic nerve, *Biochem. Pharmacol.*, 12, 1367, 1963.
62. **Zimmerman, H. and Whittaker, V. P.,** Effect of electrical stimulation on the yield and composition of synaptic vesicles from the cholinergic synapse of the electric organ of *Torpedo:* a combined biochemical electrophysiological and morphological study, *J. Neurochem.*, 22, 435, 1974.
63. **Dahlström, A., Evans, C. A. N., Häggendal, J., Heiwall, P. O., and Saunders, N.,** Rapid transport of acetylcholine in rat sciatic nerve proximal and distal to a lesion, *J. Neural Transm.*, 35, 1, 1974.
64. **Blaber, L. C. and Cuthbert, A. W.,** A sensitive method for the assay of acetylcholine, *J. Pharm. Pharmacol.*, 13, 445, 1961.
65. **Feldberg, W.,** Synthesis of acetylcholine on sympathetic and cholinergic nerve, *J. Physiol.*, 101, 432, 1943.
66. **Dahlström, A., Häggendal, J., Heilbronn, E., Heiwall, P. O., and Saunders, N.,** Proximo-distal transport of acetylcholine in perpheral cholinergic neurons, in, *Dynamics of Degeneration and Growth in Neurons,* Fuxe, K., Olson, L., and Zotterman, Y., Eds., Pergamon Press, Oxford, 1974, 275.
67. **Dahlström, A., Heiwall, P. O., and Larsson, P. A.,** Comparison between the effects of colchicine and lumicolchicine on axonal transport in rat motor neurons, *J. Neural Transm.*, 37, 305, 1975.
68. **Heiwall, P. O., Saunders, N. R., Dahlström, A., and Häggendal, J.,** The effect of local application of vinblastine or colchicine on acetylcholine accumulation in rat sciatic nerve, *Acta Physiol. Scand.*, 96, 478, 1976.
69. **Häggendal, J., Dahlström, A., and Saunders, N.,** Axonal transport and acetylcholine in rat preganglionic neurons, *Brain Res.*, 58, 494, 1973.
70. **Dahlström, A., Bööj, S., Heiwall, P. O., and Larsson, P. A.,** The effect of chronic nicotine and withdrawal on intra-neuronal dynamics of acetylcholine and related enzymes in a preganglionic neuron system of the rat, *Acta Physiol. Scand.*, 110, 13, 1980.
71. **Ulin, B., Gustavii, K., and Persson, B. A.,** Bioanalysis of picomole amounts of acetylcholine by ion pair partition chromatography applied on rat ischias nerve, *J. Pharm. Pharmacol.*, 28, 672, 1976.
72. **Waldenlind, L. and Anderson, P.,** Subcellular distribution and transport of acetylcholine in cat ventral roots and sciatic nerve, *J. Neurochem.*, 30, 1067, 1978.
73. **Dahlström, A., Bööj, S., Larsson, P. A., and McLean, W. G.,** The turnover of acetylcholine in ligated sciatic nerves of the rat, *Acta Physiol. Scand.*, 115, 493, 1977.
74. **Woodward, W. R., and Lindström, S. H.,** A potential screening technique for neurotransmitters in the CNS. Model studies in the cat spinal cord, *Brain Res.*, 137, 37, 1977.
75. **Sanes, J. B., Carlson, S. S., von Weden, R. J., and Kelly, R. B.,** Antiserum specific for motor nerve terminals in skeletal muscle, *Nature (London)*, 280, 403, 1979.
76. **Carlson, S. S. and Kelly, B.,** An antiserum specific for cholinergic synaptic vesicles from electric organ, *J. Cell. Biol.*, 87, 98, 1980.
77. **Hooper, J. E., Carlson, S. S., and Kelly, R. B.,** Antibodies to synaptic nerve vesicles purified from *Narcine* electic organ bind a subclass of mammalian nerve terminals, *J. Cell Biol.*, 87, 104, 1980.
78. **Coons, A. H.,** Flourescent antibody methods, in, *General Cytochemical Methods*, Danielli, J. F., Ed., Academic Press, New York, 1958, 399.
79. **Dahlström, A., Bööj, S., and P. A. Larsson,** Axonal transport of enzymes and transmitter organelles in cholinergic neurons, in, *Axoplasmic Transport,* Weiss, D. G., Ed., Springer-Verlag, Basel, 1982, 131.
80. **Miledi, R. and Slater, C. R.,** Electron-microscopic structure of denervated skeletal muscle, *Proc. R. Soc. London B.*, 174, 253, 1969.
81. **Dahlström, A., Larsson, P. A., Carlson, S. S., and Bööj, S.,** Localization and axonal transport of immunoreactive cholinergic organelles in rat motor neurons — an immunofluorescent study, *Neuroscience*, in press, 1984.
82. **Dahlström, A, Bööj, S., Carlson, S. S., and Larsson, P. A.,** Rapid accumulation and axonal transport of "cholinergic vesicles" in rat sciatic nerve, studied by immunohistochemistry, *Acta Physiol. Scand.*, 111, 217, 1981.
83. **Gruber, H. and Zenker, W.,** Acetylcholinesterase: histochemical differentiation between motor and sensory nerve fibres, *Brain Res.*, 51, 207, 1973.
84. **Woog, R. H. and Bennett, M. R.,** Vesicle types in injured nerve fibres, *Proc. Aust. Physiol. Pharmacol. Soc.*, 3, 89, 1972.
85. **Lundberg, J. M., Hökfelt, T., Schultzberg, M., Uvnäs-Wallenstein, K., Köhler, C., and Said, S. I.,** Occurence of vasoactive intestinal polypeptide (VIP)-like immunoreactivity in certain cholinegic neurons of the cat: evidence from combined immunohistochemistry and acetylcholine-esterase staining, *Neuroscience*, 4, 1539, 1979.

86. **Johansson, O. and Lundberg, J. M.**, Ultrastructural localization of VIP-like immunoreactivity in large dense cored vesicles of "cholinergic type" nerve terminals in cat exocrine glands, *Neuroscience*, 5, 847, 1981.
87. **Lundberg, J. M., Fried, G., Fahrenkrug, J., Holmstedt, B., Hökfelt, T., Lagercrantz, H., Lundgren, G., and Änggård, A.**, Subcellular fractionation of cat submandibular gland: comparitive studies on the distribution of acetylcholine and vasoactive intestinal polypeptide (VIP), *Neuroscience*, 6, 1001, 1981.
88. **Larsson, P. A., Goldstein, M., and Dahlström, A.**, A new methodological approach for studying axonal transport — cytofluorimetric scanning of nerves, *J. Histochem. Cytochem.*, in press, 1983.
89. **Dahlström, A., Bööj, S., Goldstein, M., and Larsson, P. A.**, Cytofluorimetric scanning: a tool for studying axonal transport in monoaminergic neurons, *Brain Res. Bull.*, 9, 61, 1982.
90. **Dahlöff, A. G., Bööj, S., Larsson, P. A., and Dahlström, A.**, Intra-axonal transport in the mammalian motor neuron and the influence of supraspinal nerve activity, *Scand. J. Rehab. Med.*, 9 (Suppl.), 27, 1983.
91. **Hökfelt, T., Elfvin, L. G., Elde, R., Schultzberg, M., Goldstein, M., and Luft, R.**, Occurence of somatostatin-like immunoreactivity in some peripheral sympathetic noradrenergic neurons, *Proc. Natl. Acad. Sci. U.S.A.*, 74, 3587, 1977.
92. **Hökfelt, T., Johansson, O., Kellerth, J. O., Ljungdahl, Å., Nilsson, G., Nygårds, A., and Pernow, B.**, Immunohistochemical distribution of substance P in, *Substance P*, von Euler, U. S. and Pernow, B., Eds., Raven Press, New York, 1977, 117.
93. **Lundberg, J. M.**, Enkephalin, substance P, VIP, somatostatin, gastrin/CCK and neurotensin in perpheral neurons, *Acta Physiol. Scand.*, 473 (Suppl.), 14, 1979.
94. **Schultzberg, M., Dreyfus, C. F., Gershon, M. D., Hökfelt, T., Elde, R. P., Nilsson, G., Said, S., and Goldstein, M.**, VIP-, enkephalin-, substance P- and somatostatin-immunoreactivity in neurons intrinsic to the intestine: immunohistochemical evidence from organotypic tissue cultures, *Brain Res.*, 155, 239, 1978.
95. **Rehfeld, J. F., Golterman, N. R., Larsson, L. I., Emson, P. C., and Lee, C. M.**, Gastrins and cholecystokinins in central and peripheral neurons, *Fed. Proc. Fed. Am. Soc. Exp. Biol.*, 38, 2325, 1979.
96. **Dockray, G. J., Gregory, R. A., Tracy, H. J., and Zhu, W.-Y.**, Transport of cholecystokinin octapeptide-like immunoreactivity towards the gut in afferent vagal fibers in cat and dog, *J. Physiol. (London)*, 314, 501, 1981.
97. **Schultzberg, M., Hökfelt, T., Nilsson, G., Terenius, L., Rehfield, J. F., Brown, M., Elde, R., Goldstein, M., and Said, S.**, Distribution of peptide and catecholamine-containing neurons in the gastrointestinal tract of rat and guinea pig: immunohistochemical studies with antisera to substance P, vasoactive intestinal polypeptide, enkephalins, somostatin, gastrin/cholecystokinin, neurotensin and dopamine-β-hydroxylase, *Neuroscience*, 5, 689, 1980.
98. **Johansson, O., Hökfelt, T., and Elde, R. P.**, Distribution of somatostatin-like immunoreactivity in the central nervous system of the adult rat, *Neuroscience*, in press, 1983.
99. **Lundberg, J. M., Hökfelt, T., Nilsson, G., Terenius, L., Rehfield, J., Elde, R., and Said, S.**, Peptide neurons in the vagus, splanchnic and sciatic nerves, *Acta Physiol. Scand.*, 104, 499, 1978.
100. **Giachetti, A. and Said, S. I.**, Axonal transport of vasoactive intestinal peptide in sciatic nerve, *Nature (London)*, 281, 574, 1979.
101. **Lundberg, J. M., Fahrenkrug, J., and Brimijoin, S.**, Characteristics of the axonal transport of vasoactive intestinal polypeptide (VI) in nerves of the cat, *Acta Physiol. Scand.*, 112, 427, 1981.
102. **Gainer, H., Sarne, Y., and Brownstein, M. J.**, Biosynthesis and axonal transport of rat neurohypophysial proteins and peptides, *J. Cell Biol.*, 73, 366, 1977.
103. **Gilbert, R. F. T., Emson, P. C., Fakrenkrug, J., Lee, C. M., Penman, E., Stakahashi, T., and Wass, J.**, Axonal transport of neuropeptides in the cervical vagus nerve of the rat, *J. Neurochem.*, 34, 108, 1980.
104. **Takahashi, T. and Otsuka, M.**, Regional distribution of substance P in the spinal cord and nerve roots of the cat and the effect of dorsal root section, *Brain Res.*, 87, 1, 1975.
105. **Hökfelt, T., Kellerth, J.-O., Nilsson, G., and Pernow, B.**, Experimental immunohistochemical studies on the localization and distribution of substance P in cat primary sensory neurons, *Brain Res.*, 100, 235, 1975.
106. **Hökfelt, T., Johansson, O., Kellerth, J.-O., Ljungdahl, Å., Nilsson, G., Nygårds, A., and Pernow, B.**, Immunohistochemical distribution of substance P, in, *Substance P, Nobel Symposium*, von Euler, U. S. and Pernow, B., Eds., Raven Press, New York, 1977, 117.
107. **Brimijoin, S., Lundberg, J. M., Brodin, E., Hökfelt, T., and Nilsson, G.**, Axonal transport of substance P in the vagus and sciatic nerves of the guinea pig, *Brain Res.*, 191, 443, 1980.
108. **Olgart, L., Gazelius, B., Brodin, E., and Nilsson, G.**, Release of substance P-like immunoreactivity from the dental pulp, *Acta Physiol. Scand.*, 101, 510, 1977.
109. **Brodin, E., Gazelius, B., Olgart, L., and Nilsson, G.**, Substance P-like immunoreactivity (SPLI) in the dental pulp: tissue concentration, release and degradation, *Acta Physiol. Scand.*, 473 (Suppl.), 142, 1979.

110. **Bill, A., Stjernschantz, J., Mandahl, A., Brodin, E., and Nilsson, G.,** Substance P: release on trigeminal nerve stimulation, effects in the eye, *Acta Physiol. Scand.*, 1979.
111. **Hinsey, J. C. and Gasser, H. S.,** The component of the dorsal root mediating vasodilation and the Sherrington contracture, *Am. J. Physiol.*, 92, 679, 1930.
112. **Lembeck, F.,** Zur Frage der zentralen Übertragung afferenter Impulse, III. Mitteilung. Das Vorkommen und die Bedeutung der Substanz P in den dorsalen Wurzeln des Rückenmarks, *Naunyn-Schmiedeb. Arch. Exp. Pathol. Pharmak.*, 219, 197, 1953.
113. **Burnstock, G., Hökfelt, T., Gershon, M. D., Iversen, L. L., Kosterlitz, H. W., and Szurszewski, J. H.,** Non-adrenergic, non-cholinergic mechanisms in the autonomic nervous system, *NRP Program Bull.*, 3, 17, 1979.
114. **Elfvin, L.-G. and Dalsgaard, C. J.,** Retrograde axonal transport of horseradish peroxidase in affernt fibers of the inferior mesenteric ganglion of the guinea pig. Identification of the cells of origin in dorsal root ganglia, *Brain Res.*, 126, 149, 1977.
115. **Rehfeld, J. F. and Lundberg, J. M.,** Cholecystokinin in feline vagal and sciatic nerves: concentration, molecular forms and transport velocity, *Brain Res.*, 275, 341, 1983.
116. **Häggendal, J. and Dahlström, A.,** The importance of axoplasmic transport of amine granules for the functions of the adrenergic neurons, *Acta Neuropathol.*, V(Suppl.), 238, 1971.
117. **Dahlström, A.,** Aminergic transmission, Introduction and short review, *Brain Res.*, 62, 441, 1973.
118. **Dahlström, A. and Häggendal, J.,** Intra-axonal transport of young amine granules: implications for nerve terminal function, in *Central Nervous System—Studies on metabolic regulation and function*, Genazzani, E. and Herken, H., Eds., Springer-Verlag, Basel, 1973, 94.
119. **Smith, A. D.,** Cellular control of the uptake, storage and release of noradrenaline in sympathetic nerves, *Biochem. Soc. Symp.*, 36, 103, 1972.
120. **DePotter, W. P., Chubb, I. W., Put, A., and De Schaepdryver, A. F.,** Facilitation of the release of noradrenaline and dopamine-β-hydroxylase at low stimulation frequencies by α-blocking agents, *Arch. Int. Pharmacodyn. Ther.*, 193, 191, 1971.
121. **Patterson, P. H. and Chunn, L. L. Y.,** Induction of acetylcholine synthesis in primary cultures of dissociated rat sympathetic neurons. I. Effects of conditioned medium, *Dev. Biol.*, 56, 263, 1977.
122. **Landis, S. C. and Keefe, D.,** Development of cholinergic sympathetic innervation of eccrine sweat glands in rat footpad, *Tenth Ann. Meet. Soc. Neurosci.*, 6, 379, 1980.
123. **Landis, S. C.,** Development changes in the neurotransmitter properties of dissociated sympathetic neurons: a cytochemical study of the effects of the medium, *Dev. Biol.*, 77, 349, 1980.
124. **Johansson, O.,** Localization of vasoactive intestinal polypeptide- and avian pancreatic polypeptide-like immunoreactivity in the Golgi apparatus of peripheral neurons, *Brain Res.*, 262, 71, 1983.
125. **Hollenbeck, P. J.,** A third front for cell motility, *Nature (London)*, 317, 17, 1985.
126. **Brady, S. T.,** A novel brain ATPase with properties expected for the fast axonal transport motor, *Nature (London)*, 317, 73, 1985.

Chapter 12

AXONAL TRANSPORT IN NERVE AND MUSCLE DISEASES

Kenneth A. Skau

TABLE OF CONTENTS

I. INTRODUCTION

The involvement of axonal transport in nerve and muscle diseases is of particular interest to both scientists and clinicians. On the one hand, an aberration of axonal transport, in general, or of a particular substance or class of substances may result in expression of clinical symptoms or in more generalized disturbances of nerve or muscle function. On the other hand, a primary nerve or muscle disease may result in secondary disturbances of axonal transport, which may produce other symptoms. To date, the studies on axonal transport in various nerve and muscle diseases have not led to a general hypothesis correlating symptomology and transport defect.

II. AXONAL TRANSPORT IN ANIMAL MUSCULAR DYSTROPHY

A. Mouse Muscular Dystrophy

A spontaneous mutation in the ReJ/129 inbred strain of mice at the Jackson Memorial Laboratories resulted in an hereditary muscle defect which , in some ways, resembles certain human muscular dystrophies.[1] Although it is generally recognized that important differences exist between the mouse disease and human dystrophies, the mutant mouse strain serves as an important model to study the biochemistry, physiology, and pathology of abnormal muscle. The initial biochemical defect has yet to be identified, and there is some controversy as to whether the disease is an expression of a primary muscle defect, a primary nerve defect, or if both nerve and muscle are independently affected.

Certain nerve abnormalities are evident in the diseased animals. Spinal roots of afflicted mice lack myelin — a condition that resembles fetal ventral roots.[2] In the peripheral nervous system axons are myelinated, but appear to have widened nodes of Ranvier and shortened internodal segments.[3] Although abnormalities of axon-Schwann cell interactions are evident in both motor and sensory nerves, the defect is greater in motor nerves.[4] These abnormalities, as well as certain alterations in the number of motor units,[5] led to the hypothesis of functional denervation in which it was proposed that murine dystrophy is neurogenic.[5-7] It is now clear that dystrophic muscles do not exhibit many of the signs associated with surgical denervation,[8,9] but the concept of a neurogenic basis of murine dystrophy has not been disproven.

This possibility, that murine dystrophy is neurogenic, led several investigators to examine axonal transport in dystrophic mice. Transport studies in mice are technically difficult, particularly in dystrophic mice which are typically one half the weight of clinically normal litter-mates. Thus, injection of radiolabeled precursors is reproducibly accomplished only with great skill. Ligation studies are also difficult because nerves are short, allowing only a short period of time for accumulation before considerable nerve damage is evident. For similar reasons in vitro accumulation studies are technically demanding.

Despite these difficulties, several investigators, using various techniques, have reported on axonal transport in dystrophic mice. Bradley and Jaros injected ^{3}H-leucine into spinal cords of dystrophic mice and used high resolution autoradiography to assess axonal transport of protein.[10] They suggested that the amount of radiolabeled protein moving at slow rates was reduced but the amount moving at fast rates (100 mm/day) was increased. This study used only a limited number of animals and compared the transport in dystrophic animals to controls of different strain of mice. Consequently, only limited conclusions can be drawn from these results.

Injection of radiolabeled precursors was also used by Komiya and Austin in a series of papers on the transport of proteins, phospholipids, and cholesterol in dystrophic mice.[11-13] These investigators reported that they observed a superfast transport rate (ca.2000 mm/day) of protein in both dystrophic and clinically normal mice, but that the amount of protein in this phase was reduced in dystrophic mice. By contrast, the conventional fast axonal transport fraction (ca. 500 mm/day) showed an increased amount of protein in dystrophic mice. Although a slow component, containing most of the radiolabeled protein, was observed in control mice, the difficulty in keeping dystrophic animals alive hampered similar studies in diseased mice. This work has been criticized on several grounds.[14] The superfast transport rate is not accepted by many investigators. This rate has not been convincingly shown to be an axonal transport and may be related to blood-borne precursor. Also, waves of radioactivity have been considered to be indistinct. Another report, in which the investigators used radioactive precursors, arrived at quite different conclusions.[15] In this study, the amount of radiolabeled protein that accumulated at a ligature was reduced in both motor and sensory nerves.

A series of papers from Brimijoin's laboratory has examined the accumulations of specific proteins at sciatic nerve ligatures.[16-18] The slowly transported enzyme choline aectyltransferase (CAT), which is restricted to cholinergic nerves,[19] was elevated in sciatic nerves of dystrophic mice, but the average velocity of transport was reduced. Accumulation of CAT at the ligature was only 52% of nondystrophic litter-mates; average velocity was 0.9 mm/day in dystrophic vs. 1.9 mm/day in controls. Furthermore, a defect of transport of dopamine-β-hydroxylase (DBH), an enzyme localized in autonomic sympathetic nerves, suggests that transport in nonmotor neurons was also affected. Basal levels of DBH in dystrophic sciatic nerve were normal, but accumulation proximal to a ligature was reduced by one third. A recent report[18] suggests that transport of acetylcholinesterase (AChE) in dystrophic sciatic nerve was also disrupted. Specifically, accumulation of the G_4 form of AChE was reduced by one third. Significantly, a major biochemical abnormality of dystrophic mouse muscle is the absence of G_4-AChE.[9] Transport of other forms of the enzyme was also reduced, but less dramatically. Skau, however, did not find an abnormality of AChE transport in dystrophic mice.[20]

All of the above studies were conducted on dystrophic mice of the ReJ/129 strain. A second dystrophic mutant mouse has been identified in the C57B1/6J strain (dy^{2J}/dy^{2J}). Only one study has reported on axonal transport in these mice.[21] Retrograde transport of nerve growth factor and serum albumin appeared to be reduced.

The significance of abnormal axonal transport in muscular dystrophy is not clear. One might speculate that an abnormal transport might lead to necrotic muscle destruction by failing to provide sufficient trophic factors. Conversely, a primary muscle defect may fail to produce factors necessary to maintain normal neuronal function, including axonal transport. The latter hypothesis seems unlikely since transport in autonomic and sensory nerves is also affected. One must also consider that both nerve and muscle might be affected independently and that defects in axonal transport are unrelated to the muscle defect. Finally, there is the possibility that reduced transport is due to reduced body temperature in dystrophic mice. Brimijoin and Jablecki[17] found identical temperature of the hind limbs of dystrophic and nonafflicted litter-mates. In unpublished results I have found that dystrophic mice have slightly, but significantly lower rectal temperature (control $= 35.21° \pm 0.16$; dystrophic $= 34.15° \pm 0.24$; $p < 0.05$). However, this temperature difference could account for no more than 10% reduction in transport.[22]

B. Chicken Muscular Dystrophy

In addition to the mouse models for muscular dystrophy, an inherited dystrophy that affects primarily fast-twitch muscle fibers has been reported in a line of chickens.[23] In fact, some investigators believe that the chicken dystrophy model is more representative of the human disease than are the mouse models. Neuronal abnormalities of the type found in dystrophic mice have not been observed in diseased chickens, although there is some indication of a defect at the motor nerve terminal.[24] Several laboratories have examined axonal transport in dystrophic chickens, but no abnormalities have been reported. DeSantis et al.[25]

found the retrograde transport of horseradish peroxidase to be normal, although they were careful not to exclude the possiblity that transport if a specific endogenous substance might be affected. Likewise, there was no abnormality in anterograde and retrograde transport of the molecular forms of AChE accumulating at the cut ends of sciatic nerves.[26] Stromska et al.[27] used local injection of ^{3}H-leucine and ^{3}H-fucose into spinal cord to label rapidly transported proteins or glycoproteins, but found no abnormality in the rate or amount of transported label, nor in the subcelluar distribution of the labeled proteins. It thus seems unlikely that an abnormality of axonal transport, per se, is involved in chicken dystrophy.

C. Hamster Muscular Dystrophy

An heredity dystrophy has also been observed in Syrian hamsters, although the characteristics of this disease differ considerably from mice.[28] The disease is relatively mild and affects not only skeletal muscle but cardiac muscle as well. The only study on axonal transport in dystrophic hamsters found no abnormalities.[29] Transport of proteins and glycoproteins in motor and sensory fibers following injection of ^{3}H-precursors into ventral horn or dorsal root ganglia, revealed no differences between dystrophic and control hamsters. Likewise, accumulation of choline acetyltransferase at a sciatic nerve ligature was not altered.

III. MISCELLANEOUS NERVE AND MUSCLE DISORDERS

A mutant mouse strain, called trembler has a Schwann cell abnormality that results in abnormal myelination patterns. One study failed to find abnormal axonal transport in these animals, suggesting that myelination abnormalities do not affect transport.[30] The wobbler mutant exhibits a progressive neural atrophy affecting primarily ventral horn cells. Bird et al. first suggested that these mice exhibited a defect in the amount, but not the rate of slow-transport.[31] However, others have found no alteration in slow-transport and only a slight increase in fast-transport.[10,32] Relatively few animals were used in these studies, thus limiting the conclusion that can be drawn.

Experimental allergic neuritis (EAN) is considered to be an animal model for Guillain-Barré Syndrome. Fast axonal transport of AChE was not abnormal in animals with EAN, but the amount of enzyme transported in the retrograde direction was reduced to 3.2% of total AChE (compared to 4.7% in controls).[33] The authors believe that this is due to a defect in turnaround of the protein at nerve terminals.

IV. AXONAL TRANSPORT IN HUMAN NEUROPATHY

Studies on the involvement of axonal transport in human nerve and muscle disorders are inherently more difficult than animal studies. Obviously, there is a reluctance to subject a patient that is already pathologic to a procedure that may further compromise neuronal function. Such powerful techniques as local injection of radiolabeled protein precursors or removal of an entire neuron are not acceptable methodologies. One is basically limited to in vitro studies of nerves that are of minor functional significance, or biopsies of cadaver tissue. In addition, these studies are often performed on nerves that have undergone considerable damage by the time the neuropathy is diagnosed.

An early report on axonal transport in diseased human sural nerves used Nomarski optics to observe the movement of particles within individual teased axons.[34] Velocities of mitochondria and spherical segments of endoplasmic reticulum were estimated in nerves of patients exhibiting a variety of neurological diseases. These investigators recognized no abnormalities of movement but admitted that their methods were biased toward large fibers that may not have been pathological.

The most extensive series of studies on axonal transport in diseased human nerve has been performed in Brimijoin's lab.[35-39] In these studies segments of human sural nerve were ligated at both ends and incubated in physiological saline at 37°C for various time periods. The transport of the enzymes DBH and AChE were assessed by cutting the nerve into 3 mm segments and assaying the segments. The average velocity of transport (V_{av}) was calculated and represents a function of true transport velocity and the fraction of enzyme that is mobile.

Not surprisingly, patients with herditary motor neuropathies exhibited no abnormalities either in basal levels or V_{av} of DBH.[36] The sural nerve contains no motor fibers. On the other hand, nerves from patients with hereditary sensory or mixed hereditary motor and sensory neuropathies exhibited reduced V_{av} of DBH. The most striking cases were patients diagnosed as having hereditary motor or sensory neuropathy, type III (Dejerine-Sottas neuropathy). In these patients basal levels of DBH were not significantly reduced, but V_{av} was greatly compromised. In at least one case no accumulation of DBH could be detected.[35] Significant reduction in axonal transport was also observed in peroneal muscular atrophy and hereditary sensory neuropathy, type II.[37]

In some cases of acquired neuropathy transport was also reduced. Patients with uremia exhibited reduced V_{av} of DBH although basal levels of the enzyme were normal.[38] One uremic patient exhibited a reduced V_{av} greater than two standard deviations from control despite a lack of symptoms of neuropathy. This suggests that transport abnormalities may precede frank neuropathy.

Both DBH and AChE transport were reduced in ten diabetic patients exhibiting neuropathies.[38] Basal levels of the enzymes were reduced, as were the number of axons. However, true velocity for DBH was reduced by two thirds and that of AChE by one half. In the most recent report from this lab[39] the stop flow technique[40] was used on a limited number of nerve biopsies from diabetic (n = 3) and control (n = 2) patients. Velocity of the moving peak of DBH activity in controls was 12 mm/hr, a velocity similar to that previously found in rabbit nerves using stop flow.[40] Diabetic nerves exhibited a velocity of 6 mm/hr. This technique offers a direct means of measuring transport. However, the potential exists that the apparent abnormal velocity might be related to a delay in recovery of the nerve from the reduced temperatures used to stop transport.[39]

Other investigators have examined axonal transport in diseased patients. Norris used Nomarski optics to estimate the movement of particles in alpha motor fibers innervating external intercostal muscles.[41] He concluded that patients with amyotrophic lateral sclerosis exhibited 30% reduction in movement of particles. Dustin and Flament-Durand suggested that abnormal accumulation of dense particles and swelling of neuronal processes in brain biopsies of patients with Alzheimer's disease may reflect reduced transport.[42] They noted that reduced microtubules and neurofilaments further suggest an axonal transport abnormality. They also speculated that reduced transport might be responsible for the abnormalities of AChE and CAT in this disease.

V. CONCLUSIONS

Studies in experimental animals, as well as in human nerves, provide strong evidence that there is an association between axonal transport and peripheral neuropathy. To date, however, it is difficult to determine whether neuropathies can arise as a consequence of abnormal axonal transport. While it seems unlikely that a generalized disturbance of transport is an explanation for nerve diseases, it remains to be determined whether breakdown of transport of selective substances or groups of substances may lead to dysfunction.

REFERENCES

1. **Michelson, A. M., Russell, E. S., and Harman, P. J.,** Dystrophia muscularis: a hereditary primary myopathy in the house mouse, *Proc. Natl. Acad. Sci. U.S.A.*, 41, 1079, 1955.
2. **Bradley, W. G. and Jenkison, M.,** Abnormalities of peripheral nerves in murine muscular dystrophy, *J. Neurol. Sci.*, 18, 227, 1973.
3. **Bradley, W. G., Jaros, E., and Jenkison, M.,** The nodes of Ranvier in the nerves of mice with muscular dystrophy, *J. Neuropathol. Exp. Neurol.*, 36, 797, 1977.
4. **Jaros, E. and Jenkison, M.,** Quantitative studies of the abnormal axon-Schwann cell relationship in the peripheral motor and sensory nerves of the dystrophic mouse, *Brain Res.*, 258, 181, 1983.
5. **Harris, J. B. and Wilson, P.,** Mechanical properties of dystrophic mouse muscle, *J. Neurol. Neurosurg. Psychiatry*, 34, 512, 1971.
6. **Baker, N., Wilson, L., Oldendorf, W., and Blahd, W. H.,** Supersensitivity to neostigmine and resistance to d-tubocurarine in mice with hereditary myopathy, *Am. J. Physiol.*, 198, 926, 1960.
7. **McComas, A. J. and Mrozek, K.,** Denervated muscle fibers in hereditary mouse dystrophy, *J. Neurol. Neurosurg. Psychiatry*, 30, 526, 1967.
8. **Law, P. K. and Atwood, H. L.,** Nonequivalence of surgical and natural denervation in dystrophic mouse muscles, *Exp. Neurol.*, 34, 200, 1972.
9. **Skau, K. A. and Brimijoin, S.,** Abnormal distribution of skeletal muscle acetylcholinesterase molecular forms in dystrophic mice, *Exp. Neurol.*, 74, 111, 1981.
10. **Bradley, W. G. and Jaros, E.,** Axoplasmic flow in axonal neuropathies. II. Axoplasmic flow in mice with motor neuron disease and muscular dystrophy, *Brain*, 96, 247, 1973.
11. **Komiya, Y. and Austin, L.,** Axoplasmic flow of protein in the sciatic nerve of normal and dystrophic mice, *Exp. Neurol.*, 43, 1, 1974.
12. **Tang, B. Y., Komiya, Y., and Austin, L.,** Axoplasmic flow of phospholipids and cholesterol in the sciatic nerve of normal and dystrophic mice, *Exp. Neurol.*, 43, 13, 1974.
13. **Kuffer, A. D., Komiya, Y., and Austin, L.,** Proteins of fast axoplasmic transport in the sciatic nerve of the dystrophic mouse, *Exp. Neurol.*, 55, 74, 1977.
14. **Brimijoin, S.,** The role of axonal transport in nerve disease in *Peripheral Neuropathy*, Dyck, P. J., Thomas, P. K., and Lambert, E. H., Eds., W. B. Saunders, Philadelphia, 1983, chap. 22.
15. **McLane, J. A. and McClure, W. O.,** Rapid axoplasmic transport in dystrophic mice, *J. Neurochem.*, 29, 825, 1977.
16. **Jablecki, C. and Brimijoin, S.,** Reduced axoplasmic transport of choline acetyltransferase activity in dystrophic mice, *Nature (London)*, 250, 151, 1974.
17. **Brimijoin, S. and Jablecki, C.,** Reduced axonal transport of dopamine-β-hydroxylase activity in dystrophic mice: Evidence for abnormality of adrenergic nerve cells, *Exp. Neurol.*, 53, 454, 1976.
18. **Brimijoin, S. and Schreiber, P. A.,** Reduced axonal transport of 10S acetylcholinesterase in dystrophic mice, *Muscle Nerve*, 5, 405, 1982.
19. **Kasa, P., Mann, S. P., and Hebb, C.,** Localization in choline acetyltransferase. Ultrastructiural localization in spinal neurones, *Nature (London)*, 226, 814, 1970.
20. **Skau, K. A.,** Distribution of acetylcholinesterase molecular forms in organs of dystrophic mice, *Soc. Neurosci. Abstr.*, 8, 331, 1982.
21. **Rutherford, P. S. and Boegman, R. J.,** Axonal transport of nerve growth factor and serum albumin in the dystrophic mouse, *Neurosci. Lett.*, 34, 209, 1982.
22. **Bisby, M. A. and Jones, D. L.,** Temperature sensitivity of axonal transport in hibernating and nonhibernating rodents, *Exp. Neurol.*, 61, 74, 1978.
23. **Asmundson, V. S., Kratzer, P. H., and Julian, L. M.,** Inherited myopathy in the chicken, *Ann. N.Y. Acad. Sci.*, 138, 49, 1966.
24. **Albuquerque, E. X. and Warnick, J. E.,** Electrophysiological observations in normal and dystrophic chicken muscles, *Science*, 172, 1260, 1971.
25. **DeSantis, M., Hoekman, T., and Limwongse, V.,** Retrograde transport of peroxidase in motor neurons innervating slow and fast muscles: absence of a difference between normal and dystrophic chickens, *Brain Res.*, 119, 454, 1977.
26. **DiGiamberardino, L., Couraud, J. Y., and Barnard, E. A.,** Normal axonal transport of acetylcholinesterase forms in peripheral nerves of dystrophic chickens, *Brain Res.*, 160, 196, 1979.
27. **Stromska, D., Ochs, S., and Muller, J.,** Rate of axoplasmic transport in the dystrophic chicken, *Exp. Neurol.*, 74, 530, 1981.
28. **Homburger, F. and Bajusz, E.,** New models of human disease in Syrian hamsters, *JAMA*, 212, 604, 1970.
29. **Boegman, R. J. and Wood, P. L.,** Axonal transport in dystrophic hamsters, *Can. J. Physiol. Pharmacol.*, 59, 202, 1981.

30. **Boegman, R. J., Aguayo, A. J., and Bray, G. M.,** Axoplasmic transport in (trembler mouse) nerves with a widespread disorder of myelination, *J. Neuropathol. Exp. Neurol.*, 36, 590, 1977.
31. **Bird, M. T., Shuttleworth, E., Koestner, A., and Reinglass, J.,** The wobbler mouse mutant: an animal model of hereditary motor system disease, *Acta Neuropathol.*, 19, 39, 1971.
32. **Bradley, W. G., Murchison, D., and Day, M. J.,** The range of velocities of axoplasmic flow. A new approach, and its application to mice with genetically inherited spinal muscular atrophy, *Brain Res.*, 35, 185, 1971.
33. **Oikarinen, R., Molnar, G. K., and Riekkinen, P. J.,** Axonal transport of cholinesterase activity in experimental allergic neuritis, *Acta Univ. Oulu A97 Biochem.*, 29, 85, 1980.
34. **Kirkpatrick, J. B. and Stern, L. Z.,** Axoplasmic flow in human sural nerve, *Arch. Neurol.*, 28, 308, 1973.
35. **Brimijoin, S., Capek, P., and Dyck, P. J.,** Axonal transport of dopamine-β-hydroxylase by human sural nerves in vitro, *Science*, 180, 1295, 1973.
36. **Brimijoin, S. and Dyck, P. J.,** Axonal transport of dopamine-β-hydroxylase by abnormal human sural nerves., in *Dynamics of Degeneration and Growth in Neurons*, Fuxe, K., Olson, L., and Zetterman, Y., Eds., Pergamon Press, Oxford, 1974, 291.
37. **Brimijoin, S. and Dyck, P. J.,** Axonal transport and peripheral nerve disease in man, in *Peripheral Neuropathies*, Caval, N. and Pozza, G., Eds., Elsevier/North Holland, Amsterdam, 1978, 139.
38. **Brimijoin, S. and Dyck, P. J.,** Axonal transport of dopamine-β-hydroxylase and acetylcholinesterase in human peripheral neuropathy, *Exp. Neurol.*, 66, 467, 1979.
39. **Brimijoin, S., Dyck, P. J., Jakobsen, J., and Lambert, E. H.,** Axonal transport in human nerve disease and in the experimental neuropathy induced by p-Bromophenylacetylurea, in *Axoplasmic Transport in Physiology and Pathology*, Weiss, D. G. and Gorio, A., Eds., Springer-Verlag, Basel, 1982, 124.
40. **Brimijoin, S.,** Stop-flow: a new technique for measuring axonal transport, and its application to the transport of dopamine-β-hydroxylase, *J. Neurobiol.*, 6, 379, 1975.
41. **Norris, F. H.,** Axon particle movement in motor neuron disease, *Trans. Am. Neurol. Assoc.*, 101, 276, 1976.
42. **Dustin, P. and Flament-Durand, J.,** Disturbances of axoplasmic transport in Alzheimer's disease, in *Axoplasmic Transport in Physiology and Pathology*, Weiss, D. G. and Gorio, A., Eds., Springer-Verlag, Basel, 1982, 131.

Chapter 13

AXONAL TRANSPORT IN EXPERIMENTAL DIABETES AND OTHER MODELS OF METABOLIC AND TOXIC NEUROPATHIES

Johannes Jakobsen and Per Sidenius

TABLE OF CONTENTS

I. DIABETIC NEUROPATHY

Peripheral nerve disease in diabetes includes a wide variety of clinical manifestations affecting somatic and autonomic functions. In its most common form the diabetic patient notices impaired sensory perception in the feet and the physical examination reveals decreased sensations of touch, pain and vibration, and weak or absent knee and ankle jerks. In some of these patients paresthesias and pain develop which characteristically are fortified in the warmth of the bed. Still others realize muscular weakness in the feet and calves. In a few cases a disabling and even life threatening state occurs especially in those having severe autonomic dysfunction.

Clinical signs of neuropathy are dependent on the duration of the metabolic derangement and develop in type I as well as in type II diabetes. After 25 years of diabetes the prevalence of loss of tendon reflexes and impaired vibration perception is close to 50%.[1-3]

Studies of peripheral nerves of patients who have died after many years of diabetes show axonal degeneration, segmental demyelination, fiber loss and thickening of the vessel wall, with narrowing of the lumen of capillaries and of smaller intraneural arteries. The etiology of diabetic neuropathy is unknown, but from histological reports it has been suggested that ischemia caused by diabetic angiopathy leads to the development of neuropathy.[4,5] Other authors have reported that intraneural vessel abnormalities are sparse in diabetic patients with peripheral neuropathy or that the degree of angiopathy is not related to the presence of signs and symptoms of nervous disease.[6-9]

Today, however, there is good evidence that metabolic factors are significant to at least some of the neuropathic manifestations. Before the era of regular insulin treatment it was reported from large series of patients that symptoms of neuropathy also occurred in unrecognized diabetics or in patients who had neglected their diabetic control.[6] Pain and cramps in the legs were the main symptoms which disappeared after a few weeks of insulin treatment. Since angiopathy is a late and irreversible manifestation of diabetes the acute neuropathic category most likely has a metabolic origin. More sophisticated evidence for the metabolic etiology comes from neurophysiologic studies of the nerve conduction velocity which reflects the functional state of the group of large myelinated fibers. Thus, motor conduction velocity in the peroneal nerve is slightly reduced soon after the establishment of the diagnosis of diabetes and, in addition, normalization takes place during treatment of hyperglycemia.[10,11] Also it has been shown that withdrawal of insulin treatment in a group of juvenile patients with short duration of diabetes leads to impairment of conduction velocity within a few days.[11]

Hyperglycemia is associated with increased activity of the polyol pathway resulting in augmented intraneural levels of sorbitol and fructose in diabetic man and in the animal models.[12,13] It has recently been demonstrated that treatment for 9 weeks with an inhibitor of the aldose reductase, the rate-limiting first enzyme of the polyol pathway improves the conduction velocity slightly in stable diabetics in whom glycemic control was constant during the period of study.[14]

Although metabolic factors exert an acute effect on peripheral nerve tissues and despite the early observation of acute symptoms of neuropathy in untreated and unrecognized diabetic patients, very little is known about the role of these acute peripheral nerve abnormalities to the peripheral neuropathy that accompany long-term diabetes. It is likely, however, that randomized clinical trials of the effect of near normalization of the metabolic state obtained by insulin pump therapy will provide new insight within the next few years to the interdependence of metabolic disturbances and long-term nervous system dysfunction.

Peripheral nerve tissues of diabetic patients can be obtained at autopsy or as biopsies at the ankle level of the sural nerve that innervates the skin of the lateral aspect of the foot.

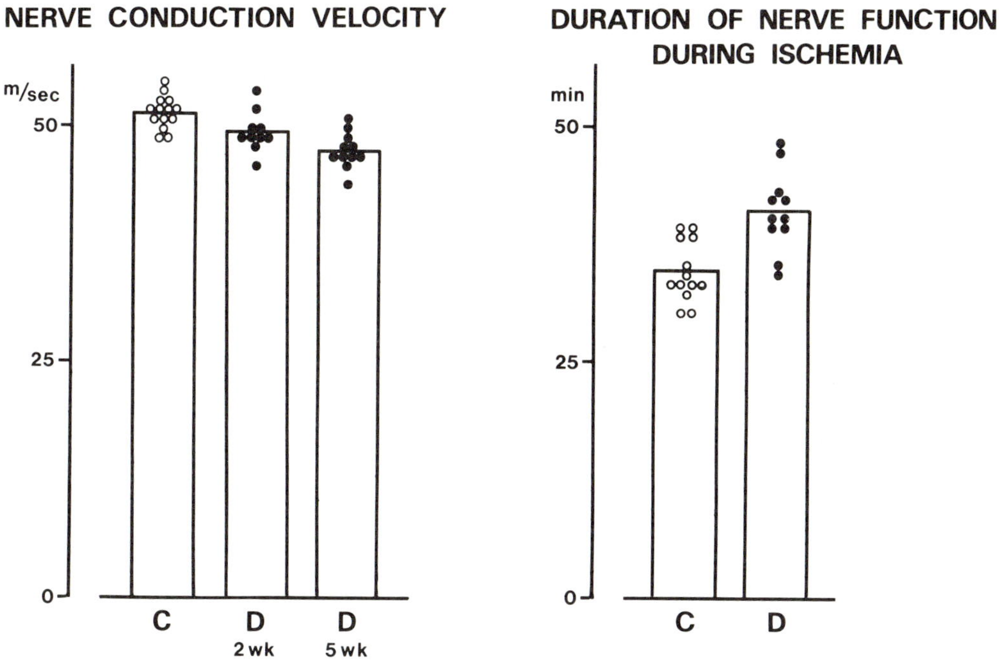

FIGURE 1. Conduction velocity and time until cessation of impulse propagation in tail nerve of controls and of streptozotocin diabetic rats with 2 and 5 weeks duration of diabetes.[18,20] All rats were of the same age at the time studied. It appears that conduction time as well as preservation time during ischemia are both prolonged.

Many unsolved biological problems of diabetic neuropathy, therefore, can at present only be approached in animal models. This is true in particular to the question of the role of axonal transport. There is one study, however, on the axonal transport of acetylcholine esterase and dopamine-beta-hydroxylase in patients with diabetic polyneuropathy.[15] The content of these two transmitter enzymes in sural nerve biopsies was reduced and so was also the calculated transport velocity.

A. Animal Models

Even though the need of an animal model for the study of diabetic neuropathy is obvious no such means exist. There are at present no convincing models that include the main features of the common type of peripheral neuropathy accompanying long-term diabetes. The characteristic myelinated fiber loss is not imitated by any of the models currently available. The most commonly applied model of type I diabetes, namely the streptozotocin rat, shares several metabolic characteristics with the human state. Hyperglycemia and ketonemia develop, hormonal patterns are very similar in the two states and angiopathy develops as well.[16] One obvious possibility to explain the discrepancy between man and model is that the duration of the state cannot be mimicked in the rat.

Early diabetes in man is associated with a slight reduction in conduction velocity and a prolonged preservation of peripheral nerve function during ischema produced by inflation of a pneumatic cuff placed proximally around an extremity.[16,17] Both neurophysiologic abnormalities are present in the streptozotocin rat after a few weeks duration of diabetes (Figure 1).[18] In short-term experiments nerve conduction drops very slightly within the very first days of diabetes and subsequently remains stable whereas conduction time improves in controls during the following weeks.[19,20] Also the increased resistance to ischemia is present within a few days following induction of diabetes. Both neurophysiologic phenomenons in the model can be normalized by insulin treatment[89] as is the case in early diabetes in man.[11,21]

A new observation that further stresses the similarities between man and the model is that treatment with aldose reductase inhibition improves the conduction time in both states.[14,22] Therefore, we have postulated that the streptozotocin rat with a few weeks duration of diabetes is a model of early peripheral nervous system abnormalities in juvenile insulin deficient diabetic man.[18]

The structure of peripheral nerve in this or in other models of early diabetes has been studied extensively within the last few years (Table 1). Nevertheless there is still some disagreement to the interpretation of the observations. One of us reported in 1976 that nerve fibers of streptozotocin rats with 4 weeks duration of diabetes had reduced myelinated nerve fiber size as compared to controls.[23] Later studies showed that control animals increased their fiber caliber during the experimental period whereas diabetic rats kept the caliber constant or even showed absolute reduction in size as compared to start values.[20,24] Similar findings have been reported later by others in the same model, in the spontaneous diabetic BB Wistar rats and in the obese hyperglycemic mouse that share hormonal patterns with type II diabetes in man.[25-27,40]

The finding of reduced size was originally described as axonal dwindling since longitudinal growth was preserved.[28] Other authors found that the best description was that of a maturational defect.[25] They emphasized that the nerve fiber failed to increase its caliber and observed a negligible retardation in skeletal growth.

We find it important to state that the reduction in fiber caliber is not due to a general maturational defect including retarded skeletal growth. Neither is it likely to be caused by malnutrition. The change in axon caliber seems rather specific and the term maturational defect to us seems misleading.

The dwindling of the axon is associated with reduced size of the perykaryon of dorsal root ganglion cells and of lower motor neurons (Table 1).[29]

Also quantitative studies of ultrastrucutural photomontages of peripheral nerve have demonstrated that besides diminished caliber of myelinated fiber experimental diabetes is associated with reduced amount of Schwann cell cytoplasm and endoneurial edema (Table 1).[30]

B. Fast Anterograde Axonal Transport

1. Studies of Enzyme Activity

In the first publication on axonal transport in diabetes Schmidt et al.[31] reported a decreased accumulation of transmitter enzymes proximal to a ligature placed for 12 hr on the sciatic nerve of streptozotocin rats who were kept hyperglycemic for 3 to 4 weeks before study. Accumulation of acetylcholinesterase (AChE) and choline acetyl transferase (CAT) activities were reduced by 22 and 41%, respectively, whereas the enzyme content remained normal in nonaccumulating nerve segments. Insulin treatment maintained for 2 weeks prior to transport study normalized the accumulation indicating that the defect is caused by the deranged metabolic state and not by a toxic effect of streptozotocin itself. AChE is a particulate enzyme and its transport rate is close to that obtained for the fast transport of labeled proteins.[32] The finding of reduced AChE accumulation, therefore, points to either a reduction in fast anterograde transport velocity, in transported amount or in fraction of enzyme in motion.

There were two later reports which failed to confirm the observation of decreased AChE accumulation in proximal collection segments of the sciatic nerve in streptozotocin diabetic rat. Jakobsen et al.[33] studied the accumulation of AChE and dopamine-beta-hydroxylase (DBH) following 6 or 19 hr periods of transport in rats kept diabetic for 1, 4, and 20 weeks. Neither of the two enzymes studied at the various intervals of diabetes duration showed abnormal activities. In the report by Mayer and Tomlinson[34] the AChE activity was slighlty and insignificantly reduced following a 24 hr collection period and 3 weeks duration of diabetes. The reduction which amounted to 7% was not explained by changes in the collection

Table 1
QUANTITATIVE DATA FROM ELECTRON MICROSCOPIC PHOTOMONTAGES[a]

Nerve Cell Body Size

		Nucleolus of LMN (μm^3)	Nucleus of LMN (μm^3)	Perikaryon of LMN (μm^3)	Perikaryon of DRG (μm^3)
Control	(n,12)	103 ± 38	2548 ± 580	33280 ± 6930	22680 ± 3460
Diabetes	(n,11)	87 ± 32	2227 ± 289	27450 ± 5090	18520 ± 3230
p-Values		NS	NS	0.05	0.02

Myelinated Nerve Fibers

		Number of fibers	Mean axon size (μm^2)	Mean myelin sheath size (μm^2)	Schwann cell cytoplasm per fiber (μm^2)
Control	(n,10)	1896 ± 238	19 ± 2	28 ± 3	5.5 ± 1.1
Diabetes	(n,10)	1798 ± 204	16 ± 2	25 ± 3	3.9 ± 0.6
p-Values		NS	<0.005	NS	<0.002

Organelles of Myelinated Axons

		Number of SER profiles per 100 μm^2 axon	SER area fraction ($\times 10^3$)	Number of filaments per 0.1 μm^2 axon	Mitochondrial area fraction ($\times 10^2$)
Control	(n,10)	10.4 ± 3.0	5.6 ± 1.8	14 ± 1	1.7 ± 0.2
Diabetes	(n,10)	12.4 ± 4.8	7.7 ± 3.9	15 ± 1	1.6 ± 0.2
p-Values		NS	NS	NS	NS

Unmyelinated Axons

		Number	Mean area (μm^2)	Number of SER profiles	SER area fraction ($\times 10^2$)
Control	(n,8)	3160 ± 734	0.47 ± 0.07	1.5 ± 0.4	5.7 ± 1.7
Diabetes	(n,10)	2864 ± 1139	0.45 ± 0.07	1.4 ± 0.4	3.4 ± 1.3
p-Values		NS	NS	NS	<0.01

Endoneurial Space

		Collagen ($\mu m^2 \times 10^3$)	Structureless endoneurium ($\mu m^2 \times 10^3$)	Vessels ($\mu m^2 \times 10^3$)	Cells (μm^2)
Control	(n,10)	17 ± 2	7 ± 4	3.3 ± 2.9	0.7 ± 0.5
Diabetes	(n,10)	20 ± 6	23 ± 8	2.2 ± 2.4	1.0 ± 0.4
p-Values		NS	<0.0001	NS	NS

Note: LMN = lower motor neurons, DRG = dorsal root ganglion cells, SER = smooth endoplasmic reticulum. Values are means and SD's.

[a] These photomontages are of cross sections of the common peroneal nerve and from light microscopic studies of dorsal root ganglion cells of the fifth lumbar root as well as of motor neurons of the fifth lumbar segment in rats with 4 weeks of streptozotocin diabetes and in age matched controls.

segments but was ascribed to minor changes in neighboring nerve segments in unconstricted contralateral nerve included for the calculation of relative accumulation.

The transport velocity of norepinephrine (NE) is probably identical to the maximal anterograde transport rate of labeled protein.[35] There are two studies on accumulation of NE in collection segments proximal to a ligature on the sciatic nerve. In the first study Giachetti[36] demonstrated reduced accumulation in obese spontaneously diabetic mice whereas Tomlinson et al.[37] found normal accumulation of NE in streptozotocin rats with a diabetes duration of even up to 6 months.

Thus, the accumulation studies of the transmitter enzymes AChE, DBH, and NE have not provided consistent evidence for changes in amount or velocity of the fast anterograde transport.

2. Studies of Labeled Proteins

The first study of fast anterograde transport profiles using radiolabeling of unidentified proteins was carried out in streptozotocin rats with 4 weeks duration of diabetes.[38] We calculated the transport velocity as the difference between a 2 and a 4 hr profile of activity in the sciatic nerves of the same animal following injection of tritiated leucine and ^{14}C-glucosamine into the fifth lumbar dorsal root ganglion. The transport velocity of the leucine labeled anterograde fast component (aFC) was identical in diabetics and controls, namely 15.8 ± 1.2 mm/hr (Figure 2). Neither were there any differences in the transport velocity of glucosamine label.

Bisby[39] studied the transport of leucine label in the same model and in the same system of sensory fibers as above. Both sciatic nerves were removed after 1.5 to 2.75 hr and then transferred to an incubation bath for further 1 hr of transport. The advantage of this design is that during the period decisive for the calculated transport velocity (the distance between transport profiles divided by the time elapsed) nerve temperature can be defined precisely. The in vitro as well as the in vivo transport rate in test nerves remained within normal limits.

Mendell et al.[40] in experimental diabetes found reduced aFC velocity in the same fiber system as used in our study mentioned above and in that of Bisby. They studied transport in the spontaneously diabetic BB Wistar rat and the diabetes duration presumably was somewhat longer than in the two previous studies. Down flow rates (DFR) for individual animals were measured from one sciatic nerve only. In controls DFR was 17.3 ± 0.9 mm/hr vs. 14.5 ± 1.0 mm/hr in nonketotic, hyperglycemic rats, 14.5 ± 1.6 mm/hr in ketotic and hyperglycemic rats and 16.3 ± 0.9 mm/hr in insulin treated diabetics.

A decrease of 21% in aFC velocity has been reported by Meiri and McLean[41] in the motor system of the streptozotocin rat with a diabetes duration of 2 weeks. Tritiated leucine was injected to the motor neurons of the spinal cord and transport velocity and time from injection until start of transport (lag-time) were obtained from shifted transport profiles of sciatic nerves. Lag-time for transport of leucine label was normal in diabetic rats, a result in accordance with the observations of Bisby[39] and those of ourselves.[38] We found, however, that lag-time for transport of glucosamine was significantly prolonged from 37 ± 17 min in controls to 49 ± 7 min in diabetic rats.[38]

Also the fast axonal transport velocity has been studied in sensory, mainly unmyelinated vagal fibers of the alloxan diabetic rabbit by Bajada et al.[42] After leucine injection into the nodose ganglion velocity was 17.1 ± 1.1 mm/hr in controls as compared to 18.1 ± 2.0 mm/hr in diabetic animals. Besides normal transport velocity in vagal fibers, in diabetes it also appears that transport velocities seem very similar in myelinated and unmyelinated nerve fibers.

The amount of label accumulating proximal to a ligature has been measured by Meiri and McLean[41] and by us[38] in sensory fibers at collection intervals ranging from 2 to 24 hr (Figure 3). No reduction in amount accumulated has been observed at any of the intervals.

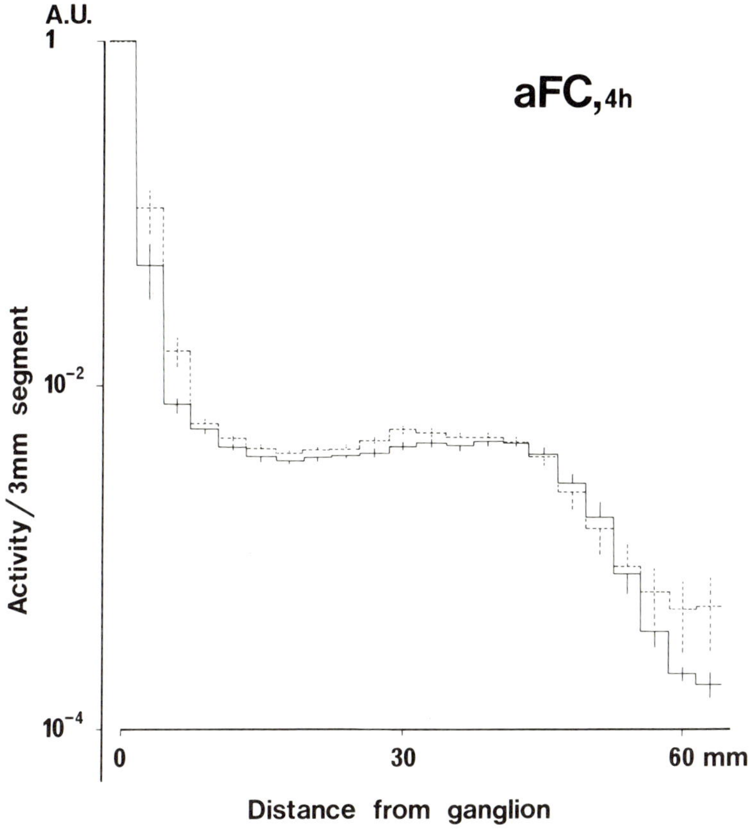

FIGURE 2. The distribution of labeled protein in the sciatic nerve 4 hr after injection of tritiated leucine into the fifth lumbar dorsal root ganglion.[38] The activity per 3 mm segment is relative to the activity in the ganglion. Values are mean ± SEM. The full drawn curve represents controls and the dashed curve age-matched, streptozotocin rats with 4 weeks of diabetes. The transport profile in diabetes is identical to that of controls.

The aFC velocity is extremely dependent on temperature. The Q_{10} is close to 2.3[32] indicating that changes of a few centigrades would be able to explain the reduction found in the two studies by Mendell et al.[40] and Meiri and McLean.[41] Streptozotocin rats with a few weeks duration of diabetes and a considerable degree of hyperglycemia lose body weight in the range of 20 to 40% as compared to controls. Also during barbiturate anesthesia diabetic rats lose body temperature faster. We have observed temperature differences of 3 to 4°C between diabetics and controls within half an hour after start of anesthesia at a room temperature of 20°C.[89]

Measures have been taken in all studies mentioned above to prevent this fallacy and rectal temperature has been measured in most of the studies. It should be considered, however, whether control of body temperature is a sufficient procedure in studies of peripheral nerve in experimental diabetes. Temperature in the extremities can differ significantly from that in the truncus as part of heat regulation. Nerve temperature, therefore, could vary considerably without suspicious changes in rectal temperature.

The minor discrepancies in transport rate observed in the various studies of experimental diabetes might of course be due to differences in the experimental design especially to degree or duration of diabetes or to the application of different models or fiber systems. More important, however, to our opinion could be the lack of sufficient temperature control.

Taken in concert most studies have shown unchanged velocity and amount of the fast

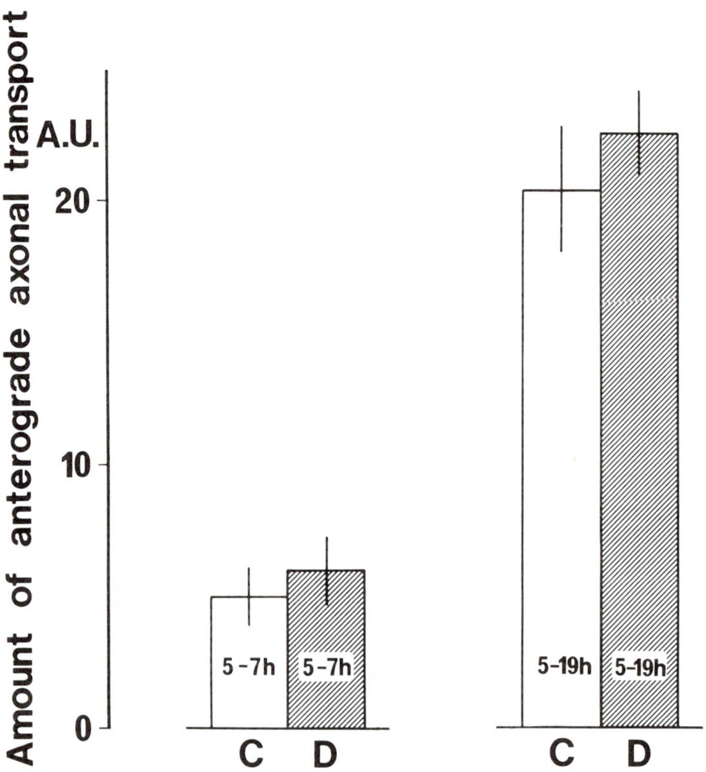

FIGURE 3. The amount of the anterograde fast component of axonal transport measured as the accumulation of labeled protein proximal to a double ligature placed on the sciatic nerve for 2 hr intervals at a specified time after injection of ^{35}S-methionine into the fifth lumbar dorsal root ganglion.[55] The values of the 5 to 7 hr interval are shown to the left. The accumulation is given in arbitrary units as mean ± SD. To the right columns indicate the summation of 2 hr intervals from 5 to 19 hr. It appears that fast anterograde accumulation is normal in diabetes.

component of the anterograde transport system in early experimental diabetes. The difficulty of sufficient temperature control implies that observations of minor slowing or accumulation should be evaluated cautiously.

C. Slow Anterograde Axonal Transport

1. Studies of Enzyme Activity

Choline acetyltransferase (CAT) is considered to be mainly soluble in the cytoplasm.[43] It accumulates very slowly at a ligature and the calculated transport velocity is close to that of labeled proteins of the slow anterograde transport.[44]

As early as in 1975, Schmidt et al.[31] reported a considerably decrease in proximal accumulation of CAT in streptozotocin rats. CAT content in nonaccumulating nerve segments was normal or slightly elevated indicating that the decrease in accumulation was due to decreased transport rather than to loss of neurons.

Mayer and Tomlinson[34] studied CAT accumulation proximal to a ligature placed for 24 hr on sciatic and vagus nerve after 1 and 3 weeks of diabetes in streptozotocin rats. CAT accumulation in the sciatic nerve was normal after 1 week but was one half following 3 weeks of diabetes. The decrease in accumulation as well as fall in motor conduction velocity was prevented in a group of diabetic rats treated with insulin during the 3 week period of study. No abnormality in proximal accumulation took place in the vagus nerve.

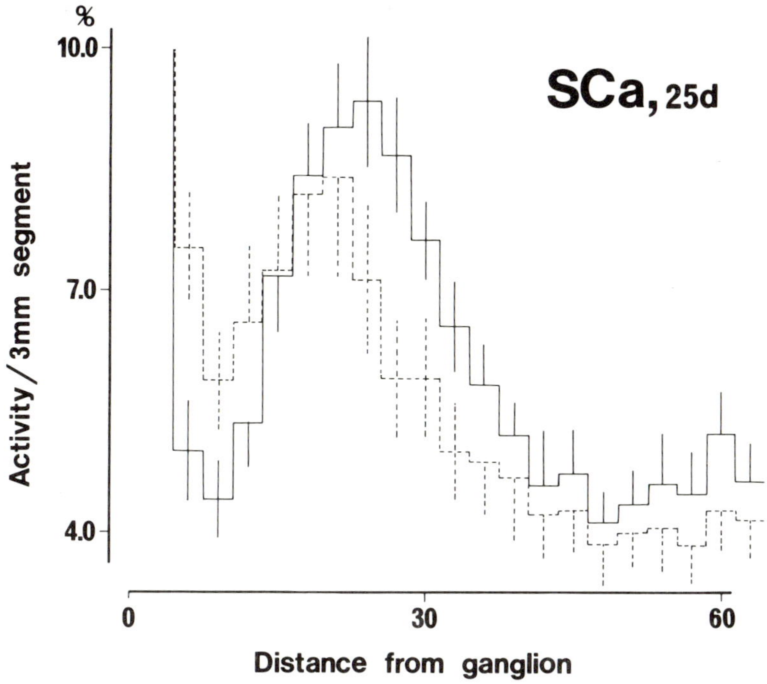

FIGURE 4. The distribution of labeled protein in the sciatic nerve 25 days after injection of tritiated proline into the fifth lumbar dorsal root ganglion.[47] The activity per 3 mm segment is relative to the activity in the ganglion. Values are mean ± SEM. The full drawn curve is a control group and the dashed curve an age-matched, diabetic group injected with streptozotocin 3 days before injection of precursor. The diabetic curve is more right-skewed and lower than that of the controls. Mean velocity and amount of activity for individual animals are calculated from the bell shaped wave of activity (Figure 5).

The peripheral nerve abnormalities in diabetes have been related to increased sorbitol levels in nerve and to lack of myoinositol. Daily administration of an aldose reductase inhibitor which impairs the conversion of glucose to sorbitol prevents the deficit in accumulation of CAT enzymes proximal to a ligature in streptozotocin diabetic rats.[45,46] Daily administration of oral myoinositol has the same effect on CAT accumulation in diabetes.[46] How myoinositol and sorbitol influence axonal transport is unknown.

In their study of transport velocity of labeled material in vagus sensory fibers of alloxan diabetic rabbits Bajada et al.[42] observed no changes in position of the mid-front of the most advanced part of the curve of radioactivity. Since the transport rate in vagus fibers is 25 to 35 mm/day the nature of the transport system is, however, uncertain.

2. Studies of Labeled Proteins

Bisby[39] in his study of axonal transport in streptozotocin diabetes in rat observed velocities of slow component a (SCa) that were slower in diabetic rats (1.10 mm/hr) than in controls (1.18 mm/hr). The difference, however, was not significant. Jakobsen and Sidenius[47] studied the slow axonal transport in sciatic nerve following injection of tritiated proline into the fifth dorsal root ganglion in the same animal model as used in our previous studies on structure, conduction velocity, and fast axonal transport. After 25 days transport velocity of SCa was significantly decreased from 1.03 ± 0.08 mm/day (mean ± SD) in controls to 0.88 ± 0.12 mm in rats who had been kept diabetic for the whole period of the study (Figure 4). After 50 days transport velocity was 0.76 ± 0.05 mm in controls versus 0.64 ± 0.05 mm in diabetics, a difference that was significant also.

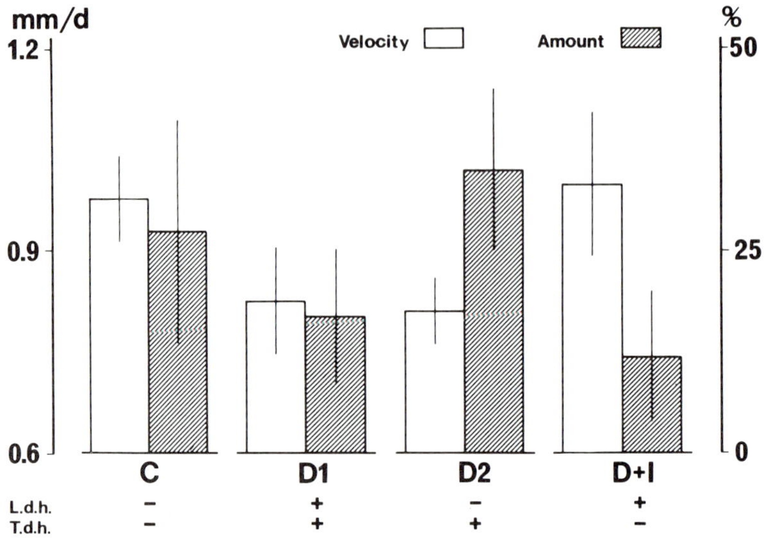

FIGURE 5. Transport parameters of SCa calculated for four experimental groups from profiles similar to those shown in Figure 4.[48] Transport velocity is indicated by the left column and transported amount by the hatched column to the right. Values are mean ± SD. C is a control group, age-matched to the following groups. D1 is a diabetic group rendered diabetic 2 days *before* the injection of the labeled amino acid in the fifth lumbar dorsal root ganglion. D2 is a group rendered diabetic 2 days *after* the injection of the labeled amino acid. D + I is a group rendered diabetic 2 days before the injection of labeled amino acid. After further 1 days insulin treatment was started and blood glucose kept close to normal for the rest of the transport period. The result for D1 shows that hyperglycemia during the whole experimental period leads to a decreased velocity as well as a decreased amount of SCa. In D2 normoglycemia during labeling of SCa is associated with a normal amount of label in transport, while the subsequent diabetic state decreases the transport velocity. D + I gives the opposite result, i.e., a decreased amount of label incorporated into SCa during hyperglycemia and normal transport velocity during the subsequent period of treatment. Abbreviations: L.d.h. = labeling during hyperglycemia; T.d.h. = transport during hyperglycemia.

The amount of material transported was calculated relatively to ganglion activity and no significant difference was found although diabetic rats transported 27 and 24% less than controls. Neither were there any significant changes in turnover of labeled ganglion protein as calculated from remaining activity 3 hr and 10, 25, and 30 days after labeling. Thus, no compensation took place in diabetic rats for the reduced transport rate.

In later studies[48] it was shown that the slowing of transport velocity of SCa also occurred in rats normoglycemic at the time of labeling and in whom diabetes was induced a few days later (Figure 5). This demonstrates that the influence on the transport is an effect on the transport mechanism itself. Studies of insulin treatment in the same report showed that the slowing in transport velocity could not only be prevented but also reversed. The amount of label carried in SCa was normal in rats who were normoglycemic when proteins were labeled whereas rats who were hyperglycemic at that time transported a decreased amount of label. This latter observation also applies to an insulin treated group with normal transport rate. It, therefore, seems that the metabolic derangement in diabetes besides its effect on the slow transport mechanism influences protein synthesis in the neuron. The delivery of proteins into SCa is decreased and the increased skewness of the curve of radioactivity points to the possibility that a fraction of proteins is affected predominantly.

The polypeptide pattern of labeled proteins in sciatic nerve of mutant diabetic mice has been studied by Vitadello et al.[49] by polyacrylamide slab gel electrophoresis. They found normal advancement of the bands corresponding to the molecular weight of tubulin and actin whereas the proteins of the neurofilament triplet were less advanced. This suggests that the neurofilaments are transported more slowly in diabetes.

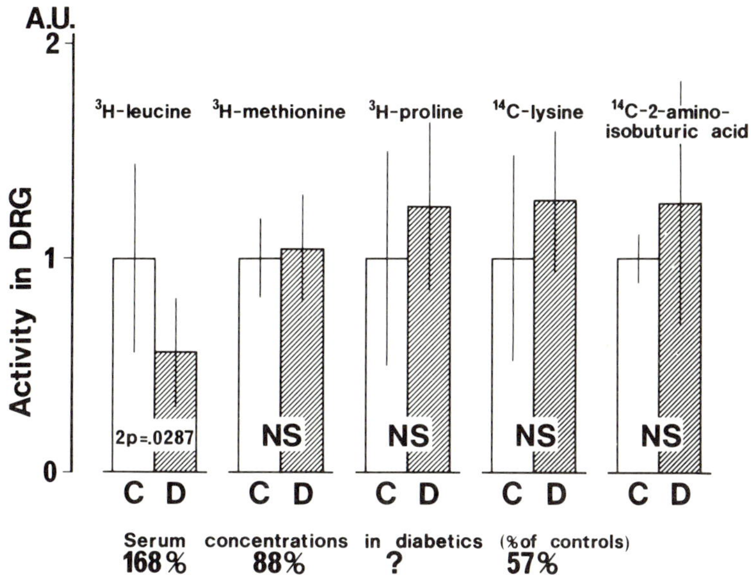

FIGURE 6. Amount of incorporated label of leucine, methionine, proline, and lysine in the fifth lumbar dorsal root ganglion 3 hr after ganglion injection in streptozotocin rats with 4 weeks duration of diabetes and in controls. Also the amount of 2-aminoisobutyric acid taken up in the ganglia is shown. It appears that the incorporated amount of label is inversely related to the serum concentrations of the respective amino acid.[50] 2-aminoisobutyric acid is not present in serum and the result obtained indicates normal uptake of amino acids into the cells of the diabetic ganglia. Values are mean ± SD.

On the background of studies of incorporation of label in neurons in vivo we suggested from an earlier report that protein synthesis in streptozotocin diabetes was impaired.[38] In the early study on fast anterograde axonal transport a significant decrease of leucine label in the dorsal root ganglion was found 2 and 4 hr following application. Later observations,[89] however, support the assumption that the decrease in amount of incorporated leucine is due to decreased specific activity in the ganglion. Thus, labeling with tritiated methionine, tritiated proline, ^{14}C-labeled lysine and ^{14}C-labeled 2-aminoisobutyric acid is associated with an unchanged amount of label in diabetes 2 and 4 hr following application (Figure 6). Furthermore, studies on amino acid concentrations in streptozotocin rats have shown considerable increase in plasma concentrations of leucine after 4 days of diabetes.[50] Since the amount of incorporated label in the ganglion varies with the type of amino acid and since the uptake of 2-aminoisobutyric acid is unchanged, we conclude that the absolute amount of leucine label is an invalid index of protein synthesis and amount of transport in experimental diabetes.

D. Retrograde Axonal Transport

Retrograde axonal transport in experimental diabetes has been studied by means of ligation techniques combined with the use of labeled proteins and endogenous enzymes. These techniques estimate the flux of material transported but allow no distinction between transport velocity and density of transported material. Sidenius and Jakobsen[38] studied accumulation of label following dorsal root ganglion injection of tritiated leucine and ^{14}C-labeled glucosamine in rats with 4 weeks of streptozotocin diabetes. The accumulation of glucosamine label was increased proximal to a ligature following a 24 hr collection period whereas a decreased accumulation was found distal to a ligature. In contrast no differences in accu-

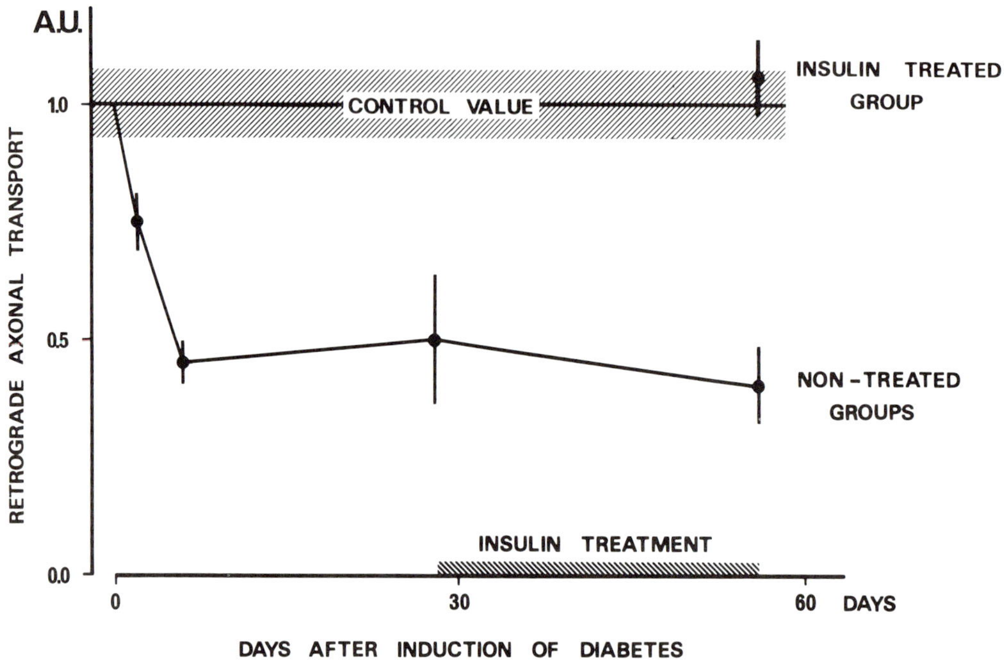

FIGURE 7. Amount of fucose label in the segment distal to the double ligature on the sciatic nerve representing the collection interval 19 to 21 hr following injection into the fifth lumbar dorsal root ganglion of streptozotocin rats with 2, 7, 28, and 60 day of diabetes as well as in insulin treated diabetic rats and in controls (hatched area).[33,52] The figure shows that the retrograde accumulation is reduced early in diabetes and remains impaired throughout the period of study. Values are mean ± SEM.

mulation of leucine label at proximal or distal sites occurred. It was suggested that an abnormality of the retrograde fast component (rFC) of axonal transport was the most likely explanation of this observation. Accumulations obtained by the use of long collection intervals as those used in our study of glucosamine label reflect the combined effect of anterograde and retrograde transport.

The time course of accumulation distal to a ligature following pulse labeling of the neurons with radioactive amino acids and the optimal collection intervals have been mapped for rat sensory and motor fibers of the sciatic nerve.[51] To test the suggestion that rFC was influenced in diabetes distal accumulation was studied during short collection periods of 2 hr at optimal intervals following application of label.[52] We found a normal proximal accumulation of fucose label but distal accumulation was reduced. The short collection intervals allow better separation of anterograde and retrograde fluxes and a decrease in the retrograde flux of glycoprotein label was assumed to be associated with the diabetic state.

Later studies[53] have shown that the decrease in distal accumulation of fucose label during short collection intervals can be demonstrated as early as 24 to 48 hr after the establishment of a hyperglycemic state and that distal accumulation of methionine label remains normal.

To further elucidate the abnormality in retrograde transport of labeled glycoproteins, transmitter enzymes and nerve growth factor (NGF) were compared in the same study.[33] Again distal accumulation of fucose label was decreased at intervals of 1, 4, and 8 weeks after induction of diabetes and could be reversed by insulin treatment (Figure 7). Distal accumulation of DBH and AChE, however, remained normal even in rats with 4 months duration of diabetes. This finding indicates normal capacity of the transport system in diabetes and, therefore, the explanation of the decrease in retrograde flux is likely to be an altered metabolism of glycoproteins rather than an impaired transport mechanism.

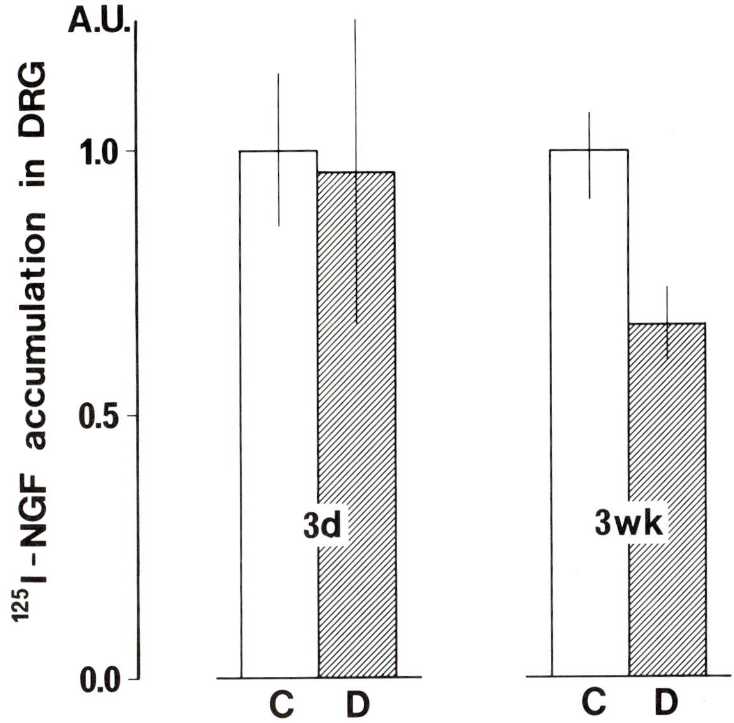

FIGURE 8. Amount of ^{125}I-NGF (nerve growth factor) accumulated in the fifth lumbar dorsal root ganglion 24 hr following subcutaneous application into a footpad of streptozotocin rats with 3 day and 3 week duration of diabetes and in controls.[33] Accumulation is normal at the early interval but decreased after 3 weeks of diabetes. Values are mean and ± SD.

The study of retrograde transport of labeled NGF showed, however, that additional factors are likely to be involved. NGF injected subcutaneously in the footpad of the hind legs of rats accumulated to a lesser degree distal to a ligature in the mid-thigh portion of the sciatic nerve as well as in the dorsal root ganglion of unligated nerves (Figure 8).

A marked decrease of radio labeled NGF has also been demonstrated in the superior mesenteric ganglion of the sympathetic nervous system of rats with a diabetes duration ranging from 2 days to 10 months,[54] whereas the superior cervical ganglion accumulated NGF normally. These abnormalities are consistent with a change in uptake of macromolecules at the nerve terminal level or of a change in retrograde transport.

That an altered metabolism of glycoproteins is not the sole abnormality of rFC in experimental diabetes has been demonstrated in studies of nerve crushed 1 day prior to or at the time of protein labeling.[55] In this situation a delay and a decrease in distal accumulation of protein label as well as of glycoprotein label took place in collection segments placed proximal to the crush, which functionally acts as new nerve terminals.

Taken in concert experimental diabetes is associated with abnormalities in retrograde build up of macromolecules, an observation that might be explained partly by changes in glycoprotein metabolism. Studies of NGF transport as well as of the effect of a nerve crush on transport point to an additional defect of the nerve terminal level or to a decrease in the amount of material transported retrogradely.

We have advanced a hypothesis to connect the various abnormalitities in structure, nerve conduction, and axonal transport in early experimental diabetes.[53] The earliest abnormality detected is that of the decreased build up of retrogradely transported glycoproteins which possibly reflects a defect of the nerve terminals.

The change in delivery of retrogradely transported material to the nerve cell body is the signal that resets the protein synthesis leading to impaired production of protein in SCa (neurofilaments) and to slowing of the slow anterograde transport velocity. The consequence is dwindling of the axons and slowing of nerve conduction.

II. AXONAL TRANSPORT IN OTHER METABOLIC CONDITIONS AND TOXIC NEUROPATHIES

The classification of peripheral neuropathies in man is based on clinical, histologic, and neurophysiologic criteria. The features decisive for diagnosis of these conditions still are the symptoms of the patient and the distribution of neurological signs of deranged nerve function. The comprehension of peripheral nerve disease is rather limited and mainly based on the erratic histologic decision whether degeneration of the neuron itself or the Schwann cell predominates. In general, most of current knowledge has a purely descriptive character, biochemical insight is the exception and pathophysiologic considerations are rather speculative. However, the study of axonal transport has improved this epistemologic poverty. However, e.g. for the demonstration of the slowing of the anterograde axonal transport of neurofilaments and for the retrograde alterations of protein transport associated with early diabetes. Hoping to further this process we have started a systematic study of axonal transport abnormalities in the various models of neurotoxic and metabolic disorders (Figure 9).

A. Other Metabolic Neuropathies

1. The Galactose Fed Rat

Another and commonly used model of the metabolic nerve changes in diabetes is the galactose fed rat. Galactose, an epimere of glucose, accumulates in peripheral nerves of the rat if fed in large amounts. Galactose is treated analogous with glucose by the polyol pathway and as in diabetes sugar alcohols accumulate in nerve tissues. The content of galactiol is increased whereas the amount of *myo*-inositol is decreased.[56,57] A metabolic model simulating the effects of high glucose undisturbed by the hormonal and vascular changes characteristic of diabetes, therefore, is at hand.

The early structural and functional changes are similar to those of experimental diabetes. Axonal dwindling accompanied by an endoneurial edema and a corresponding decrease of the nerve conduction velocity have been demonstrated in several studies.[58-60] To elucidate the role of axonal transport in this model we investigated the axonal transport of isotope labeled endogenous proteins during the first 4 weeks of galactose feeding.[61] The age of the animals was 24 weeks as in our studies on diabetes. We found that the velocity of SCa, measured with a transport period of 35 days starting 4 days after the start of galactose feeding, was decreased like in diabetes. The retrograde FC was investigated in previous crushed nerves as well as in intact nerves by the double ligature technique.[51] For the crushed nerves changes were as in streptozotocin diabetes: a decreased amount of retrogradely transported fucose-labeled glycoproteins as well as of methionine-labeled proteins, a decrease probably reflecting a delayed reversal of FC at the crush. In the intact nerves no changes were found after 5 days of galactose feeding. However, at 28 days the magnitude of retrograde transport had declined by 35% for both methionine- and fucose-labeled material. For aFC no differences between galactose fed rats and controls were found. Thus, streptozotocin diabetic and galactose fed rats share in common a decreased nerve conduction velocity by a magnitude which can be explained by the accompanying axonal caliber reduction, a decreased transport velocity of SCa and decreases in rFC, despite normal aFC.

However, there are also differences between the galactose model and diabetes. The galactiol content of the nerves is approximately three times as high as the sorbitol content of the diabetic nerves and correspondingly the amount of endoneurial edema is much larger.[56,60]

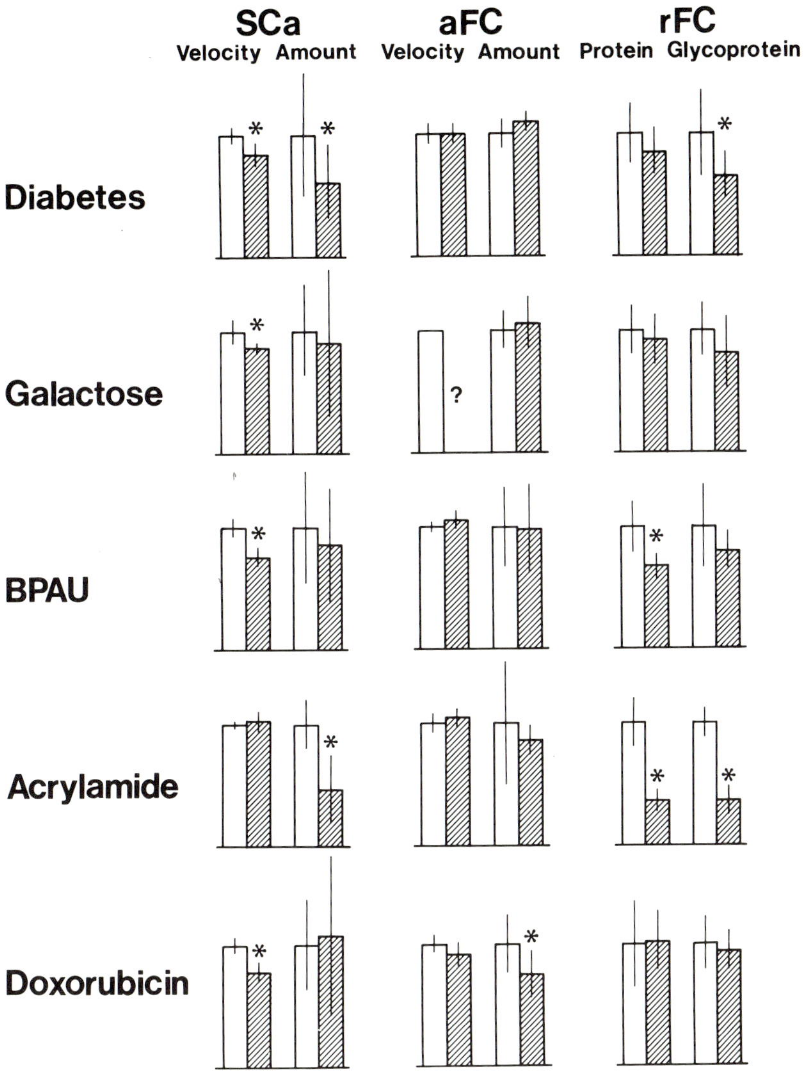

FIGURE 9. The figure shows selected axonal transport parameters for streptozotocin diabetes[38,47,48,52] and galactose fed rats[61] as well as for BPAU,[70] acrylamide,[72,77] and doxorubicin intoxication. The values are mean ± SD and in each instance the columns of test group and the respective controls are opposed. The values are normalized relative to the mean of the controls. Asterisks indicate statistically significant differences. While galactose gives rise to an axonal transport spectrum similar to diabetes none of the other experimental conditions share in common a pattern of transport abnormalities.

The endoneurial fluid pressure has been measured in both models and while no changes are seen in diabetes galactose feeding gives rise to an increased pressure after several months.[58,62] Furthermore, severe changes of Schwann cells and demyelination develops after 2 years galactose feeding, a feature which has not been documented in experimental diabetes.[63]

In contrast, the paradoxically increased resistance to ischemia seen in diabetes is not found in the galactose fed rats (unpublished observation). This fact seems to rule out the possibility that the endoneurial edema is a necessary precondition for this phenomenon.

The studies on the galactose fed rat and the demonstrated similarity with the early changes of diabetes suggest that current studies on the biochemical consequences of hyperglycemia, i.e., the increase of the polyol pathway and the impairment of *myo*-inositol metabolism are among the most promising lines of study at the present.

2. Hypoglycemia

Greene and Winegrad[64,65] studied the energy metabolism of endoneurial preparations of peripheral nerves and demonstrated that the peripheral nerve is almost exclusively dependent on glucose utilization. However, nervous tissue seems to adapt itself to the current level of glucose, so it might be inferred that short-term fluctuations in glucose level are especially harmful for the nerve.[66] In badly controlled diabetes, thus, the prevalent high blood glucose could make the nerve susceptible to sudden drops in glucose level, e.g., the nocturnal hypoglycemia which is rather common in long-term diabetic patients. We investigated the possibility of inducing a peripheral neuropathy by insulin treatment in rats. In a group of nondiabetic rats treated with high doses of insulin to induce severe prolonged hypoglycemia ($<$2 mmol/1), it was found that axonal degeneration depended upon the severity of hypoglycemia.[67] To investigate this further aFC was studied during profound hypoglycemia. Acute hypoglycemia causes a significant fall in the amount of fast-transported protein though the transport velocity is unchanged. If the rats were preconditioned by 4 days of insulin treatment immediately before the axonal transport experiment the same level of hypoglycemia was not associated by any changes of the measured parameters of aFC.[89] Whether the decrease in aFC is responsible for the degeneration seen in our structural study remains to be clarified.[67] Nevertheless, our finding corroborate the idea that complicating factors in the course and treatment of diabetes may play a role in the pathogenesis of diabetic neuropathy.

B. Toxic Neuropathies

Recent studies on axonal transport after administration of neurotoxic substances have changed the current view on the role of transport abnormalities in the development of neuropathies.[68] As representative of substances leading to distal axonopathies hexacarbons, zinc pyridinethione (ZPT), p-bromophenylacetylurea (BPAU), and acrylamide have been intensively investigated by axonal transport techniques. β,β′-iminodipropionitrile (IDPN) is held to be the model of proximal axonopathies and doxorubicin (Adriamycin®) the model of sensory neuronopathies.

1. Distal Axonopathies

A distal axonopathy is classified as such when pathology is selectively distal and progresses towards the nerve cell body. Preterminal focal swellings containing tubulo-vesicular material and neurofilaments appear proximal to the nodes of Ranvier preceding the fiber breakdown.[69] The exact nature and natural history of these accumulations has not been agreed upon.

For BPAU and acrylamide an early and substantial decrease of the retrograde axonal transport of proteins has been demonstrated.[70-72] The decrease is seen after subclinical doses before signs of neuropathy develop, and is correlated to the clinical scoring of the neuropathy.[72] A similar decrease has also been demonstrated in clinically affected rats intoxicated by ZPT and 2,5-hexanedione.[71,73] The decrease could be due either to a distal block of anterograde FC, an impaired reversal of FC in the nerve terminals, a decreased capacity for retrograde FC or a decreased velocity of rFC. Evidence from studies on acrylamide favors an impairment of anterograde and retrograde FC preterminally. Chrétien et al.[74] showed by radioautography that fast transported material was retained focally in pilings of tubulovesicular profiles, smooth endoplasmic reticulum, dense core vescicles and mitochondria. In later studies, the axonal transport of the different molecular forms of acetylcholinesterase was studies in acrylamide intoxicated chickens.[75] The content of the fast transported forms was increased while the anterogradely transported amount was decreased. The retrograde transport was unaffected as well as the transport and content of the slowly transported forms of the enzyme. This supports a concept of preterminal inhibition of aFC. However, Miller et al.[76] injected labeled nerve growth factor (NGF) into the front foot pad of rats and measured the subsequent accumulation in the spinal ganglia C4-C7, and found a decreased accumulation after single doses of acrylamide. The decrease was seen prior to detectable alterations in peripheral nerve function. Thus, rFC seems to be affected as well.

As to aFC in the more proximal part of the axon the results are somewhat conflicting. However, recent quantitative studies with proper temperature control have failed to find any slowing of aFC.[77]

As to the slow components enzyme studies have reported unchanged amounts of transported enzyme activity in acrylamide and BPAU intoxication.[70,75] In studies using isotope techniques, however, treatment with BPAU and acrylamide gives rise to disturbances of transport. In BPAU neuropathy the transport velocity of a normal amount of SCa was decreased.[70] The decrease was present only in severely affected animals. In contrast, administration of acrylamide gave the opposite result: a decreased amount of radioactive label transported by an unchanged velocity.[77] The decrease in amount was seen in severely disabled rats only. The transport differences obtained between studies using either enzyme or isotope techniques could indicate that different types of proteins are affected differently as in diabetes. Though the changes in slow axonal transport have been studied in the proximal part of the axon, where degenerative changes presumably are minor, they are nevertheless late phenomenons appearing at a time when the animals have been disabled for several days.

Sterman[78,79] investigated the nerve cell body reaction to 2,5-hexanedione and acrylamide with morphological techniques. He found reorganization of the Nissl substance and nuclear eccentricity at a time when distal degeneration was sparse. While the cell body changes might be reactions to the events taking place in the distal part of the axon the findings stress that the nerve cell should be considered *in toto* in order to understand the development of distal nerve damage.

It was mentioned above that the changes in SCa transport were different in BPAU and acrylamide axonopathy. The same is, in fact, true for rFC. In BPAU the decrease is primarily in the early arriving material indicating a delayed retrograde transport, though a decrease in the total amount takes place also.[70] During the early phase of acrylamide intoxication no delay or decrease of the early arriving material is seen while the total amount of proteins returned during the first 24 hours is reduced considerably.[72] It appears that the distal axonopathies by no means are a homogenous entity when the pathophysiology of different neurotoxic states is considered.

Apart from the differences in the specters of axonal transport abnormalities (Figure 9) distal axonopathies are characterized by accumulation of subcellular material in the preterminal regions, while diabetes is accompanied by fiber dwindling and a corresponding slowing of nerve conduction, phenomenons so far not demonstrated in distal axonopathies. Even though distal axonopathy, as pointed out, is a rather broad category for the description of damage to the nerve cell the abnormalities of axonal transport in early experimental diabetes are not easily encompassed.

2. Proximal Axonopathies

Proximal axonopathies are defined as disorders where the initial pathology is localized to the proximal part of the axon. The IDPN neuropathy is the only well described model of this entity. Structurally, neurofilaments accumulate in large amounts in the proximal axon later followed by atrophy of the rest of the fiber.[80] The SCa transport of neurofilament is halted in the proximal part of the axon[81] whereas tubulin and other slowly transported proteins are less affected by IDPN treatment.[82] The anterograde FC transport of labeled proteins as well as the retrograde axonal transport investigated by distal uptake and subsequent transport of labeled tetanus toxin was found to be normal.[81] Recently it has been reported that small amounts of IDPN lead to distal swellings of the nerve fibers.[83] This finding has been taken as an indication of a more profound similarity than previously suspected between the distal and the proximal axonopathies. Though these kinds of simplification of our concepts are tempting the possibility of pathological mimicry should be borne in mind. A tentative description of the pathophysiology in IDPN neuropathy could be a normal synthesis of proteins in the nerve cell body followed by a segregation of the neurofilaments and relegation to the periphery in the proximal axon leaving the other transport components essentially undisturbed.[84] The consequence could be proximal swelling and distal atrophy of the axon.

The lack of axonal swellings in diabetes and the lack of disturbances of retrograde transport in IDPN intoxication imply that the proximal axonopathies do not include the diabetic nervous abnormalities.

3. Neuronopathies

The category neuronopathy is used to designate conditions with primary damage to the nerve cell body. Doxorubicin which gives rise to a sensory neruonopathy is a convenient model for studying this class of neuropathies. Structurally, the earliest changes appear in the nucleus, followed by loss of Nissl substance and vacuolation of the perikaryon, eventually leading to phagocytosis, formation of nodules of Nageotte and loss of the axon.[85] The mechanism of the degeneration is thought to be an interference of doxorubicin with DNA and a subsequent impairement of the protein synthesis in the nerve cell body.[86] The subsequent degeneration of the axon has been claimed to progress in a proximodistal way, but no quantitative data to support the claim have been offered.[87]

The anterograde FC in this model was investigated by Sahenk and Mendell[88] 2 days after the administration of 10 mg/kg. The transport velocity was found unchanged while the time of perikaryal processing before start of fast axonal transport was prolonged. Using a toxic dose of 6 mg/kg we did not find any prolongation of lag-time 10 days after administration of doxorubicin.[90] However, the amount of fast-transported material was reduced. Approximately 10 days is the time needed before symptoms of the disorder start to develop. The relationship, however, between the decrease of anterograde FC and the axonal degeneration remains to be clarified. The retrograde axonal transport was investigated 2 days after administration of 6 mg/kg of doxorubicin, and no change occurred. Investigation of the SCa transport showed that the velocity was decreased in a dose dependent manner. Even a small dose of 2 mg/kg which caused no fiber degeneration after 4 weeks was associated with a decrease of velocity. Following single doses of doxorubicin the decrease in transport velocity was correlated to the number of degenerated fibers. It should be mentioned that the decrease associated with the first signs of fiber degeneration was larger than the decrease seen in experimental diabetes where no degeneration has been demonstrated. Apparently, there were no changes in the amounts of incorporated and transported isotopes, but as mentioned above the isotope technique is not suitable for estimating protein synthesis.

The decrease in SCa is common for diabetes and the doxorubin neuronopathy. The retrograde transport is affected in experimental diabetes, whereas the anterograde FC remains undisturbed. Since the opposite changes in transport take place in the case of doxorubicin neuropathy, the two conditions seem rather unlike.

III. CONCLUSION

The early nervous abnormalities of diabetes share features in common with the various toxic neuropathies though no one displays the same spectrum of abnormalities. Grouping in accordance with the traditional classification based on fiber pathology only seems not possible. On the other hand, the different studies reviewed here emphasize the importance of axonal transport in the endeavor to understand the development of neuropathies.

Among the distal axonopathies caused by BPAU, ZPT, 2,5-hexanedione and acrylamide the former two states display a delayed and decreased retrograde axonal transport while the latter two intoxications only lead to decrease in flux. The proximal nerve fiber swellings of

IDPN intoxication offer no hindrance to the fast-transport components, whereas the distal nerve fiber swellings caused by acrylamide seem to be traps for the same material. Finally, the decreased velocity of SCa in experimental diabetes, IDPN, BPAU, and doxorubicin develops in conditions characterized by significant dissimilarities of symptomatology and pathology.

We presume that future experimental work using different neurotoxic substances as dissecting tools to study the significance of the various axonal transport abnormalities will help clarify the pathophysiology underlying peripheral neuropathies.

REFERENCES

1. **Pirart, J.,** Diabète et complications dégénératives. Présentation d'une étude prospective portant sur 4400 cas observés entre 1947 et 1973. Première partie, *Diabete Metab.*, 3, 97, 1977.
2. **Pirart, J.,** Diabète et complications dégénératives. Présentation d'une étude prospective portant sur 4400 cas observés entre 1947 et 1973. Deuxième partie, *Diabete Metab.*, 3, 173, 1977.
3. **Pirart, J.,** Diabète et complications dégénératives. Présentation d'une étude prospective portant sur 4400 cas observés entre 1947 et 1973. Troisième et dernière partie, *Diabete Metab.*, 3, 245, 1977.
4. **Woltman, H. W. and Wilder, R. M.,** Diabetes mellitus. Pathologic changes in the spinal cord and peripheral nerves, *Arch. Intern. Med.*, 44, 576, 1929.
5. **Fagerberg, S.-E.,** Diabetic neuropathy. A clinical and histological study on the significance of vascular affections, *Acta Med. Scand.*, Suppl. 164, 345, 1959.
6. **Jordan, W. R.,** Neuritic manifestations in diabetes mellitus, *Arch. Intern. Med.*, 57, 307, 1936.
7. **Thomas, P. K. and Lascelles, R. G.,** The pathology of diabetic neuropathy, *Q. J. Med. N. S.*, 35, 489, 1966.
8. **Reske-Nielsen, E. and Lundbæk, K.,** Pathological changes in the central and peripheral nervous system of young long-term diabetics, *Diabetologia*, 4, 34, 1968.
9. **Chopra, J. S., Hurwitz, L. J., and Montgomery, D. A. D.,** The pathogenesis of sural nerve changes in diabetes mellitus, *Brain*, 92, 391, 1969.
10. **Gregersen, G.,** Diabetic neuropathy: influence of age, sex, metabolic control and duration of diabetes on motor conduction velocity, *Neurology*, 17, 972, 1967.
11. **Gregersen, G.,** Variations, in motor conduction velocity produced by acute changes of the metabolic state in diabetic patients, *Diabetologia*, 4, 273, 1968.
12. **Gabbay, K. H. and O'Sullivan, J. B.,** The sorbitol pathway. Enzyme localization and content in normal and diabetic nerve and cord, *Diabetes*, 17, 239, 1968.
13. **Ward, J. D., Baker, R. W. R., and Davis, B. H.,** Effect of blood sugar control on the accumulation of sorbitol and fructose in nervous tissues, *Diabetes*, 21, 1173, 1972.
14. **Judzewitsch, R. G., Jaspan, J. B., Polonsky, K. S., Weinberg, C. R., Halter, J. B., Halar, E., Pfeifer, M. A., Vukadinovic, C., Bernstein, L., Schneider, M., Liang, K.-Y., Gabbay, K. H., Rubenstein, A. H., and Porte, D., Jr.,** Aldose reductase inhibition improves nerve conduction velocity in diabetic patients, *New Eng. J. Med.*, 308, 119, 1983.
15. **Brimijoin, S. and Dyck, P. J.,** Axonal transport of dopamine-β-hydroxylase and acetylcholinesterase in human peripheral neuropathy, *Exp. Neurol.*, 66, 467, 1979.
16. **Rasch, R.,** Prevention of diabetic glomerulopathy in streptozotocin diabetic rats by insulin treatment. Glomerular basement membrane thickness, *Diabetologia*, 16, 319, 1979.
17. **Steiness, I.,** Vibratory perception in diabetics during arrested blood flow to the limb, *Acta Med. Scand.*, 163, 195, 1959.
18. **Jakobsen, J.,** Early peripheral nerve abnormalities in diabetes: an animal model, *Horm. Metab. Res.*, Suppl., 30, 1980.
19. **Gabbay, K. H.,** Role of sorbitol pathway in neuropathy, in *Vascular and Neurological Changes in Early Diabetes. Advances in Metabolic Disorders*, Suppl. 2, Camerini-Dávalos, R. A. and Cole, H. S., Eds., Academic Press, New York, 1973, 417.
20. **Jakobsen, J.,** Early and preventable changes of peripheral nerve structure and function in insulin-deficient diabetic rats, *J. Neurol. Neurosurg. Psychiatry*, 42, 509, 1979.
21. **Gregersen, G.,** A study of the peripheral nerves in diabetic subjects during ischaemia, *J. Neurol. Neurosurg. Psychiatry*, 31, 175, 1968.

22. **Yue, D. K., Hanwell, M. A., Satchell, P. M., and Turtle, J. R.,** The effect of aldose reductase inhibition on motor nerve conduction velocity in diabetic rats, *Diabetes,* 31, 789, 1982.
23. **Jakobsen, J.,** Axonal dwindling in early experimental diabetes. I. A study of cross sectioned nerves, *Diabetologia,* 12, 539, 1976.
24. **Sugimura, K., Windebank, A. J., Natarajan, V., Lambert, E. H., Schmid, H. H. O., and Dyck, P. J.,** Interstitial hyperosmolarity may cause axis cylinder shrinkage in streptozotocin diabetic nerve, *J. Neuropathol. Exp. Neurol.,* 39, 710, 1980.
25. **Sharma, A. K., Bajaba, S., and Thomas, P. K.,** Influence of streptozotocin-induced diabetes on myelinated nerve fibre maturation and on body growth in the rat, *Acta Neuropathol.,* 53, 257, 1981.
26. **Robertson, D. M. and Sima, A. A. F.,** Diabetic neuropathy in the mutant mouse (C57BL/ks(db/db)). A morphometric study, *Diabetes,* 29, 60, 1980.
27. **Sharma, A. K., Thomas, P. K., Gabriel, G., Stolinski, C., Dockery, P., and Hollins, G. W.,** Peripheral nerve abnormalities in the diabetic mutant mouse, *Diabetes,* 32, 1152, 1983.
28. **Jakobsen, J.,** Axonal dwindling in early experimental diabetes. II. A study of isolated nerve fibers, *Diabetologia,* 12, 547, 1976.
29. **Sidenius, P. and Jakobsen, J.,** Reduced perikaryal volume of lower motor and primary sensory neurons in early experimental diabetes, *Diabetes,* 29, 182, 1980.
30. **Jakobsen, J.,** Peripheral nerves in early experimental diabetes. Expansion of the endoneurial space as a cause of increased water content, *Diabetologia,* 14, 113, 1978.
31. **Schmidt, R. E., Matschinsky, D. A., Godtfrey, A. D., Williams, A. D., and McDougal, D. B., Jr.,** Fast and slow axoplasmic flow in sciatic nerve of diabetic rat, *Diabetes,* 24, 1081, 1975.
32. **Grafstein, B. and Forman, D. S.,** Intracellular transport in neurons, *Physiol. Rev.,* 60, 1167, 1980.
33. **Jakobsen, J., Brimijoin, S., Skau, K., Sidenius, P., and Wells, D.,** Retrograde axonal transport of transmitter enzymes, fucose-labeled protein and nerve growth factor in streptozotocin-diabetic rats, *Diabetes,* 30, 797, 1981.
34. **Mayer, J. H. and Tomlinson, D. R.,** Axonal transport of cholinergic transmitter enzymes in vagus and sciatic nerves of rats with acute experimental diabetes mellitus; correlation with motor nerve conduction velocity and effects of insulin, *Neuroscience,* 9, 951, 1983.
35. **Livett, B. G., Geffen, L. B., and Austin, L.,** Proximo-distal transport of ^{14} C-noradrenaline and protein in sympathetic nerves, *J. Neurochem.,* 15, 931, 1968.
36. **Giachetti, A.,** Axoplasmic transport of noradrenaline in the sciatic nerves of spontaneously diabetic mice, *Diabetologia,* 16, 191, 1979.
37. **Tomlinson, D. R., Gillon, K. R. W., and Smith, M. G.,** Axonal transport of noradrenaline and noradrenergic transmission in rats with streptozotocin-induced diabetes, *Diabetologia,* 22, 199, 1982.
38. **Sidenius, P. and Jakobsen, J.,** Axonal transport in early experimental diabetes, *Brain Res.,* 173, 315, 1979.
39. **Bisby, M. A.,** Axonal transport of labeled protein and regeneration rate in nerves of streptozotocin-diabetic rats, *Exp. Neurol.,* 69, 74, 1980.
40. **Mendell, J. R., Sahenk, Z., Warmolts, J. R., Marshall, J. K., and Thibert, P.,** The spontaneously diabetic BB Wistar rat. Morphologic and physiologic studies of peripheral nerve, *J. Neurol. Sci.,* 52, 103, 1981.
41. **Meiri, K. F. and McLean, W. G.,** Axonal transport of protein in motor fibres of experimentally diabetic rats — fast anterograde transport, *Brain Res.,* 238, 77, 1982.
42. **Bajada, S., Sharma, A. K., and Thomas, P. K.,** Axoplasmic transport in vagal afferent fibres in normal and alloxan-diabetic rabbits, *J. Neurol. Sci.,* 47, 365, 1980.
43. **Fonnum, F.,** Surface charge of choline acetyltransferase from different species, *J. Neurochem.,* 17, 1095, 1970.
44. **Saunders, N. R., Dziegielewska, K., Haggendal, C. J., and Dahlström, A. B.,** Slow accumulation of choline acetyltransferase in crushed sciatic nerves of the rat, *J. Neurobiol.,* 4, 95, 1973.
45. **Tomlinson, D. R., Holmes, P. R., and Mayer, J. H.,** Reversal, by treatment with an aldose reductase inhibitor, of impaired axonal transport and motor nerve conduction velocity in experimental diabetes, *Neurosci. Lett.,* 31, 189, 1982.
46. **Mayer, J. H. and Tomlinson, D. R.,** Prevention of defects of axonal transport and nerve conduction velocity by oral administration of myo-inositol or an aldose reductase inhibitor in streptozotocin-diabetic rats, *Diabetologia,* 25, 433, 1983.
47. **Jakobsen, J. and Sidenius, P.,** Decreased axonal transport of structural proteins in streptozotocin diabetic rats, *J. Clin. Invest.,* 66, 292, 1980.
48. **Sidenius, P. and Jakobsen, J.,** Reversibility and preventability of the decrease in slow axonal transport velocity in experimental diabetes, *Diabetes,* 31, 689, 1982.
49. **Vitadello, M., Filliatreau, G., Dupont, J. L., Hassig, R., Gorio, A., and DiGiamberdino, L.,** Altered axonal transport of cyoskeletal proteins in the mutant diabetic mouse, *J. Neurochem.,* 45, 860, 1985.

50. **Frayser, R. and Buse, M. G.,** Branched chain amino acid metabolism in the retina of diabetic rats, *Diabetologia,* 14, 171, 1978.
51. **Bisby, M. A. and Bulger, V. T.,** Reversal of axonal transport at a nerve crush, *J. Neurochem.,* 29, 313, 1977.
52. **Jakobsen, J. and Sidenius, P.,** Decreased axonal flux of retrogradely transported glycoproteins in early experimental diabetes, *J. Neurochem.,* 33, 1055, 1979.
53. **Sidenius, P. and Jakobsen, J.,** Retrograde axonal transport. A possible role in the development of neuropathy, *Diabetologia,* 20, 110, 1981.
54. **Schmidt, R. E., Modert, C. W., Yip, H. K., and Johnson, E. M., Jr.,** Retrograde axonal transport of intravenously administered ^{125}I-nerve growth factor in rats with streptozotocin-induced diabetes, *Diabetes,* 32, 654, 1983.
55. **Sidenius, P. and Jakobsen, J.,** Impaired retrograde axonal transport from a nerve crush in streptozotocin diabetic rats, *Diabetologia,* 19, 222, 1980.
56. **Stewart, M. A., Sherman, W. R., Kurien, M. M., Moonsammy, G. I., and Wisgerhof, M.,** Polyol accumulations in nervous tissue of rats with experimental diabetes and galactosaemia, *J. Neurochem.,* 14, 1057, 1967.
57. **Gabbay, K. H. and Snider, J. J.,** Nerve conduction defect in galactose-fed rats, *Diabetes,* 21, 295, 1972.
58. **Myers, R. R., Costello, M. L., and Powell, H. C.,** Increased endoneurial fluid pressure in galactose neuropathy, *Muscle Nerve,* 2, 299, 1979.
59. **Sharma, A. K., Thomas, P. K., and Baker, R. W. R.,** Peripheral nerve abnormalities related to galactose administration in rats, *J. Neurol. Neurosurg. Psychiatry,* 39, 794, 1976.
60. **Malmgren, L. T., Jakobsen, J., and Olsson, Y.,** Permeability of blood-nerve barrier in galactose-fed rats, *Exp. Neurol.,* 66, 758, 1979.
61. **Sidenius, P. and Jakobsen, J.,** Axonal transport in rats after galactose feeding, *Diabetologia,* 19, 229, 1980.
62. **Powell, H. C., Costello, M. L., and Myers, R. R.,** Endoneurial fluid pressure in experimental models of diabetic neuropathy, *J. Neuropathol. Exp. Neurol.,* 40, 613, 1981.
63. **Powell, H. C. and Myers, R. R.,** Schwann cell changes and demyelination in chronic galactose neuropathy, *Muscle Nerve,* 6, 218, 1983.
64. **Greene, D. A. and Winegrad, A. I.,** In vitro studies of the substrates for energy production and the effects of insulin on glucose utilization in the neural components of peripheral nerve, *Diabetes,* 28, 878, 1979.
65. **Greene, D. A. and Winegrad, A. I.,** Effects of acute experimental diabetes on composite energy metabolism in peripheral nerve axons and Schwann cells, *Diabetes,* 30, 967, 1981.
66. **Gjedde, A. and Crone, C.,** Blood-brain glucose transfer: repression in chronic hyperglycemia, *Science,* 214, 456, 1981.
67. **Sidenius, P. and Jakobsen, J.,** Peripheral neuropathy in rats induced by insulin treatment, *Diabetes,* 32, 383, 1983.
68. **Jakobsen, J., Brimijoin, S., and Sidenius, P.,** Axonal transport in neuropathy, *Muscle Nerve,* 6, 164, 1983.
69. **Schaumburg, H. H., Wiesniewski, H. M., and Spencer, P. S.,** Ultrastructural studies of the dying-back process. I. Peripheral nerve terminal and axon degeneration in systemic acrylamide intoxication, *J. Neuropathol. Exp. Neurol.,* 33, 260, 1974.
70. **Jakobsen, J. and Brimijoin, S.,** Axonal transport of enzymes and labeled proteins in experimental axonopathy induced by p-bromophenyl-acetylurea, *Brain Res.,* 229, 103, 1981.
71. **Sahenk, Z. and Mendell, J. R.,** Acrylamide and 2,5-hexanedione neuropathies: abnormal bidirectional transport rates in distal axons, *Brain Res.,* 219, 397, 1981.
72. **Jakobsen, J. and Sidenius, P.,** Early and dose-dependent decrease of retrograde axonal transport in acrylamide intoxicated rats, *J. Neurochem.,* 40, 447, 1983.
73. **Sahenk, Z. and Mendell, J. R.,** Axoplasmic transport in zinc pyridinethione neuropathy: evidence for an abnormality in distal turn-around, *Brain Res.,* 186, 343, 1980.
74. **Chrétien, M., Patey, G., Souyri, F., and Droz, B.,** "Acrylamide-induced" neuropathy and impairment of axonal transport of proteins. II. Abnormal accumulation of smooth endoplasmic reticulum at sites of focal retention of fast transported proteins. Electron microscope radioautographic study, *Brain Res.,* 205, 15, 1981.
75. **Couraud, J. Y., Di Giamberardino, L., Chrétien, M., Souri, F., and Fardeau, M.,** Acrylamide neuropathy and changes in the axonal transport and muscular content of the molecular forms of acetylcholinesterase, *Muscle Nerve,* 5, 302, 1982.
76. **Miller, M. S., Miller, M. J., Burks, T. F., and Sipes, I. G.,** Altered retrograde axonal transport of nerve growth factor after single and repeated doses of acrylamide in the rat, *Toxicol. Appl. Pharmacol.,* 69, 96, 1983.

77. **Sidenius, P. and Jakobsen, J.,** Anterograde axonal transport in rats during intoxication with acrylamide, *J. Neurochem.*, 40, 697, 1983.
78. **Sterman, A. B.,** Cell body remodeling during dying-back axonopathy: DRG changes during advanced disease, *J. Neuropathol. Exp. Neurol.*, 41, 400, 1982.
79. **Sterman, A. B.,** Altered sensory ganglia in acrylamide neuropathy. Quantitative evidence of neuronal reorganization, *J. Neuropathol. Exp. Neurol.*, 42, 166, 1983.
80. **Clark, A., Griffin, J. W., and Price, D. L.,** The axonal pathology in chronic β,β′-iminodipropionitrile intoxication, *J. Neuropathol. Exp. Neurol.*, 39, 42, 1980.
81. **Griffin, J. W., Hoffman, P. N., Clark, A. W., Carroll, P. T., and Price, D. L.,** Slow axonal transport of neurofilament proteins: impairment by β,β-iminodipropionitrile, *Science*, 202, 633, 1978.
82. **Papasozomenos, S. C., Yoon, M., Crane, R., Autilio-Gambetti, L., and Gambetti, P.,** Redistribution of proteins of fast axonal transport following administration of β,β′-iminodipropionitrile: a quantitative autoradiographic study, *J. Cell Biol.*, 95, 672, 1982.
83. **Griffin, J. W., Gold, B. G., Cork, L. C., Price, D. L., and Lowndes, H. E.,** IDPN neuropathy in the cat. Coexistence of proximal and distal axonal swellings, *Neuropathol. Appl. Neurobiol.*, 8, 351, 1982.
84. **Griffin, J. W., Fahnestock, K. E., Price, D. L., and Hoffman, P. N.,** Microtubule-neurofilament segregation produced by β,β′-iminodipropionitrile: evidence for the association of fast axonal transport with microtubules, *J. Neurosci.*, 3, 557, 1983.
85. **Cho, E. S., Spencer, P. S., Jortner, B. S., and Schaumburg, H. H.,** A single intravenous injection of doxorubicin (adriamycin^R) induces sensory neuronopathy in rats, *Neurotoxicology*, 1, 583, 1980.
86. **Pigram, W. J., Fuller, W., and Hamilton, L. D.,** Stereochemistry of intercalation — interaction of daunomycin with DNA, *Nature New Biol.*, 235, 17, 1972.
87. **Mendell, J. R. and Sahenk, Z.,** Interference of neuronal processing and axoplasmic transport by toxic chemicals, in *Experimental and Clinical Neurotoxicology*, Williams & Wilkins, Baltimore, 1980, 139.
88. **Sahenk, Z. and Mendell, J. R.,** Analysis of fast axoplasmic transport in nerve ligation and adriamycin-induced neuronal perikaryon lesions, *Brain Res.*, 171, 41, 1979.
89. **Jakobsen, J. and Sidenius, P.,** unpublished results.
90. **Sidenius, P.,** The effect of doxorubicin on slow and fast components of the axonal transport system in rats, *Brain*, in press, 1986.

Chapter 14

AXOPLASMIC TRANSPORT OF NEUROPEPTIDES IN PATHOLOGICAL SYSTEMS

C. G. Rasool

TABLE OF CONTENTS

I. INTRODUCTION

Several different peptides are widely but precisely distributed in central, peripheral, and autonomic nervous systems. Although their functional importance in many of their sites remains to be understood, their occurrence in specific cells strongly suggests that they serve as neurotransmitters or as neuromodulators.[1,2,3] At present, at least 33 peptides have been described in neurons and nerve terminals within mammalian CNS, other than those related to endocrine function which may be considered as neurotransmitters or neuromodulators. Although their discovery has added a totally new dimension to our analysis of neural activity, the inaccessibility and diffuse localization of central peptidergic pathways have made it difficult to identify many peptides as neurotransmitters. However, the presence of several different peptides in the peripheral nervous system has made it possible to carry out experiments necessary to determine whether these peptides can be designated as transmitters. This group of putative neurotransmitters in peripheral neurons includes somatostatin, substance P, vasoactive intestinal polypeptide, and neurotensin.[2,4,5,6] If a peptide is released as a transmitter or a modulator substance from a nerve terminal, it would be expected that its terminal store would be maintained by axoplasmic transport of the peptide. Evidence for axoplasmic transport of certain neuropeptides has been provided by accumulation of immunoreactive peptides proximal to the site of peripheral nerve ligation or crush.[4,6,7] (For review see References 2 and 5.)

It is not my intention to discuss the evidence for and function of peptides as transmitters or modulators. The purpose of the present review is to present a progress report on recent studies on axoplasmic transport of somatostatin and substance P, particularly somatostatin in pathological conditions. There are numerous qualitative studies available on the axoplasmic transport of peptides under normal conditions, particularly somatostatin, and substance P.[4,5,6,7,8] However, very few studies are available on axoplasmic transport of peptides in pathological conditions. Recently, we have completed studies on axoplasmic transport of substance P and somatostatin in various metabolic conditions. In this article, in addition to our studies, I will discuss other available data on transport of neuropeptides.

II. AXOPLASMIC TRANSPORT OF NEUROPEPTIDES

In most of the studies, the ligation technique has been used to determine the quantitative aspect of axoplasmic transport of neuropeptides in normal axons and in various altered metabolic states. In spite of species differences and disagreement on absolute numbers, it is generally accepted that the mobile fraction of substance P is transported with an average velocity of 240 mm/day and this lies in a range considered to represent fast transport.[4,7,8,9] Although somatostatin has not been subjected to rigorous quantitative transport studies, the available data suggest that the mobile fraction of this peptide also flows at a rapid velocity.[4,6]

A. Capsaicin and the Neuropeptides, Substance P, and Somatostatin

Administration of capsaicin, a homovanillic acid derivative, depletes substance P from primary sensory neurons and causes a prolonged increase in the thermal and chemical pain threshold of adult rats.[10,11] These observations, together with the demonstration that capsaicin treatment reduces the substance P level in the dorsal horn and has no effect on the activities of glutamic acid decarboxylase and choline acetyltransferase,[12] led to the suggestion that the neurotoxic action of capsaicin is specific to substance P neurons of spinal ganglia and appeared to support the hypothesis of an involvement of substance P in pain mechanisms.[13] However, subsequent work with capsaicin has shown that it causes indiscriminate damage to all unmyelinated sensory fibers. Also, capsaicin injures, in addition to substance P neurons, other peptidergic sensory neurons including those containing somatostatin, vasoactive intestinal polypeptide, and neurotensin.[7,14] Studies from our own laboratory on the effect of capsaicin on substance P and somatostatin in spinal dorsal horn confirm these results.[60] Recently, Gamse et al.[7] have studied the effect of capsaicin on the axoplasmic transport of substance P and somatostatin in peripheral nerve. The local application of capsaicin led to a complete block of axoplasmic transport of substance P and somatostatin with no effect on the transport of noradrenaline and acetylcholinesterase. These results, taken together with previous observations,[14] suggest that administration of capsaicin may also affect axoplasmic transport of other peptides such as neurotensin, vasoactive intestinal polypeptide, or cholecystokinin-like peptide. Thus, although capsaicin cannot be regarded as a selective neurotoxin for substance P-containing neurons, it seems to be specific to peptidergic sensory neurons. Therefore, the interesting pattern of capsaicin specificity (e.g., its lack of effect on the transport of noradrenaline or acetylcholinesterase) will help us to understand the functional role of peptidergic primary sensory neurons.

B. Acrylamide and Neuropeptide Transport

Acrylamide is known to produce axonal-type dying-back neuropathy. (For review, see Reference 15.) The studies on axoplasmic transport in acrylamide-induced neuropathy provided rather conflicting results. Summer et al.[16] reported a small reduction in the rate of fast transport without affecting slow-transport. Using the ligation technique, Rasool and Bradley[17] reported an almost 50% reduction in the absolute transport rate of acetylcholinesterase enzymes, which represent a group of proteins being transported at a fast velocity of 400 to 500 mm/24 hr.[18,19,20] Our recent studies on the axoplasmic transport of somatostatin and substance P in sciatic nerves of acrylamide-treated rats indicated a reduction of 50% in the accumulation of these peptides proximal to a ligature. Although the precise mechanisms of acrylamide action on axoplasmic transport are not clear, the present observations taken together with previous studies suggest that acrylamide treatment affects the fast axoplasmic transport system in general and, unlike capsaicin treatment, is not specific to peptidergic sensory neurons.

C. Streptozotocin-Induced Diabetes and Neuropeptide Transport

Streptozotocin-induced diabetes is a model for many features of both acute and chronic spontaneous diabetes in man[21,22] and in rat.[23] Several investigators have reported normal transport rates of radiolabeled proteins in acute streptozotocin-induced diabetes.[24,25] Recently, we have studied the axoplasmic transport of somatostatin in streptozotocin-treated rat sciatic nerve. Diabetes mellitus was produced by intravenous injection of streptozotocin at a dose of 50 mg/kg b.w. The drug was dissolved in 0.9% NaCl and acidified with citric acid to pH 4.5, immediately before injection. Diabetic status was evaluated by daily measurement of body weight, urine volume, urinary glucose, and ketones. Axoplasmic transport of somatostatin was studied by the single ligation technique. As previously reported,[6] the rate of accumulation of somatostatin proximal to the ligature was linear up to at least 8 hr post-ligation. In the case of diabetic rats, the rate of accumulation of somatostatin was indistinguishable from that in the control group (79.7 ± 9.3 vs. 87.9 ± 15.3 mm/24 hr $\pm$ SEM). The finding that acute diabetes does not modify transport of somatostatin, a peptide with fast velocity, confirms a number of other studies that indicate normal anterograde fast transport rates.[24-28]

D. Hypothyroidism and Neuropeptide Transport

Severe impairment of neural function occurs with some frequency in adult patients with hypothyroidism. These changes include disturbed higher functions which can progress to coma,[29-31] slowing of the dominant rhythm of cerebral cortical electroencephalograms,[32,33] reduced rate of peripheral nerve conduction,[34,35] and peripheral neuropathy.[34-37] The mech-

Table 1
AXOPLASMIC TRANSPORT OF IR-SOMATOSTATIN, PLASMA T_3 AND Ca^{++} LEVELS IN NORMAL AND VARIOUS METABOLIC STATES

	Amount in contralateral control nerve segment (pg/3 mm)	Apparent transport rate (mm/24 hr)	Absolute transport rate (mm/24 hr)	Plasma T_3 (ng/dℓ)	Plasma Ca^{++} (mg/dℓ)	Ref.
Normal	13.7 ± 2.5	87.1 ± 15.3	435	54.8 ± 6.3	9.9 ± 0.1	18
Thyroidectomized	13.5 ± 1.8	54.6 ± 3.7[a]	273[a]	19.7 ± 2.5	8.9 ± 0.2	10
Parathyroidectomized	11.8 ± 2.1	85.3 ± 11.1[b]	426[b]	51.1 ± 3.7	6.3 ± 0.9	12

Note: For the calculation of axoplasmic transport parameters, see Reference 6. Absolute transport rates were calculated on the basis of 20% mobile fraction of neuropeptide. Plasma Ca^{++} level was determined by atomic absorption spectrophotometric method and T_3 levels were measured according to the method of Nejad et al.[61] The values are the means ± SE for the number of animals shown in parentheses and unpaired student's *t*-test was used for statistical evaluation.

[a] $p < 0.05$ (transport rate in normal).
[b] $p \geq 0.3$ (transport rate in normal).

anism underlying these changes have not been established. Adult brain, unlike most other tissues (with the exception of spleen and testes), shows no change in oxidative metabolism in states of hyper- or hypothyroidism,[38-40] and thyroxine does not stimulate the biosynthesis of proteins in a system containing adult brain mitochondria, unlike those containing liver or immature brain mitochondria.[41] This seeming metabolic inertness of the adult brain in response to thyroid hormone is in marked contrast to the striking effects of thyroid hormone deficiency on the developing mammalian nervous system, which include poor cerebral growth, impaired differentiation of axons, dendrites,[42] and nerve terminals,[43] and reduced synthesis of proteins,[44-46] enzymes,[47] and receptors.[48] Brain cytosol receptors for T_3 decrease during maturation.[49]

At present no data are available on the effect of thyroidectomy on the axonplasmic transport system. We have considered the hypothesis that hypothyroidism might impair neuronal function by interfering with axoplasmic transport. Anterograde axoplasmic transport is responsible for translocation of structural proteins, enzymes, neurotransmitters, and other subcellular elements from sites of synthesis in the cell body to axonal and dendritic terminals.[50] It is conceivable that impaired delivery of one or more transported components could interfere with nerve activity in hypothyroidism. We have utilized measurement of somatostatin transport in the sciatic nerve of the rat as a model system for elucidation of effects of thyroid hormone deficiency on axoplasmic transport.

Axoplasmic transport of somatostatin was studied in thyroidectomized, parathyroidectomized, and normal control rats. The accumulation of peptide proximal to ligature was found linear in normal and experimental rats up to 8 hr post-ligation. However, the rate of accumulation was significantly less in thyroidectomized animals than in normal and parathyroidectomized animals (Table 1). As shown in Table 1, the plasma thyroid, T_3, level was significantly less in thyroidectomized rats than in normal control and parathyroidectomized rats. The plasma Ca^{++} level was less in parathyroidectomized animals than in the normal group and no change was observed in its level in hypothyroid animals. These observations signify the importance of thyroid hormone, T_3, in the axoplasmic transport of somatostatin. Our finding of reduced accumulation of peptide could be explained by an increase in degradation of the peptide within the nerve. Dupont and colleagues[51] reported that somatostatin-degrading activity by hypothalamic extracts in the rat is increased by propylthiouracil-induced hypothyroidism, but that of cerebral cortex extracts is unchanged. Effects of thyroid deficiency on peripheral nerve somatostatin degradation are unknown. Information about the effects of thyroid status on somatostatin secretion in other parts of the brain is limited. Berelowitz et al.[52] reported that hypothyroidism reduces the somatostatin content of hypothalamus and also the secretory response to depolarization in vitro. Since these workers also noted a decrease in thalamic somatostatin as well, albeit a modest one, it may be that somatostatin secretion by most brain regions is reduced in the hypothyroid state. On the other hand, Fernandez-Durango et al.[53] found no change in hypothalamic somatostatin concentration after thyroidectomy.

The effect of thryoid hormone on oxidative metabolism must be considered in interpreting the finding of reduced accumulation of somatostatin in ligated nerve. As summarized by Ochs and Hollingsworth[54,55] and Grafstein and Formar[50,56] the fast-transport rate in peripheral nerve is dependent upon oxidative metabolism, which is, in turn, dependent upon adequate concentrations of ATP, presumably generated by mitochondria present in perikarya and transported in peripheral nerve, and by a critical concentration of calcium ion.[57] Since oxidative metabolism is reportedly unaffected in adult brain by the hypothyroid state,[38-40] thyroxine deficiency conceivably could influence axoplasmic transport by modifying ATP generation independent of oxidation or by altering the rate of synthesis of any of the individual elements responsible for assembling the transport system (such as microfilaments, actin, calmodulin, the endoplasmic reticulum, or other membrane-bound elements). The role of these organelles in the transport system is summarized by Sabatini and Kreibich.[58]

The role of impaired synthesis and/or transport in the pathogenesis of hypothyroid neuropathy must be considered. It is probable that reduced somatostatin accumulation is not the cause of the impaired neuronal function, but rather a consequence of impaired neuronal function. Smith[59] has summarized this problem by stating, "One cannot say whether reduced synthesis and transport of proteins cause any human neuropathy.... Nevertheless, a block of transport, regardless of its origin, could prevent materials that are essential for neurotransmission from reaching nerve terminals and could interfere with the repair and renewal of the axolemma. The result would be a failure of nerve function and even axonal degeneration."

At present no data are available on the effect of thyroid hormone on the axoplasmic transport of other peptides, enzymes, and proteins. It is, therefore, not possible to determine whether the effect of thryoidectomy is specific to the transport of neuropeptides or if thyroidectomy generally affects the axoplasmic transport systems, both fast and slow, of other substances as well.

III. SUMMARY

At present, only a few studies on the axoplasmic transport of neuropeptides in pathological states have been completed.

1. The study of axoplasmic transport in hypothyroidism indicating a reduction in somatostatin transport is very interesting. However, further studies have to be performed to find out the specificity of thyroid hormone in affecting transport of different peptides.
2. The effect of acrylamide on axoplasmic transport of neuropeptides and other proteins seems to be a general phenomenon, suggesting its effect on all substances flowing at fast velocity.
3. The finding of normal axoplasmic transport of somatostatin in streptozotocin-induced diabetes taken together with studies on axoplasmic transport of other substances suggests that streptozotocin-induced diabetes does not impair the fast axoplasmic transport system.

4. The effect of capsaicin on axoplasmic transport seems to be specific to peptidergic sensory neurons with no effect on post-ganglionic noradrenergic and cholinergic neurons. These various studies open up new possibilities in our understanding of the functional significance of peptidergic sensory neurons.

ACKNOWLEDGMENTS

The experimental work was carried out at Dr. Seymour Reichlin's laboratory at Tufts-New England Medical Center and supported by USPHS grant AM 16684. I am grateful to Dr. S. Reichlin for support and stimulating discussion.

I thank Drs. Dennis J. Selkoe, Kenneth S. Kosik, and Lawrence K. Duffy for useful comments on this manuscript, and Johanne Khan for typing the manuscript.

REFERENCES

1. **Emson, P. C.,** Peptides as neurotransmitter candidates in the mammalian CNS, *Prog. Neurobiol.*, 13, 61, 1979.
2. **Iversen, L. L.,** Nonopioid neuropeptides in mammalian CNS, *Ann. Rev. Pharmacol. Toxicol.*, 23, 1, 1983.
3. **Otsuka, M. and Konishi, S.,** Substance P — the first peptide neurotransmitter?, *Trends NeuroSci.*, 6, 317, 1983.
4. **Gilbert, R. F. T., Emson, P. C., Fahrenkrug, J., Lee, C. M., Penman, E., and Wass, J.,** Axoplasmic transport of neuropeptides in the cervical vagus nerve of rat, *J. Neurochem.*, 34, 208, 1980.
5. **Lundberg, J. M., Hökfelt, T., Anggard, A., Uvnas-Wallenstein, K., Brimijoins, Brodin, E., and Fahrenkrug, J.,** Peripheral peptide neurons: distribution, axonal transport, and some aspects of possible function, in *Neural Peptides and Neural Communication*, Costa, E. and Trabucchi, M., Eds., Raven Press, New York, 1980, 25.
6. **Rasool, C. G., Schwartz, A. L., Bollinger, J. A., Reichlin, S., and Bradley, W. G.,** Immunoreactive somatostatin distribution and axoplasmic transport in rat peripheral nerve, *Endocrinology*, 108, 966, 1981.
7. **Gamse, R., Petsche, U., Lembeck, F., and Jancso, G.,** Capsaicin applied to peripheral nerve inhibits axoplasmic transport of substance P and somatostatin, *Brain Res.*, 239, 447, 1982.
8. **Brimijoin, S., Lundberg, J., Brodin, E., Hökfelt, T., and Nilsson, G.,** Axonal transport of substance P in the vagus and sciatic nerve of guinea pig, *Brain Res.*, 191, 443, 1980.
9. **Gamse, R., Lembeck, F., and Cuello, A. C.,** Substance P in the vagus nerve — immunochemical and immunohistochemical evidence for axoplasmic transport, *Naunyn-Schmiedebergs Arch. Pharmacol.*, 306, 37, 1979.
10. **Jessell, T. M., Iversen, L. L., and Cuello, A. C.,** Capsaicin-induced depletion of substance P from primary sensory neurons, *Brain Res.*, 152, 182, 1978.
11. **Yaksh, T., Farb, D., Leeman, S., and Jessell, T. M.,** Intrathecal capsaicin depletes substance P in the rat spinal cord and produces prolonged thermal analgesia, *Science*, 206, 481.
12. **Nagy, J. I., Vincent, S. R., Staines, W. M., Fibiger, H. C., Reisine, T. D., and Yamamura, H. J.,** Neurotoxic action of capsaicin on spinal P neurons, *Brain Res.*, 186, 435, 1980.
13. **Nagy, J. I.,** Capsaicin: chemical probe for sensory neuron mechanism, in *Handbook of Psychopharmacology*, Iversen, L. L., Iversen, S. D., and Snyder, S. H., Eds., Plenum Press, New York, 1982, 185.
14. **Jancso, G., Hökfelt, T., Lundberg, J. M., Kiraly, E., Halasz, N., Nilsson, G., Terenius, L., Rehfeld, J., Steinbusch, H., Verhofstad, A., Elde, R., Said, S., and Brown, M.,** Immunohistochemical studies on the effect of capsaicin on spinal and medullary peptide and monoamine neurons using antisera to substance P, gastrin/CCK, somatostatin, VIP, enkephalin, neuro-tensin and 5-hydroxytryptamine, *J. Neurocytol.*, 10, 963, 1981.
15. **Summer, A.,** Physiology of dying-back neuropathies, in *Physiology and Pathobiology of Axons*, Waxman, S. G., Ed., Raven Press, New York, 1978, 349.
16. **Summer, A. J., Pleasure, J. D., and Ciesielka, K.,** Slowing of fast transport in acrylamide neuropathy, *J. Neuropathol. Exp. Neurol.*, 35, 319, 1976.
17. **Rasool, C. G. and Bradley, W. G.,** Studies on axoplasmic transport of individual proteins: 1. Acetylcholinesterase (AChE) in acrylamide neuropathy, *J. Neurochem.*, 31, 419, 1978.
18. **Lasek, R. J.,** Axoplasmic transport of labeled proteins in rat ventral motoneurons, *Exp. Neurol.*, 21, 41, 1968.

19. **Bray, J. J. and Austin, L.,** Axoplasmic transport of ^{14}C proteins at two rates in chicken sciatic nerve, *Brain Res.*, 12, 230, 1969.
20. **Sjöstrand, J. and Karlsson, J.-O.,** Axoplasmic transport in the optic nerve and tract of the rabbit: a biochemical and radioautographic study, *J. Neurochem.*, 16, 833, 1969.
21. **Thomas, P. R. and Eliasson, S. G.,** *Diabetic Neuropathy, Peripheral Neuropathy,* Vol. 2, Dick, P. J., Thomas, P. R., and Lambert, E. H., Eds., W. B. Saunders, Philadelphia, 1975, 956.
22. **Bestetti, G., Rossi, G. L., and Zemp, C.,** Changes in peripheral nerves of rats, four months after induction of streptozotocin diabetes: a qualitative and quantitative study, *Acta Neuropathol.*, 54, 129, 1981.
23. **Jefferys, J. G. R. and Brismar, T.,** Analysis of peripheral nerve function in streptozotocin diabetic rats, *J. Neurol. Sci.*, 48, 435, 1980.
24. **Sidenius, P. and Jakobsen, J.,** Axonal transport in early experimental diabetes, *Brain Res.*, 173, 315, 1979.
25. **Bisby, M. A.,** Axonal transport of labeled protein and regeneration rate in nerves of streptozotocin-diabetic rats, *Exp. Neurol.*, 69, 74, 1980.
26. **Sidenius, P. and Jakobsen, J.,** Impaired retrograde axonal transport from a nerve crush in streptozotocin diabetic rats, *Diabetologia*, 19, 222, 1980.
27. **Sidenius, P. and Jakobsen, J.,** Retrograde axonal transport: a possible role in the development of neuropathy, *Diabetologia*, 20, 110, 1981.
28. **Jakobsen, J., Brimijoin, S., Skau, K., Sidenius, P., and Wells, D.,** Retrograde axonal transport of transmitter enzymes, sucrose-labelled protein and nerve growth factor in streptozotocin-diabetes rats, *Diabetes*, 30, 797, 1981.
29. **Scheinberg, P., Stead, E. A., Brannon, E. S., and Warren, J. V.,** Correlative observations on cerebral metabolism and caridan output in myxedema, *J. Clin. Invest.*, 29, 1139, 1950.
30. **Sensebach, W., Madison, L., Eisenberg, S., and Ochs, L.,** Cerebral circulation and metabolism in hyperthyroidism and myxedema, *J. Clin. Invest.*, 33, 1434, 1954.
31. **Browning, T. B., Atkins, R. W., and Weiner, H.,** Cerebral metabolic disturbances in hypothyroidism, *Arch. Intern. Med.*, 93, 938, 1954.
32. **Ross, D. A. and Schwab, R. S.,** Cortical alpha rhythm in thyroid disorders, *Endocrinology*, 25, 75, 1939.
33. **Browning, T. B., Atkins, R. W., and Weiner, H.,** Cerebral metabolic disturbances in hypothyroidism: clinical and electroencephalographic studies of psychosis of myxedema and hypothyroidism, *Arch. Intern. Med.*, 93, 938, 1954.
34. **Fincham, R. W. and Cape, C. A.,** Neuropathy in myxedema: a study of sensory nerve conduction in the upper extremities, *Arch. Neurol.*, 19, 464, 1968.
35. **Dyck, P. J. and Lambert, E. H.,** Polyneuropathy associated with hypothyroidism, *J. Neuropath. Exp. Neurol.*, 29, 631, 1970.
36. **Crevasse, I. F. and Logue, R. B.,** Peripheral neuropathy in myxedema, *Ann. Intern. Med.*, 50, 1933, 1959.
37. **Nickel, S. N., Frame, B., Bebin, J., Tourtellotte, W. W., Parker, J. A., and Hughes, B. R.,** Myxedema neuropathy and myopathy: a clinical and pathologic study, *Neurology*, 11, 125, 1961.
38. **Barker, S. B. and Klitgaard, H. M.,** Metabolism of tissue excised from thyroxine-injected rats, *Am. J. Physiol.*, 170, 81, 1952.
39. **Barker, S. B. and Schwartz, H. S.,** Further studies on metabolism of tissues from thyroxine-injected rats, *Proc. Soc. Exp. Biol. Med.*, 83, 500, 1953.
40. **Oppenheimer, J. H.,** Thyroid hormone action at cellular level, *Science*, 203, 971, 1979.
41. **Klee, C. B. and Sokoloff, L.,** Mitochondrial differences in mature and immature brain: influence on rate of amino acid incorporation into protein and responses to thyroxine, *J. Neurochem.*, 11, 709, 1964.
42. **Legrand, J.,** The variation of the response of the cerebellum to the morphogenic action of the thyroid in the rat, as a function of the age, *Arch. Anat. Microsc. Morphol. Exp.*, 56, 291, 1967.
43. **Balazs, R., Kovacs, S., and Cocks, W. A.,** Biochemical effects of thyroid deficiency on the developing brain, *J. Neurochem.*, 15, 1335, 1968.
44. **Krawiec, L., Garcia-Argiz, C. A., Gomez, C. J., and Pasquini, J. M.,** Hormone regulation of brain development. III. Effects of triiodothyronine and growth hormone on the biochemical changes in the cerebral cortex and cerebellum of neonatally thyroidectomized rats, *Brain Res.*, 15, 209, 1969.
45. **Balazs, R., Cocks, W. A., Eayrs, J. T., and Kovacs, S.,** Biochemical effects of thyroid hormones on the developing brain, in *Hormones in Development*, Hamburgh, M. and Barrington, E. J. W., Eds., Appleton-Century-Crofts, New York, 1971, 357.
46. **Geel, S. E. and Timiras, T. S.,** The role of thyroid and growth hormones on RNA metabolism in the immature brain, in *Hormones in Development*, Hamburgh, M. and Barrington, E. J. W., Eds., Appleton-Century-Crofts, New York, 1971, 391.
47. **Bera, H. and Chatterjee, G. C.,** Effect of thyroid deficiency on the level of several enzymes in rat brain, *J. Neurochem.*, 245, 859, 1975.

48. **Smith, R. M., Patel, R. J., Kingsbury, A. E., Hunt, A., and Balazs, R.,** Effects of thyroid state on brain development: Beta-adrenergic receptors and '5-nucleotidase activity, *Brain Res.*, 198, 375, 1980.
49. **Geel, S. E.,** Development-related changes of triiodothyronine binding to cytosol receptors, *Nature (London)*, 269, 428, 3977.
50. **Grafstein, B. and Forman, D.,** Intracellular transport in neurons, *Physiol. Rev.*, 60, 1167, 1980.
51. **Dupont, A., Merand, Y., and Barden, N.,** Effect of propylthiouracil and thyroxine on the inactivation of somatostatin by rat hypothalamus, *Life Sci.*, 23, 2007, 1978.
52. **Berelowitz, M., Kiyoshi, M., Harris, S., and Frohman, L. A.,** The effect of alterations in the pituitary-thyroid axis on hypothalamic content and in vitro release of somatostatin-like immunoreactivity, *Endocrinology*, 107, 24, 1980.
53. **Fernandez-Durango, R., Arimura, A., Fishback, J., and Schally, A. V.,** Hypothalamic somatostatin and LH-RH after hypophysectomy in hyper- or hypothyroidism and during anesthesia in rats, *Proc. Soc. Exp. Biol. Med.*, 157, 235, 1978.
54. **Ochs, S.,** Local supply of energy to the fast axoplasmic transport mechanism, *Proc. Natl. Acad. Sci. U.S.A.*, 68, 1279, 1971.
55. **Ochs, S. and Hollingsworth, D.,** Dependence of fast axoplasmic transport in nerve on oxidative metabolism, *J. Neurochem.*, 18, 107, 1971.
56. **Grafstein, B.,** Axonal transport: the intracellular traffic of the neuron, in *Handbook of Physiology; The Nervous System*, Sec. I, Vol. I, American Physiology Society, Bethesda, Md., 2977, 691.
57. **Chan, S. Y.,** The requirement for calcium ions and the effect of other ions on axoplasmic transportation in mammalian nerve, *J. Physiol.*, 301, 477, 1980.
58. **Sabatini, D. D. and Kreibich, G.,** Functional specialization of membrane-bound ribosomes in eukaryotic cells, in *The Enzymes of Biological Membranes*, Vol. 2, Mortonosi, A., Ed., Plenum Press, New York, 1976, 531.
59. **Smith, B. H.,** Pathophysiology of axonal transport, *Neurosci. Res. Prog. Bull.*, 20, 89, 1981.
60. **Rasool, C. G. et al.,** unpublished results.
61. **Nejad et al.,** Measurement of plasma and tissue triiodothyronine concentration in rat by radioimmunoassay, *Endocrinology*, 96, 773, 1975.

Chapter 15

CHEMICAL NEUROTOXINS AND DISRUPTION OF THE AXONAL TRANSPORT SYSTEM

Mohammad I. Sabri

TABLE OF CONTENTS

I. INTRODUCTION

Agents which exhibit their primary toxic action on the nervous system are referred to as "neurotoxins". The term neurotoxins include biological toxins, such as curari, diphtheria toxin, tetanous toxin, saxitoxin etc., and a large number of new chemicals and drugs that are introduced every year. There is increasing evidence that many neurological disorders might be related to abnormalities in the axoplasmic transport. Several excellent reviews[1-8] and books[9-11] are currently available on axoplasmic transport and its relation to neuronal integrity and pathology.[12-16] The purpose of this chapter is to review the properties of selected chemical neurotoxins known to disrupt the axonal transport system.

Neurotoxic agents may disrupt the axonal transport system in many ways.[17-21] Some agents (i.e., doxorubicin) may do so by binding to nuclear DNA thereby reducing RNA synthesis. Others (methyl mercury) by interfering with protein synthesis in the perikaryon. Neurotoxic agents may also disrupt cytoskeletal elements in the axon (i.e., colchicine, IDPN) inhibit enzymes or bind to coenzymes of energy metabolism and thereby block energy dependent axonal transport system.

Axonal transport is essential for the maintenance of neuronal (axonal) integrity. It is therefore not surprising that abnormalities in the axonal transport system are often thought to be the likely cause of axonal disease. However, a cause-and-effect relationship between axonal transport defect and axonal degeneration has been difficult to establish. Since axonal degeneration is a hallmark of several human neurodegenerative diseases, chemical neurotoxins have become important biological tools to study the mechanism of neuronal degeneration.

II. ROLE OF THE AXONAL TRANSPORT SYSTEM

Materials required for the maintenance of nerve fibers are synthesized in the neuronal perikaryon and transported along the axon (anterograde transport) to sites of utilization. Materials are also transported from the axon and its terminals toward the perikaryon (retrograde transport).[22] Presumably, an equilibrium exists between anterograde and retrograde transport to maintain normal function of the neuron. Axoplasmic transport has been shown to have at least three anterograde and two retrograde transport systems,[7,9,23] however, only three major transport systems are considered here.

A. Fast Anterograde Transport

Fast-transported materials[7,9] are mainly particulate consisting of smooth endoplasmic reticulum, synaptic vesicles, and components of the plasma membrane. In addition, membrane-bound proteins and enzymes, glycoproteins, and lipids associated with membranes, free amino acids, sugars, and Ca^{++} are known to undergo fast anterograde axonal transport.[7,9] Acetyl choline esterase[24] and dopamine beta-hydroxylase[25] are often used as markers of fast axonal transport. In mammalian sciatic nerve, fast-transport has a velocity close to 400 mm/day.[26] While, in the optic system, rate of about 250 mm/day have been reported.[7]

Although transport is not directly affected by protein-synthesis inhibitors, blocking synthesis in neuronal perikarya reduces the amounts of materials transported along the axon.[7,9,27] Fast axonal transport is dependent on oxidative metabolism,[9] temperature, and intracellular Ca^{++} concentration, but is independent of the length, diameter and electrical activity of the axon or extracellular concentration of Na^{+} or K^{+}.[7,9] Fast-transport requires a continuous supply of ATP from oxidative phosphorylation along the nerve fiber[9] and is rapidly blocked in vitro by treating the nerve with cyanide, dinitrophenol, sodium azide, iodoacetate, or anoxia.[9,28] Axonal transport is therefore critically dependent on the energy supplies in nerve fibers.[29] By using video-enhanced microscopy, a new method of studying fast axonal transport in the axoplasm,[30,31] Brady and associates[32] have shown that DNP-induced inhibition of particle movement is reversed by the addition of exogenous ATP in extruded axoplasm. Agents that disrupt microtubules are potent inhibitors of fast axonal transport, although some investigators have suggested that microtubules are not required for transport to occur.[7,9,10]

Several chemicals that block transport induce distal axonal degeneration, commonly known as central-peripheral distal (dying-back) axonopathy (CPDA).[1,13,14] CPDA is associated with blockade of either the fast and/or slow axonal transport systems (Table 1), but the mechanism by which these agents disrupt axonal transport is unknown. One idea is that certain CPDA neurotoxins act directly on the axon and deplete energy supplies necessary for axonal transport.[6,14,33]

B. Slow Anterograde Transport

After neuronal perikarya are labeled with protein precursors, a wave of radioactivity traveling at a slow rate of 0.1 to 4.0 mm/day[23,34,35] appears in the axon long after the peak of fast axonal transport has reached the terminals. As much as 50% of the labeled proteins, including cytoskeletal elements (neurofilaments, neurotubules, and actin filaments) and certain high molecular weight soluble proteins (enzymes of glycolysis) are carried by slow-transport.[7,35] The slow axonal transport system has two well defined components: slow component a (SCa), which moves at 0.1 to 1.0 mm/day and carries cytoskeletal elements (neurofilaments and nerurotubules), and slow component b (SCb) which moves at 2 to 4 mm/day carrying high molecular weight soluble protein, particularly the enzymes of intermediary metabolism.[35] Approximately 75% of the label in SCa is associated with tubulin and the triplet proteins of 10-nm neurofilaments.[7,34] The composition of SCb includes neuron-specific enolase (NSE), aldolase, glyceraldehyde-3-phosphate dehydrogenase (GAPDH), creatine phosphokinase, clathrin, calmodulin, actin, and possibly lactic dehydrogenase (LDH).[35,36] Several toxic agents block selectively the slow component of axoplasmic transport (Table 1). However, 2,5-hexanedione (2,5-HD) and its 3-methyl derivative have been reported to accelerate the rate of transport in the rat optic system.[37,38] The mechanism of action of these agents on slow transport is not understood.

C. Retrograde Transport

Materials carried in the axon by fast and slow anterograde transport have been shown to turnaround at terminals and proceed retrogradely toward the perikaryon.[22,39] Approximately 50% of the fast anterogradely transported protein turns around and returns toward the perikaryon by retrograde transport, at a rate somewhat less than that of fast anterograde transport.[29,39,40] The movement of fast anterogradely and retrogradely transported particles within the axon has also been observed microscopically.[8,41-43]

Like fast anterograde transport, retrograde transport is sensitive to many chemicals and drugs (see Table 1). For example, colchicine or vinblastine block the transport of horseradish peroxidase (HRP).[44] Toxic agents which interfere with energy metabolism generally are also potent blockers of retrograde transport.[45] Several investigators have suggested that abnormalities in the retrograde axonal transport system play an important role in the pathogenesis of the distal axonopathy induced by *para*-bromophenylacetylurea,[46,47] zinc pyridinethione,[13,48] 2,5-hexanedione (2,5-HD),[49] and acrylamide.[49,50-52] A recent study by Forman et al.[53] of the giant motor axon of lobsters has shown a selective inhibition of retrograde transport by erythro-9-[3-(2-hydroxyanonyl)] adenine (EHNA). It would be of interest to determine whether the inhibition of retrograde transport by EHNA leads to distal nerve fiber degeneration in mammals.

III. EFFECT OF SELECTED NEUROTOXIC CHEMICALS ON THE AXONAL TRANSPORT SYSTEM

A number of chemically unrelated toxic compounds cause nerve fiber degeneration and

Table1
NEUROTOXINS AFFECTING AXONAL TRANSPORT SYSTEM

Axonal transport	Rate	Components	Destination	Role	Neurotoxin	Ref.
Fast anterograde	200—400 mm/day	Smooth endoplasmic reticulum, synaptic vesicles acetylcholine esterase, dopamine-B-hydroxylase, Ca^{2+}, gangliosides	Axolemma and nerve terminals	Renewal of axolemma and synaptic constituents, including vesicle membranes	Acrylamide	49, 92, 93, 120, 121
					DFP[g]	123—127
					MnBK[a]	119,120,144
					2,5-HD[b]	68
					Adriamycin	49, 69, 70
					Aluminum	153
					Methyl-Hg	165
					Alcohol	168,169 170
Slow anterograde	0.1—4.0 mm/day	Microtubules, neurofilaments, microfilaments, glycolytic enzymes	Axon	Renewal of axoplasm; provides substrate for other transport systems	IDPN[c]	135, 136
					DMHD[d]	83[a]
					Aluminum	166, 167
					2,5-HD	37
					Acrylamide	92, 119—121
					3-MHD[e]	38
					CS_2[f]	161
					Alcohol	170
					DFP[g]	119, 144
Retrograde	> 250 mm/day	Smooth endoplasmic reticulum, nerve growth factor, (acetyl choline esterase), dopamine-B-hydroxylase, gangliosides, phospholipids	Cell body	Recycling of endogenous materials	BPAU[h]	47, 154, 155
					Zinc pyridinethione	149
					Acrylamide	47, 49—52, 127
					EHNA[i]	53
					Disulfiram	160

[a] Methyl *n*-butyl ketone.
[b] 2,5-Hexanedione.
[c] B,B′-Iminodipropionitrile.
[d] 3,4-Dimethyl-2,5-hexanedione.

[e] 3-Methyl-2,5-hexanedione.
[f] Carbon disulfide.
[g] Diisopropyl phosphofluoridate.
[h] p-Bromophenylacetalurea.
[i] Erythro-9-[3-(2-hydroxynonyl)] adenine.

Modified from Donald L. Price and John W. Griffin, *Experimental and Clinical Neurotoxicology*, Williams and Wilkins Baltimore. With permission.

inhibit fast and/or slow axonal transport. Whether the axonal transport defect is the cause or merely a reflection of axonal degeneration is under investigation in various laboratories. Irrespective of their site of action, toxic agents which alter the axonal transport system have been discussed below.

A. Spindle Inhibitors and Vinca Alkaloids

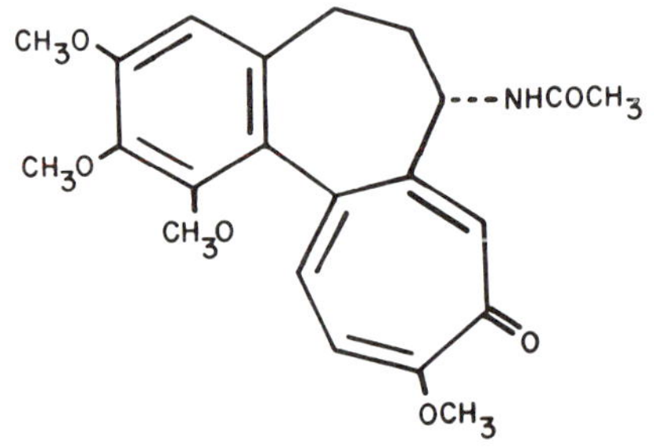

Colchicine

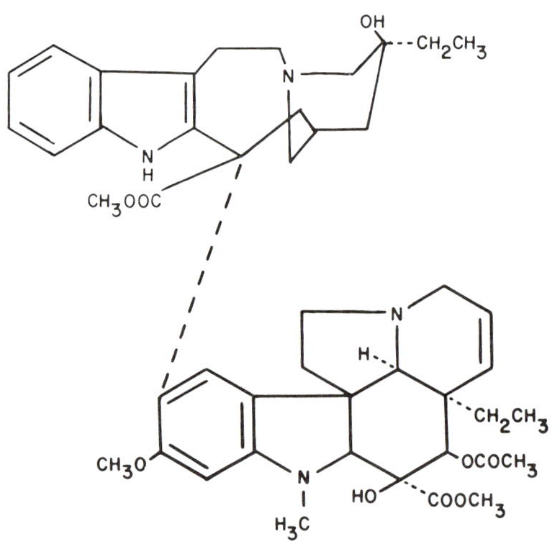

Vinblastin

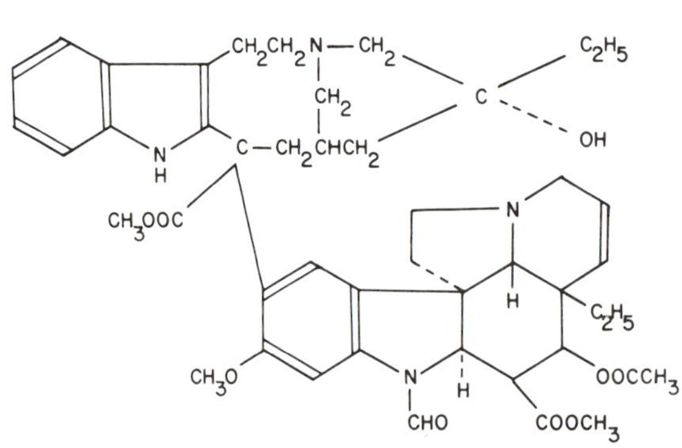

Vincristine

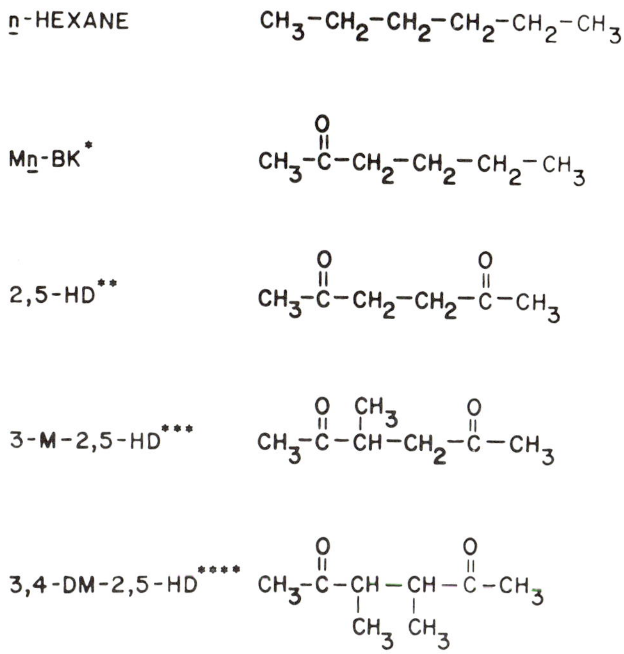

FIGURE 1. Neurotoxic hexacarbons known to produce distal axonal degeneration. Both hexane and M*n*BK are metabolized to 2,5-HD in vivo. * Methyl-*n*-butyl ketone; ** 2,5-hexanedione; *** 3-methyl-2,5-hexanedione; **** 3,4-dimethyl-2,5-hexanedione.

Spindle inhibitors such as colchicine and vinca alkaloids (e.g., vinblastine and vincristine) were probably the first neurotoxic compounds shown to disrupt axonal transport and cause axonal degeneration.[9,13,54] The precise effect of these agents on microtubular function in the nerve fiber is still unknown. It is widely accepted that these agents bind to tubulin and cause disassembly of microtubules, important cytoskeletal elements for axonal transport.[7,9] On the other hand, evidence has been presented that microtubules are not required for fast axonal transport.[58,59]

Vinca alkaloids and colchicine have been widely used in studies of axonal transport in a variety of systems.[13] These agents cause blockade of fast and slow axonal transport in rabbit vagus nerve.[9] Maytansine, which has antimitotic properties similar to that of colchicine, also causes microtubular diassembly and binds selectively to brain tubulin.[55] Several investigators have studied the effect of tubulin-binding agents on axonal transport and have suggested a role of intact microtubules in the mechanism of axonal transport.[56,57] There is increasing evidence that microtubules and neurofilaments are normally present in interconnected form in axoplasm and that this association is vital for nerve-fiber integrity. However, systemic intoxication with a number of mitotic spindle inhibitors causes an increase only in the number of neurofilaments and not of neurotubules. Colchicine also disrupts microtubules in cultured neurons and subsequently causes an aggregation of neurofilaments.

There are also reports which show little or no decrease in the number of microtubules under conditions of axonal transport blockade.[60] Taxol, an experimental antitumor compound isolated from *Taxus brevifolia* has been shown to stabilize microtubules and promote their assembly.[61] The consequences for axonal transport have yet to be investigated.

B. Hexacarbon Compounds

n-Hexane and its metabolites, methyl *n*-butyl ketone (M*n*BK), 2,5-HD, and its 3,4-dimethyl derivative, DMHD (Figure 1), produce a common pattern of axonal degeneration

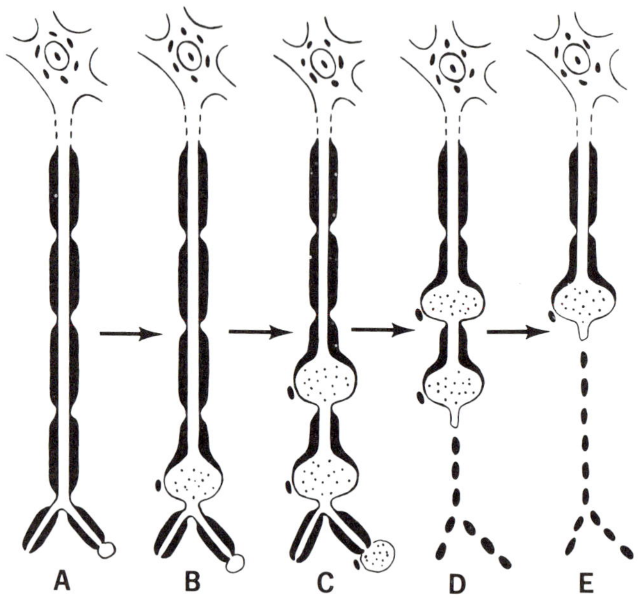

FIGURE 2. Distal (dying-back) axonopathy produced by neurotoxic hexacarbon compounds such as M*n*BK and 2,5-HD. (A) Normal axon. (B) During chronic intoxication, transported materials (stippling) accumulate, causing giant axonal swellings and myelin retraction to develop on the proximal sides of nodes of Ranvier in the distal portion of the fiber. (C) Paranodal swellings appear at more proximal loci with time, as transport failure develops at these sites, while (D,E) the distal portion below an axonal swelling undergoes wallerian-like degeneration. (From Spencer, et al., *Ann. Neurol.*, 5, 501, 1979. With permission.)

in the central and peripheral nervous system, referred to as gamma-diketone neuropathy.[62] Both *n*-hexane and M*n*BK are biotransformed, in vivo, to the neurotoxic diketone 2,5-HD which persists in the body.[63] The neurotoxic properties of this and other long-chain aliphatic diketones are directly linked to the gamma-spacing of the carbonyl groups.[64]

Recent studies have shown that DMHD is considerably more neurotoxic than the parent 2,5-HD.[65] The cardinal initial feature of hexacarbon neuropathy is the appearance of giant axonal swellings filled with 10-nm neurofilaments on the proximal sides of nodes of Ranvier in distal myelinated nerve fibers.[15] The accumulation of neurofilaments presumably reflects a local abnormality of the triplet proteins of SCa. As the disease progresses, axonal swellings appear in more proximal portions of the nerve and the distal region undergoes Wallerian-like degeneration (Figure 2). This distal-proximal degeneration of nerve fibers is known as dying-back disease or simply central peripheral distal axonopathy (CPDA).[66] CPDA is the most common cause of retrograde neuronal degeneration and is produced by exposure to a variety of toxic chemicals such as carbon disulfide, organophosphorus esters, and para-bromophenyl acetyl urea.[67]

Mendell and associates[68] studied fast axonal transport system in M*n*BK-exposed animals. These investigators observed the accumulation of 10-nm neurofilaments and other cell organelles in axonal swellings in association with a significant decrease in fast-transport velocity; the drop in the mean rate of transport correlated with the severity of hexacarbon neuropathy (Figure 3). Morphological studies revealed that, while the number of 10-nm neurofilaments was abnormally increased, microtubules in axonal swellings remained unchanged although their distribution in the axon was altered. In a subsequent study, Sahenk and Mendell[49] reported a decrease in bidirectional transport (both anterograde and retrograde)

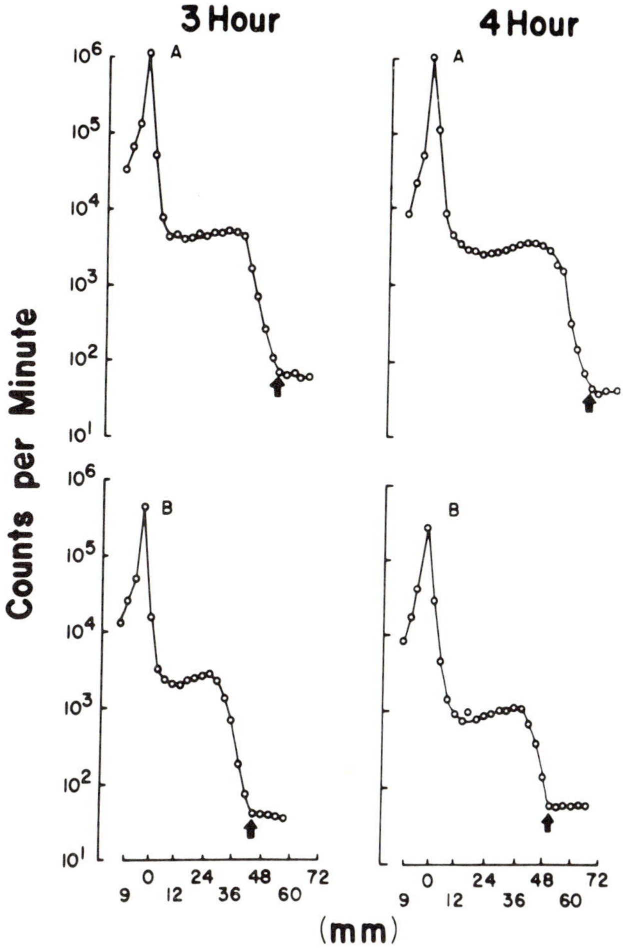

FIGURE 3. Transport profiles at 3 and 4 hr after injection of (^{3}H)-leucine into the DRG. The activity present in 3-mm segments of nerve roots, DRG, and sciatic nerve is given on the ordinate in logarithmic divisions in counts per minute. (A) Represents normal controls and (B) severely paralyzed rats. (From Mendell, J. R., et al., *Brain Res.*, 133, 107, 1977. With permission.)

with no change in distal turnaround at the nerve terminals. In another study, 2,5-hexanediol blocked fast-transported radioactivity in axonal swellings proximal to nodes of Ranvier, an observation consistent with a defect in fast-transport in regions already displaying defects of slow neurofilament transport.[69] Griffin and associates[70] observed local neurofilament accumulation within 2 hr of intraneural injection of 2,5-HD, but fast axonal transport was unchanged. These studies led to the conclusion that a fast-transport defect is not a primary lesion in hexacarbon neuropathy but secondary to cytoskeletal reorganization.

The biochemical mechanism of hexacarbon neuropathy is not understood, although toxic inhibition of selected enzymes in glycolysis is associated with nerve fiber degeneration. Spencer and associates[33] suggested that hexacarbon compounds inhibit key glycolytic enzymes along nerve fibers, thereby restricting energy supply in the axon. Subsequent studies showed that 2,5-HD inhibited glycolytic enzymes, GAPDH, phosphofructokinase (PFK), and enolase (but not LDH, transketolase and succinic dehydrogenase).[71-74] Reduced ATP levels were also observed in rat sciatic nerves treated in vitro with 2,5-HD. ATP levels were restored close to normal by treating the nerve with sodium pyruvate. These findings led to the suggestion that hexacarbons inhibit the axonal transport system by inactivating the glycolytic enzymes (i.e., GAPDH) necessary for driving the transport system along the axon. Glycolytic enzymes are intrinsic components of axoplasm and are transported as a component of slow axonal transport.[35,36] Hexacarbon-induced inhibition of GAPDH activity and the reduction in oxygen consumption suggest that oxidative metabolism in the nerve may be impaired in hexacarbon neuropathy.[75,76]

The mechanism by which hexacarbons inhibit certain enzymes is not understood. Sabri et al.[71,72] suggested that glycolytic enzyme inhibition by 2,5-HD might be linked to its reactivity with sulfhydryl groups essential for enzyme activity. Earlier it was suggested that 2,5-HD has the potential to react with proteins to form Schiff-bases.[77,78] This idea was further explored by Graham and Abou-Donia[79] who showed that 2,5-HD caused an irreversible inhibition of GAPDH activity and forms an orange chromophore (Schiff-base) with a number of proteins. Graham[80] postulated a cross-linkage between 2,5-HD and the ϵ-amino group of neurofilament proteins (e.g., lysine) and suggested a progressive chemical reaction which would denature axonal neurofilaments during their transport along the axon and thereby cause neurofilament accumulation. Graham and associates[65,80] believe that cross-linking of neurofilaments is the molecular event which precipitates axonal transport blockade, leading to axonal degeneration in hexacarbon neuropathy. Recent studies have implicated neurofilaments as the target in 2,5-HD[81,83] and DMHD[65,82,83a] neuropathy. Neurofilament accumulation, however, is not a specific event in gamma diketone neuropathy since a variety of chemically unrelated neurotoxic chemicals (i.e., carbon disulfide, organophosphorous compounds, para-bromophenyl acetyl urea) also produces neurofilamentous accumulation.

A two to threefold increase in the rate of transport of neurofilaments has been reported in the optic system of rats treated with 2,5-HD.[37] Although this is unexpected, it may account for the massive increase of 10-nm neurofilaments in axonal swellings in the distal axon. 3-Methyl-2,5-HD, which has a greater neurotoxic potency than the parent 2,5-HD, also accelerated the rate of transport of neurofilaments in the optic system of rats.[38] A defect in sterol synthesis has also been reported as the biochemical lesion in hexacarbon neuropathy.[84,85] The significance of these reports in relation to the pathogenesis of hexacarbon neuropathy is unknown, since biochemical studies were carried out with animals already exhibiting signs of neuropathy. Future studies should focus on biochemical changes before animals display clinical and/or morphological signs of neurotoxicity.

C. Acrylamide

$$CH_2{=}CH{-}\overset{\overset{\displaystyle O}{\|}}{C}{-}NH_2$$

Acrylamide is one of the most extensively studied neurotoxic compounds (for review see References 86 to 89). The neurotoxic potnetial of acrylamide was first recognized among workers engaged in the production of this chemical for the polymer industry. Early signs of acrylamide neuropathy are sensory loss in the fingers followed by hind limb weakness and paralysis.[62,67] Laboratory animals treated repeatedly with acrylamide also develop peripheral neuropathy which is characterized by hind limb weakness.[90] The pathological hallmark of neuropathy is a slowly evolving distal, retrograde axonal degeneration, similar to that seen with hexacarbons.[90,91] Axons display abnormally large numbers of 10-nm neurofilaments and smooth vesicular profiles,[90,91] and the nerve terminals accumulate tubulovesicular structures as well as dense and multivesicular bodies.[92,93] Neuronal perikaryon display some changes but their degeneration is rarely seen in acrylamide neuropathy.[94-96]

1. Molecular Mechanisms of Acrylamide Neurotoxicity

The mechanism underlying acrylamide-induced CPDA is unknown. Pharmacokinetic studies show that radiolabeled acrylamide is rapidly distributed throughout the body, metabolized and readily excreted in the urine.[97] Small quantities of free acrylamide are detectable in peripheral nerves[97,98] but a large amount of radioactivity is found in the blood.[97,99]

Biochemical studies have focused on the action of acrylamide on protein synthesis,[100,101] glycolytic enzyme inhibition[102-105] and on its sulfhydryl reactivity.[99,106,107] Acrylamide was also reported to reduce leucine incorporation into proteins.[101,108,109] Since *N,N'*-methylene-*bis*-acrylamide (a noneurotoxic compound) also inhibits protein synthesis, it is held that protein-synthesis inhibition by acrylamide is unlikely to precipitate neuropathy.[101] Inhibition of many sulfhydryl enzymes (i.e., alcohol dehydrogenase,[110] phosphofructokinase (PFK), and GAPDH) occurs when crystalline or tissue enzymes are incubated with acrylamide in vitro. Inhibition of GAPDH and PFK can be prevented by prior addition of the sulfhydryl protectant, dithiothreitol.[102-105] Addition of glutathione also prevented the cytotoxic effects of acrylamide in chick ganglia cultures.[111] Neurochemical studies have shown that GAPDH activity in sciatic nerves is selectively reduced while normal levels are present in brain and liver of animals with acrylamide neuropathy.[105] Howland[104] demonstrated that both GAPDH and neuron specific enolase (NSE) were inhibited by acrylamide and most significantly, that the inhibition of NSE was higher distally, the site of axonal degeneration. However, since the enzyme assays were done in nerves from animals already exhibiting clinical signs of acrylamide neuropathy, a reduction in NSE activity may simply be a reflection of distal axonal degeneration. Although the inhibition of GAPDH and NSE presumably depresses ATP generation in the axon,[14] direct measurement of ATP levels in peripheral nerves might be uninterpretable because ATP is also present in the proliferating Schwann cells and fibroblasts in degenerating nerves. Support for the thesis that the inhibition of glycolytic enzymes is causally associated with axonal degeneration is obtained by studies in which animals were partially protected from the neurotoxic effects of acrylamide by dietary sodium pyruvate, an alternative substrate beyond the position of putative enzyme blockade for the generation of ATP[112,113] and by pyridoxal phosphate, a co-enzyme involved in energy transformation.[114] Supplementing the diet with Vitamin A or E has no effect on the development of acrylamide neuropathy in rats.[106] Some other biochemical effects of acrylamide which are related to its neurotoxic property have been explored. For example, interference with pyridine nucleotide function,[111,115] receptor binding,[116,117] and neurotransmission[118] have been studied.

2. Effect of Acrylamide on Axonal Transport

Several studies have examined the efficacy of the axonal transport system in animals intoxicated with acrylamide. Pleasure and colleagues[119] studied the slow axonal transport of ^{3}H-labeled proteins in normal and acrylamide-treated cats, and suggested that the blockade of slow axonal transport might underlie distal axonal degeneration in acrylamide neuropathy. Bradley and Williams[120] studied both the fast and slow components of anterograde axonal transport from dorsal root ganglia along the proximal regions of the sciatic nerve of cats with mild to moderate signs of peripheral neuropathy. Contrary to the earlier study, these investigators found no defect in slow axonal transport, although there was a small reduction in the velocity of the crest but not in the wave front of fast axonal transport. Subsequent studies reported normal rates of slow-transport along proximal axons in acrylamide neuropathy.[92,121] Gold et al.[122] have reported a modest but significant decrease in the transport of slow component proteins following a single injection of acrylamide in rats. These authors suggest that inhibition of slow-transport may lead to a proximal accumulation of neurofilaments with concomitant alteration in axonal caliber.

Rasool and Bradley[123] reported a 50% decrease in the rate transport of acetylcholinesterase, an enzyme which is carried by the fast axonal transport system. Fast anterograde transport of acetylcholinesterase (form A12) was reduced by 60% in the sciatic nerves of chickens treated with acrylamide.[124] Using (^{14}C)-leucine as the precursor, Wier et al.[125] observed a 68% inhibition of fast-transport of proteins in cats exposed to acrylamide. Examination of individual nerve fibers of acrylamide-treated chicks revealed multifocal blockade of fast-transport radiolabel in swollen paranodes and internodes filled with 10-nm neurofilaments and other axoplasmic organelles.[93,126]

Sahenk, and Mendell[49] observed that the distal turnaround of fast-transported materials was unchanged in acrylamide-treated animals, but the rate of retrograde transport was reduced along the distal portion of the axon. Similar conclusions were also drawn by Jakobsen et al.[47,50] that alteration in retrograde transport plays an important role in the development of many axonopathies including that produced by acrylamide. Sidenius and Jakobsen[127] found no change in either fast or slow anterograde axonal transport. Miller et al.[51,52] have shown that single injections of acrylamide produce a dose-dependent decrease in the retrograde transport of ^{125}I-labeled NGF and tetanus toxin. These investigators found a marked attenuation of the induction of ornithine decarboxylase activity and uridine incorporation in dorsal root ganglion by single and repeated injections of acrylamide in transected peripheral nerve,[128] and suggested that acrylamide slowed the repair process in the neuron and that this may be the leading cause of nerve fiber degeneration in acrylamide neuropathy.

In summary, acrylamide is capable of inhibiting both anterograde and retrograde axonal transport but whether these defects are the result or the cause of axonal degeneration has yet to be determined.[52]

D. B,B′-iminodipropionitrile (IDPN)

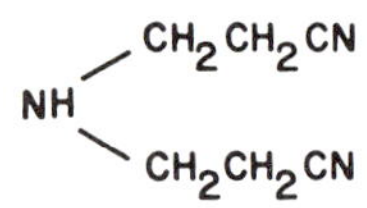

IDPN is a synthetic neurotoxic compound. Single intraperitoneal injection or repeated administration of IDPN produces proximal axonal swellings and progressive atrophy of the distal axon in anterior horn cells, brain stem, red nuclei, and Purkinje cells. Chou and Hartman[129] were the first to observe neurofilamentous accumulation in the proximal axon. More recently, axonal swellings have been observed in the distal sciatic nerve of IDPN-treated cats. Several investigators have examined the neurochemical correlates of IDPN neurotoxicity.[129-131] The most significant biochemical change was increased RNA synthesis in the spinal ganglia,[131] however, changes in amino acid incorporation were not seen in the intoxicated animals.[132] Another study reported increased succino oxidase activity in reticular neurons in IDPN-treated rats.[133]

Since neurofilament accumulation is the hallmark of IDPN intoxication, recent work has been focused on the transport of 10-nm neurofilaments. Axonal transport initially reported to be normal in IDPN-treated animals.[134] Recent work has shown blockade of slow axonal transport selectively.[135,136] Subtle changes in fast anterograde transport have also been reported.[138] Griffin et al.[135] analyzed nerves from rats intoxicated with IDPN and found blockade of slow-transport of neurofilament triplate proteins. The slow-transport of all proteins in SCa was impaired but the effect on neurofilaments was greatest (Figure 4). It was suggested by Griffin and Price[137] that the primary blockade of neurofilaments contributed to the slowing of other components of slow transport (i.e., actin and tubulin). This proximal accumulation of neurofilaments and distal axonal atrophy and its association with slow-transport blockade has provided an excellent model in which altered axonal transport has been directly implicated in the pathogenesis of axonal disease. The precise biochemical lesion which accounts for the inhibition of neurofilament transport is not understood.

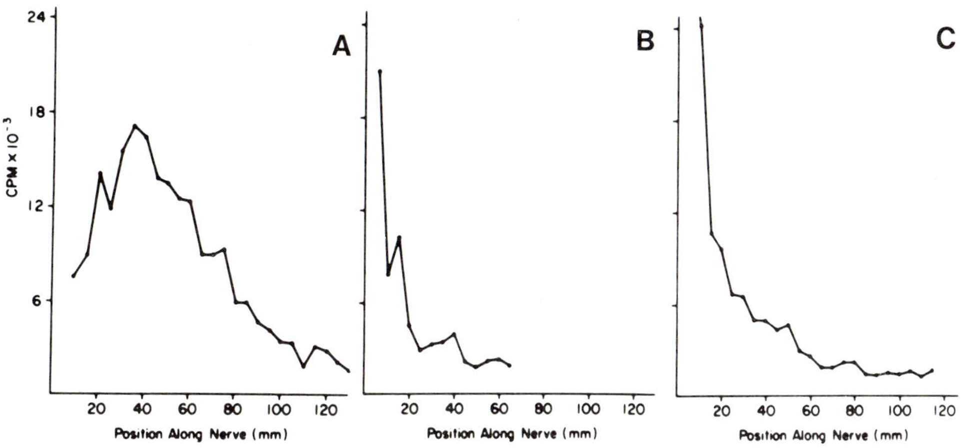

FIGURE 4. Plots of the distribution of slowly transported radioactivity in rat sciatic nerves 3 weeks after injection of labeled amino acids — (^{3}H)-leucine or (^{35}S)-methionine — into the lumbar ventral horns. (A) Plot of a normal root. (B) Plot of a root taken from a rat injected with B,B′-iminodipropionitrile (IDPN) the day after intraspinal injection of labeled amino acid. (C) Plot of a root taken from a rat intoxicated with IDPN for 9 months prior to the axonal transport study. Note, in contrast to the normal slow transport profile in A, the failure of the main slow component peak to move beyond the initial 5 to 10 mm of the nerve in B and C. (From Griffin, J. W., et al., *Science*, 202, 633, 1978. With permission.)

Griffin et al. reported that a defect in slow axonal transport may be preceded by cytoskeletal reorganization in the axon.[138] It was demonstrated that treatment with IDPN given systemically or injected directly into rat sciatic nerve results in a displacement of neurofilaments to the periphery of the axon with neurotubules and mitochondria moved toward the center.[139] It was postulated that IDPN disturbed the interaction between neurofilaments and neurotubules and that this change might underlie the defect in slow transport. These findings also suggest that neurofilaments are not necessary for fast axonal transport since normal transport was observed in IDPN-treated animals.

E. Organophosphorus Compounds

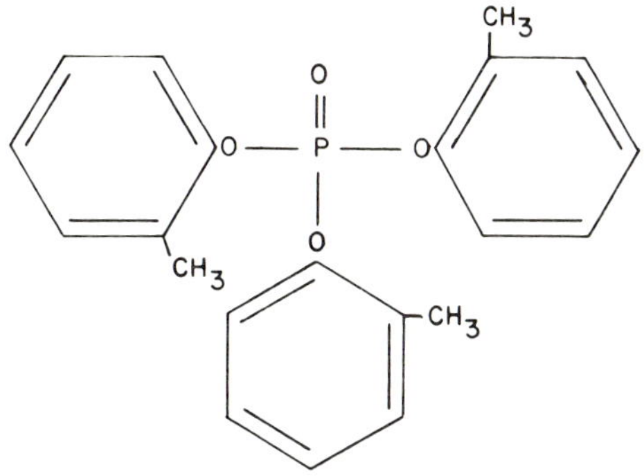

Certain organophosphorus (OP) also induce axonal degeneration of the CPDA type (for review see References 140 and 141). This neurotoxic property does not involve the inhibition of acetylcholinesterase, but requires phosphorylation and aging of the phosphorylated protein known as Neuropathy Target Esterase (formerly Neurotoxic Esterase).[142,143]

Pleasure et al.[119] found no change in slow axonal transport in the sensory and motor nerve fibers of cats treated with TOCP. Fast-transport was not studied by these investigators. James and Austin[144] investigated the effect of diisopropyl phosphofluoridate (DFP) on both fast- and slow-transport of protein in peripheral nerve of chickens. DFP had no effect on the distribution of radioactivity in the sciatic nerve carried by either fast or slow axonal transport. Bradley and Williams[120] compared fast- and slow-transport of radioactivity following ^{3}H-leucine injection in the L7 dorsal root ganglion of control and TOCP-treated cats. These authors observed a reduction in the rate of movement of the crest of fast-transport in TOCP-treated animals, however, the advancing front of radioactivity remained unchanged. Although these authors observed a reduction in the velocity of slow-transport in TOCP neuropathy, they concluded that a defect in axonal transport is unlikely to cause axonal degeneration in TOCP neuropathy.[120]

F. Zinc Pyridinethione

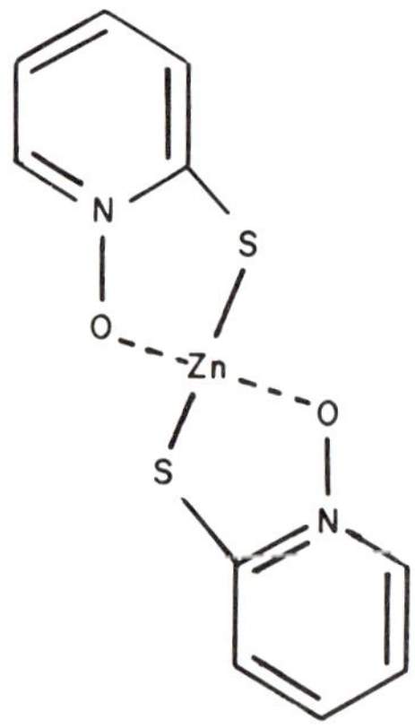

Zinc pyridinethione (ZPT) is the water-insoluble salt of I-hydroxy-2(IH)-pyridinethione. ZPT is a broad spectrum antifungal and antibacterial compound used in certain antidandruff hair shampoos.[145] In rodents, ZPT or its parent compound produces CPDA and hind limb paralysis. The mechanism of action of ZPT is not understood. ZPT is capable of inactivating certain enzymes by chelating metal cofactors which might account for its biological activity.[145] It has also been suggested that pyridinethione could act as an antimetabolite of nicotinic acid.[145] The parent compound of ZPT inhibits glycolytic enzymes[146] which may account for the reduced levels of intracellular ATP levels in fungi.[147] The pathogenesis of ZPT neuropathy is not understood but it appears to be independent of zinc since NaPT also displays neurotoxic properties. Neuropathological studies show that ZPT administration causes accumulation of tubulovesicular profiles in motor nerve terminals followed by distal axonal degeneration.[145,148] Studies of fast axonal transport in ZPT neuropathy, carried out by Sahenk and Mendell,[149] suggest that accumulation of membranous materials may be related to an abnormality in the turnaround of fast-transported materials in the distal axon: this accounts for the decrease of transported materials in retrograde transport.[145]

G. *p*-Bromophenylacetylurea

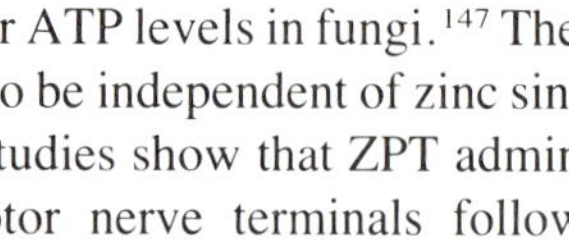

p-Bromophenylacetylurea (BPAU) is a synthetic crystalline compound sparingly soluble in water.[154] A single dose of BPAU produces hind limb weakness after 7 to 8 days leading to paralysis. BPAU causes distal axonal degeneration in the central and peripheral nervous system (CPDA). Distal regions of the axon show the presence of tubulomembranous debris, which suggests a defect in the fast-transport system. Neuronal perikarya reveal little or no change. The mechanism of action of BPAU is unknown. Incorporation of labeled amino acid into protein decreases in dorsal root ganglia of BPAU-treated animals, but whether this effect is related to the pathogenesis of axonal disease is unknown.

The accumulation of tubulomembranous structures in distal axons in BPAU neuropathy is similar to that seen in cases where accumulation of smooth endoplasmic reticulum is produced by ligation or cold-block. Jakobsen and Brimijoin[155] have investigated axonal transport alterations in BPAU-treated animals and found that BPAU neuropathy is not associated with a change in fast anterograde axonal transport of proteins; however, there was a reduction in the retrograde transport system due to a defect in the distal turn around of membranous organelles, similar to that seen in zinc pyridinethione neuropathy.[149] Jakobsen et al.[47] believe that changes in retrograde axonal transport play an important role not only in BPAU neuropathy but also in the development of many toxic neuropathies.

H. Disulfiram and Carbon Disulfide

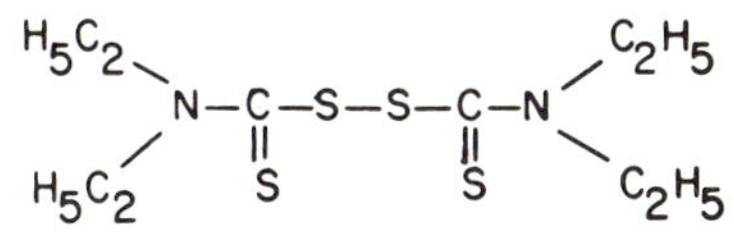

Disulfiram (tetraethylthiuram disulfide) is used in the treatment of chronic alcoholic patients. Disulfiram is neurotoxic and known to produce peripheral neuropathy.[156] It is catabolized to carbon disulfide (CS_2) in the body. Neurotoxicity studies in experimental animals and humans have shown that carbon disulfide induces a typical distal axonopathy and accumulation of neurofilaments.[157] Systemically injected labeled carbon disulfide has been shown to bind to neurofilament protein,[158] although it is unknown whether this is selective. Anzil[159] studied alterations in axonal transport in disulfiram neuropathy in the rat and reported a reduction in retrograde transport. Anzil et al.[160] have also reported that topical application of disulfiram reduces retrograde transport of horseradish peroxidase to approximately 50% of its original amount.

Pappolla et al.[161] have recently reported an increase in the rate of transport of neurofilaments in carbon disulfide neuropathy. This phenomenon, it was suggested, may account for the accumulation of neurofilaments in the distal part of the optic axon. Carbon disulfide inhibits key enzymes in glycolysis, uncouples oxidative phosphorylation,[162,164] and reduces ATP levels.

I. Doxorubicin

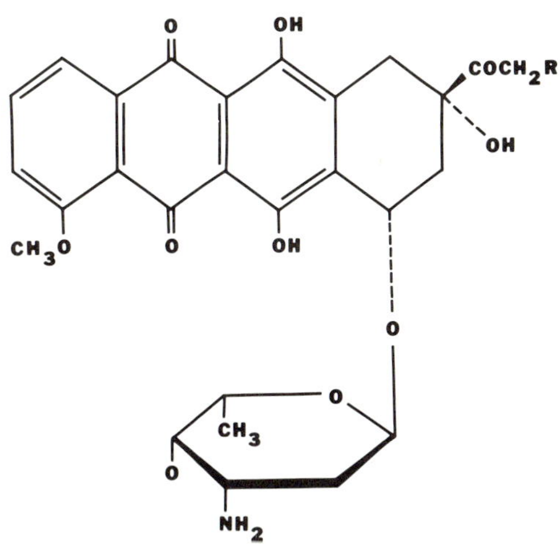

Doxorubicin, commercially known as Adriamycin®, is a glycosidic anthracycline antibiotic currently being used in the treatment of cancer (for review see Reference 150). The drug is rapidly distributed to various tissues including dorsal root ganglia following intravenous injection and forms a complex with deoxyribonucleic acid, and blocks RNA synthesis. Changes in ganglion cells have been observed in cancer patients treated with this drug. However, no specific neurological abnormality was observed in these patients. Philips and associates,[151] and Cho et al.[152] studied neurological and pathological effects, respectively, of doxorubicin in rats. Morphological examination revealed alterations in the lumbosacral dorsal root ganglia. Sahenk and Mendell[153] studied fast axonal transport in rats treated with adriamycin. These investigators observed little change in the rate of fast-transport with doxorubicin although the time of perikaryal processing of (^{3}H)-leucine by dorsal root ganglia was increased. Further intoxication with doxorubicin causes breakdown of the neuronal perikaryon and axonal degeneration in a proximo-distal or dying-forward manner.

J. Other Neurotoxins

1. Aluminum

Liwnicz et al.[165] investigated axoplasmic transport in rabbits intoxicated with Al^{+3} and found no inhibitory effect on fast axonal transport. Injections of $AlCl_3$ into the lumbar spinal cord or at the midline of the medulla causes a selective blockade of neurofilaments.[166,167] It is believed that a selective defect in the transport of NF might be the underlying cause of Al^{+3} neurotoxicity which is characterized by massive accumulation of neurofilaments within the perikarya and axon.

2. Erythro-9-[3-(2-Hydroxynonyl)] Adenine (EHNA)

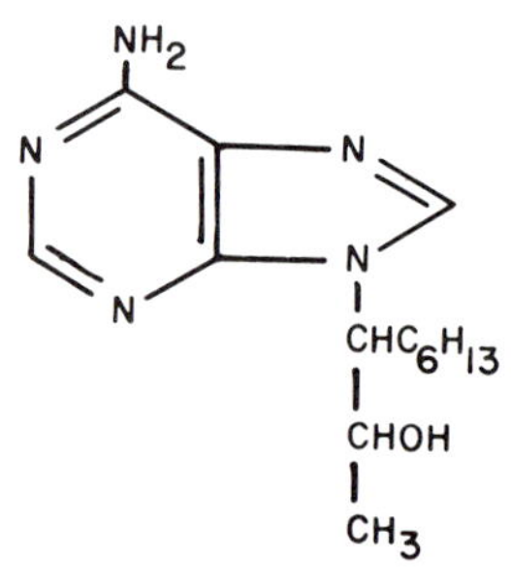

Erythro-9-[3-(2-hydroxynonyl)] adenine (EHNA) selectively reduces the velocity of retrograde but not anterograde transport of subcellular particles in the lobster.[53] The inhibitory effect of EHNA on retrograde transport may be related to its action on dynein and protein carboxymethylation. EHNA might be useful in controlling the direction of fast axonal transport.

3. Methylmercury

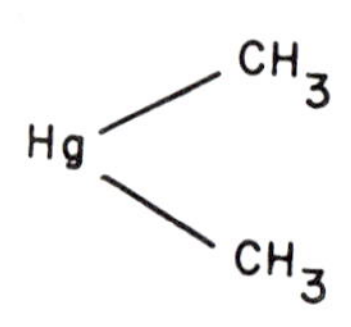

Increased rates of fast-transport were seen in chronic methylmercury intoxication, however, when locally injected, methylmercury inhibited fast axonal transport in the rat.[168] In an earlier report, Abe et al.[169] found that methylmercury blocked fast-transport as well as depolymerized reassembled microtubules.

4. Ethanol

$$CH_3 - CH_2 - OH$$

A significant increase in fast anterograde transport of acetyl cholinesterase and no change in slow-transport of cholineacetyltransferase has been reported in sciatic nerves of alcohol-exposed rats.[170] The relationship between alteration in fast-transport in sciatic nerves and a mild distal axonal degeneration in the ventral caudal nerve of rats chronically treated with ethanol is not understood.

IV. CONCLUSION

This chapter reviews available information on the effect of selected neurotoxic compounds on axoplasmic transport. Most of the compounds discussed are chemcially unrelated and appear to have a selective action. For example, some neurotoxins block only slow-transport (IDPN) and have no effect on fast-transport. Certain neurotoxins interfere only with the fast retrograde transport system and have no significant effect on anterograde transport.[53] Axonal transport abnormalities coupled with biochemical assays can provide clues for specific sites and levels at which a given neurotoxin acts. An understanding of the mechanism of action of neurotoxic chemicals could provide a basis for a rational treatment of neurotoxic diseases.

ACKNOWLEDGMENTS

The author is grateful to Dr. Peter S. Spencer, Director, Institute of Neurotoxicology for his interest, encouragement, and suggestions in organizing this manuscript. The author appreciates the invaluable help of Mrs. Gloria Warkenthien, Ms. Patricia A. Vacchelli, and Mrs. Monica Bischoff-Fenton in the preparation of this manuscript. Finanacial support was provided by OH00851 and NS19611.

REFERENCES

1. **Spencer, P. S.,** Neurotoxic chemicals as probes of cellular mechansims of neuromuscular diseases, in *Current Topics in Nerve and Muscle Research,* Aguayo, A. and Karpati, G., Eds., North Holland, Amsterdam, 1978, 274.
2. **Jonsson, G.,** Chemical neurotoxins as denervation tools in neurobiology, *Ann. Rev. Neurosci.,* 3, 169, 1980.
3. **Kosik, K. S. and Selkoe, D. J.,** Experimental models of neurofilamentous pathology, in, Neurofilaments, Marotta, C. A., Ed., University of Minnesota Press, Minneapolis, Miss., 1983, 155.
4. **Griffin, J. W., Price, D. L., and Hoffman, P. N.,** Neurotoxic probes of the axonal cytoskeleton, *Trends Neurosci.,* 6, 490, 1983.
5. **Schwartz, J. H.,** Axonal transport: components, mechanisms, and specificity, *Ann. Rev. Neurosci.,* 2, 467, 1979.
6. **Lasek, R. J.,** Axonal transport: a dynamic view of neuronal structures, *Trends Neurosci.,* 3, 87, 1980.
7. **Grafstein, B. and Forman, D. S.,** Intracellular transport in neurons, *Physiol. Rev.,* 60, 1168, 1980.
8. **Forman, D. S.,** New approaches to the study of the mechanism of fast axonal transport, *Trends Neurosci.,* 7, 112, 1983.
9. **Ochs, S.,** *Axoplasmic Transport and Its Relation to Other Nerve Functions,* John Wiley & Sons, New York, 1982.
10. **Weiss, D. G., Ed.** *Axoplasmic Transport,* Springer-Verlag, Basel, 1982.
11. **Weiss, D. C. and Gorio, A., Eds.,** *Axoplasmic Transport in Physiology and Pathology,* Springer-Verlag, Basel, 1982.

12. **Price, D. L. and Griffin, J. W.,** Neurons and ensheathing cells as targets of disease process, in, *Experimental and Clinical Neurotoxicology*, Spencer, P. S. and Schaumburg, H. H., Eds., Williams & Wilkins, Baltimore, 1980, 2.
13. **Mendell, J. R. and Sahenk, Z.,** Interference of neuronal processing and axoplasmic transport by toxic chemicals, in, *Experimental and Clinical Neurotoxicology*, Spencer, P. S. and Schaumburg, H. H., Eds., Williams & Wilkins, Baltimore, 1980, 139.
14. **Sabri, M. I. and Spencer, P. S.,** Toxic distal axonopathy: biochemical studies and hypothetical mechanisms, in, *Experimental and Clinical Neurotoxicology*, Spencer, P. S. and Schaumburg, H. H., Eds., Williams & Wilkins, Baltimore, 1980, 206.
15. **Spencer, P. S., Schaumburg, H. H., Sabri, M. I., and Veronesi, B.,** The enlarging view of hexacarbon neurotoxicity, *CRC Crit. Rev. Toxicol.*, 7, 279, 1980.
16. **Spencer, P. S., Miller, M. S., Ross, S. M., Schwab, B. W., and Sabri, M. I.,** Biochemical mechanism underlying primary degeneration of axons, in, *Handbook of Neurochemistry*, Lajtha, A., Ed., Plenum Press, New York, 1985, 31.
17. **Schoental, R. and Cavanagh, J. B.,** Mechanisms involved in the ''dying-back'' process — an hypothesis implicating co-enzymes, *Neuropathol. Appl. Neurobiol.*, 3, 145, 1977.
18. **Cavanagh, J. B.,** The significance of the process in experimental and human neurological disease, *Int. Rev. Exp. Pathol.*, 3, 219, 1964.
19. **Cavanagh, J. B.,** Peripheral neuropathy caused by chemical agents, *CRC Crit. Rev. Toxicol.*, 2, 365, 1973.
20. **Spencer, P. S. and Schaumburg, H. H.,** A review of acrylamide neurotoxicity. II. Experimental animal neurotoxicity and pathologic mechanisms, *Can. J. Neurol. Sci.*, 1, 152, 1974.
21. **Sterman, A. B.,** Cell body remodelling during dying-back axonopathy: DRG changes during advanced disease, *J. Neuropathol. Exp. Neurol.*, 41, 400, 1982.
22. **Kristensson, K.,** Retrograde transport of macromolecules in axons, *Ann. Rev. Pharmacol. Toxicol.*, 18, 97, 1978.
23. **Willard, M., Cowan, W. M., and Vagelos, P. R.,** The polypeptide composition of intra-axonally transported proteins: evidence for four transport velocities, *Proc. Natl. Acad. Sci. U.S.A.*, 71, 2183, 1974.
24. **Ranish, N. and Ochs, S.,** Fast axoplasmic transport of acetylcholinesterase in mammalian nerve fibers, *J. Neurochem.*, 19, 2641, 1972.
25. **Brimijoin, S.,** Stop-flow: a new technique for measuring axonal transport, and its application to the transport of dopamine-B-hydroxylase, *J. Neurobiol.*, 8, 251, 1975.
26. **Ochs, S., Sabri, M. I., and Johnson, J.,** Fast transport system of material in mammalian nerve fibers, *Science*, 163, 686, 1969.
27. **Ochs, S., Sabri, M. I., and Ranish, N.,** Somal site of synthesis of fast transported materials in mammalian nerve fibers, *J. Neurobiol.*, 1, 329, 1970.
28. **Ochs, S. and Smith, C.,** Fast axoplasmic transport in mammalian nerve *in vitro* after block of glycolysis with iodoacetic acid, *J. Neurochem.*, 18, 833, 1971.
29. **Sabri, M. I. and Ochs, S.,** Relation of ATP and creatine phosphate to fast axoplasmic transport in mammalian nerve, *J. Neurochem.*, 19, 2821, 1972.
30. **Allen, R. D., Allen, N. S., and Travis, J. L.,** Video-enhanced contrast, differential interference contrast (AVEC-DIC) microscopy: a new method capable of analyzing microtubule-related motility in the reticulopodial network of allogromia laticollaris, *Cell Motil.*, 1, 291, 1981.
31. **Allen, R. D., Metuzals, J., Tasaki, I., Brady, S. T., and Gilbert, S. P.,** Fast axonal transport in squid giant axon, *Science*, 218, 1127, 1982.
32. **Brady, S. T., Lasek, R. J., and Allen, R. D.,** Fast axonal transport in extruded axoplasm from squid giant axon, *Science*, 218, 1129, 1982.
33. **Spencer, P. S., Sabri, M. I., Schaumburg, H. H., and Moore, C. L.,** Does a defect of energy metabolism in the nerve fiber underlie axonal degeneration in polyneuropathies?, *Ann. Neurol.*, 5, 501, 1979.
34. **Hoffman, P. N. and Lasek, R. J.,** The slow component of axonal transport, *J. Cell Biol.*, 66, 351, 1975.
35. **Brady, S. T. and Lasek, R. J.,** The slow components of axonal transport: movements, compositions and organization, in, *Axoplasmic Transport*, Weiss, D. G., Ed., Springer-Verlag, Basel, 1982, 206.
36. **Brady, S. T. and Lasek, R. J.,** Nerve-specific enolase and creatine phosphokinase in axonal transport: soluble proteins and the axoplasmic matrix, *Cell*, 23, 515, 1981.
37. **Monaco, S., Crane, R., Autilio-Gambetti, L., and Gambetti, P.,** Transport of neurofilaments is accelerated in 2,5-hexanedione axonopathy, *Am. Assoc. Neuropathol.*, 59, 76, 1983.
38. **Monaco, S., Wongmongkolrit, T., Sayre, L., Autilio-Gambetti, L., and Gambetti, P.,** Giant axonopathy in 3-methyl-2,5-hexanedione (3-M-2,5-HD). Effect on morphology and slow axonal transport, *Am. Assoc. Neuropathol.*, 60, 33, 1984.
39. **Bisby, M. A.,** Retrograde axonal transport of endogenous proteins, in, *Axoplasmic Transport*, Weiss, D. G., Ed., Springer-Verlag, Basel, 1982, 193.

40. **Brimijoin, S. and Helland, L.,** Rapid retrograde transport of dopamine B-hydroxylase as examined by the stop flow technique, *Brain Res.*, 102, 217, 1976.
41. **Kirkpatrick, J. B., Bray, J. J., and Palmer, S. M.,** Visualization of axoplasmic flow *in vitro* by Nomarski microscopy. Comparison to rapid flow of radioactive proteins, *Brain Res.*, 43, 1, 1972.
42. **Leestma, J. E.,** Velocity measurement of particulate neuroplasmic flow in organized mammalian CNS tissue cultures, *J. Neurobiol.*, 7, 173, 1976.
43. **Forman, D. S., Padjen, A. L., and Siggins, G. R.,** Axonal transport of organelles visualized by light microscopy: cinemicro-graphic and computer analysis, *Brain Res.*, 136, 197, 1977.
44. **Bunt, A. H. and Lund, R. D.,** Vinblastine-induced bolckage of orthograde and retrograde axonal transport of protein in retinal ganglion cells, *Exp. Neurol.*, 45, 288, 1974.
45. **Edstrom, A. and Hanson, M.,** Retrograde axonal transport of protein *in vitro* in frog sciatic nerves, *Brain Res.*, 61, 311, 1973.
46. **Brimijoin, S., Dyck, P. J., Jakobsen, J., and Lambert, E. H.,** Axonal transport in human nerve disease and in the experimental neuropathy induced by p-bromophenylacetylurea, in, *Axoplasmic Transport in Physiology and Pathology,* Weiss, D. G. and Gorio, A., Eds., Springer-Verlag, Basel, 1982, 124.
47. **Jakobsen, J., Brimijoin, S., and Sidenius, P.,** Axonal transport in neuropathy, *Muscle Nerve,* 6, 164, 1983.
48. **Sahenk, Z. and Mendell, J. R.,** Axoplasmic transport in zinc pyridinethione neuropathy: Evidence for an abnormality in distal turn around, *Brain Res.*, 186, 343, 1980.
49. **Sahenk, Z. and Mendell, J. R.,** Acrylamide and 2,5-hexanedione neuropathies: abnormal bidirectional transport rates in distal axons, *Brain Res.*, 219, 397, 1981.
50. **Jakobsen, J. and Sidenius, P.,** Early and dose-dependent decrease of retrograde axonal transport in acrylamide-intoxicated rats, *J. Neurochem.*, 40, 447, 1983.
51. **Miller, M. S., Miller, M. J., Burks, T. F., and Sipes, I. G.,** Altered retrograde axonal transport of nerve growth factor after single and repeated doses of acrylamide in the rat, *Toxicol. Appl. Pharmacol.*, 69, 96, 1983.
52. **Miller, M. S. and Spencer, P. S.,** Single doses of acrylamide reduce retrograde transport velocity, *J. Neurochem.*, 43, 1401, 1984.
53. **Forman, D. S., Brown, K. J., and Promersberger, M. E.,** Selective inhibition of retrograde axonal transport by erythro-9-(3-(2-hydroxynonyl)) adenine (EHNA), *Soc. Neurosci. Abstr.*, 9, 147, 1983.
54. **Iqbal, Z. and Ochs, S.,** Uptake of vinca alkaloids into mammalian nerve and its subcellular components, *J. Neurochem.*, 34, 59, 1980.
55. **Madelbaum-Shavit, F., Wolpert-DeFilippes, M. R., and Johns, D. G.,** Binding of maytansine to rat brain tubulin, *Biochem. Biophys. Res. Commun.*, 12, 47, 1976.
56. **Schmitt, F. O.,** Fibrous proteins — neuronal organelles, *Proc. Natl. Acad. Sci. U.S.A.*, 60, 1092, 1968.
57. **Ochs, S.,** Fast transport of materials in mammalian nerve fibers, *Science,* 176, 252, 1972.
58. **Brady, S., Crothers, S. D., Nosal, C., and McClure, W. O.,** Fast axonal transport in the presence of high Ca^{2+}: evidence that microtubules are not required, *Proc. Natl. Acad. Sci.*, 77, 5909, 1980.
59. **Brady, S.,** Microtubules and the mechanism of fast axonal transport, in, *Axoplasmic Flow,* Weiss, D. G., Ed., Springer-Verlag, Basel 1982, 301.
60. **Dustin, P.,** *Microtubules,* Springer-Verlag, Basel, 1978.
61. **Schiff, P. B. and Horwitz, S. B.,** Taxol stabilizes microtubules in mouse fibroblast cells, *Proc. Natl. Acad. Sci. U.S.A.*, 77, 1561, 1980.
62. **Schaumburg, H. H., Spencer, P. S., and Thomas, P. K.,** *Disorders of Peripheral Nerves,* F. A. Davis, Philadelphia, 1983.
63. **DiVincenzo, G. D., Hamilton, M. L., Kaplan, C. J., and Dedinas, J.,** Metabolic fate and disposition of ^{14}C-labeled methyl *n*-butyl ketone in the rat, *Toxicol. Appl. Pharmacol.*, 41, 547, 1976.
64. **O'Donoghue, J. J. and Krasavage, W. J.,** Identification and characterization of methyl *n*-butyl ketone neurotoxicity in laboratory animals, in, *Experimental and Clinical Neurotoxicology,* Spencer, P. S. and Schaumburg, H. H., Ed., Williams & Wilkins, Baltimore, Md., 1980, 856.
65. **Anthony, D. C., Giangaspero, F., and Graham, D. G.,** The spatio-temporal pattern of the axonopathy associated with the neurotoxicity of 3,4-dimethyl-2,5-hexanedione in the rat, *J. Neuropathol. Exp. Neurol.*, 42, 548, 1983.
66. **Spencer, P. S. and Schaumburg, H. H.,** Central peripheral distal axonopathy — the pathology of dying-back polyneuropathies, in, *Progress in Neuropathology,* Vol. 3., Zimmerman, H. M., Ed., Grune & Stratton, New York, 1976, 253.
67. **Spencer, P. S. and Schaumburg, H. H., Eds.,** *Experimental and Clinical Neurotoxicology,* Williams & Wilkins, Baltimore Md., 1980.
68. **Mendell, J. R., Sahenk, Z., Saida, K., Weiss, H. S., Savage, R., and Couri, D.,** Alterations of fast axoplasmic transport in experimental methyl *n*-butyl ketone neuropathy, *Brain Res.*, 133, 107, 1977.

69. **Spencer, P. S. and Griffin, J. W.,** Disruption of axoplasmic transport by neurotoxic agents. The 2,5-hexanedione model, in, *Axoplasmic Transport in Physiology and Pathology,* Weiss, D. G. and Gorio, A., Eds., Springer-Verlag, Basel, 1982, 92.
70. **Griffin, J. W., Fahnstock, R. E., and Price, D. L.,** Fast axonal transport through focal regions of cytoskeletal segregation produced by chemical neurotoxins, *Neurology,* 32, A214, 1982.
71. **Sabri, M. I., Moore, C. L., and Spencer, P. S.,** Studies on the biochemical basis of distal axonopathies. I. Inhibition of glycolysis by neurotoxic hexacarbon compounds, *J. Neurochem.,* 32, 683, 1979.
72. **Sabri, M. I., Ederle, K., Holdsworth, C. E., and Spencer, P. S.,** Studies on the biochemical basis of distal axonopathies. II. Specific inhibition of fructose-6-phosphate kinase by 2,5-hexanedione and methyl *n*-butyl ketone, *Neurotoxicology,* 1, 285, 1979.
73. **Sabri, M. I.,** *In vitro* effect of *n*-hexane and its metabolites on selected enzymes in glycolysis, pentose phosphate pathway and citric acid cycle, *Brain Res.,* 297, 145, 1984.
74. **Howland, R. D., Vyas, I. L., Lowndes, H. W., and Argentieri, T. M.,** The etiology of toxic peripheral neuropathies: *in vitro* effects of acrylamide and 2,5-hexanedione on brain enolase and other glycolytic enzymes, *Brain Res.,* 202, 131, 1980.
75. **Couri, D. and Nachtman, J. P.,** Biochemical and biophysical studies of 2,5-hexanedione neuropathy, *Neurotoxicology,* 1, 269, 1979.
76. **Couri, D. and Milks, M.,** Toxicity and metabolism of the neurotoxic hexacarbons *n*-hexane, 2-hexanone, and 2,5-hexanedione, *Ann. Rev. Pharmacol. Toxicol.,* 22, 145, 1982.
77. **Scheontal, R. and Cavanagh, J. B.,** Mechanisms involved in the ''dying-back'' process — an hypothesis implicating co-enzymes, *Neuropathol. Appl. Neurobiol.,* 3, 145, 1977.
78. **Savolainen, H.,** Some aspects of the mechanisms by which industrial solvents produce neurotoxic effects, *Chem. Biol. Interact.,* 18, 1, 1977.
79. **Graham, D. G. and Abou-Donia, M. B.,** Studies of the molecular pathogenesis of hexane neuropathy. I. Evaluation of the inhibition of glyceraldehyde-3-phosphate dehydrogenase by 2,5-hexanedione, *J. Toxicol. Environ. Health,* 6, 621, 1980.
80. **Graham, D. G.,** Hexane neuropathy: a proposal for pathogenesis of a hazard of occupational exposure and inhalant abuse, *Chem. Biol. Interact.,* 32, 339, 1980.
81. **DeCaprio, A. P., Olajos, E. S., and Weber, P.,** Convalent binding of a neurotoxic *n*-hexane metabolite: conversion of primary amines to substituted pyrrole adducts by 2,5-hexanedione, *Toxicol. Appl. Pharmacol.,* 65, 440, 1982.
82. **Anthony, D. C., Szaka-Quin, G. Y., and Graham, D. G.,** Stable proteins as targets in diketone neuropathy, *Am Assoc. Neuropathol.,* 305, 1984.
83. **Durham, H. D., Pena, S. D. J., and Carpenter, S.,** The neurotoxins 2,5-hexanedione and acrylamide promote aggregation of intermediate filaments in cultured fibroblasts, *Muscle Nerve,* 6, 631, 1983.
83a. **Griffin, J. W., Anthony, D. C., Fahnestock, K., Hoffman, P. N., and Graham, D. G.,** 3,4-Dimethyl-2,5-hexanedione impairs axonal transport of neurofilament proteins, *Neurosci. Abstr.,* 9, 147, 1983.
84. **Gillies, P. J., Norton, R. M., and Bus, J. S.,** Effect of 2,5-hexanedione on lipid biosynthesis in sciatic nerve and brain of the rat, *Toxicol. Appl. Pharmacol.,* 54, 210, 1980.
85. **Gillies, P. J., Norton, R. M., White, E. L., and Bus, J. S.,** Inhibition of sciatic nerve sterologenesis in hexacarbon-induced distal axonopathy in the rat, *Toxicol. Appl. Pharmacol.,* 54, 217, 1980.
86. **Spencer, P. S. and Schaumburg, H. H.,** A review of acrylamide neurotoxicity. I. Properties, uses and human exposure, *Can. J. Neurol. Sci.,* 1, 143, 1974.
87. **Spencer, P. S. and Schaumburg, H. H.,** A review of acrylamide neurotoxicity. II. Experimental animal neurotoxicity and pathologic mechanisms, *Can. J. Neurol. Sci.,* 1, 152, 1974.
88. **LeQuesne, P. M.,** Acrylamide, in, *Experimental and Clinical Neurotoxicology,* Spencer, P. S. and Schaumburg, H. H., Eds., Williams & Wilkins, Baltimore, 1980, 309.
89. **Sabri, M. I.,** Mechanism of action of acrylamide on the nervous system, *Biol. Memoirs.* 8, 16, 1983.
90. **Schaumburg, H. H., Wisniewski, H. M., and Spencer, P. S.,** Ultrastructural studies of the dying-back process. I. Peripheral nerve terminal and axon degeneration in systemic acrylamide intoxication, *J. Neuropathol. Exp. Neurol.,* 33, 260, 1974.
91. **Prineas, J.,** The pathogenesis of dying-back polyneuropathies. II. An ultrastructural study of experimental acrylamide intoxication in the cat, *J. Neuropathol. Exp. Neurol.,* 28, 598, 1969.
92. **Griffin, J. W., Price, D. L., and Drachman, D. B.,** Impaired axonal regeneration in acrylamide intoxication, *J. Neurobiol.,* 8, 355, 1977.
93. **Chretien, M., Patey, G., Souyri, F., and Droz, B.,** Acrylamide-induced neuropathy and impairment of axonal transport of proteins. II. Abnormal accumulation of smooth endoplasmic reticulum at sites of focal retention of fast transported protein. Electron microscope radioautographic study, *Brain Res.,* 205, 15, 1981.
94. **Sterman, A. B.,** Acrylamide induces early morphologic reorganization of the neuronal cell body, *Neurology,* 32, 1023, 1982.

95. **Sterman, A. B.,** Altered sensory ganglia in acrylamide neuropathy. Quantitative evidence of neuronal reorganization, *Exp. Neurol.*, 42, 166, 1983.
96. **Sterman, A. B.,** Acrylamide induced remodelling of perikarya in rat superior cervical ganglia, *Neuropathol. Appl. Neurobiol.*, 3, 221, 1984.
97. **Miller, M. J., Carter, D. E., and Sipes, I. G.,** Pharmacokinetics of acrylamide in Fisher-334 rats, *Toxicol. Appl. Pharmacol.*, 63, 36, 1982.
98. **Poole, C. F., Sye, W.-F., Zlatkis, A., and Spencer, P. S.,** Determination of acrylamide in nerve tissue homogenates by electron-capture gas chromatography, *J. Chromatography*, 217, 239, 1981.
99. **Hashimoto, K. and Aldridge, W. N.,** Studies on acrylamide, a neurotoxic agent, *Biochem. Pharmacol.*, 19, 2581, 1970.
100. **Hashimoto, K. and Ando, K.,** Alteration of amino acid incorporation into proteins of the nervous system *in vitro* after administration of acrylamide to rats, *Biochem. Pharmacol.*, 22, 1057, 1973.
101. **Schotman, P., Gipon, L., Jennekens, F. G. I., and Gispen, W. H.,** Polyneuropathies and CNS protein metabolism III. Changes in protein synthesis rate induced by acrylamide intoxication, *J. Neuropathol. Exp. Neurol.*, 37, 820, 1978.
102. **Sabri, M. I. and Spencer, P. S.,** Inhibition of glyceraldehyde-3-phosphate dehydrogenase and other glycolytic enzymes by acrylamide, *Neurosci. Lett.*, Suppl. 5, 455, 1980.
103. **Howland, R. D., Vyas, I. L., Lowndes, H. E., and Argentieri, T. M.,** The etiology of toxic peripheral neuropathies: *in vitro* effects of acrylamide and 2,5-hexanedione on brain enolase and other glycolytic enzymes, *Brain Res.*, 202, 131, 1980.
104. **Howland, R. D.,** The etiology of acrylamide neuropathy: enolase, phosphofructokinase, and glyceraldehyde-3-phosphate dehydrogenase activities in peripheral nerve, spinal cord, brain and skeletal muscle of acrylamide intoxicated cats, *Toxicol. Appl. Pharmacol.*, 60, 324, 1981.
105. **Sabri, M. I.,** *In vitro* and *in vivo* inhibition of glycolytic enzymes by acrylamide, *Neurochem. Pathol.*, 1, 179, 1983.
106. **Edwards, P. M.,** Neurotoxicity of acrylamide and its analogues and effects of these analogues and other agents on acrylamide neuropathy, *Br. J. Ind. Med.*, 32, 31, 1975.
107. **Dixit, R., Mukhtar, H., Seth, P. K., and Krishna, Murti, C. R.,** Conjugation of acrylamide with glutathione catalyzed by glutathione-s-transferase of rat liver and brain, *Biochem. Pharmacol.*, 30, 1739, 1981.
108. **Kaplan, M. L. and Murphy, S. D.,** Effect of acrylamide on rotarod performance and sciatic nerve B-glucuronidase activity of rats, *Toxicol. Appl. Pharmacol.*, 22, 259, 1972.
109. **Tanii, H. and Hashimoto, K.,** Neurotoxicity of acrylamide and related compounds in rats. Effects on rotarod performance, morphology of nerves and neurotubulin, *Arch. Toxicol.*, 54, 203, 1983.
110. **Dixit, R., Mukhtar, H., and Seth, P. K.,** *In vitro* inhibition of alcohol dehydrogenase by acrylamide: Interaction with enzyme — SH groups, *Toxicol. Lett.*, 1, 487, 1981.
111. **Sharma, R. P. and Obersteiner, E. J.,** Acrylamide cytotoxicity in chick ganglia cultures, *Toxicol. Appl. Pharmacol.*, 42, 149, 1977.
112. **Sabri, M. I., Dairman, W., Juhasz, L., Bischoff, M. C., and Spencer, P. S.,** Is acrylamide neurotoxicity pyruvate sensitive?, *Trans. Am. Soc. Neurochem.*, 12, 155, 1981.
113. **Dairman, W., Sabri, M. I., Bischoff-Fenton, M., Juhasz, L., Ng, T., and Spencer, P. S.,** Dietary pyruvate retards development of acrylamide neuropathy in rats, *Muscle Nerve*, submitted.
114. **Loeb, A. L. and Anderson, R. J.,** Antagonism of acrylamide by supplementation with vitamin B6, *Neurotoxicology*, 2, 625, 1981.
115. **Johnson, E. C. and Murphy, S. D.,** Effect of acrylamide intoxication on pyridine nucleotide concentrations and functions in rat cerebral cortex, *Biochem. Pharmacol.*, 26, 2151, 1977.
116. **Agrawal, A. K., Seth, P. K., Squibb, R. E., Tilson, H. A., Uphouse, L. L., and Bondy, S. C.,** Neurotransmitter receptors in brain regions of acrylamide-treated rats. I. Effects of a single exposure to acrylamide, *Pharmacol. Biochem. Behav.*, 14, 527, 1981.
117. **Agrawal, A. K., Squibb, R. E., and Bondy, S. C.,** The effects of acrylamide treatment upon dopamine receptor, *Toxicol. Appl. Pharmacol.*, 58, 89, 1981.
118. **Rafales, L. S., Lasley, S. M., Greenland, R. D., and Mandybur, T.,** Effects of acrylamide on locomotion and central monoamine function in the rat, *Pharmacol. Biochem. Behav.*, 9, 635, 1983.
119. **Pleasure, D. E., Mishler, K. C., and Engel, W. K.,** Axonal transport of proteins in experimental neuropathies, *Science*, 166, 524, 1969.
120. **Bradley, G. and Williams, M. H.,** Axoplasmic flow in axonal neuropathies. I. Axoplasmic flow in cats with toxic neuropathies, *Brain*, 96, 235, 1973.
121. **Sumner, A., Pleasure, D., and Cresielka, K.,** Slowing of fast axoplasmic transport in acrylamide neuropathy, *J. Neuropathol. Exp. Neurol.*, 35, 319, 1976.
122. **Gold, B. G., Griffin, J. W., and Price, D. L.,** Slow axonal transport in acrylamide neuropathy: different abnormalities produced by single dose and continuous administration, *J. Neurosci.*, 5, 1755, 1985.

123. **Rasool, C. G. and Bradley, W. G.,** Studies on axoplasmic transport of individual proteins: 1-acetylcholinesterase (AChE) in acrylamide neuropathy, *J. Neurochem.*, 31, 419, 1978.
124. **Couraud, J. Y., Di Giamberardino, L., Chretien, M., Souyri, F., and Fardeau, M.,** Acrylamide neuropathy and changes in the axonal transport and muscular content of the molecular forms of acetylcholinesterase, *Muscle Nerve*, 5, 302, 1982.
125. **Wier, R. L., Blaubiger, G., and Chase, T. N.,** Inhibition of fast axoplasmic transport by acrylamide, *Environ. Res.*, 17, 251, 1978.
126. **Souyri, F., Chretien, M., and Droz, B.,** Acrylamide-induced neuropathy and impairment of axonal transport of proteins. I. Multifocal retention of fast transported proteins at the periphery of axons as revealed by light microscope radioautography, *Brain Res.*, 205, 1, 1981.
127. **Sidenius, P. and Jakobsen, J.,** Anterograde axonal transport in rats during intoxication with acrylamide, *J. Neurochem.*, 40, 697, 1983.
128. **Miller, M. S. and Spencer, P. S.** Inhibition of acrylamide of increased ornithine decarboxylase activity and RNA synthesis in dorsal root ganglion following sciatic nerve transection, *J. Neurochem.*, 43, 1401, 1984.
129. **Chou, S. M. and Hartman, H. A.,** Electron microscopy of focal neuroaxonal lesion produced by β,β′-iminodipropionitrile (IDPN) in rats, *Acta Neuropathol.*, 4, 590, 1965.
130. **Hartmann, H. A. and Shively, M. C.,** RNA of nerve cell bodies and axons after β,β′-iminodiproprionitrile, *Acta Neuropathol.*, 11, 275, 1968.
131. **Kitiyara, A., Shively, M. C., and Hartmann, H. A.,** Synthesis and fate of RNA in neurons from mice given β,β′-iminodipropionitrile, *Exp. Neurol.*, 54, 403, 1977.
132. **Chou, S. M. and Klein, R. A.,** Autoradiographic studies of protein turnover in motoneurons of IDPN-treated rats, *Acta Neuropathol.*, 22, 183, 1972.
133. **Hartmann, H. A. and Scoleier, D.,** Succinoxidase activity in reticular neurons of hyperactive rats, *Acta Neuropathol.*, 16, 154, 1970.
134. **Goldberg, M. A.,** Protein turnover and axonal flow in B,B′-iminodipropionitrile induced axonal dystrophy, *J. Neurochem.*, 28, 259, 1977.
135. **Griffin, J. W., Hoffman, P. N., Clark, A. W., Carroll, P. T., and Price, D. L.,** Slow axonal transport of neurofilament proteins: impairment by B,B′-iminodipropionitrile administration, *Science*, 202, 633, 1978.
136. **Yokoyama, K., Tsukita, S., Ishikawa, H., and Kurokawa, M.,** Early changes in the neuronal cytoskeleton caused by B,B′-iminodipropionitrile: selective impairment of neurofilament polypeptide, *Biomed. Res.*, 1, 537, 1980.
137. **Griffin, J. W. and Price, D. L.,** Proximal axonopathies induced by toxic chemicals, in, *Experimental and Clinical Neurotoxicology*, Spencer, P. S. and Schaumburg, H. H., Eds. Williams & Wilkins, Baltimore, Md., 1980, 161.
138. **Griffin, J. W., Hoffman, P. N., and Price, D. L.,** Axonal transport in B,B′-iminodiproprionitrile neuropathy, in, *Axoplasmic Transport in Physiology and Pathology*, Weiss, D. G., and Gorio, A., Eds., Springer-Verlag, Basel, 1982, 109.
139. **Papasozomenos, S. Ch., Autilio-Gambetti, L., and Gambetti, P.,** Reorganization of axoplasmic organelles following B,B′-iminodiproprionitrile administration, *J. Cell Biol.*, 91, 866, 1981.
140. **Davis, C. S. and Richardson, R. J.,** Organophosphorous compounds, in, *Experimental and Clinical Neurotoxicology*, Spencer, P. S. and Schaumburg, H. H., Eds., Williams & Wilkins, Baltimore, Md., 1980, 527.
141. **Ecobichon, D. J.,** Organophosphorous ester insecticides, in, *Pesticides and Neurological Diseases*, Ecobichon, D. J. and Joy, R. J., Eds., CRC Press, Boca Raton, Fla., 1982, 151.
142. **Johnson, M. K.,** The delayed neuropathy caused by some organophosphorous esters: mechanism and challenge, *CRC Crit. Rev. Toxicol.*, 3, 289, 1975.
143. **Johnson, M. K.,** *Reviews in Biochemical Toxicology*, Hodgson, E., Bend, J., and Philpot, R. M., Eds., Elsevier-North Holland, Amsterdam, 1982, 141.
144. **James, K. A. C. and Austin, L.,** The effect of DFP on axonal transport of proteins in chicken sciatic nerve, *Brain. Res.*, 13, 192, 1970.
145. **Sahenk, Z. and Mendell, J. R.,** Zinc pyridinethione, in, *Experimental and Clinical Neurotoxicology*, Spencer, P. S. and Schaumburg, H. H., Eds., Williams & Wilkins, Baltimore, Md., 1980, 578.
146. **Sabri, M. I. and Spencer, P. S.,** On the mechanism of action of sodium pyridinethione on the nervous system, *Neurosci. Abstr.*, 5, 86, 1979.
147. **Chandler, C. J. and Segel, I. H.,** Mechanism of the antimicrobial action of pyrithione: effect on membrane transport, ATP levels and protein synthesis, *Antimicrob. Agents Chemother.*, 14, 60, 1978.
148. **Sahenk, Z. and Mendell, J. R.,** Ultrastructural study of zinc pyridinethione-induced peripheral neuropathy, *J. Neuropathol. Exp. Neurol.*, 38, 532, 1979.
149. **Sahenk, Z. and Mendell, J. R.,** Axoplasmic transport in zinc pyridinethione neuropathy: evidence for an abnormality in distal turn around, *Brain Res.*, 186, 343, 1980.

150. **Cho, E.-S., Spencer, P. S., and Jortner, B. S.,** Doxorubicin, in, *Experimental and Clinical Neurotoxicology,* Spencer, P. S. and Schaumburg, H. H., Williams & Wilkins, Baltimore, Md., 1980, 430.
151. **Philips, F. S., Gilladoga, A., Marquardt, H., Sternberg, S. S., and Vidal, P. M.,** Some observations on the toxicity of adriamycin, *Cancer Chemother. Rep.,* 6, 177, 1975.
152. **Cho, E.-S., Jortner, B. S., Schaumburg, H. H., and Spencer, P. S.,** A single intravenous injection of doxorubicin (Adriamycin^R) induces sensory neuronopathy in rats, *Neurotoxicology,* 1, 583, 1980.
153. **Sahenk, Z. and Mendell, J. R.,** Analysis of fast axoplasmic transport in nerve ligation and adriamycin-induced neuronal perikaryon lesions, *Brain Res.,* 171, 41, 1979.
154. **Brimijoin, S., Dyck, P. J., Jakobsen, J., and Lambert, E. H.,** Axonal transport in human nerve disease and in the experimental neuropathy induced by p-bromophenylacetylurea, in *Axoplasmic Transport in Physiology and Pathology,* Weiss, D. G. and Gorio, A., Eds., Springer-Verlag, Basel, 1982, 124.
155. **Jakobsen, J. and Brimijoin, S.,** Axonal transport of enzymes and labeled proteins in experimental axonopathy induced by p-bromo-phenylacetalurea, *Brain Res.,* 229, 103, 1981.
156. **Sterman, A. B. and Schaumburg, H. H.,** Neurotoxicity of selected drugs, in, *Experimental and Clinical Neurotoxicology,* Spencer, P. S. and Schaumburg, H. H., Eds., Williams & Wilkins, Baltimore, Md., 1980, 593.
157. **Seppalainen, A. M. and Haltia, M.,** Carbon disulfide, in, *Experimental and Clinical Neurotoxicology,* Spencer, P. S. and Schaumburg, H. H., Eds., Williams & Wilkins, Baltimore, Md., 1980, 356.
158. **Savolainen, H., Lehtonen, E., and Vainio, H.,** CS_2 binding to rat spinal neurofilaments, *Acta Neuropathol.,* 37, 219, 1977.
159. **Anzil, A. P.,** Selected aspects of experimental disulfiram neuromyopathy, in *Advances in Neurotoxicology,* Mango, L., Ed., Pergamon Press, New York, 1980, 359.
160. **Anzil, A. P., Isenberg, G., and Kreutzberg, G. W.,** Disulfiram-induced impairment of horseradish peroxidase retrograde transport in rat sciatic nerve, in *Axoplasmic Transport in Physiology and Pathology,* Weiss, D. G. and Gorio, A., Eds., Springer-Verlag, Basel, 1982, 117.
161. **Pappolla, M., Monaco, S., Weiss, H., Miller, C., Sahenk, Z., Autilio-Gambetti, L., and Gambetti, P.,** Slow axonal transport in carbon disulfide (CS_2) giant axonopathy, *Am. Assoc. Neuropathol. Abstr.,* 60, 305, 1984.
162. **Sidorowicz, W., Zatonski, W., Andrzejak, R., and Smolek, R.,** The effect of CS_2 on red cell metabolism, *Acta Biol. Med. Ger. Band,* 36, 645, 1977.
163. **Sabri, M. I. and Spencer, P. S.,** Inhibition of glycolysis by chemically unrelated industrial neurotoxins which produce polyneuropathies, *Proc. Int. Soc. Neurochem.,* 7, 564, 1979.
164. **Tarkowski, S. and Sobczak, H.,** Oxidation and phosphorylation process in brain mitochondria of rats exposed to carbon disulfide, *J. Neurochem.,* 18, 177, 1971.
165. **Liwnicz, B. H., Kristensson, H. M., Wisniewski, H. M., Shelanski, M. L., and Terry, R. D.,** Observations on axoplasmic transport in rabbits with aluminum-induced neurofibrillary tangles, *Brain Res.,* 80, 413, 1974.
166. **Troncoso, J. C., Griffin, J. W., Hoffman, P. N., Hess-Kozlow, K. M., Blum, J. R., and Price, D. L.,** Slow axonal transport of neurofilament proteins in aluminum, *Soc. Neurosci. Abstr.,* 9, 1190, 1983.
167. **Bizzi, A., Autilio-Gambetti, L., Crane, R., and Gambetti, P.,** Effect of aluminum on neurofilament transport and morphology of peripheral axons, *Soc. Neurosci. Abstr.,* 9, 1191, 1983.
168. **Wakabayashi, M., Araki, K., and Takahashi, Y.,** Increased rate of fast axonal transport in methyl mercury induced neuropathy, *Brain Res.,* 117, 524, 1976.
169. **Abe, T., Haga, T., and Kurokawa, M.,** Blockage of axoplasmic transport and depolymerization of reassembled microtubules by methyl mercury, *Brain Res.,* 86, 504, 1975.
170. **Bosch, E. P., Pelham, R. W., Rasool, C. G., Chatterjce, A., Lash, R. W., Brown, L., Munsat, T. L., and Bradley, W. G.,** Animal models of alcoholic neuropathy: morphologic, electrophysiologic, and biochemical findings, *Muscle Nerve,* 2, 133, 1979.

Chapter 16

THE RELEVANCE OF AXOPLASMIC TRANSPORT IN NEUROTROPHIC FUNCTIONS

Hugo L. Fernandez and J. Alejandro Donoso

TABLE OF CONTENTS

I. INTRODUCTION

The idea that nerve cells exert trophic (nutritive) influences on themselves, as well as on the target cells they innervate, has a long history extending back to the mid-19th century.[1,2] By the turn of the 20th century it was believed that such influences were mediated by substances produced in the neuronal cell body and translocated along the axon to their cellular targets.[3-5] This view, connecting trophic influences with the process now known as axoplasmic transport, became increasingly popular in the early 1930s. For example, Parker and Paine,[6] describing experiments on nerve regeneration, stated

> " . . . secondary or Wallerian degeneration may be regarded as the result of an interruption of a flow of substance which emanating from the nucleated part of the neuron, passes out through its processes including the terminals. The interruption of this flow would give rise to a progressive rather than a simultaneous degenerative change . . . "

During the subsequent 2 decades the above concept remained relatively unexplored and partly eclipsed by electrophysiological studies that were aimed primarily at understanding the conduction of nerve impulses and the consequent activation of the innervated cells. In the years that followed, however, the experimental demonstration of axoplasmic transport[7,8] and the accumulation of evidence for neurotrophic effects not necessarily mediated by nerve impulses,[9-11] led to renewed interest in the possible relationship between both phenomena.

At present, it is well established that aside from impulse conduction, transmitter release, neuroendocrine secretion, and other functions, nerve cells play a vital role in the long-term control of their differentiated characteristics, i.e., they exert a *propriotrophic* influence in support of their own structural and functional integrity. In this regard, there is considerable evidence linking axoplasmic transport to the renewal of axonal constituents, as well as to the supply of substances for neuronal growth and regeneration.[12] It is also generally accepted that nerve cells have an *exterotrophic* function related to the maintenance of the various cells they innervate. This is dramatically illustrated by the loss of normal structure and operation of effector and receptor cells when their respective nerves are transected. For instance, following denervation skeletal muscles show decreased resting membrane potentials,[13] increased extrajunctional sensitivity to acetylcholine (ACh),[14] loss of junctional acetylcholinesterase (AChE),[15] and various metabolic alterations which eventually lead to their complete degeneration.[16] In turn, when sensory structures such as taste buds,[17] muscle spindle receptors,[18] and fish lateral line organs,[19] are deprived of their nerve supply they also undergo a series of degenerative changes. Since some of the innervated or target cells also influence the functional state of the corresponding neurons,[20] the aforementioned supportive and maintenance functions should be viewed as bidirectional processes.

While the phenomenology of long-term cellular interactions has recently burgeoned, the molecular mechanisms underlying trophic events are only beginning to be appreciated. In this context, certain experimental findings which suggest a role for axoplasmic transport in trophic functions may attain particular importance. Several extensive reviews have dealt separately with the process of axoplasmic transport[12,21-23] and with the general problems of nerve-target cell interactions.[11,16,20,24,25] In this chapter, we present a summarized account of studies on *exterotrophic* neuronal functions which support the idea that axoplasmic transport is an important mediator of trans-synaptic target cell maintenance. Although a large number of cellular actions, including several which occur during growth and differentiation, have been considered within the framework of trophic phenomena, we will emphasize only those which exist in adult tissues. In doing so, we often refer to the cellular interactions of motor nerves and skeletal muscles, but it must be kept in mind that the mechanisms of neurotrophic regulation in one system cannot a priori be extrapolated to other systems.

II. NEURONAL FEATURES INVOLVED IN TROPHIC PHENOMENA

Historically there has been considerable controversy over which target cell characteristics are under direct trophic control and which nerve factors are involved.[24] One viewpoint has been that nerve electrical activity and synaptic transmission are directly implicated.[26] Another has ascribed trophic effects to unidentified neurogenic substances carried by axoplasmic transport,[16] whereas a third view has favored the participation of nerve-target cell contact.[24] One of the few experimental models in which at least some of the above mechanisms can be separated is given by the neurotrophic maintenance of taste buds in mammalian species. These structures are organized around the sensory nerve terminal and undergo a constant turnover with full cellular renewal over a period of several days.[27] Taste buds actually disappear when their nerve supply is interrupted, but they are replaced after reinnervation.[17,28] This replacement takes place upon reinnervation with gustatory sensory nerves (glossopharyngeal, vagus, and chorda tympani), but not with motor (hypoglossal) or nongustatory sensory nerves (lingual and auriculotemporal).[28] Futhermore, the failure of neural receptor function and the degeneration of taste buds induced by nerve transection occurs later when there is a longer distance between the site of axotomy and the receptor organs.[29,30] These observations, together with the fact that sensory electrical information is expressed in the central nervous system whereas neural maintenance of taste buds is exerted in the periphery, support the existence of a trophic effect which is independent of nerve electrical activity.[16]

Unfortunately, the aforegoing conclusion cannot be freely extrapolated to other types of cellular arrangements such as the mammalian skeletal neuromuscular system. Thus far, it has not been possible to unambiguously determine whether certain typical muscle denervation changes result from depletion of ACh, absence of putative trophic substances, muscle inactivity, and/or alterations in the structural organization of the motor endplates.[31] Part of the above complications arise from the fact that many different neuromuscular preparations have been used and often less than adequate markers have been chosen to evaluate the muscle's response to the different neural factors (e.g., trophic substances, ACh, nerve impulse activity, etc.) involved. An additional problem in this area has been the lack of knowledge concerning whether a single or several muscle cell characteristics are directly regulated by one or a combination of neural factors. In any case, such characteristics may be involved in the triggering of a sequence of physiological events which, although not specifically influenced by a trophic factor(s), would nontheless parallel changes specifically regulated by the nerve. Clearly, further studies are needed to sort out the relative contributions of all these factors in the maintenance of functional, structural, and chemical, properties of muscle. Special attention should be given to the possible interrelationship between the various factors, as well as to other influences (e.g., hormonal) which may modify the operation of trophic mechanisms. Finally, however, it should be noted that despite the number and nature of the proposed factors, most may ultimately depend upon an intact axoplasmic transport system.

III. EVIDENCE SUPPORTING AXOPLASMIC TRANSPORT AS A TROPHIC MEDIATOR

A classical illustration of neurotrophic influences which exist in the absence of nerve impulse activity is the finding that denervation changes develop sooner if the nerve is cut close to, rather than far from, the innervated structure. In the case of the neuromuscular system, Luco and Eyzaguirre[9] were the first to show that the time of onset of denervation-induced fibrillation and ACh supersensitivity occurred with a latency which was directly related to the length of nerve that remained attached to the muscle after nerve transection. This outstanding observation was later corroborated[33] and extended to other denervation-induced changes in skeletal muscle such as: loss of glycogen;[10] increase in proteolytic activity;[33] appearance of tetrodotoxin-resistant action potentials;[34] decrease in endplate cholinesterase (ChE),[35] AChE,[36] and 16S AChE;[37] increase in cyclic AMP concentration;[38] decrease in resting membrane potential;[13] decrease in miniature endplate potential activity;[39] and, decrease in muscle phosphatidylethanolamine.[40] Those findings clearly pointed to the importance of the amount of nerve in the maintenance of various muscle properties. Since nerve-evoked muscle activity is abolished at the time of axotomy regardless of nerve stump length, such results could not be attributed solely to muscle disuse. Indeed, it was suggested that motor nerves provide regulatory substances which maintain certain muscle features for more extended periods of time when the degenerating distal nerve stump is longer. These substances would be conveyed by axoplasmic transport,[22] a process which continues in axonal segments that have been separated from their respective cell bodies.[41,42] Based on this assumption, the rates of movement of putative regulatory substances were estimated in the range of 4 to 500 mm/day depending on the type of neuromuscular preparation studied and the particular muscle feature measured.[12,22,25,36,43,44] Other investigators have claimed that the observed nerve stump length-dependent muscle changes may be due to structural and/or biochemical alterations resulting from or parallel to nerve terminal degeneration.[45-47] Both possibilities, one involving the lack of trophic substances and the other the presence of products of nerve degeneration, are not mutually exclusive. It must be considered, however, that in some preparations the denervation-induced muscle changes occurred earlier than complete nerve terminal degeneration.[37] In addition, experiments described below show that denervation-like changes in muscle can take place in the absence of nerve terminal degeneration.

Further support for the involvement of axoplasmic transport in trophic functions was given by experiments in which this process was selectively blocked with pharmacological agents. In those studies the antimicrotubular drugs colchicine or vinblastine, which are known to block axoplasmic transport without interfering with action potential conduction,[22] were locally applied to motor nerves either by intraneural injection or by means of silastic cuffs. As a result of such treatments skeletal muscles showed a number of denervation-like changes including: muscle membrane depolarization, tetrodotoxin-resistant action potentials and extrajunctional sensitivity to ACh;[48,49] fibrillation;[50,51] quantitative changes in muscle soluble proteins;[52] decrease in oxidative and glycolytic enzymes;[53] and, decrease in motor endplate AChE.[54] Most of the muscle alterations observed were found to be reversible and to occur later than those caused by short nerve stump denervation.[50-54] Furthermore, in contrast to the effects of denervation, they were measured in the absence of any detectable effect on: nerve impluse conduction;[48,51,55] spontaneous and evoked transmitter release;[48,49] neuromuscular transmission;[56,57] miniature endplate potentials;[48,49] muscle contraction properties including maximum isometric twitch tension, tetanic tension, as well as contraction and half-relaxation times;[49,51,53] rate of in vitro release of motor endplate AChE;[53] and, neuromuscular junction ultrastructure.[48,53] All these findings provide compelling evidence for the involvement of axoplasmic transport in the trans-synaptic maintenance of a number of important muscle features, and one may thus infer the existence of some trophic or regulatory substance(s) which is translocated along the axon and supplied to the post-neural cell.

Other investigators have challenged the above conclusion, but in so doing they have overlooked several of the results and important technical details described in the aforementioned studies.[48-54] In this connection, it has been claimed that subepineural injection of colchicine not only causes denervation-like changes in the muscle innervated by the treated nerve, but also in the contralateral control muscle.[58] Therefore, one argument has been that colchicine or vinblastine applied locally to motor nerves may gain entry into the circulation thereby affecting the muscles directly.[59,60] This interpretation, however, neglects the possibility that under the latter conditions colchicine may have blocked axoplasmic transport not only in the experimental nerve but also in the intramuscular nerve fibers of the contralateral control muscle. Even more important, perhaps, is that in the previously mentioned studies, in which a 20-fold lower concentration of colchicine was used, it was clearly shown that: the drug remained at least temporarily localized within a discrete segment of nerve around the injection site and in no case reached the muscle in detectable amounts;[50,51] the time of onset of the colchicine-induced changes was directly related to the distance between the site of injection and the experimental muscle;[51,53] there was no measurable systemic effect of the drug on the absolute activities of various enzymes in the control muscles;[53,54] and the ultrastructure of the neuromuscular junction was not affected by drug treatment.[53] Another criticism has questioned the mode of administration of the axoplasmic transport blocking agents. Specifically, it has been argued that both subepineural injections and application of drug-impregnated silastic cuffs may cause physical damage to the nerve, thereby causing partial denervation of the muscle.[58] In the case of local injections, such an interpretation is inconsistent with results showing that control injections with solvent alone usually fail to produce any of the denervation-like changes detected after application of the drug.[50-54] Also, in experiments using cat hypoglossal nerve-geniohyoid muscle preparations, surgical damage to the nerve was avoided by injecting the drugs into the main nerve trunk and, therefore, the agents reached the parallel lateral branch innervating the experimental muscle only through diffusion.[53] Regarding silastic cuffs, it has been indicated that in certain instances this method may produce muscle denervation-like changes by mechanical compression of the nerve.[58] In turn, this compression effect would lead to blockage of nerve impulse conduction and eventually to nerve degeneration. Taken at face value the experimental evidence presented to support this view is at best conflicting, especially if one considers that silastic cuffs free of drugs have not been found to induce any muscle degenerative changes.[55,61] These questions have recently been analyzed in great detail and it was concluded that such objections do not apply to results obtained in properly controlled experiments.[22,55,62] Accordingly, it is now well established that muscle denervation-like changes induced by local application of colchicine or vinblastine to motor nerves are brought about by the direct blocking action of these drugs on axoplasmic transport.

In the above discussion, it has become evident that combined results from the long-short nerve stump and axoplasmic transport blockade experiments provide strong evidence for the existence of neurogenic substances which are transported along the motor nerve axons, and subsequently regulate certain features of the adult muscle cell. The question remains, then, as to the precise nature of the substances which participate in trophic interactions.

IV. THE SEARCH FOR NEUROTROPHIC SUBSTANCES

Identification of neural substances influencing target cell properties has been approached mainly by attempting artificial replacement of materials which are missing from noninnervated embryonic cells maintained in tissue culture. For skeletal muscle cells, it has been indicated that crude soluble protein extracts from brain, spinal cord, and nerve, as well as neural explants, stimulate morphological development, protein synthesis, and enhance total AChE activity of the cells in culture.[63-68] Similar extracts have also been shown to affect the number and clustering of ACh receptors,[69,70] tetrodotoxin-sensitive spike potentials,[71] and electrical properties of the cultured muscle cells.[72,73] The biologically active principle in some of the crude extracts has been partially purified and characterized. For example, a myotrophic protein, sciatin, isolated from chicken sciatic nerve has been found to be an acidic 86,400 daltons glycoprotein with biochemical properties closely resembling those of transferrin.[70,74,75] Thus, it was suggested that trophic regulation of certain membrane and metabolic features of muscle is mediated by soluble nerve substances which are proteinaceous in nature.[63-74] The difficulty with this interpretation lies in determining whether the effects of crude or purified nerve extracts are caused by their direct action on certain muscle properties or by promoting growth and differentiation of the cultured cells.[26]

Another approach used to identify possible trophic substances has been to show that nervous tissue extracts or neural explants affect certain properties of adult denervated muscles in vitro. In fact, addition of neurogenic substances to skeletal muscles in organ culture is known to effectively delay or partly prevent the effects of denervation on glucose-6-phosphate dehydrogenase,[76] cyclic AMP,[77] and ChE.[78,79] When the latter was evaluated it was found that crude soluble homogenates from sensory ganglia, brachial nerve, spinal cord, and brain, retarded the denervation-induced decrease of ChE in whole tricep muscle of the newt. This effect was abolished by heat inactivation of the extracts and it was not detected when homogenates of kidney, oviduct, spleen, and liver, were tested.[78,81] The importance of these studies is somewhat lessened, however, because neural regulation of junctional vs. nonjunctional ChE appears to involve different mechanisms.[82,83] More recently, however, the effects of nerve extracts on endplate ChE activity have been assessed independently from those on nonendplate ChE in rat extensor digitorum longus muscles maintained in organ culture.[84,85] Cultures supplemented with soluble extracts from brain, spinal cord, sciatic nerve, and innervated diaphragm muscle demonstrated a dose-dependent increase in total endplate ChE activity and ACh sensitivity. Similar results were obtained with substances which accumulate proximal to a sciatic nerve transection and also with substances that are released from neuromuscular preparations upon nerve stimulation.[85] In turn, extracts from liver, spleen, denervated or noninnervated diaphragm muscle, and boiled innervated muscle, were all found to be ineffective. Interpretation of these findings may seem complex in that the ChE contained by the nervous tissue extracts, or that released from nerve upon stimulation,[86] could have contributed to the enhancement seen in muscle enzymatic activity.[85] Our own experiments, however, have indicated that specific AChE in comparable nerve extracts cannot entirely account for the substantial increase detected in endplate AChE activity.[87] Furthermore, such extracts prevented the denervation-induced increase in the rate of release of AChE from the muscle endplates.[87] Thus, taken together, the above results suggest that endplate ChE and AChE activities may be at least partially maintained by a neurogenic substance(s) supplied to the muscle.[54,81,84]

While this conclusion is quite reasonable, the supporting experimental evidence may still be considered circumstantial.[26] This is partly because of the inherent drawbacks in measuring ChE or AChE activities in muscle endplate regions which unavoidably include some noninnervated tissues and blood components that are also known to contain such enzymes.[37,88] In this regard, measurement of the heaviest asymmetric molecular form of AChE, characterized by a 16S sedimentation coefficient, provides a more specific assay for putative trophic substances.[89,90] This enzyme is not present in blood,[88] but it is highly concentrated at neuromuscular junctions[91] where it is found in association through ionic interactions with the basal lamina.[90,92,93] At this location, its activity decreases drastically with motor nerve transection,[91] in a nerve stump length-dependent fashion,[37] but returns to control levels upon reinnervation.[94] More recently, we have shown that proteinaceous soluble extracts from rat obturator or sciatic nerves partly prevent, in a dose-related manner, the denervation-induced decay of 16S AChE activity in anterior gracilis muscles maintained in organ culture.[87] The nerve extracts were devoid of 16S AChE itself, their effect was eliminated when they were trypsinized or boiled, and similar homogenates of brain, liver, nonendplate muscle regions, or blood serum, were all ineffective.[95] In addition, it has also been shown that nerve explants[96] and a basic neuronal protein with a molecular weight of 35,000[97] also retard the decrease of 16S AChE in denervated newt muscles in organ culture.

More recently we have found that extracts derived from an obturator nerve segment proximal to a ligature placed on the nerve 3 days earlier, are significantly more effective in preventing the decay of endplate 16S AChE in gracilis muscles. Inasmuch as substances

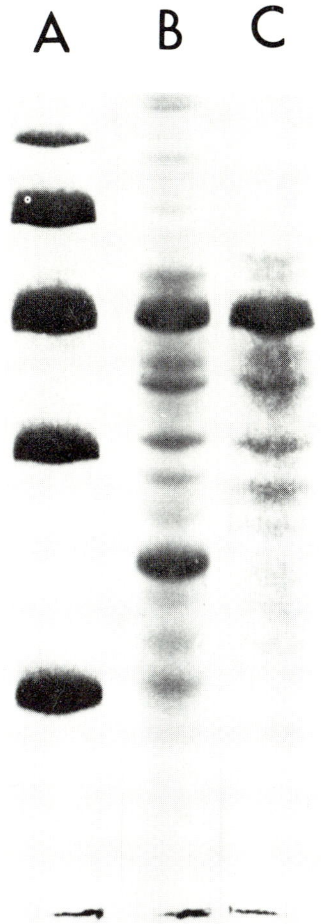

FIGURE 1. Sodium dodecylsulfate polyacrilamide electrophoresis gels (5.6%) intentionally overloaded with crude (B; 20 μg protein) or clean (C; 35 μg protein) rat obturator nerve extracts and stained with coomassie blue. Molecular weight markers (A) are from top to bottom β-galactosidase (130,000), phosphorylase-a (94,000), bovine serum albumin (68,000), aldolase (40,000), and ribonuclease (13,700). Crude nerve extract was prepared as previously reported[87] and the clean extract was obtained by loading 7.5 mg protein of crude extract onto cynogen bromide-activated Sepharose-4B columns with coupled rabbit anti-rat serum antibodies (60% recovery). The proteins eluted from the columns (detected by monitoring absorbance at 280 nm) were pooled, dialyzed overnight against distilled water, lyophilized, and stored at −70°C until used.

conveyed by axoplasmic transport accumulate in ligated nerves in the region just proximal to the ligature,[12,22] the substance(s) involved in the maintenance of 16S AChE would appear to be transported along the axons. This contention is in agreement with a previous report showing that the major protein components in such extracts are carried by axoplasmic transport.[96] Additional experiments with the obturator nerve soluble homogenates indicate that they include at least 20 to 22 protein bands, some of which represent blood serum contaminants. After removal of these contaminants by affinity chromatography in Sepharose-4B columns, containing rat serum antibodies, at least 7 to 9 protein bands persist (Figure 1). Double immunodiffusion and immunoelectrophoresis tests also indicate that in contrast with the crude obturator nerve extract, rat serum and serum albumin, the pooled protein fractions remaining after chromatography do not cross-react with rat serum antibodies (Figure 2). More importantly, the protein extract free of serum contaminants (clean extract) is as

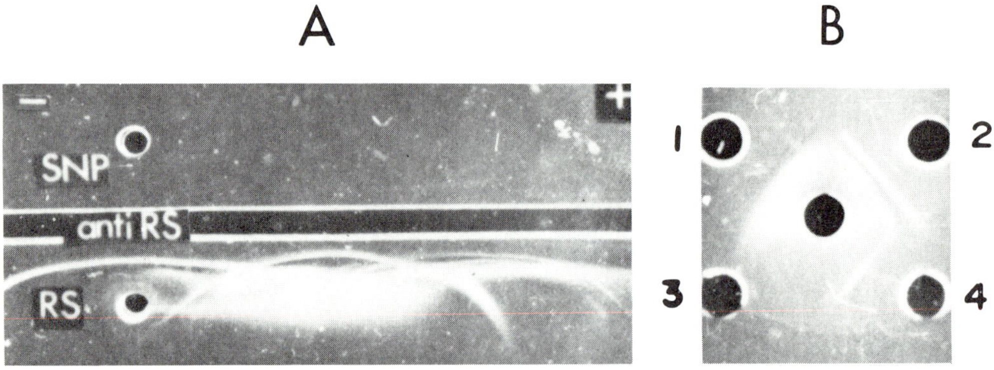

FIGURE 2. Immunoelectrophoresis (A) and double immunodiffusion (B) of rabbit anti-rat serum (anti RS, trough, and center well) against: clean soluble nerve extract (SNP, 2 μg protein, upper well in A and well 3 in B); rat serum (RS, 0.2 μg protein, lower well in A and well 1 in B); crude extract (1 μg protein, well 2 in B); and rat serum albumin (1 μg protein, well 4 in B).

Table 1
EFFECTS OF CRUDE AND CLEAN OBTURATOR NERVE EXTRACTS[a]

Experimental conditions	Protein (mg/200μℓ)	16S AChE activity (mU)	16S AChE activity (% of control)	Nerve extract[b] specific activity
Control (12)	0	31.1 ± 2.1	100	0
Crude extract (10)	0.54	45.1 ± 2.5[c]	145	26
Clean extract (12)	0.32	46.1 ± 2.8[c]	148	46

[a] Denervated rat anterior gracilis muscles were maintained in organ culture for 48 hr in the presence or absence of either crude or clean soluble obturator nerve extracts. Number of experiments are in parentheses and values are mean ± SEM. The 1 mU of AChE activity[87] is 1 mol ^{3}H-acetate formed per min × 10^{-3}.

[b] Nerve extract specific activity = (experimental mU-control mU)/mg of nerve extract protein.

[c] Denotes highly significant difference ($p << 0.001$) from corresponding controls.

effective as the crude nerve extract in partly preventing the decrease in 16S AChE resulting from denervation (Table 1).

To address the question as to whether such extracts affect the promotion and/or prevent the degradation of 16S AChE, gracilis muscles were cultured for 2 days in the presence of nerve extracts, after a previous 4-day in vivo denervation period. Results showed that both the crude and the clean obturator nerve extracts significantly increased 16S AChE activity not only over corresponding control levels, but also over the values obtained for denervated muscles which were kept in the animals for 4 and 6 days (Table 2). Since after 4 days of denervation in vivo endplate 16S AChE is known to account for only 10 to 15% of control activity,[37] our findings suggest that the nerve extracts may promote the formation of 16S AChE. Although other interpretations are possible, our conclusion is entirely compatible with reports indicating that 16S AChE is formed in embryonic skeletal muscle in tissue culture only upon induction by neuronal elements[99] and it is produced in smooth muscle, where it is not normally present, upon reinnervation by cholinergic motor fibers.[100]

The converging lines of evidence described above clearly point to the existence of neurogenic proteinaceous substances encoding trophic information for the maintenance of se-

Table 2
NERVE EXTRACT-INDUCED RECOVERY OF 16S AChE IN DENERVATED MUSCLE[a]

Experimental conditions	16S AChE activity (mU)	16S AChE activity (% of control)	Nerve extract specific activity
In vivo			
Control (18)	80 ± 9.0	100	—
Denervated, 4d (10)	13 ± 2.0	16	—
Denervated, 6d (8)	9 ± 0.5	11	—
In organ culture			
Control (10)	8 ± 0.6	100	0
Crude extract (10) (0.54 mg/200 μℓ)	30 ± 2.7	375	41
Clean extract (10) (0.32 mg/200 μℓ)	35 ± 2.0	437	84

[a] Denervated rat anterior gracilis were maintained either in vivo or in organ culture for 2 days (after a 4-day denervation period in vivo) in the presence or absence of obturator nerve extracts. Number of experiments are in parentheses and values are mean ± SEM. Units of AChE activity and nerve extract specific activities are as defined in footnotes of Table 1.

lected features of adult skeletal muscles. Such evidence, however, is still indirect and further studies are needed to accomplish a more precise biochemical characterization of putative trophic substances, as well as to elucidate their mechanisms of action. Such studies should consider that trophic effects in the neuromuscular system may depend on the relative biological activities of stimulatory vs. inhibitory neurogenic substances and on the physiological interplay between these substances and other neurotrophic factors. In addition, attempts should be made to demonstrate that putative trophic substances can specifically reproduce the normal trophic operation of nerves.

V. CRITERIA FOR NEUROTROPHIC SUBSTANCES

The aforementioned findings leave a number of unanswered questions and illustrate the complexities involved in the identification of trophic substances. In general, various partially purified substances extracted from nervous tissues have been claimed to exhibit trophic capabilities.[24,28,64,101] As yet, however, there is little agreement concerning their biochemical characteristics and distinction from growth factors (also referred to as secondary hormones such as nerve growth factor[102] and neurotrophin[97]). In this context, it seems worthwhile to establish certain guidelines to be followed when exploring the existence of potential trophic substances in a particular cellular system. By analogy with the criteria determined for neurotransmitters,[103,104] the establishment of some rules for trophic materials should provide a useful framework for studies attempting their identification. Hence, we propose the following criteria for neurotrophic substances in the adult mammalian motor nerve-skeletal muscle system. Although these guidelines have been designed specifically for the neuromuscular system, they may be extended to other adult cellular systems with minor modifications.

A neurogenic substance, other than the neurotransmitter, which is suspected as a trophic factor for the regulation of skeletal muscle should:

1. Be present only, or in greater proportion, within motor neurons
2. Be synthesized in the motor neuron's cell body

3. Be conveyed by axoplasmic transport
4. Be released at the nerve endings and supplied to the muscle
5. Interact with a molecular target or receptor located in the postsynaptic membrane and/or within the muscle cell
6. Exert a dose-dependent regulatory action on one or several definable metabolic, functional, and/or structural properties of muscle

As a corollary to these criteria one would at least expect a trophic substance to: (1) be absent, or reduced in concentration, in denervated muscle after complete nerve terminal degeneration; (2) be depleted from nerve terminals upon selective blockage of axoplasmic transport or its synthesis, a result which should lead to the onset of muscle denervation-like changes; and (3) prevent or reverse the muscle alterations previously elicited by its absence.

Thus far, partially purified and even crude nerve extracts have already been shown to fulfill some of the above criteria. It must be noted, however, that final confirmation of the trophic nature of a neurogenic substance(s) will be possible only after its purification and biochemical characterization are achieved.

VI. NEUROTROPHIC MECHANISMS

While in recent years a considerable amount of information has been gathered on the neural regulation of certain skeletal muscle properties, knowledge gained on the molecular mechanisms behind trophic phenomena has been meager. As previously mentioned, one obvious problem has been to determine to what extent one or more neural factors are directly involved in such mechanisms. In addition, one particular pitfall that must be avoided is the unjustified attempt to demonstrate that one particular factor is indeed the only essential regulatory neural component. For example, it has been reported that complete blockage of ACh transmission by α-bungarotoxin (α-BTX), with no impairment of fast axoplasmic transport, causes a fall in muscle membrane resting potential and an increase in extrajunctional ACh receptors.[106,107] Since these muscle changes were found to be identical to those induced by surgical denervation it was concluded that "...there seems to be no need to postulate trophic mediators other than ACh to account for the neural regulation of the muscle properties studied in this investigation".[107]

To maintain this conclusion, however, one clearly must disregard previous studies showing that selective inhibition of axoplasmic transport, in the absence of any effect on spontaneous and evoked transmitter release, also leads to decreased muscle membrane resting potentials and increased ACh sensitivity.[48,55,62] These two sets of results, independently supporting the preeminence of either ACh or other neural substance in the trophic control of muscle membrane properties, are difficult to reconcile. It must be noted, however, that in those experiments with α-BTX[107] the transport of an unlabeled or even a small amount of labeled trophic substance may have been blocked without being detected. In addition, the drug may have impaired the movement of substances carried at rates slower than those of fast axoplasmic transport and/or it could have affected the release of substances other than ACh from the nerve terminals. The latter possibility is in agreement with a report indicating that aside from its action on postsynaptic ACh receptors, α-BTX also shows presynaptic binding.[108] Other alternative explanations may involve direct α-BTX-induced inactivation of a trophic substance distinct from ACh or an interaction with the receptors for such a substance. In view of these, and others,[22,62] considerations it is apparent that the results obtained by inhibiting ACh transmission with α-BTX, or other means, do not exclude the likely participation of substances other than ACh in the trophic regulation of certain muscle features. Likewise, since the outcome of selective blockage of axoplasmic transport is not identical to that of muscle denervation,[48,52,53] the participation of factors other than transported trophic substances cannot be disregarded.

At present no definitive conclusion can be made on the mechanisms underlying neurotrophic control of even a single muscle property. As often pointed out,[16,20,22,25] it seems that a variety of factors acting together or in opposition may subserve the complex regulation of muscle functions. Factors such as substances conveyed by axoplasmic transport, ACh, nerve-evoked muscle activity, hormones, etc. are all capable of exerting a certain trophic influence, the extent and importance of which may depend on the particular species and neuromuscular preparation studied. In this context, it is important to determine the relative contribution of each of these factors in a single preparation in which an adequate marker of the muscle's reactions to such agents can be quantified. Since more than one factor may simultaneously influence the same or different myogenic properties, it seems equally important to elucidate, as has been recently proposed,[109] whether muscle cells have a common target mechanism for all such factors. In this regard, it is possible that all neurotrophic influences act on a "second messenger system" (e.g., adenyl cyclase-cyclic AMP system) which translates such influences into signals that directly regulate various muscle features.[109]

VII. SUMMARY AND CONCLUSIONS

At present, it is well established that nerve cells exert long-term *propiotrophic* influences related to the preservation of their own structure and function, as well as *exterotrophic* effects concerned with the maintenance of the various target cells they innervate. Results from studies involving the long-short nerve stump experimental paradigm and the drug induced selective blockage of axoplasmic transport, have clearly shown that such transport is an important mediator of trophic phenomena. This implies that substances synthesized in the neuronal cell body and conveyed by axoplasmic transport are essential neuronal components of trophic interactions. That this may be the case, at least for exterotrophic functions, is supported by experiments showing that neurogenic proteinaceous substances have innervation-like effects on target cells deprived of their nerve supply. Thus far, the purification and biochemical characterization of such putative trophic substances have not been completely achieved. In this regard, in order to establish a framework for future studies, we have proposed a set of criteria which we consider necessary characteristics of a neural substance(s) thought to encode trophic information.

Aside from substances carried by axoplasmic transport, other factors have been claimed to participate in trophic mechanisms. For example, nerve electrical activity, synaptic transmission, and nerve-target cell contact, have all been proposed as trophic mediators. The relative contribution of these factors in the maintenance of a single or several target cell characteristics has not yet been determined and may depend upon the type of preparation considered. Current information suggests that a number of factors acting together or in opposition may be involved in the apparently complex mechanisms underlying neurotrophic functions. Among other aspects, the interrelationship between the aforementioned factors and the possible existence of a common target mechanism for all of them remain to be elucidated.

ACKNOWLEDGMENTS

This work was supported by the Medical Research Service of the Veterans Administration, the Muscular Dystrophy Association of America, and the Amyotrophic Lateral Sclerosis Research Center of the University of Kansas Medical Center.

REFERENCES

1. **Magendie, M.,** De l'influence de la cinquieme paire des nerfs sur la nutrition et les fonctions de l'oeil, *J. Physiol. Exp. Pathol.*, 4, 176, 1824.
2. **Waller, A. V.,** A new method for the study of the nervous system, *London J. Med.*, 43, 609, 1852.
3. **Scott, F. H.,** On the relation of nerve cell to fatigue of their nerve fibers, *J. Physiol. (London)*, 34, 145, 1906.
4. **Ramon y Cajal, S.,** *Studies on Degeneration and Regeneration of the Nervous System*, Vol. 1, Oxford University Press, Cambridge, 1928, 77.
5. **Gerard, R. W.,** Nerve metabolism, *Physiol. Rev.*, 12, 469, 1932.
6. **Parker, G. H. and Paine, V. L.,** Progressive nerve degeneration and its rate in the lateral-line nerve of the catfish, *Am. J. Anat.*, 54, 1, 1934.
7. **Weiss, P. and Hiscoe, H. B.,** Experiments on the mechanisms of nerve growth, *J. Exp. Zool.*, 107, 315, 1948.
8. **Droz, B. and Leblond, C. P.,** Migration of proteins along the axons of the sciatic nerve, *Science*, 137, 1047, 1962.
9. **Luco, J. V. and Eyzaguirre, C.,** Fibrillation and hypersensitivity to ACh in denervated muscle: effect of length of degenerating nerve fibers, *J. Neurophysiol.*, 18, 65, 1955.
10. **Gutmann, E., Vodička, Z., and Zelená, J.,** Changes in cross-striated muscle after nerve section depending on length of peripheral nerve stump, *Physiol. Bohemoslov.*, 4, 181, 1955.
11. **Miledi, R.,** An influence of nerve not mediated by impulses, in *The Effect of Use and Disuse of Neuromuscular Functions*, Gutmann, E. and Hnìk, P., Eds., Publishing House of the Czechoslovak Academy of Sciences, Prague, 1963, 35.
12. **Grafstein, B. and Forman, D. S.,** Intracellular transport in neurons, *Physiol. Rev.*, 60, 1167, 1980.
13. **Albuquerque, E. X., Schuh, F. T., and Kauffman, F. C.,** Early membrane depolarization of the fast mammalian muscle after denervation, *Pfluegers Arch.*, 328, 36, 1971.
14. **Axelsson, J. and Thesleff, S.,** A study of supersensitivity in denervated mammalian skeletal muscle, *J. Physiol. (London)*, 147, 178, 1959.
15. **Guth, L., Albers, R. W., and Brown, W. C.,** Quantitative changes in cholinesterase activity of denervated muscle fibers and sole plates, *Exp. Neurol.*, 10, 236, 1964.
16. **Guth, L.,** "Trophic" influences of nerve on muscle, *Physiol. Rev.*, 48, 645, 1968.
17. **Olmsted, J. M. D.,** The nerve as a formative influence in the development of taste buds, *J. Comp. Neurol.*, 31, 465, 1920.
18. **Tower, S. S.,** Atrophy and degeneration in the muscle spindle, *Brain*, 55, 77, 1932.
19. **Parker, G. H.,** On the trophic impulse so-called, its rate and nature, *Am. Nat.*, 66, 147, 1932.
20. **Smith, B. H. and Kreutzberg, G. W.,** Neuron-target cell interactions, *Neurosci. Res. Prog. Bull.*, 14, 211, 1976.
21. **Thoenen, H. and Kreutzberg, G. W.,** The role of fast transport in the nervous system, *Neurosci. Res. Prog. Bull.*, 20, 1, 1981.
22. **Ochs, S.,** *Axoplasmic Transport & Its Relation to Other Nerve Functions*, 1st ed., John Wiley & Sons, New York, 1982, 1.
23. **Weiss, D. G.,** *Axoplasmic Transport*, 1st ed., Springer-Verlag, Basel, 1982, 1192.
24. **Harris, A. J.,** Inductive functions of the nervous system, *Ann. Rev. Physiol.*, 36, 251, 1974.
25. **Rosenthal, J.,** Trophic interactions of neurons, in *Handbook of Physiology*, Vol. 1, Brookhart, M. and Mountcastle, V. B., Eds., American Physiology Society, Bethesda, Md., 1977, 775.
26. **Drachman, D. B.,** Trophic interactions between nerves and muscle: the role of cholinergic transmission (including usage) and other factors, in *Biology of Cholinergic Function*, 1st ed., Goldberg, A. M. and Hanin, I., Eds., Raven Press, New York, 1976, 161.
27. **Beidler, L. M.,** Dynamics of taste cells, in *Olfaction and Taste*, Vol. 1, Zotterman, W., Ed., Pergamon Press, Oxford, 1963, 133.
28. **Guth, L.,** "Trophic" effects of vertebrate neurons, *Neurosci. Res. Prog. Bull.*, 7, 1, 1969.
29. **Torrey, T. W.,** The relation of taste buds to their nerve fibers, *J. Comp. Neurol.*, 59, 203, 1934.
30. **Donoso, J. A. and Zapata, P.,** Effects of denervation and decentralization upon taste buds, *Experientia*, 32, 591, 1976.
31. **Gutmann, E.,** Problems in differentiating trophic relationships between nerve and muscle cells, in *Motor Innervation of Muscle*, Thesleff, S., Ed., Academic Press, New York, 1976, 323.
32. **Salafsky, B, Bell, J., and Prewitt, M. A.,** Development of fibrillation potentials in denervated fast and slow skeletal muscle, *Am. J. Physiol.*, 215, 637, 1968.
33. **Hájek, I., Gutmann, E., and Syrovy, I.,** Proteolytic activity in denervated and reinnervated muscle, *Physiol. Bohemoslov.*, 13, 32, 1964.
34. **Thesleff, S.,** Physiological effect of denervation of muscle, *Ann. N.Y. Acad. Sci.*, 228, 89, 1974.

35. **Davey, B. and Younkin, S. G.,** Effect of nerve stump length on cholinesterase in denervated rat diaphragm, *Exp. Neurol.*, 59, 168, 1978.
36. **Ranish, N. A., Dettbarn, W. D., and Wecker, L.,** Nerve stump length-dependent loss of acetylcholinesterase activity in endplate regions of rat diaphragm, *Brain Res.*, 191, 379, 1980.
37. **Fernandez, H. L., Duell, M. J., and Festoff, B. W.,** Neurotrophic control of 16 S acetylcholinesterase at the vertebrate neuromuscular junction, *J. Neurobiol.*, 10, 441, 1979.
38. **Carlsen, R. C.,** The possible role of cyclic AMP in the neurotrophic control of skeletal muscle, *J. Physiol. (London)*, 247, 343, 1975.
39. **Miledi, R. and Slater, C. R.,** On the degeneration of rat neuromuscular junctions after nerve section, *J. Physiol. (London)*, 207, 507, 1970.
40. **Fernandez, H. L., Dorman, R. V., and Festoff, B. W.,** Neurotrophic control of skeletal muscle phospholipids, *Muscle Nerve*, 2, 118, 1979.
41. **Edström, J. and Emmelin, N.,** Movement of choline acetyltransferase in axons disconnected from their cell bodies, *J. Physiol. (London)*, 216, 247, 1971.
42. **Ochs, S. and Ranish, N.,** Characteristics of the fast transport system in mammalian nerve fibers, *J. Neurobiol.*, 1, 247, 1969.
43. **Harris, J. B. and Thesleff, S.,** Nerve stump length and membrane changes in denervated skeletal muscle, *Nature New Biol.*, 236, 60, 1972.
44. **Heslop, J. P.,** Axonal flow and fast transport in nerves, *Adv. Comp. Physiol. Biochem.*, 6, 75, 1975.
45. **Lømo, T.,** The role of activity in the control of membrane and contractile properties of skeletal muscle, in *Motor Innervation of Muscle*, Thesleff, S., Ed., Academic Press, New York, 1976, 289.
46. **Jones, R. and Vrbová, G.,** Two factors responsible for the development of denervation hypersensitivity, *J. Physiol. (London)*, 236, 517, 1974.
47. **Vrbová, G.,** Induction of an extrajunctional chemosensitive area in intact innervated muscle fibers, *J. Physiol. (London)*, 191, 20, 1967.
48. **Albuquerque, E. X., Warnick, J. E., Tasse, J. R., and Sansone, F. M.,** Effects of vinblastine and colchicine on neural regulation of the fast and slow skeletal muscle of the rat, *Exp. Neurol.*, 37, 607, 1972.
49. **Hofmann, W. W. and Thesleff, S.,** Studies on the trophic influence of nerve on skeletal muscle, *Eur. J. Pharmacol.*, 20, 256, 1972.
50. **Fernandez, H. L.,** Neurotrophic effects in relation to axoplasmic transport, *Acta Physiol. Lat. Am.*, 23, 189, 1973.
51. **Fernandez, H. L. and Ramirez, B. U.,** Muscle fibrillation induced by blockage of axoplasmic transport in motor nerves, *Brain Res.*, 79, 385, 1974.
52. **Fernandez, H. L. and Ramirez, B. U.,** Neurotrophic effects: axoplasmic transport involvement in the regulation of skeletal muscle soluble proteins, *Neurosci. Lett.*, 2, 211, 1976.
53. **Inestrosa, N. C. and Fernandez, H. L.,** Muscle enzymatic changes induced by blockage of axoplasmic transport, *J. Neurophysiol.*, 39, 1236, 1976.
54. **Fernandez, H. L. and Inestrosa, N. C.,** Role of axoplasmic transport in neurotrophic regulation of muscle endplate acetylcholinesterase, *Nature (London)*, 262, 55, 1976.
55. **Warnick, J. E., Albuquerque, E. X., and Guth, L.,** The demonstration of neurotrophic function by application of colchicine or vinblastine to the peripheral nerve, *Exp. Neurol.*, 57, 622, 1977.
56. **Hofmann, N. W. and Peacock, J. H.,** Post-junctional changes induced by partial interruption of axoplasmic flow in motor nerves, *Exp. Neurol.*, 41, 345, 1973.
57. **Hofmann, N. W., Struppler, A., Weindl, A., and Velho, F.,** Neuromuscular transmission with colchicine-treated nerves, *Brain Res.*, 49, 208, 1973.
58. **Cangiano, A. and Fried, J. A.,** The production of denervation-like changes in rat muscle by colchicine, without interference with axonal transport or muscle activity, *J. Physiol. (London)*, 265, 63, 1977.
59. **Cangiano, A. and Fried, J. A.,** Neurotrophic control of skeletal muscle of the rat, *J. Physiol. (London)*, 239, 31, 1974.
60. **Lømo, T.,** Neurotrophic control of colchicine effects on muscle?, *Nature (London)*, 249, 473, 1974.
61. **Sellin, L. C. and McArdle, J. J.,** Colchicine blocks neurotrophic regulation of the resting membrane potential in reinnervating skeletal muscle, *Exp. Neurol.*, 55, 483, 1977.
62. **Guth, L. and Albuquerque, E. X.,** The neurotrophic regulation of resting membrane potential and extrajunctional acetylcholine sensitivity in mammalian skeletal muscle, in *Muscle Regeneration*, 1st ed., Mauro, A., Ed., Raven Press, New York, 1979, 405.
63. **Oh, T. H., Johnson, D. D., and Kim, S. U.,** Neurotrophic effect of isolated chick embryo muscle in culture, *Science*, 178, 1298, 1972.
64. **Oh, T. H.,** Neurotrophic effect: characterization of the nerve extract that stimulates muscle development in culture, *Exp. Neurol.*, 46, 432, 1975.
65. **Oh, T. H.,** Neurotrophic effect of sciatic nerve extract on muscle development in culture, *Exp. Neurol.*, 50, 376, 1976.

66. **Markelonis, G. J. and Oh, T. H.,** A protein fraction from peripheral nerve having neurotrophic effects on skeletal muscle cells in culture, *Exp. Neurol.*, 58, 285, 1978.
67. **Stevens, W. F., Hooisma, J., Slaf, D. W., and Meeter, E.,** Neurotrophic influence in tissue culture, *Prog. Brain Res.*, 48, 31, 1978.
68. **Hooisma, J., Slaf, D. W., Meeter, E., and Stevens, W. F.,** Trophic support by neural explants of cultured muscle fibers, *Exp. Neurol.*, 62, 628, 1978.
69. **Podleski, T. R., Axelrod, D., Ravdin, P., Greenberg, I., Johnson, M. M., and Salpeter, M. M.,** Nerve extract induces increase and redistribution of acetylcholine receptors on cloned muscle cells, *Proc. Natl. Acad. Sci. U.S.A.*, 75, 2035, 1978.
70. **Markelonis, G. J., Oh, T. H., Eldefrawi, M. E., and Guth, L.,** Sciatin: a myotrophic protein increases the number of acetylcholine receptors and receptor clusters in cultured skeletal muscle, *Dev. Biol.*, 89, 353, 1982.
71. **Cano, M., Suzuki, N., and Ojima, H.,** Neurotrophic effect of nerve extract on development of tetrodotoxin-sensitive spike potential in culture, *J. Cell. Physiol.*, 99, 327, 1979.
72. **Engelhordt, J. K., Ishikawa, K., Mori, J., and Shimabukuro, Y.,** Neurotrophic effect on the electrical properties of cultured muscle produced by conditioned medium from spinal cord explants, *Brain Res.*, 128, 243, 1977.
73. **Hasegawa, S. and Kuromi, H.,** Ventral part of spinal cord contains neurotrophic factor for the action potential of cultured muscle, *Brain Res.*, 157, 153, 1978.
74. **Markelonis, G. J. and Oh, T. H.,** Purification of sciatin using affinity chromatography on concanavalin-a agarose, *J. Neurochem.*, 37, 95, 1981.
75. **Markelonis, G. J., Bradshaw, R. A., Oh, T. H., Johnson, J. L., and Bates, O. J.,** Sciatin is a tranferrin-like polypeptide, *J. Neurochem.*, 39, 315, 1982.
76. **Robbins, N.,** Neurotrophic regulation of rat muscle glucose-6-phosphate dehydrogenase in vitro, *Brain Res.*, 225, 387, 1981.
77. **Lentz, T. L.,** Effect of denervation on muscle cyclic AMP in the newt, *Exp. Neurol.*, 49, 716, 1975.
78. **Lentz, T. L.,** Nerve trophic function: in vitro assay of effects of nerve tissue on muscle cholinesterase activity, *Science*, 171, 187, 1971.
79. **Lentz, T. L.,** Development of the neuromuscular junction. III. Degeneration of motor endplates after denervation and maintenance *in vitro* by nerve explants, *J. Cell Biol.*, 55, 93, 1972.
80. **Lentz, T. L.,** Effect of brain extracts on cholinesterase activity of cultured skeletal muscle, *Exp. Neurol.*, 45, 520, 1974.
81. **Lentz, T. L.,** Neurotrophic regulation at the neuromuscular junction, *Ann. N.Y. Acad. Sci.*, 228, 323, 1974.
82. **Inestrosa, N. C., Ramirez, B. U., and Fernandez, H. L.,** Effect of denervation and of axoplasmic transport blockage on the *in vitro* release of muscle endplate acetylcholinesterase, *J. Neurochem.*, 28, 941, 1977.
83. **Walker, C. R. and Wilson, B. W.,** Control of acetylcholinesterase by contractile activity of cultured muscle cells, *Nature (London)*, 256, 215, 1975.
84. **Davey, B., Younkin, L. H., and Younkin, S. G.,** Neural control of skeletal muscle cholinesterase: a study using organ-cultured rat muscle, *J. Physiol. (London)*, 288, 501, 1979.
85. **Younkin, S. G., Bret, R. S., Davey, B., and Younkin, L. H.,** Substances moved by axonal transport and released by nerve stimulation have an innervation-like effect on muscle, *Science*, 200, 1292, 1978.
86. **Skau, K. A. and Brimijoin, S.,** Release of acetylcholinesterase from rat hemidiaphragm preparation stimulated through the phrenic nerve, *Nature (London)*, 275, 224, 1978.
87. **Fernandez, H. L., Patterson, M. R., and Duell, M. J.,** Neurotrophic control of 16 S acetylcholinesterase from mammalian skeletal muscle in organ culture, *J. Neurobiol.*, 11, 557, 1980.
88. **Fernandez, H. L., Duell, M. J., and Festoff, B. W.,** Cellular distribution of 16 S acetylcholinesterase, *J. Neurochem.*, 32, 581, 1979.
89. **Massoulié, J., Bon, S., and Vigny, M.,** The polymorphism of cholinesterase in vertebrates, *Neurochem. Int.*, 2, 161, 1980.
90. **Massoulié, J. and Bon, S.,** The molecular forms of cholinesterase and acetylcholinesterase in vertebrates, *Ann. Rev. Neurosci.*, 5, 57, 1982.
91. **Hall, Z. W.,** Multiple forms of acetylcholinesterase and their distribution in endplate and nonendplate regions of rat diaphragm muscle, *J. Neurobiol.*, 4, 343, 1973.
92. **McMahan, U. J., Sanes, J. R., and Marshall, L. M.,** Cholinesterase is associated with the basal lamina at the neuromuscular junction, *Nature (London)*, 271, 172, 1978.
93. **Lwebuqa-Mukasa, J. S., Lappi, S., and Taylor, P.,** Molecular forms of acetylcholinesterase from *Torpedo californica:* their relationship to synaptic membranes, *Biochemistry*, 15, 1425, 1976.
94. **Vigny, M., Koenig, J., and Rieger, F.,** The motor endplate specific form of acetylcholinesterase: appearance during embryogenesis and reinnervation of rat muscle, *J. Neurochem.*, 27, 1347, 1976.
95. **Fernandez, H. L.,** Regulation of nerve-target cell interactions, *Sci. Proc. Univ. Kansas*, 1, 147, 1983.

96. **Rathbone, M. P., Vickers, J. D., and Logan, D. M.,** Neural regulation of cholinesterase in newt skeletal muscle. III. Regulation in vitro by proteins from chick embryo brains, *J. Exp. Zool.*, 210, 463, 1979.
97. **Lentz, T. L., Addis, J. S., and Chester, J.,** Partial purification and characterization of a nerve trophic factor regulating muscle acetylcholinesterase activity, *Exp. Neurol.*, 73, 542, 1981.
98. **Festoff, B. W., Duell, M. J., and Fernandez, H. L.,** Trophic effects of axonally transported proteins on muscle cells in culture, in *Maturation of Neurotransmission*, 1st ed., Vernadakis, A. and Filogamo, G., Eds., S. Karger, Basel, 1978, 152.
99. **Koening, J. and Vigny, M.,** Neural induction of the 16 S acetylcholinesterase in muscle cell cultures, *Nature (London)*, 271, 75, 1978.
100. **Inestrosa, N. C., Mendez, B., and Luco, J. V.,** Acetylcholinesterase like that of skeletal muscle in smooth muscle reinnervated by a motor nerve, *Nature (London)*, 280, 504, 1979.
101. **Gospodarowicz, D. and Moran, J. S.,** Growth factors in mammalian cell culture, *Ann. Rev. Biochem.*, 45, 531, 1976.
102. **Bradshaw, R. A.,** Nerve growth factor, *Ann. Rev. Biochem.*, 47, 191, 1978.
103. **Orrego, F.,** Criteria for the identification of central neurotransmitters, and their application to studies with some nerve tissue preparations in vitro, *Neuroscience*, 4, 1037, 1979.
104. **Auerbach, A. A.,** Transmitter release at chemical synapses, in *Structure and Function of Synapses*, 1st ed., Pappas, G. D. and Purpura, D. P., Eds., Raven Press, New York, 1972, 137.
105. **Drachman, D. B., Pestronk, A., and Stanley, E. F.,** Neurotrophic interaction between nerves and muscles: role of acetylcholine, in *Diseases of the Motor Unit*, Schotland, E. L., Ed., John Wiley & Sons, New York, 1982, 107.
106. **Drachman, D. B.,** The role of acetylcholine as a neurotrophic transmitter, in *Trophic Functions of the Neuron*, Drachman, D. B., Ed., New York Academy Science, New York, 1974, 160.
107. **Drachman, D. B., Stanley, E. F., Pestronk, A., Griffin, J. W., and Price, D. L.,** Neurotrophic regulation of two properties of skeletal muscle by impulse-dependent and spontaneous acetylcholine transmission, *J. Neurosci.*, 2, 232, 1982.
108. **Lentz, T. L., Mazurkiewicz, J. E., and Rosenthal, J.,** Cytochemical localization of acetylcholine receptors at the neuromuscular junction by means of horseradish peroxidase-labeled α-bungarotoxin, *Brain Res.*, 132, 423, 1977.
109. **Thesleff, S.,** Studies on neurotrophic regulation of skeletal muscle, in *The Cholinergic Synapse*, Tuček, S., Ed., *Prog. Brain Res.*, 49, 385, 1979.

Chapter 17

REGENERATION STUDIES IN RELATION TO TRANSPORT

Lawrence Austin

TABLE OF CONTENTS

I. INTRODUCTION

The nervous system is capable of regeneration as are other organs. There are, however, a number of important differences between neuronal regeneration and that of other cell types in both the regrowth capacity and the biochemical changes which occur during the regenerative process. The major difference is related to two well-known characteristics of the neuron; the inability to divide after maturation and the asymmetry of the cell. When the liver regenerates it does so mainly using an accelerated process of cell division but in the nervous system the neuron must replace the axon which is a major part of itself. The one exception is the neuron of the olfactory epithelium.[1] Thus, one part of a cell, typically 20 to 50 nm in diameter is called on to produce replacement components involved in the elongation of the cell by distances which may be meters long. These components are quite varied in chemical and physical structures but, in the main, they may be subdivided into various classes; membranes of the axon, cytoskeletal elements, mitochondria, lysosomes, vesicles, and a variety of diffusible enzymes required for energy and neurotransmitter production.

Faced with such a formidable task it is hardly surprising that successful regeneration of nerves leading to correct reinnervation, at least in mammals, is frequently not achieved. In the central nervous system (CNS) successful reinnervation is extremely rare; that is not to say that the CNS neurons are incapable of regeneration, but functional return is usually unattainable.

Regeneration of nerve has attracted much attention and because of the need to supply the variety of components to the elongation axon, the role of axoplasmic transport in these axons has been extensively studied. The site for biosynthesis for these components is the cell body and it will be necessary to discuss some aspects of these synthetic processs before considering the involvement of axoplasmic transport.

II. PERIKARYON CHANGES DURING AXON REGENERATION

Certain obvious changes in the neuronal cell body were observed by Nissl[2] after injury to peripheral nerve. These may occur to a greater or lesser degree and are characterized by a dissolution of the Nissl substance (now known to be largely composed of rough endoplasmic reticulum (RER)), a swelling of the cell body and a movement of the nucleus towards the periphery of the cell.[3] There is often an increase in the size of the nucleolus. These changes occur at various times after nerve injury and they vary with the type of nerve and the species of animal. The onset can vary from 1 to 7 days after injury and be either very obvious or barely observable. These morphological changes are usually referred to as chromatolysis. The axon reaction, defined as the biochemical changes which are set in train at the commencement of the regenerative process, are not necessarily dependent on an obvious chromatolytic response. Thus, injury to the vagus nerve of the rabbit leads to a very pronounced chromatolytic response in the nodose ganglion while in the rat this response is less obvious. Nevertheless the biochemical changes in the rat are very striking.[4] No chromatolytic response is seen in frog spinal neurons at low temperatures, after sciatic nerve section but the axon reaction proceeds albeit at a somewhat reduced rate.[5] Furthermore, the cell body reaction may increase with the severity of the nerve injury.[6] The axon reaction is sustained for much longer periods in young animals even when experimental procedures are employed to prevent successful regrowth of the axon.[7] This enhanced ability of nerve regeneration in the young has been known to neurosurgeons for many years as success of human nerve repair generally declines with age.

The morphological and biochemical changes suggest two very different approaches to the study of the initiation and progression of regeneration. On the one hand, the delay in response to injury, to some extent dependent on the distance between the cell body and the injury site, has led to a search for a chromatolytic signal.[8,9] This hypothetical molecule, generally assumed to be a protein, is often thought to be carried to the cell body and therefore presumably to the nucleus, by retrograde axonal transport. It is also considered to originate at the injury site or alternatively, it is transported from the distal regions of the axon to the cell body but, due to injury, its supply is interrupted. Furthermore, it is important to determine what initiates these changes; is it a signal from the periphery carried by axonal transport or is it due to some other controlling factor such as an altered ion flux in the cell body? If a signal molecule does exist it would be expected to reach the cell body by retrograde axonal transport and therefore this transport would be an important factor in the initiation of the regenerative process. If, on the other hand, an altered ionic composition of the cell body cytoplasm results from the loss of axoplasm at the site of injury,[10] factors other than retrograde axonal transport would be involved.

The second approach is to study changes brought about in the cell body as a result of injury. Changes to the RER suggest a reorganization of the membrane-bound ribosomes while those in the nucleolus point to an alteration in gene expression. Consequently there have been many studies undertaken to investigate the metabolism of RNA; (1) as an indication of changes in gene expression, i.e., messenger RNA (mRNA) changes, (2) to determine more precisely the turnover and replacement of ribosomes, i.e., ribosomal RNAs (rRNA), and (3) to establish, by indirect means, whether there is any alteration in protein assembly, i.e., transfer RNA (tRNA). Initial studies concentrated on the total RNA of the cell. The results were conflicting due to the crude methodology of the time and the failure to take into account that there may be a very significant increase in volume of the cytoplasm of the cell body. Thus, there have been reports of a decrease in RNA,[11,12] an increase in RNA content,[13-17] and no change in RNA content.[4] The involvement of RNA in the axon reaction has been well established, particularly since a block in RNA production by the neuron, effected by administration of actinomycin D, abolishes the axon reaction.[18-20] However, it is not the changes in total RNA that are important but rather, the types of RNA which change.

III. RIBONUCLEIC ACID METABOLISM DURING REGENERATION

There are a number of different RNA molecules in the eukaryotic cell but 75 to 80% of the total RNA is contributed by the rRNA components. It follows that any major change in cellular RNA would almost certainly be due to altered ribosome status. The constancy of base ratios[15] supports this view as do various observations on single neurons undergoing the axon reaction.[8,21,22] Very significant variations in metabolism of rRNA species 28S and 18S RNA have been demonstrated which commence within 24 hr of axon injury.[23-26] Even earlier changes are seen in metabolic responses of rapidly turning over (presumed messenger) RNA.[27] These appear within a few hours of the injury and suggest not only an early alteration in protein synthesis but also a very rapid response to the altered circumstances of the neuron. Production of tubulin mRNA has been shown to increase markedly in the regenerating goldfish optic system 10 days after injury.[28]

The production of transfer RNA is also increased, following a time course similar to that of rRNA.[23,25,26] This is the only stable RNA species which is carried along the nerve by axoplasmic transport[29-33] and the reported increase in amount of tRNA transported during goldfish optic nerve regeneration may simply be a result of the increase in cellular production.[34]

The neuron contains relatively high levels of stable small molecular weight RNAs (smwRNA). For some of these the role in metabolic control is known; 2 smwRNAs respond by way of increased synthesis after nerve injury.[35] Of these the 7S molecule, smwL is involved in translational events at the ribosome and protein transport within the cell through the SER[36,37] while smwD is implicated both in messenger RNA splicing and preribosomal RNA processing.[38-40]

The time course of response of these various RNA species varies within the one system and therefore there must be multiple control sites operating during the change from metabolic maintenance to regrowth. For example, 28S, 18S, and 5S RNA are all components of the ribosome but they respond differently to axotomy, probably due to the differing locations on the genome.[35] Furthermore, there are increases in activity of processing enzymes, carbomoyl phosphate synthetase and acid phosphatase[4] as well as ribonuclease or its endogenous inhibitor.[41] Since the most likely effect of the elusive signal for chromatolysis, assumed to be carried to the cell body by axoplasmic transport, is to alter the protein synthetic machinery which so intimately is connected with RNA and its metabolism for its continuation, there would seem to be a need for a number of signals because of the numerous metabolic controls involved. Evidence for the existence of at least two signals operating at different points in time, has been presented.[26]

IV. PROTEIN SYNTHETIC RESPONSES TO AXOTOMY

When mouse sciatic nerve is cut there is no obvious chromatolytic response in the ventral horn cells but a marked rise in protein content occurs, confined to the cytoplasm.[14] Single motor neurons were injected by Lux et al.[42] with radioactively labeled glycine and they subsequently showed that electrical stimulation of these cells increased the incorporation rate by about 100%. This increased cellular protein was mainly exported to the axon, chiefly by fast axoplasmic transport. The origin of this extra synthetic activity was shown to be at the rough endoplasmic reticular level and its movement into the axon has been followed using electron microscope radioautography.[43,44]

The increased protein synthesis may, in part, be due to an increased amino acid uptake into the cell.[45,46] If this is so it may mean that there is a need for yet another signal operating at the level of the neuronal cell body plasma membrane. These uptake studies have been made using the goldfish retinal ganglion cells, a tissue which has been widely used to study protein synthesis and protein transport during regeneration. Although many interesting observations have been obtained using this model it must be remembered that it may not necessarily represent regenerating mammalian systems. In the regenerating CNS neurons there is a very obvious increase in synthesis of tubulin.[47-49] The increased protein synthesis in the cell body does not appear to be due to some feedback mechanism from the damaged axon resulting from protein utilization since blockage of axonal transport by vinblastine did not alter the synthetic rate in Aplysia neurons.[50]

V. AXOPLASMIC TRANSPORT OF PROTEINS

The major function of protein transport is to supply axonal components to maintain an equilibrium along the whole length of the axon which is in a continually dynamic state due to varying rates of turnover of its constituent parts. This equilibrium will be upset if the neuron is called upon to initiate active regrowth as a result of injury.

The measurement of axoplasmic transport in regenerating nerves has been used to determine the success of the regenerative process and its extent. Ochs and Barnes[51] cut the L7 ventral root in the cat and joined it to the cut end of the dorsal root on the other side. Flow studies were carried out over a period of a year to determine regeneration. Accumulation of labeled protein at the growth cones as well as its distribution along the nerve was measured and the extent of regeneration as determined by this method was found to correspond to that measured by the pinch test.[52]

Many studies have been carried out to examine alterations in axoplasmic transport during nerve regeneration and to equate these changes with the requirements for regrowth. These changes vary very significantly from one system to another and, unfortunately, there have been many generalized conclusions made from a single experimental system that do not apply to regeneration in general.

A. Transport Rates

There is a significant rise in both the rate of transport and amount of slowly transported protein in the regenerating goldfish optic nerve; these remain elevated for some time after reinnervation has taken place. There is a lesser rise in the rate of the fast component of axonal transport but a very large increase, of 3- to 5-fold, in amount of protein delivered at this rate, particularly at the time that reinnervation has taken place.[53-55] In the frog sciatic, a peripheral nerve, the amount of transported protein increases over control in the regenerating system[56] but no increase was found in the dorsal roots.[57] In both these studies the transported proteins were examined using two-dimensional gel electrophoresis but little change in protein patterns was seen.

Nerves which are preganglionic to the chick ciliary ganglion are affected by section of the postganglion axons. Such an injury results in the late arrival of slowly transported protein, a reduction in material carried at the fast rate and an increased acetylcholinesterase (AChE) activity.[58]

Responses in rates in mammals are quite variable. Ochs found no variation in either rate or amount of protein transported at the fast rate in cat sensory nerves during nerve regeneration.[59] This seems to be quite general for sensory nerves, at least for those examined by measuring the outflow of labeled proteins after isotopically labeled amino acids are deposited in a dorsal root ganglion. No very significant change in either rate or amount of fast-transport protein has been observed in sensory neurons in the rat and this seems to apply whether the peripheral or central branches are studied.[60-62] There is, however, a higher rate of overall regeneration in the peripheral branch.[62] This is not the case with motor axons where, although there is no rate change, rapidly transported proteins increase very significantly in amount.[63] There appears to be a similar increase in amount of fast-transport protein in regrowing fascial neurons of the hamster, particularly in the adult.[64]

The hypoglossal nerve is another system which has been used for regeneration studies. In the guinea pig the amount of slowly transported protein is increased with no change in rate[65] while an increase in both rate and amount have been observed in the rabbit.[66] The amount of fast-transported protein, but not the rate, is also increased in the rabbit hypoglossal.[67] The opposite situation is found for the regenerating rabbit vagus nerve. Here both rate and amount of slowly transported protein is less than controls whereas the amount only of fast-transport protein is reduced.[66,67] In motor axons of the rat sciatic, some cytoskeleton proteins are reduced in amount while others are increased.[68]

It can be seen that both the amount of transported protein and the rate of transport can vary considerably between species and within species (Table 1). This may be due to a number of factors, not all of which have been considered by those studying these effects. When the axon is damaged there may be a rapid out-pouring of axoplasm from the site of injury[20] and in some nerves there may be an overall loss of axonal volume which may be as high as 60% of the initial volume.[65] Clearly, such large losses of axoplasm could very significantly affect the amount of protein transported and perhaps also give a misleading value for the measured rate. An additional factor is that in most of the studies quoted above, total protein flow was measured. If some proteins are exported in increased amounts and others are reduced, the total amount transported may vary with the needs of the particular neuron. This seems to be so when sensory fibers are compared with motor axons in the sciatic nerve. In the regenerating sensory fibers the overall amount of rapidly transported proteins were found to be no different from controls when examined using an in vitro flow approach but when

Table 1
SOME VARIATION IN RATE OF PROTEIN TRANSPORT AND AMOUNT TRANSPORTED DURING NERVE REGENERATION

System	Rate Slow transport	Amount	Ref.
Goldfish optic nerve	Increase		53
		Increase	54
Toad optic nerve		Increase	82
Rabbit hypoglossal nerve	Increase		66
		Increase	65
Rabbit vagus nerve	Decrease	Decrease	66
Rat sciatic nerve	No change (tubulin neurofilaments, actin)	Decrease (tubulin, neurofilaments) Increase (actin)	68
	Fast transport		
Goldfish optic nerve	Increase		53
		Increase	55
	Increase (also contralateral)		55
Toad optic nerve		Increase	82
		Increase, GAPS, 24, 43, 50	
		Decrease, 190 K protein	
Frog sciatic nerve		Increase — some proteins, others unchanged	56
Frog, dorsal roots		No change	56
Chick, ciliary preganglionic fibers		Decrease	58
Cat sciatic nerve		No change 6—165d	59
Rat sciatic nerve	No change	Increase	63
		Some proteins, increase, some decrease	69
Rat dorsal roots		No change	60, 61
Hamster facial neurons		Increase in many proteins	64
Rat hypoglossal nerve	No change	Increase	65
Rabbit hypoglossal nerve	No change	Increase	67
		Increase, GAPS. 23 and 43	83
Rabbit vagus nerve	No change	Decrease	67
Rabbit optic nerve		No change, GAPS. 23 and 43	83

these proteins were separated by polyacrylamide gel electrophoresis some were found to be increased while others were reduced.[69]

B. The Proteins Transported

The functioning axon and its terminals require a continual supply of membrane components, cytoskeletal elements, and enzymes for various functions. The study of the supply of these has been the cause of much research activity in recent years; in addition, their alteration from normal during the regenerating state has been followed.

The cytoskeletal elements are composed of microtubules, neurofilaments, and actin fibers. They make up the bulk, but not all, of the proteins which are traditionally classified as slow flow proteins.[70-72] In the regenerating goldfish visual system there is a marked increase in transport of tubulin, both in a soluble and particulate form.[47,73] There is an increase in proteins of 68 to 70,000 daltons and the appearance of a 45,000 daltons protein which is barely discernible in the control. In the regenerating rat sciatic nerve, on the other hand, the amount of tubulin transported is reduced, as are the triplet proteins which make up the neurofilaments.[68] There is an increased transport of actin. Sutherland[74] counted the density of microtubules in the regenerating hypoglossal nerve and found no change in numbers compared to controls. In the goldfish visual system there is an increased rate of synthesis of tubulin together with an increase in tubulin mRNA.[28] It may be that in the mammalian peripheral motor nerve the turnover of tubulin in the regenerating axon is reduced, thus leading to an inadequate structuring of the cytoskeletal system resulting from a reduced tubulin synthesis.

Proteins transported at the fast rate have been studied in a number of ways in regenerating systems. McEwan et al.[71] used proteins labeled by incorporation of tritium-labeled glucosamine or fucose into glycoproteins, proteins which are only associated with the fast-transport phase. This approach has been used by others in subsequent investigations to identify this fast phase. Griffin et al.[75] examined the distribution of glycoproteins along the regenerating nerve, using EM radioautography to locate in the axon tritiated fucose incorporated into the glycoprotein in the cell body. They identified the position of the regenerating tip with [^{35}S] incorporated into transported protein. As early as the first day the newly synthesized glycoproteins were inserted in the membranes of the newly formed sprouts and they appeared along the length of the axon. Once the glycoproteins are located in the membrane they appear to remain quite stable; very little redistribution takes place. Almost all of the radioactivity is associated with the axolemma, as might be expected for glycoprotein molecules. These membrane glycoproteins of newly formed axon are not identical to those of the established axon membrane as they are rather smaller in size.[76] In the mammalian optic pathway the production and perhaps transport of glycoproteins is more complex in that there are two waves, one appearing rapidly, within hours of isotope deposition while the second takes up to a week to be transported. This is probably reflecting the turnover of different membrane structures and, in fact, the composition of the proteins differ between the two pools.[77] This complication of pools of proteins with differing processing and turnover times is often ignored. Multiple pools have been demonstrated for axonally transported proteins in Aplysia neurons.[78] Here there are at least two such pools with half-lives of approximately 1 hr and 40 hr respectively. Specific proteins appear to be able to enter either pool; their transport times are governed by passage through the Golgi and the proximity of the Golgi to the transport site. Thus, it would be unreasonable to expect that any specific protein might behave in the same manner in systems that differ as widely as the mammalian peripheral nervous system and the amphibian CNS.

C. Growth Associated Proteins

Evidence for the existence of specific proteins in the optic system of the goldfish has been presented.[79] There are four such proteins, identified using two-dimensional gel electrophoresis, which are highly concentrated in the intact optic nerve and all fall dramatically in amount in the denervated tectum but rise on reinnervation, suggesting that they are transported to the tectum by axoplasmic transport. In both amphibian and mammalian nerves, a number of proteins dramatically increase when the nerve regenerates after injury.[80] These have been named growth associated proteins (GAPs) by the authors.

When the toad optic nerve is crushed it can regenerate. During the regenerative phase there is an increase in the amount of a number of proteins transported including actin, α- and β-tubulins.[81] There is also a more impressive increase in a small number of proteins which are barely discernible in the intact nerve. Of these, three have attracted the most interest. They are all membrane-bound, transported at the fast rate and increase in labeling over the first week. They remain elevated for about 4 weeks, i.e., presumably until tectal reinnervation is complete.[81] They are known as GAP-50, GAP-43, and GAP-24, where the numbers refer to molecular weight in kilodaltons.[82] GAP-50, which is a fucosylated glycoprotein, is transported together with GAP-43 to the growing tip whereas GAP-24 is distributed evenly along to axon and is probably a component of the axolemma. This latter GAP-24 is more transient than the other two in that its half-life is only 4 to 6 hr compared with about 1 day for GAP-50 and 2 to 3 days for GAP-43.

It is interesting that similar proteins occur in the mammalian nervous system.[83] Retinal ganglion cells of the neonatal rabbit transport GAP-43 at the fast rate during the early stages of development. The amount produced falls off precipitately with further development but there is no return when the rabbit optic nerve is crushed. Unlike the toad, optic nerve regeneration in the rabbit does not occur. GAP-43 and another GAP protein, GAP-23, is found in the rabbit peripheral (hypoglossal) nerve. Here, unlike the optic nerve, the hypoglossal nerve can regenerate and when induced to do so by nerve crush, it produces and transports significantly elevated levels of GAP-43. On the other hand, GAP-23 is present in readily detectable amounts in the intact hypoglossal nerve and no change in this level was found in regeneration.[83] These findings have led the authors to suggest that these proteins are concerned in the regenerative process, perhaps at more than one stage and that the failure of mammalian CNS neurons to regenerate may be related, at least in part, to their inability to alter their gene expression to produce the appropriate GAP proteins. A more detailed study of these proteins and others produced by degenerating nerves[84] throw more light on the important question of failure of CNS regeneration.

D. Specific Proteins

The two-dimensional gel electrophoresis technique of O'Farrell[85] whereby proteins are separated by iso-electric focusing followed by molecular sieving, has allowed for a very high degree of resolution of protein mixtures. This approach was first used on regenerating nerve by Hall et al.[86] who separated proteins in the rat superior cervical ganglion after axotomy. Although a number of changes were observed, these were confined to about 5 proteins; another 30 showed minor changes. The only changes seen were with respect to the amount of protein present and no new protein spots were observed. As to the cytoskeletal proteins, there was a small increase in the level of actin but no concomitant change in tubulin.

Others have examined the types of protein involved in axoplasmic transport, both in normal and regenerating nerves, using either single or two dimensional gel electrophoresis or by tissue subfractionation. In regenerating goldfish olfactory nerves there are three types of nerve fiber which regenerate at 5.8, 2.1, and 0.8 mm/day.[87] Fast protein transport in these regenerating nerves showed little difference from those of intact nerve except that, on subfractionation there was found to be a higher proportion of protein in the lighter membrane fractions than in intact nerve. It may be inappropriate to compare these regenerating nerves with other types of central or peripheral systems since the olfactory nerves regenerate from dividing neurons, unlike those elsewhere in the nervous system.[1]

Regenerating rat sciatic motor fibers have been solubilized in sodium dodecyl sulfate (SDS) and the proteins separated in one-dimensional gels.[88] Changes in amounts of proteins were seen, commencing 3 days after injury and reaching a maximum in 7 to 14 days. In particular, three were measured, one with a molecular weight of 13,000 daltons fell while two others of 18,000 and 66,000 daltons rose in amount. These returned to normal levels at about the time of reinnervation. Later, two other proteins were studied, of 22,000 and 18,000 daltons.[89] The rates of these two rose during regeneration and remained high when regeneration, although proceeding, was prevented from forming a synaptic connection. This elevated ratio was measured over 98 days and may be compared with a similar situation in which altered mRNA production is also elevated for prolonged periods (at least 93 days) in the regenerating fibers of neurons in the nodose ganglion when they are misdirected by myelin tubes into the thorax and therefore are unable to find their appropriate innervating sites in the abdomen.[27,90]

Changes in protein patterns have been observed in nerves of fish and amphibia. A comparison of intact and regenerating goldfish optic nerve was made by crushing one nerve and subsequently labeling one eye with [^{3}H] and the other with [^{14}C]-labeled proline.[91] Both nerves were removed from fish at various intervals after crush and proteins extracted from the combined nerve mass using SDS. Electrophoretic patterns showed that labeling varied over a threefold range, reaching a maximum at 8 days. This increased labeling was observed in a number of proteins of size 24,000, 27,000, 44,000, 110,000, 140,000, and 210,000 daltons.

The transported proteins which were increased in the nerve were also increased in the perikaryon. In regenerating frog sensory axons of the sciatic nerve, increases of certain proteins of the fast transport phase, identified by two-dimensional gel electrophoresis, were found not only in the regenerating axons but also in the cell bodies as well as the uninjured proximal branch. No changes were observed in either tubulin or actin; there seems to be little or no change in these two proteins in regenerating sensory fibers.

The fate of rapidly transported proteins is somewhat different in regenerating sensory and motor axons, in that, in the motor fibers they are mainly laid down at the region of the growth cones, whereas in sensory fibers they are distributed along the length of the growing axon.[92]

E. Membrane Transport

It is becoming increasingly obvious that fast axonal transport is a system whereby materials, packaged in vesicles or vacuoles, are carried to the axon, the axolemma, and the axon terminals. This is hardly surprising since membranes are largely preformed and transported as integral membrane structures to their specific locations in vesicle form.[93] In addition, proteins and enzymes required for membrane repair[94] will also presumably be transported. This membrane packaging has been demonstrated in regenerating nerves.[43,95] In a study of material transported both by the orthograde and retrograde route, Griffin et al.[96] showed that all of the transported material which arrives at the reactive axonal swelling of nerves ligated 2 to 24 hr before labeling were to be found in vesicles and tubules of the SER. A detailed autoradiographic study of the regenerating rabbit hypoglossal was made by Tessler et al.[97] Less than 10% of labeled glycoprotein, a membrane-bound protein, was incorporated into the axolemma 6 hr after injury but by 18 hr this had increased to close to 100%. At both time periods, about 30% of the label was to be found in the growth cones. The authors concluded that the axon elongates by continual insertion of membrane into the axolemma and that the glycoproteins, as part of this new membrane are inserted along its entire length. There appears to be a constant ratio between the inserted glycoprotein and the transported glycoprotein, implying that the two are in equilibrium.

F. Reversal of Protein Transport

Bray et al.[98] showed that newly synthesized protein, when transported down an axon, returns by retrograde flow if it meets an obstruction. This turnaround has been further studied in recent years.[99,100] In the intact nerve there seems to be little difference between the proteins which return to the cell body and those which leave it.[101] In the goldfish optic nerve, glycoproteins are transported from the ganglion cell at 80 mm/day and return at 36 mm/day with a turnaround time of 1 to 2 hr. There is a very large increase, to about fivefold in the amount returning after 2 weeks postnerve crush.[102] Particle-specific enzymes also undergo this turnaround process. Acetylcholinesterase, associated with the SER, acid phosphatase with lysosomes and monoamine oxidase with mitochondria all turnaround in rat sciatic nerve at an injury site.[103] These studies revealed no evidence to suggest that retrogradely transported protein undergoes turnaround to become part of the orthogradely transported protein.

G. Retrograde Transport; Does it Signal the Cell Body?

The retrograde transport of proteins has been studied from a number of points of view. One of these is to test the hypothesis that the neuronal metabolic responses to injury are triggered by the arrival of some trophic molecule from the periphery, the presence of which results from the axonal injury. It is well established that such a retrograde transport operates and it is responsible for the reversal of flow which is a consequence of turnaround. This retrograde transport is frequently followed by introducing a foreign protein, commonly horseradish peroxidase (HRP), in the region of the axon terminals or at the site of injury. HRP must be applied to the injured nerve within a short time. If left for more than 30 to 60 min, the degree of uptake is markedly reduced. After uptake it is rapidly cleared from the cytoplasm and incorporated into vesicular and tubular organelles and, as such, transported to the cell body.[104,105] When applied at later stages of regeneration, the HRP is taken up only by growth cones, not by the growing axon.[106] Use has been made of the delay in re-establishment of retrograde transport to show that cholinergic fibers respond more quickly than sympathetic fibers after injury.[107]

The timing of appearance of the retrogradely transported protein at the cell body has been compared with the onset of chromatolysis.[108] This comparison may have little meaning as, has been pointed out above, there are a number of demonstrable metabolic changes which appear in the neuronal cell body long before the morphological changes which are collectively known as chromatolysis and, the chromatolytic response does not necessarily bear any relationship to successful regeneration.

In the regenerating rabbit hypoglossal nerve the appearance of retrogradely transported protein in the cell body is reduced, when compared to a control nerve, for periods up to 26 days but over later periods it exceeds the control level.[109] Ultimately the two levels correspond. The arrival of HRP at the neuronal cell body after application to the injured sciatic nerve precedes the first stages of chromatolysis.[110,111] There was no correlation with distance of injury from the cell body or between the species used which were rats, mice, and rabbits. The time of arrival from the point of injury of the facial nerve was as little as 6 hr, very much earlier than any observed chromatolytic response.[110]

Only a few studies have been made into the types of protein carried by retrograde transport during regeneration. In the regenerating rabbit vagus nerve the amount of protein transported is larger than in controls and this elevated level persists for several weeks. Both generally labeled proteins and specifically labeled glycoproteins are elevated.[112] When various exogenous macromolecules are applied to the crushed hypoglossal nerve some, but not all proteins, are taken into the nerve and transported suggesting some specificity in the process.[113]

Denburg[114] used HRP conjugated to wheat germ agglutinin to identify regenerating neurons that had regrown into the cockroach leg muscle. With this approach he was able to identify the neurons which successfully innervated the muscle since other neurons, which did not innervate the muscle were eliminated. He concluded that the reinnervation was not dependent on material supplied along the path of the nerve or by the muscle. These results, if extrapolated to more complex life forms, would limit the source of any trophic molecule.

Axoplasmic transport may be blocked by chemical as well as mechanical means. When colchicine was applied to an avian ciliary nerve, changes were observed in the RNA metabolism in the ciliary ganglion and there was a chromatolytic response.[115] This might suggest that the chromatolysis signal is produced by blockage of supply of some substance to the cell body which is normally carried by retrograde transport. Contrary to this conclusion, axotomy of the hypoglossal nerve leads to increased energy demands in the hypoglossal nucleus, as shown by increased uptake of 2-deoxyglucose and this response is abolished if colchicine is placed on the nerve on the proximal side of the injury.[116] In the same system (hypoglossal) it has been shown that colchicine blocks retrograde transport of HRP from the tongue, suggesting that the signal originates at the site of injury. These results need to be treated with some caution since, although colchicine blocks both orthograde and retrograde transport it has other actions, the most damaging of which is its ability to destroy retinal ganglion cells.[117,118] Since the hypoglossal nucleus is located in the CNS it would be unwise to make direct comparisons between this system and a peripheral ganglion in an avian species.

H. Neurotransmitter Metabolizing Enzymes

It is frequently assumed that when an axon is damaged all metabolic effort will be directed into the production of components for axonal regrowth and that synthesis of other proteins, primarily the enzymes involved in neurotransmitter synthesis and degradation, will be either reduced or turned-off. This is frequently, but not always the case. Most studies have concentrated on the cholinergic system, partly to determine whether the principle enzymes concerned with acetylcholine production, choline acetyltransferase (CAT), and with acetylcholine destruction, acetylcholinesterase (AChE) are reduced, and partly to use these enzymes to measure the extent of regeneration. This has been stimulated by the demonstration that acetylcholine is transported in both intact and regenerating axons.[119] Furthermore, it was thought that CAT is a component of slow-transport and AChE of fast-transport.[120,121] This is not strictly correct as AChE has at least four molecular forms ranging in size from 5S to 20S and the two larger isomeric forms of 11.5S and 20S are transported at much faster rates than the smaller components.[122]

In the early stages of nerve regeneration there is a fall in both CAT and AChE activity in the regenerating nerve.[123-126] All four AChE isoenzymes were reduced in activity 4 to 7 days after cutting the chick sciatic nerve, the large, more rapidly transported proteins being the more severely affected.[127] The enzyme levels increase at a later stage of regeneration to levels equal to or above normal.[120,123-125] The recovery is dependent on the type of injury. When crushed or frozen this is more rapid than when ligatured. Little recovery is seen after the nerve is cut.[126]

Sympathetic fibers appear to recover their ability to transport both transmitters and their enzymes more quickly than do cholinergic nerves.[128,129]

Adenylate cyclase is transported in normal and regenerating frog sciatic nerve.[130,131] Here there is transport of the enzyme during the early regenerative phase but this declines within 24 hr, suggesting a role for cyclic AMP in the initiation of the early phase of the regeneration process.

Most of these studies are carried out by measurement of accumulation of the enzyme at a constriction placed on the nerve above the site of injury some time after the injury has occurred. The times used vary from hours to weeks. This approach may not necessarily measure the actual situation in the regenerating nerve since the constriction is, of necessity, a second injury. It has been shown that a nerve responds in an accelerated fashion when injured a second time shortly after the initial trauma.[132-134] Thus, the responses measured may be altered by the constriction which is applied to the nerve to allow enzyme accumulation.

VI. SCHWANN CELL INFLUENCE ON REGENERATION

Fluids from Schwann cell cultures stimulate neurite outgrowth in cultured neurons.[135] This finding has led to a number of investigations to determine whether there is some trophic influence from the nerve sheath cells which stimulates axonal regeneration. Many of these approaches have utilized cuffs of various materials which are placed at the proximal and distal ends of a cut nerve. Axons can regrow over a distance of up to 10 mm under these circumstances[136] and they grow preferentially in a chamber with a distal portion of the same nerve in it.[137] This cuff may be concentrating trophic molecules or in some way reduce the occurrence of scar tissue, or both.[138] When the system is modified to provide a semipermeable membrane which excludes molecules larger than about 50,000 daltons, there is a significant improvement in regeneration.[139] It has recently been shown that at least some of these effects may be due to a protein of 57,000 daltons which is synthesized by Schwann cells in response to injury.[84] The synthesis occurs in cells in the distal stump, is produced in relatively large quantities and is regulated to commence and subsequently increase some days to weeks after injury. This protein is also produced in rat optic nerve. There is no evidence available to indicate where this protein may act. It is possible that it is taken up by the nerve at the site of injury and conveyed by retrograde transport to the cell body. Alternatively it may place some role in guidance or viability of the sprouting axons.

Studies on the influence of Schwann cells on axonal regrowth have been undertaken to clarify the enigma of the inability of CNS neurons, at least in higher organisms, to regenerate. When a piece of peripheral nerve is inserted in the CNS, axonal outgrowths from CNS cells can enter the perineural tubes and elongate over long distances. This does not happen in the CNS itself.[140] The reason for the ability of the peripheral axon tubes to support growth is unknown; it is probably not directly related to the provision of the 57 kD protein described by Skene and Shooter[84] since rabbit optic tracts also produce this protein on injury but no regeneration occurs.

VII. NONPROTEIN TRANSPORT

A. Phospholipids

It is widely accepted that fast axoplasmic transport represents components packaged in vesicles or some form of membrane-enclosed package. Since membrane-bound proteins such as glycoproteins are always transported at the fast rate, it is to be expected that phospholipids which are also entirely membrane components would be transported at the rapid rate. This has been shown to be so in a number of studies, both in intact and regenerating nerves, but rate measurements are complicated by exchange between molecules in transit and those in membranes, both in the axon and the myelin sheath.[141,142] The measurement of axoplasmic transport is also complicated by differences in apparent turnover which is influenced by metabolic synthesis and degradation as well as membrane exchange and inter- and intracellular movement.[143] The choice of label may also influence the results; this may be in the glycerol moiety, the phosphate component, the base, or a fatty acid. Subsequent metabolism of these fractions as well as events such as phospholipid exchange will influence the interpretation of the apparent movement within the axon.

In the regenerating rat sciatic nerve there is an increase transport of phospholipids, particularly phosphatidylcholine.[144-146] The labeled phospholipid builds up at the crush site but is also elevated along the segment of the nerve proximal to the injury.[145] These authors have also shown that the response to injury, as measured by elevated phospholipid output is very rapid and the time delay indicates that a retrogradely transported signal would have a transport rate of at least 140 mm/hr. The response is blocked by administration of KCl at the crush site at about the time of injury but it is initiated by the injection of either local anaesthetic or tetrodotoxin.

When [^{14}C] choline was used as the label only phosphatidylcholine and sphingomyelin were labeled in intact and regenerating nerves but elevated levels of phospholipid were found in regenerating nerves.[147] Radioactive labeling of glycerol or fatty acids also demonstrated that regenerating nerves incorporate rapidly transported phospholipids in greater amounts than control nerves up to 14 days after injury; most of the label was found to be in phosphatidylcholine and phosphatidylethanolamine.[148]

B. Ribonucleic Acid

The difficulties inherent in studying axoplasmic transport of ribonucleic acid (RNA) were recognized early in the study of transportable substances.[149-152] The major problem concerned the diffusion of precursor molecules both in the axoplasm and periaxonal spaces as well as incorporation into supporting cells, complicated by turnover in these cells and further diffusion and reincorporation of the products of degradation. In later studies use was made of the virtually exclusive location of certain small molecular weight RNAs in the nucleus of the neuron to assess transport of various RNA species relative to these nuclear located molecules.[29,30,32] This approach showed that there is a discrete transport of 4S RNA with strong evidence to show that ribosomal RNA is not transported. The system used was the chick visual system and it was further demonstrated that a similar pattern of 4S RNA transport operates in the adult chicken sciatic nerve.[30] Thus, axoplasmic transport of 4S RNA occurs in normal, intact nerves of higher species. This 4S RNA is largely, if not entirely, transfer RNA (tRNA).[33] It is a discrete transport of most, if not all tRNAs; another soluble small molecular weight RNA species (species L or 7S RNA) is not transported, suggesting that there is some control over the transport system or its initiation and that the tRNA has a continuing role in the axon or at its terminal.

Evidence has been presented for increased amounts of transported 4S RNA in the regenerating goldfish optic system.[153,155] It should be remembered that there is a very large increase in the amount of transported tubulin in this system during regeneration but this is not found in regenerating nerves in higher animals. When RNA synthesis in the retinal ganglion cells was inhibited by injection of 3 deoxyadenosine, the labeled 4S RNA appearing in the tectum was reduced, but so was the tectal-labeled ribosomal RNA.[156,157] Polyamines which associate with ribonucleic acids such as putrescine, spermidine and spermine are also transported in regenerating goldfish optic nerves.[158-161]

The role of axonally transported tRNA is not known but it has been suggested[33] that it may be needed for post-translational protein modification at the synapse. It may play a similar role in the growth cone.

VIII. SOME FACTORS RELEVANT TO AXOPLASMIC TRANSPORT IN REGENERATING NERVE

The axon, like extensions of other cells, contains a cytoplasm interlaced with a complex of membranes, fibrous and filamentous structures, a variety of vesicles, mitochondria, and other membrane-enclosed organelles. These all must be replaced if the regrowing axon is to regain function and the manner in which the system of axoplasmic transport carries out this task is relevant to this discussion. It became apparent a number of years ago that the bulk of the slowly transported material is composed of structural elements of the axoplasm whereas rapidly transported components are mainly particulate organelles (for review see Reference 162).

There have been many investigations into the role of the cytoskeletal elements in axoplasmic transport since the demonstration that tubulin is a component of slowly transported protein,[70] some into changing patterns in regenerating nerve and others into the role of these structures in the transport process. Since this role would influence transport in a regenerating nerve it will be briefly reviewed here.

Weiss and his collaborators considered that axoplasm moves as a viscous column at a very slow rate[163] but is modified to allow for a separate cannicular system within this column to account for rapid movement of less viscous liquid channels.[164] Schmidt[165] proposed a model, analogous to the sliding filament model of muscle contraction in which vesicles undergo a mechanochemical coupling with microtubules, using an ATPase controlled bridge (Figure 1a). A number of variations have been proposed to this model. In particular it has been extended by Ochs to incorporate the need for energy (presumed ATP) to maintain fast-transport and for the regulation of calcium levels in the axoplasm (Figure 1b).[166] These models have lead to the search for ATPase associated either with the tubulins or microtubule-associated proteins (MAPs) as well as cross-bridges, equivalent to dynein arms in microtubule containing flagellae. Since ATPases of varying functions are widespread throughout tissues,

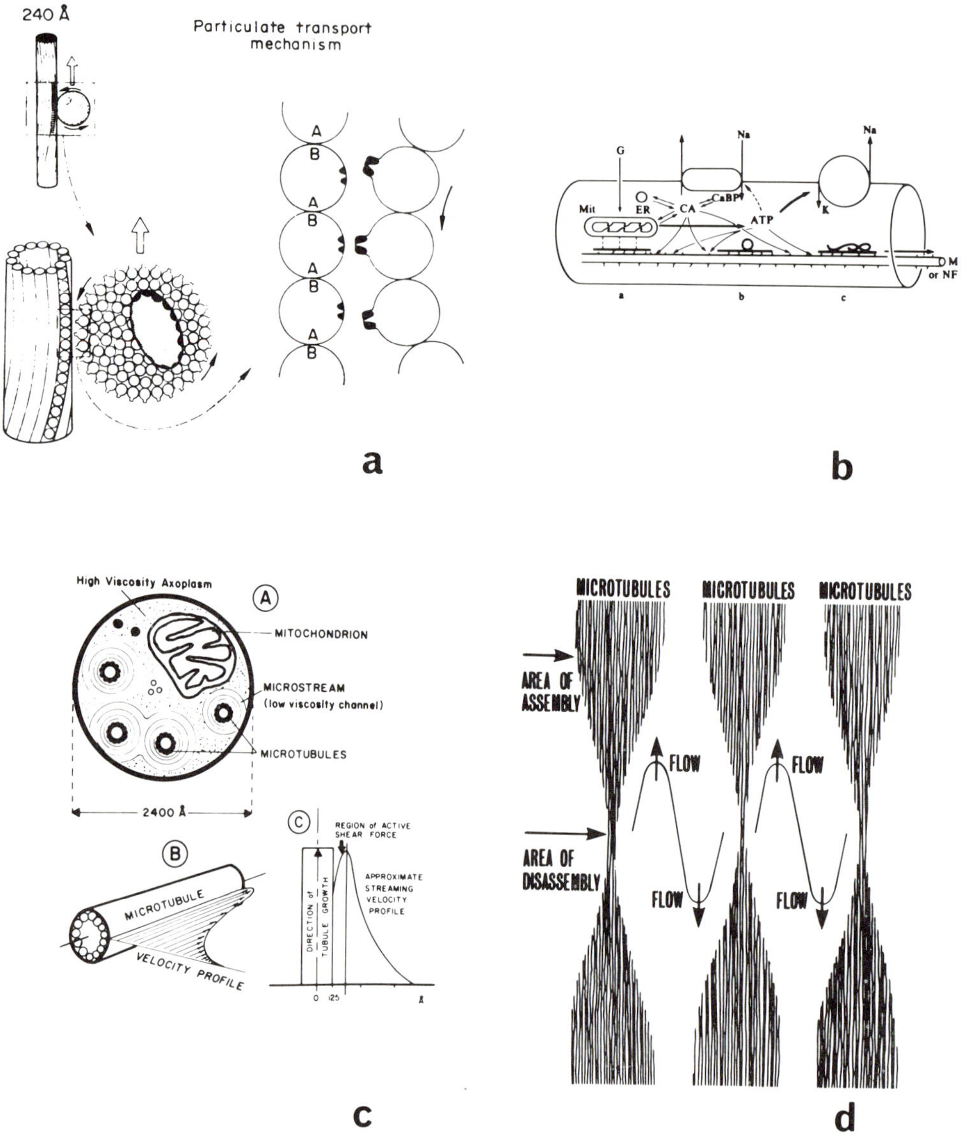

FIGURE 1. (a) The sliding-vesicle model of fast axoplasmic transport. Vesicles are shown interacting with microtubules in a manner analogous to the sliding-filament system operating in skeletal muscle. (From Schmitt, F. O., *Neurosci. J. Res. Prog. Bull*, 6, 119, 1968. With permission.) (b) The transport filament hypothesis. This model takes into account the energy needs of axoplasmic transport. Glucose (G) supplies the metabolic fuel for ATP production which provides energy for movement of transport filaments which are postulated to carry attached components such as mitochondria (a), soluble proteins (b), and particulate structures (c). (From Ochs, S., *Basic Neurochemistry*, Little Brown, Boston, 1981. With permission.) (c) The microstream hypothesis. Regions of low viscosity are postulated in the vicinity of microtubules, induced by a sheer force generating system which is based on ATPase activity. Rapid transport is envisaged as the movement of components in these microstreams. Slower rates of transport result from absorption to and desorbtion from membranes. (From Gross, G. W., *Adv. Neurol.*, 12, 283, 1975. With permission.) (d) The structured flow hypothesis. Microtubules assemble and disassemble throughout the axoplasm and this process leads to the control of the rates of structured flow in the axoplasm. The assembly-disassembly process would be dependent on energy provided either as GTP or ATP and local concentrations of free calcium. Structured flows operate in two directions and would allow for both anterograde and retrograde transport. (From Comper, W. D., et al., *Neurochem. Res.*, 8, 943, 1983. With permission.)

no conclusive evidence has been presented which eliminates the possibility of contamination of the tubulin preparations by small amounts of other cellular material. Much attention has been paid to the central role of microtubules in axoplasmic transport, mainly because of the many demonstrations that colchicine and Vinca alkaloids, which interfere with the microtubule assembly process, all block both fast- and slow-transport.[162] In addition, morphological evidence suggests an association of subcellular particles with microtubules in axons.[167,168] This analogy of particle movement in axons with filament movement in muscle was rejected by Gross who proposed that a chromatographic-like system operates in which absortion and desorbtion to membranes takes place with the whole process driven by a zone of low viscosity in the region of the microtubules;[169] this low viscosity zone would result from ATP splitting. Thus, a microstreaming of cytoplasm would carry particles along the axon (Figure 1c).

The involvement of microtubules in axonal transport is not universally accepted. Rat sensory fibers when incubated in vitro in a high calcium medium continue to maintain fast axoplasmic transport but there is a complete loss of microtubular structures within a few hours.[170]

Droz et al. consider that the smooth endoplasmic reticulum (SER),[171] which is present as a complex membranous structure in axoplasm, is the network which guides particles along the axons in both orthograde and retrograde directions. This view is supported by Tsukita and Ishikawa[172] but extended to suggest that the SER is itself moving and continuously budding-off pieces of membrane. They further extended this hypothesis to involve both the axolemma and microtubule channels which associate with the SER. On the other hand, horseradish peroxidase, when injected into the chick tectum and taken up into optic nerve shows no association with the SER.[173]

The importance of the axolemma in a force generating system, operating through submembranous structures such as actin have been emphasized by Willard and colleagues.[174,175] The enzyme DNAase I complexes with actin and when introduced intracellularly, it blocks axoplasm transport.[176,177] Thus, one of the current concepts of the driving force for axoplasmic transport has swung back to the muscle analogy but this time invoking myosin-like interactions with actin filaments rather than with microtubules.

Although the axolemma may have some indirect role to play in axoplasmic transport it does not appear to be intimately involved since ordered streaming of axoplasmic structures continues for hours even in axoplasm extruded from its axon.[178] These authors state that "understanding of cellular phenomena at the molecular level requires model systems that permit direct analysis of the cellular elements . . . and that can be maintained long enough to conduct the manipulations". Such a model system has been described by Preston et al. and adapted to the particular situation of axoplasmic transport.[179] In this paradigm it is pointed out that macromolecular streaming sets up and continues spontaneously when there is a concentration differential of the macromolecule in a medium containing other macromolecular elements. Such a situation is continuously occurring in an axon (and other cytoplasm) due to synthesis and degradation of the components where the synthetic and degradative sites are remote from each other. The rate of movement is influenced by viscosity and this viscosity flux would be expected to occur in the axoplasm due to the assembly-disassembly process of microtubules (Figure 1d). The model also relates to that of muscle contraction but only with respect to the influence of calcium on the microtubule assembly-disassembly through the presence of a sarcoplasmic reticulum-like meshwork, which could control the free calcium levels. Such a structure has been shown to exist in nonmuscle cells including those in brain.[180,181] In this model energy is required in the microtubule assembly-disassembly and the microtubules themselves do not direct the flow process but simply affect the rate of flow. This would lead to a microstreaming through the axoplasm somewhat in the manner proposed by Gross.[169] Both anterograde and retrograde streaming would occur.

The above considerations have relevance to the regenerating axon in that the renewal process requires the provision of cytoskeletal elements, membrane units, probably prepackaged for axonal elongation,[93] as well as a variety of organelles, enzymes, and enzyme substrates. The sequence in which these elements are provided will determine the efficacy of supply by axoplasmic transport if any of these are deeply involved in the transport process. It may be that microtubules, actin filaments, SER, and even the microtrabecular structure[182] are involved but to differing degrees in different axonal types. This may explain, at least in part why there are such confusing differences between regenerating nerves of various types with respect to transport of particular proteins such as the tubulins.

IX. CENTRAL NERVOUS SYSTEM REGENERATION AND AXOPLASMIC TRANSPORT

Successful regeneration of any part of the mammlian central nervous system (CNS) is a very rare event but in lower orders of animals regeneration is frequently successful. A variety of explanations have been advanced to account for the inability of the mammalian CNS neurons to elongate axons after damage, back to their original sites of innervation. These include an innate incapacity for regeneration in the CNS neuron but this is clearly no longer tenable since evidence has been presented to show regeneration processes proceeding when the neuronal environment is suitably altered.[140,183] Other impediments to regeneration that have been proposed include:

1. Physical barriers such as the formation of scar tissue or myelin and glial debris
2. Inappropriate formation of synapses
3. An autoimmune response which inhibits regrowth
4. An alteration in perivascular permeability
5. A role for growth factors which are missing or in insufficient supply in the CNS (see References 184 to 186)

Of these, the only ones which may involve axoplasmic transport are (1) and (5) above.

A. Physical Barriers

Insertion of degenerated peripheral nerve into a damaged region of the CNS can provide an environment in which central neurons can elongate.[140] Furthermore, Schwann cells in degenerating nerve are induced to synthesize a specific protein which may enhance the regenerative capacity of the peripheral nerve.[84] Although there is clearly a great deal of work needed in this area, it could mean that the Schwann cells are capable of producing growth factors which either act directly on the growth cone or are taken up at the tip and retrogradely transported to the cell body to exert some beneficial action on it. There may even be some specificity; i.e., the correct axon regenerates more readily into the Schwann cell tube.

B. Growth Factors

The demonstration of alterations in gene expression to stimulate production of certain growth associated proteins after nerve injury has stimulated the view that these GAPs are required to either initiate or sustain regrowth.[81-83] The proteins are produced in regenerating peripheral nerves[83] and regenerating CNS nerves but not nonregenerating CNS nerves.[82,83] Nothing is known of the role of these proteins. They are membrane-associated and transported at the fast rate. They probably have differing roles since GAP-43 is preferentially transported to the growth cone whereas GAP-24 is laid down along the length of the axolemma. Thus, there may be a role for these proteins in directing or aligning newly laid-down membrane in the actively growing axon.

The regenerating toad optic nerve transports, in addition to the GAPs, significantly elevated amounts of other proteins and, in particular, α and β tubulin and actin.[81] A rather similar situation has been observed in the regenerating goldfish optic nerve,[47] where elevated levels of cytoskeletal proteins were also found. The rate of transport of slow components has been known to be approximately equivalent to that of regeneration. This may simply mean that provision of the cytoskeleton is the rate-limiting step in regeneration.

C. Stimulation of Regeneration by Conditioning Lesions

When the goldfish optic nerve is damaged twice the rate of regrowth is increased after the second injury. The first injury is therefore a conditioning lesion.[132] Similar stimulations to growth are found in mammalian peripheral nerve.[133] The reason for this effect is not known but it may be the result of some cell body change which has already been effected by the first lesion. This change may be related to the reorganization of the ribosomal population of the cell body as a result of injury.[26,28] This process is completed in a two phase fashion, probably reflecting changes in processing of 28S and 18S RNA initially and cytoplasmic assembly at a later stage.[35] With the altered protein synthetic machinery in place the cell would be in a more favorable position to respond to a second lesion.

X. CONCLUSION

Axoplasmic transport performs a number of functions in the regenerative process; in particular it may be important in the mechanism through which the cell body is informed that an injury has occurred and also in the provision of materials to the damaged axon so that repair can be effected. In this latter regard, the two most important factors are the components of the cytoskeleton and the vesicles which convey preformed membranes to the redeveloping axon. It is unfortunate that much of the information currently available about the details of transport of these two classes of axonal structures has been obtained from very diverse life forms. There have been relatively few attempts to correlate this information between, on the one hand, the CNS of fish or amphibia and on the other, the peripheral system of mammals.

Investigations are needed to define more precisely the order of rearrangement of the cytoskeleton because of its importance in controlling axoplasmic transport and its various roles in maintaining the functioning axon. Furthermore, other systems which control numerous axonal events need further study. In particular, calcium sequestering proteins such as calmodulin are probably needed to ensure a correct sequence of axonal development.

The growth associated proteins have only been investigated in a few systems. They are potentially important in either the initiation or regulation of regeneration and, indeed, they may be crucial factors in that lack of these proteins may be a fundamental cause of non-regeneration in the CNS. Other specific proteins produced by myelin forming cells may also have a key role in regeneration of nerve. It is, however, possible that they belong to a group of proteins, known as heat shock proteins, which respond to abnormal physiological conditions such as elevated temperature and drug treatment.[187] These proteins, which are formed in response to stress, are found in a wide range of organisms. The proteins formed by Schwann cells may also be an expression of stress response.

ACKNOWLEDGMENTS

This work was supported by the Australian Research Grants Scheme. The author is grateful to Dr. P. L. Jeffrey and Dr. D. Bates for critical comments.

REFERENCES

1. **Rochel, S. and Margolis, F. L.,** The response of ornithine decarboxylose during neuronal degeneration and regeneration in olfactory epithelium, *J. Neurochem.*, 35, 850, 1980.
2. **Nissl, F.,** Über die veräderungen, der ganglienzellen am faciliskern des kanichens nach aureissung des nerven, *Allg. Z. Psychiatr.*, 48, 197, 1892.
3. **Leiberman, A. R.,** Some factors affecting retrograde neuronal responses to axonal lesions, in *Essays on the Nervous System*, Ballairs, R. and Gray, E. C., Eds., Clarendon Press, Oxford, 1974, 71.
4. **Kaye, P. L., Jeffrey, P. L., and Austin, L.,** Bilateral enzymic changes in rat nodose ganglia following unilateral cervical vagotomy, *Neuroscience*, 3, 1231, 1978.
5. **Carlsen, R. C., Kiff, J., and Ryugo, K.,** Suppression of the cell body response in axotomized frog spinal neurons does not prevent initiation of nerve regeneration, *Brain Res.*, 234, 11, 1982.
6. **Sreide, A. J.,** Variations in the axon reaction after different types of nerve lesion. Light and electron microscopic studies on the facial nucleus of the rat, *Acta Anat.*, 110, 173, 1981.
7. **Jorgensen, D. and Dyck, P. J.,** Axonal underdevelopment from axotomy in kittens, *J. Neuropathol. Exp. Neurol.*, 38, 571, 1979.
8. **Watson, W. E.,** Observations on the nucleolar and total cell body nucleic acid of injured nerve cells, *J. Physiol.*, 196, 655, 1968.
9. **Cragg, B. G.,** What is the signal for chromatolysis?, *Brain Res.*, 23, 1, 1970.
10. **Weiss, P. A.,** Neuronal dynamics and axonal flow. V. The semi-solid state of the moving axonal column, *Proc. Natl. Acad. Sci. U.S.A.*, 69, 620, 1972.
11. **Hydén, H.,** Protein metabolism in the nerve cell during growth and function, *Acta Physiol. Scand.*, 6 (Suppl. 17), 1, 1943.
12. **Gersh, I. and Bodian, D.,** Some chemical mechanisms in chromatolysis, *J. Cell. Comp. Physiol.*, 1, 253, 1943.
13. **Brattgård, S-O., Edström, J-E., and Hydén, H.,** The chemical changes in regenerating neurons, *J. Neurochem.*, 1, 316, 1957.
14. **Edström, J-E.,** Ribonucleic acid changes in the motor neurones of the frog during axon regeneration, *J. Neurochem.*, 5, 43, 1959.
15. **Lison, M. P.,** Changes in the protein content of the spinal motor neurons of the mouse after severance of the axon, *Nature (London)*, 196, 837, 1962.
16. **Lambert, B. and Daneholt, B.,** A quantitative and qualitative study of RNA in regenerating hypoglossal neurons of rabbits, in *Macromolecules and the Function of the Neuron*, Lodin, Z. and Rose, S. P. R., Eds., Elsevier/North Holland, Amsterdam, 1968, 334.
17. **Haddad, A., Incif, S., and Cruz, A. R.,** Synthesis of RNA in neurons of the hypoglossal nerve nucleus after section of the axon, in mice, *J. Neurochem.*, 16, 865, 1969.
18. **Torvik, A. and Heding, A.,** Effect of actinomycin D on retrograde nerve cell reaction. Further observations, *Acta Neuropathol.*, 14, 62, 1969.
19. **Watson, W. E.,** The binding of actinomycin D to the nuclei of axotomised neurones, *Brain Res.*, 65, 317, 1974.
20. **Torvik, A. and Skjörten, F.,** The effect of Actinomycin D upon normal neurons and retrograde nerve cell reaction, *J. Neurocytol.*, 3, 87, 1974.
21. **Kinderman, N. B. and La Velle, A.,** Ultrastructural changes in the developing nucleolus following axotomy, *Brain Res.*, 108, 237, 1976.
22. **Scott, T. M. and Foote, J.,** Factors involved in the control of RNA synthesis during regeneration of the optic nerve in the frog, *Neurosci. Lett.*, 29, 189, 1982.
23. **Kaye, P. L., Gunning, P. W., and Austin, L.,** *In vivo* synthesis of stable RNA within the rat nodose ganglion following vagotomy, *J. Neurochem.*, 28, 1241, 1977.
24. **Burrell, H. R., Dokas, L. A., and Agranoff, B. W.,** RNA metabolism in the goldfish retina during optic nerve regeneration, *J. Neurochem.*, 31, 289, 1978.
25. **Langford, C. J., Scheffer, J. W., Jeffrey, P. L., and Austin, L.,** The *in vitro* synthesis of RNA within the rat nodose ganglion following vagotomy, *J. Neurochem.*, 34, 531, 1980.
26. **Bates, D., Sirkos, S., and Austin, L.,** Increased *in vitro* labelling of stable RNA within the rat nodose ganglion following abdominal vagotomy, *Neurochem. Res.*, 8, 363, 1983.
27. **Gunning, P. W., Kaye, P. L., and Austin, L.,** *In vivo* synthesis of rapidly-labelled RNA within the rat nodose ganglion following vagotomy, *J. Neurochem.*, 28, 1245, 1977.
28. **Burrell, H. R., Heathcock, A. M., Water, R. D., and Agranoff, B.,** Increased tubulin messenger RNA in the goldfish retina during optic nerve regeneration, *Brain Res.*, 168, 628, 1979.
29. **Austin, L., Gunning, P. W., Jeffrey, P. L., and Por, S.,** tRNA, its axonal transport and fate, in *Mechanisms, Regulation and Special Function of Protein Synthesis in the Brain*, Roberts, S., Lajtha, A., and Gispen, W. H., Eds., Elsevier/North Holland, Amsterdam, 1977, 71.

30. **Por, S., Gunning, P. W., Jeffrey, P. L., and Austin, L.,** Axonal transport of 4S RNA in the chick optic system, *Neurochem. Res.*, 3, 411, 1978.
31. **Por, S. B., Komiya, Y., McGregor, A., Jeffrey, P. L., Gunning, P. W., and Austin, L.,** The axoplasmic transport of 4S RNA within the sciatic nerve of the chicken, *Neurosci. Lett.*, 8, 165, 1978.
32. **Gunning, P. W., Por, S. B., Langford, C. J., Scheffer, J., Austin, L., and Jeffrey, P. L.,** The direct measurement of the axoplasmic transport of individual RNA species transfer but not ribosomal RNA is transported, *J. Neurochem.*, 32, 1737, 1979.
33. **Scheffer, J. W., Howe, N., Gunning, P. W., and Austin, L.,** The axoplasmic transport of transfer RNA in the chick optic system, *J. Neurochem.*, in press, 1983.
34. **Lindquist, T. D., Ingoglia, N. A., and Gould, R. M.,** 4S RNA is transported axonally in normal and regenerating axons of the sciatic nerves of rats, *Brain Res.*, 230, 181, 1981.
35. **Moses, E. K., Langford, C. J., and Austin, L.,** Small molecular weight RNAs: altered metabolism in regenerating nerve, *Biochem. Int.*, 5, 177, 1982.
36. **Gunning, P. W., Beguin, P., Shooter, E. M., Austin, L., and Jeffrey, P. L.,** Characterization of the association of two small molecular weight RNAs with eukaryotic polysomes, *J. Biol. Chem.*, 256, 6670, 1981.
37. **Walter, P. and Blobel, G.,** Signal recognition particle contains a 7S RNA essential for protein translocation across the endoplasmic reticulum, *Nature (London)*, 299, 691, 1982.
38. **Lerner, M. R., Boyle, J. A., Mount, S. M., Wolin, S. L., and Steitz, J. A.,** Are snRNPs involved in splicing?, *Nature (London)*, 283, 220, 1980.
39. **Rogers, J. and Wall, R.,** A mechanism for RNA splicing, *Proc. Natl. Acad. Sci. U.S.A.*, 77, 1877, 1980.
40. **Rao, M. S., Blackstone, M., and Busch, H.,** Effects of Ul nuclear RNA on translation of mRNA, *Biochemistry*, 16, 2756, 1977.
41. **Bates, D. J.,** Changes in RNA metabolism during nerve regeneration, *J. Neurochem.*, 41, S66, 1983.
42. **Lux, H. D., Schubert, P., Krutzberg, G. W., and Globus, A.,** Excitation and axonal flow: autoradiographic study on motoneurons intracellularly injected with a ^{3}H amino acid, *Exp. Brain Res.*, 10, 197, 1970.
43. **Lentz, T. L.,** Distribution of leucine ^{3}H during axoplasmic transport within regenerating neurons as determined by electron-microscope radioautography, *J. Cell Biol.*, 52, 719, 1972.
44. **McLoon, L. K. and La Velle, A.,** Tritiated leucine incorporation in the developing hamster facial nucleus with injury: a liquid scintillation study, *Brain Res.*, 227, 237, 1981.
45. **Denburg, J. L. and Hood, N. A.,** Protein synthesis in regenerating motor neurons in the cockroach, *Brain Res.*, 125, 227, 1977.
46. **Whitnall, M. H. and Grafstein, B.,** The relationship between extracellular amino acids and protein synthesis is altered during axonal regeneration, *Brain Res.*, 220, 362, 1981.
47. **Heacock, A. M. and Agranoff, B. W.,** Enhanced labelling of a retinal protein during regeneration of optic nerve in goldfish, *Proc. Natl. Acad. Sci. U.S.A.*, 73, 828, 1976.
48. **Giulian, D., Des Reuisseaux, H., and Cowburn, D.,** Biosynthesis and intra-axonal transport of proteins during neuronal regeneration, *J. Biol. Chem.*, 255, 6494, 1980.
49. **Birks, C. and Cronly-Dillon, J.,** Reciprocal interactions between regenerating optic fibres and tectal cells modulate tubulin synthesis in goldfish retina and tectum, *J. Physiol.*, 307, 22, 1980.
50. **Berry, R. W.,** Synthesis of axonally transported proteins following inhibition of transport, *Brain Res.*, 159, 427, 1978.
51. **Ochs, S. and Barnes, C. D.,** Regeneration of ventral root fibers into dorsal roots shown by axoplasmic flow, *Brain Res.*, 15, 600, 1969.
52. **Forman, D. S. and Berenberg, R. A.,** Regeneration of motor axons in the rat sciatic nerve studied by labelling with axonally transported radioactive proteins, *Brain Res.*, 156, 213, 1978.
53. **Grafstein, B. and Murray, M.,** Transport of protein in goldfish optic nerve during regeneration, *Exp. Neurol.*, 25, 494, 1969.
54. **Grafstein, B.,** Role of slow axonal transport in nerve regeneration, *Acta Neuropathol.*, Suppl. V, 144, 1971.
55. **McQuarrie, I. Q. and Grafstein, B.,** Protein synthesis and fast axonal transport in regenerating goldfish retinal ganglion cells, *Brain Res.*, 235, 213, 1982.
56. **Perry, G. W. and Wilson, D. L.,** Protein synthesis and axonal transport during nerve regeneration, *J. Neurochem.*, 37, 1203, 1981.
57. **Perry, G. W., Krayanek, S. R., and Wilson, D. L.,** Protein synthesis and rapid axonal transport during regrowth of doral root axons, *J. Neurochem.*, 40, 1590, 1983.
58. **Kolnig, H. L. and Droz, B.,** Effects of nerve section on protein metabolism of ganglion cells and preganglionic nerve endings, *Acta Neuropathol.*, Suppl. V, 119, 1971.
59. **Ochs, S.,** Fast axoplasmic transport in the fibres of chromatolysed neurones, *J. Physiol.*, 255, 249, 1976.
60. **Bisby, M. A.,** Fast axonal transport of labelled protein in sensory axons during regeneration, *Exp. Neurol.*, 61, 281, 1978.

61. **Bisby, M. A.,** Axonal transport in the central axon of sensory neurons during regeneration of their peripheral axon, *Neurosci. Lett.*, 21, 7, 1981.
62. **Komiya, Y.,** Axonal regeneration in bifurcating axons of rat dorsal root ganglion cells, *Exp. Neurol.*, 73, 824, 1981.
63. **Griffin, J. W., Drackman, D. B., and Price, D. L.,** Fast axonal transport in motor nerve regeneration, *J. Neurobiol.*, 7, 355, 1976.
64. **Griffith, A. and La Velle, A.,** Developmental protein changes in normal and chromatolytic facial nerve nuclear regions, *Exp. Neurol.*, 33, 360, 1971.
65. **Kreutzberg, G. W. and Schubert, P.,** Changes in axonal flow during regeneration of mammalian motor nerves, *Acta Neuropathol.*, Suppl. 5, 70, 1971.
66. **Frizell, M. and Sjöstrand, J.,** The axonal transport of [^{3}H] fucose labelled glycoproteins in normal and regenerating peripheral nerves, *Brain Res.*, 81, 267, 1974.
67. **Frizell, M. and Sjöstrand, J.,** The axonal transport of [^{3}H] fucose labelled glycoproteins in normal and regenerating peripheral nerves, *Brain Res.*, 78, 109, 1974.
68. **Hoffman, P. N. and Lasek, R. J.,** Axonal transport of the cytoskeleton in regenerating motor neurons: constancy and change, *Brain Res.*, 202, 317, 1980.
69. **Theiber, R. F. and McClure, W. O.,** Rapid axoplasmic transport of proteins in regenerating sensory nerve fibres, *J. Neurochem.*, 31, 433, 1978.
70. **James, K. A. C., Bray, J. J., and Morgan, I. G.,** The effect of colchicine on the transport of axonal protein in the chicken, *Biochem. J.*, 117, 769, 1970.
71. **McEwan, B. S., Forman, D. S., and Grafstein, B.,** Components of fast and slow axonal transport in the goldfish optic nerve, *J. Neurobiol.*, 2, 361, 1971.
72. **Black, M. M. and Lasek, R. J.,** Axonal transport of actin: slow component b is the principle source of actin for the axon, *Brain Res.*, 171, 401, 1979.
73. **Heacock, A. M. and Agranoff, B. W.,** Protein synthesis and transport in the regenerating goldfish visual system, *Neurochem. Res.*, 7, 771, 1982.
74. **Sutherland, F. I.,** Controlled variation of microtubule number in neuronal processes during remodelling *in vivo, Exp. Brain Res.*, 20, 311, 1974.
75. **Griffin, J. W., Price, D. L., Drachman, D. B., and Morris, J.,** Incorporation of axonally transported glycoproteins into axolemma during nerve regeneration, *J. Cell Biol.*, 88, 205, 1981.
76. **Cole, G. J. and Elam, J. S.,** Axonal transport of glycoproteins in regenerating olfactory nerve: enhanced glycopeptide concanavalin A binding, *Brain Res.*, 222, 437, 1981.
77. **Goodrum, J. F. and Morell, P.,** Axonal transport, deposition and metabolic turnover of glycoproteins in the rat optic pathway, *J. Neurochem.*, 328, 696, 1982.
78. **Berry, R. W.,** Evidence for multiple somatic pools of individual axonally transport proteins, *J. Cell Biol.*, 87, 379, 1980.
79. **Quitschke, W., Francis, A., and Schechter, N.,** Electrophoretic analysis of specific proteins in the regenerating goldfish retinotectal pathway, *Brain Res.*, 201, 347, 1980.
80. **Levine, J., Skene, P., and Willard, M.,** GAPs and fodrin. Novel axonally transported proteins, *Trends Neurosci.*, 4, 273, 1981.
81. **Skene, J. H. and Willard, M.,** Changes in axonally transported proteins during axonal regeneration in toad retinal ganglion cells, *J. Cell Biol.*, 89, 86, 1981.
82. **Skene, J. H. P. and Willard, M.,** Characteristics of growth-associated polypeptides in regenerating toad retinal ganglion cell axons, *J. Neurosci.*, 1, 419, 1981.
83. **Skene, J. H. P. and Willard, M.,** Axonally transported proteins associated with axon growth in rabbit central and peripheral nervouse systems, *J. Cell Biol.*, 89, 96, 1981.
84. **Skene, J. H. P. and Shooter, E. M.,** Denervated sheath cells secrete a new protein after nerve injury, *Proc. Natl. Acad. Sci. U.S.A.*, 80, 4169, 1983.
85. **O'Farrell, P. H.,** High resolution two dimensional electrophoresis of proteins, *J. Biol. Chem.*, 250, 4007, 1975.
86. **Hall, M. E., Wilson, D. L., and Stone, G. C.,** Changes in synthesis of specific proteins following axotomy: detection with two-dimensional gel electrophoresis, *J. Neurobiol.*, 9, 353, 1978.
87. **Cancalon, P. and Elam, J. S.,** Rate of movement and composition of rapidly transported proteins in regenerating olfactory nerve, *J. Neurochem.*, 35, 889, 1980.
88. **Bisby, M. A.,** Changes in the composition of labelled protein transported in motor axons during their regeneration, *J. Neurobiol.*, 11, 435, 1980.
89. **Bisby, M. A.,** Prolonged alterations in composition of fast transported protein in axons prevented from regenerating after injury, *J. Neurobiol.*, 13, 377, 1982.
90. **Evans, D. H. L. and Murray, J. G.,** Histological and functional studies on the fibre composition of the vagus nerve of the rabbit, *J. Anat.*, 88, 320, 1954.
91. **Benowitz, L. I., Shashoua, V. E., and Yoon, M. G.,** Specific changes in rapidly transported proteins during regeneration of the goldfish optic nerve, *J. Neurosci.*, 1, 300, 1981.

92. **Bisby, M. A.,** Differences in incorporation of axonally transported protein in regenerating motor and sensory axons, *Exp. Neurol.*, 65, 680, 1979.
93. **Wickner, W.,** The assembly of proteins into biological membranes: the membrane trigger hypothesis, *Ann. Rev. Biochem.*, 48, 23, 1979.
94. **Dawson, R. M. C.,** The exchange of phospholipids between cell membranes, *Sub Cell. Biochem.*, 2, 69, 1973.
95. **Matthews, M. R.,** Evidence suggesting translocation of small dense-cored vesicles from the cell bodies to regenerating tips of adrenergic neurons, *J. Anat.*, 111, 508, 1972.
96. **Griffin, J. W., Price, D. L., Engel, W. K., and Drachman, D. B.,** The pathogenesis of reactive axonal swellings: role of axonal transport, *J. Neuropathol. Exp. Neurol.*, 36, 214, 1977.
97. **Tessler, A., Autiliogambetti, L., and Gambetti, P.,** Axonal growth during regeneration — quantitative autoradiographic study, *J. Cell Biol.*, 87, 197, 1980.
98. **Bray, J. J., Kon, C. M., and Breckenridge, B. M. E.,** Reversed polarity of rapid axonal transport in chicken motoneurones, *Brain Res.*, 33, 560, 1971.
99. **Bisby, M. A. and Bulgar, V. T.,** Reversal of axonal transport at a nerve crush, *J. Neurochem.*, 29, 313, 1977.
100. **Bulgar, V. T. and Bisby, M. A.,** Reversal of axonal transport in regenerating nerves, *J. Neurochem.*, 31, 1411, 1978.
101. **Bisby, M. A.,** Reversal of axonal transport: similarity of proteins transported in anterograde and retrograde directions, *J. Neurochem.*, 36, 741, 1981.
102. **Whitnall, M. H., Currie, J. R., and Grafstein, B.,** Bi-directional axonal transport of glycoproteins in goldfish optic nerve, *Exp. Neurol.*, 75, 191, 1982.
103. **Schmidt, R. E., Yu, M. J. C., and McDougall, D. B.,** Turnover of axoplasmic transport of selected particle-specific enzymes at an injury in control and DFP-treated rats, *J. Neurochem.*, 35, 641, 1980.
104. **Kristensson, K. and Olsson, Y.,** Retrograde transport of horseradish peroxidase in transected axons. III. Entry into injury axons and subsequent localization in perikaryon, *Brain Res.*, 115, 201, 1976.
105. **Olsson, T. and Kristensson, K.,** Uptake and retrograde axonal transport of horseradish peroxidase in normal and axotomized motor neurons during postnatal development, *Neuropathol. Appl. Neurobiol.*, 5, 377, 1979.
106. **Sparrow, J. R. and Kiernan, J. A.,** Uptake and retrograde transport of proteins by regenerating axons, *Acta Neuropathol.*, 47, 39, 1979.
107. **Ohshiro, S., Fujiwara, M., Sakakibara, Y., and Osumi, Y.,** Retrograde axonal transport of horseradish peroxidase in regenerating sciatic nerve of the rat, *Exp. Neurol.*, 65, 489, 1979.
108. **Olsson, T. P., Forsberg, I., and Kristensson, K.,** Uptake and retrograde axonal transport of horseradish peroxidase in regenerating facial motor neurons of mouse, *J. Neurocytol.*, 7, 323, 1978.
109. **Kristensson, K. and Sjöstrand, J.,** Retrograde transport of protein tracer in the rabbit hypoglossal nerve during regeneration, *Brain Res.*, 45, 175, 1972.
110. **Kristensson, K. and Olsson, Y.,** Retrograde transport of horseradish peroxidase in transected axons. I. Time relationships between transport and induction of chromatolysis, *Brain Res.*, 79, 101, 1974.
111. **Kristensson, K. and Olsson, Y.,** Retrograde transport of horseradish peroxidase in transected axons. II. Relations between rate of transfer from the site of injury to the perikaryon and onset of chromatolysis, *J. Neurocytol.*, 4, 653, 1975.
112. **Frizell, M., McLean, W. G., and Sjöstrand, J.,** Retrograde axonal transport of rapidly migrating labelled proteins and glycoproteins in regenerating peripheral nerves, *J. Neurochem.*, 27, 191, 1976.
113. **Malmgren, L., Olsson, Y., Olsson, T., and Kristensson, K.,** Uptake and retrograde axonal transport of various exogenous macromolecules in normal and crushed hypoglossal nerves, *Brain Res.*, 153, 477, 1978.
114. **Denburg, J. L.,** Elimination of inappropriate axonal branches of regenerating cockroach motor neurons as detected by the retrograde transport of horseradish peroxidase conjugated wheat germ agglutinin, *Brain Res.*, 248, 1, 1982.
115. **Pilar, G. and Landmasser, L.,** Axotomy mimicked by localized colchicine application, *Science*, 177, 1116, 1972.
116. **Singer, P. A., Mehler, S., and Fernandez, H. L.,** Blockade of retrograde axonal transport delays the onset of metabolic and morphologic changes induced by axotomy, *J. Neurosci.*, 2, 1299, 1982.
117. **Morelli, M., Porceddu, M. L., and Di Chiara, G.,** Loss of striatal neurons after local micro-injection of colchicine, *Neurosci. Lett.*, 16, 131, 1980.
118. **Morgan, I. G.,** Intraocular colchicine selectively destroys immature ganglion cells in chicken retina, *Neurosci. Lett.*, 24, 255, 1981.
119. **Evans, C. A. N. and Saunders, N. R.,** The distribution of acetylcholine in normal and in regenerating nerves, *J. Physiol.*, 192, 79, 1967.
120. **Kanish, N., Ochs, S., and Barnes, C. D.,** Regeneration of ventral root axons into dorsal roots as shown by increased AChE activity, *J. Neurobiol.*, 3, 245, 1972.

121. **Tûcek, S.,** Transport of cholineacyltransferase and acetylcholinesterase in the central stump and isolated segments of a peripheral nerve, *Brain Res.*, 86, 259, 1975.
122. **Courand, J. Y. and Di Giamberardino, L.,** Axonal transport of the molecular forms of acetylcholinesterase in chick sciatic nerve, *J. Neurochem.*, 35, 1053, 1980.
123. **Frizell, M. and Sjöstrand, J.,** Transport of proteins, glycoproteins and cholinergic enzymes in regenerating hypoglossal neurons, *J. Neurochem.*, 22, 845, 1974.
124. **O'Brien, R. A.,** Axonal transport of acetylcholine, choline acetyltransferase and cholinesterase in regenerating peripheral nerve, *J. Physiol.*, 282, 91, 1978.
125. **Schmidt, R. E. and McDougal, D. B.,** Axonal transport of selected particle-specific enzymes in rat sciatic nerve *in vivo* and its response to injury, *J. Neurochem.*, 30, 527, 1978.
126. **Heiwall, P. O., Dahlström, A., Larsson, P. A., and Bloloj, S.,** The intra-axonal transport of acetylcholine and cholinergic enzymes in rat sciatic nerve during regeneration of various types of axonal trauma, *J. Neurobiol.*, 10, 119, 1979.
127. **Courand, J. Y., Di Giamberardino, L., Hassig, R., and Mira, J. C.,** Axonal transport of the molecular form of acetylcholinesterase in developing and regenerating sciatic nerve, *Exp. Neurol.*, 80, 94, 1983.
128. **Ohshiro, S., Fujiwara, M., and Osumi, Y.,** Axonal transport of norepinephrine and choline acetyltransferase in regenerating sciatic nerve of the rat, *Exp. Neurol.*, 62, 159, 1978.
129. **Kessler, J. A. and Black, I. B.,** The role of axoplasmic transport in the regulation of enzyme activity in sympathetic ganglia of adult rats, *Brain Res.*, 171, 415, 1979.
130. **Carlsen, R. C.,** Axonal transport of adenylate cyclase activity in normal and axotomized frog sciatic nerve, *Brain Res.*, 232, 413, 1982.
131. **Carlsen, R. C. and Anderson, L. J.,** Sub-cellular distribution of axonally transported adenylate cyclase: effect of nerve construction and comparison with acetylcholinesterase, *J. Neurochem.*, 39, 1467, 1982.
132. **Landreth, G. E. and Agranoff, B. W.,** Explant culture of adult goldfish retina: effect of prior optic nerve crush, *Brain Res.*, 118, 299, 1976.
133. **McQuarrie, I. G., Grafstein, B., and Gersham, M. D.,** Axonal regeneration in the rat sciatic nerve: effect of a conditioning lesion and of dbc AMP, *Brain Res.*, 132, 443, 1977.
134. **McQuarrie, I. G. and Grafstein, B.,** Protein synthesis and axonal transport in goldfish retinal ganglion cells during regeneration accelerated by a conditioning lesion, *Brain Res.*, 251, 25, 1982.
135. **Varon, S., Skaper, S. D., and Manthorpe, M.,** Trophic activities for dorsal root and sympathetic ganglionic neurons in media conditioned by Schwann and other peripheral cells, *Dev. Brain Res.*, 1, 73, 1981.
136. **Lindberg, G., Dahlin, L. B., Danielsen, N., Gelberman, R. H., Longo, F. M., Powell, H. O., and Varon, S.,** Nerve regeneration in silicone chambers: influence of gap length and of distal stump components, *Exp. Neurol.*, 76, 361, 1982.
137. **Politis, M. J., Ederle, K., and Spencer, P. S.,** Tropism in nerve regeneration *in vivo*. Attraction of regenerating axons by diffusible factors derived from cells in distal nerve stumps of transected peripheral nerves, *Brain Res.*, 253, 1, 1982.
138. **Molander, H., Olsson, Y., Engkvist, O., Bowald, S., and Eriksson, I.,** Regeneration of peripheral nerve through a polyglactin tube, *Muscle Nerve*, 5, 54, 1982.
139. **Uzman, B. G. and Villegas, G. M.,** Mouse sciatic nerve regeneration through semi-permeable tubes: a quantitative model, *J. Neurosci. Res.*, 9, 325, 1983.
140. **David, S. and Aguayo, A.,** Axonal elongation into peripheral nervous system "Bridges" after central nervous system injury in adult rats, *Science*, 214, 931, 1981.
141. **Haley, J. E. and Ledeen, R. W.,** Incorporation of axonally transported substances into myelin lipids, *J. Neurochem.*, 32, 735, 1979.
142. **Haley, J. E., Tirri, L. J., and Ledeen, R. W.,** Axonal transport of lipids in the rabbit optic system, *J. Neurochem.*, 32, 727, 1979.
143. **Toews, A. D. and Morell, P.,** Turnover of axonally transported phospholipids in nerve endings of retinal ganglion cells, *J. Neurochem.*, 37, 1316, 1981.
144. **Dziegielewska, K. M., Evans, C. A. N., and Saunders, N. R.,** Effect of axotomy on rapid axonal transport of phospholipids in peripheral nerve of the rat, *J. Physiol.*, 263, 200, 1976.
145. **Dziegielewska, K. M., Evans, C. A. N., and Saunders, N. R.,** Rapid effect of nerve injury upon axonal transport of phospholipids, *J. Physiol.*, 304, 83, 1980.
146. **Gould, R. M., Spivack, W. D., Sinatra, R. S., Linquist, T. D., and Ingoglia, N. A.,** Axonal transport of choline lipids in normal and regenerating rat sciatic nerve, *J. Neurochem.*, 39, 1569, 1982.
147. **Alberghina, M., Viola, M., and Giuffrida, A. M.,** Rapid axonal transport of glycerophospholipids in regenerating hypoglossal nerve of the rabbit, *J. Neurochem.*, 40, 25, 1983.
148. **Alberghina, M., Moschella, F., Viola, M., Brancati, V., Micali, G., and Giuffrida, A. M.,** Changes in rapid transport of phospholipids in the rat sciatic nerve during axonal regeneration, *J. Neurochem.*, 40, 32, 1983.

149. **Austin, L., Bray, J. J., and Young, R. J.,** Transport of proteins and ribonucleic acid along nerve axons, *J. Neurochem.*, 13, 1267, 1966.
150. **Miani, N., Di Girolama, A., and Di Girolamo, M.,** Sedimentation characteristics of axonal RNA in rabbit, *J. Neurochem.*, 13, 755, 1966.
151. **Bray, J. J. and Austin, L.,** Flow of protein and ribonucleic acid in peripheral nerve, *J. Neurochem.*, 15, 731, 1968.
152. **Peterson, J. A., Bray, J. J., and Austin, L.,** An autoradiographic study of the flow of protein and RNA along peripheral nerve, *J. Neurochem.*, 15, 741, 1968.
153. **Ingoglia, N. A. and Tuliszewski, T.,** Transfer RNA may be axonally transported during regeneration of goldfish optic nerves, *Brain Res.*, 112, 371, 1976.
154. **Gambetti, P., Ingoglia, N. A., Autilio-Gambetti, L., and Weis, P.,** Distribution of [^{3}H] RNA in goldfish optic tectum following intraocular or intracranial injection of [^{3}H] uridine. Evidence of axonal migration of RNA in regenerating optic fibers, *Brain Res.*, 154, 285, 1978.
155. **Ingoglia, N. A.,** 4S RNA in regenerating optic axons of goldfish, *J. Neurosci*, 2, 331, 1982.
156. **Ingoglia, N.,** The effect of intraocular injections of cordycepin on retinal RNA synthesis and on RNA axonally transported during regeneration of the optic nerve of the goldfish, *J. Neurochem.*, 30, 1029, 1978.
157. **Ingoglia, N. A. and Sharma, C.,** The effect of inhibition of axonal RNA transport on the restoration of retinotectal projections in regenerating optic nerves of goldfish, *Brain Res.*, 156, 141, 1978.
158. **Ingoglia, N. A., Sturman, J. A., and Eisner, R. A.,** Axonal transport of putrescine, spermidine and spermine in normal and regenerating goldfish optic nerves, *Brain Res.*, 130, 433, 1977.
159. **Ingoglia, N. A. and Sturman, J. A.,** Axonal transport of putrescine and the polyamine, *Neurochem. Res.*, 5, 913, 1980.
160. **Ingoglia, N. A., Sharma, S. C., Pilchman, J., Baranowski, K., and Sturman, J. A.,** Axonal transport and transcellular transfer of nucleotides and polyamines in intact and regenerating optic nerves of goldfish: speculation on the axonal regulation of periaxonal cell metabolism, *J. Neurosci.*, 2, 1412, 1982.
161. **Ingoglia, N. A., Sturman, J. A., Jaggard, P., and Perez, C.,** Association of spermine and 4S RNA during axonal transport in regenerating optic nerves of goldfish, *Brain Res.*, 238, 341, 1982.
162. **Jeffrey, P. L. and Austin, L.,** Axoplasmic transport, *Prog. Neurobiol.*, 2, 105, 1973.
163. **Weiss, P. and Hiscoe, H. B.,** Experiments on the mechanism of nerve growth, *J. Exp. Zool.*, 107, 315, 1948.
164. **Biondi, R. J., Levy, M. J., and Weiss, P. A.,** An engineering study of the peristaltic drive of axonal flow, *Proc. Natl. Acad. Sci. U.S.A.*, 69, 1732, 1972.
165. **Schmitt, F. O.,** The molecular biology of neuronal fibrous proteins, *Neurosci. Res. Prog. Bull.*, 6, 119, 1968.
166. **Ochs, S.,** Axoplasmic transport, in *Basic Neurochemistry*, Siegel, G. J., Albers, R. W., Agranoff, B. W., and Katzman, R., Eds., Little Brown, Boston, 1981, 425.
167. **Martinez, A. J. and Friede, R. L.,** Accumulation of axoplasmic organelles in smaller nerve fibres, *Brain Res.*, 19, 183, 1970.
168. **Griffin, J. W., Fehnestock, K. E., Price, D. L., and Hoffman, P. N.,** Microtubule-neurofilament segregation produced by B,B-iminodiproprinonitrile: evidence for the association of fast axonal transport with microtubules, *J. Neurosci.*, 3, 557, 1983.
169. **Gross, G. W.,** The microstream concept of axoplasmic and dendritic transport, *Adv. Neurol.*, 12, 283, 1975.
170. **Brady, S. R., Crothers, S. D., Nosal, C., and McClure, W. O.,** Fast axonal transport in the presence of high Ca^{2+}: evidence that microtubules are not required, *Proc. Natl. Acad. Sci. U.S.A.*, 77, 5909, 1980.
171. **Droz, B., Rambourg, A., and Koenig, H. L.,** The smooth endoplasmic reticulum: structure and role in the renewal of axonal membrane and synaptic vesicles by fast axonal transport, *Brain Res.*, 93, 1, 1975.
172. **Tsukita, S. and Ishikawa, H.,** Morphological evidence for the involvement of the smooth endoplasmic reticulum in axonal transport, *Brain Res.*, 174, 315, 1979.
173. **Levail, J. H., Rapisardi, S., and Sugino, I. K.,** Evidence against the smooth endoplasmic reticulum as a continuous channel for the retrograde axonal transport of horseradish peroxidase, *Brain Res.*, 191, 3, 1980.
174. **Willard, M., Wiseman, M., Levine, J., and Skene, P.,** Axonal transport of actin in rabbit retinal ganglion cells, *J. Cell Biol.*, 81, 581, 1979.
175. **Levine, J. and Willard, M.,** Fodrin: axonally transported polypeptides associated with the internal periphery of many cells, *J. Cell Biol.*, 90, 631, 1981.
176. **Isenberg, G., Schubert, P., and Kreutzberg, G. W.,** Experimental approach to test the role of actin in axonal transport, *Brain Res.*, 194, 588, 1980.
177. **Goldberg, D. J., Harris, D. A., Lubit, B. W., and Scherantz, J. H.,** Analysis of fast axonal transport by intracellular injection of potentially inhibitory macromolecules: evidence for a possible role of actin filaments, *Proc. Natl. Acad. Sci. U.S.A.*, 12, 7448, 1980.

178. **Brady, S. T., Lasek, R. J., and Allen, R. D.,** Fast axonal transport in extruded axoplasm from squid giant axon, *Science,* 218, 1129, 1982.
179. **Comper, W. D., Preston, B. N., and Austin, L.,** A paradigm for axonal transport, *Neurochem. Res.,* 8, 943, 1983.
180. **Trotta, E. E. and De Meis, L.,** ATP-dependent calcium accumulation in brain microsomes. Enhancement by phosphate and oxalate, *Biochim. Biophys. Acta,* 394, 239, 1975.
181. **Trotta, E. E. and De Meis, L.,** Adenosine-5-triphosphate-orthophosphate exchange catalyzed by the Ca^{++}-transport ATPase of brain, *J. Biol. Chem.,* 253, 7821, 1978.
182. **Stearns, M. E.,** Role of the microtrabecular lattice in axoplasmic transport, *Vision Res.,* 21, 71, 1981.
183. **Björklund, A. and Stenevi, V.,** Regeneration of monoaminergic and cholinergic neurons in the mammalian central nervous system, *Physiol. Rev.,* 59, 62, 1979.
184. **Kiernan, J. A.,** Hypotheses concerned with axonal regeneration in the mammalian nervous system, *Biol. Rev.,* 54, 155, 1979.
185. **Marx, J. L.,** Regeneration in the central nervous system, *Science,* 209, 378, 1980.
186. **Robertson, M.,** The unregenerate nervous system, *Nature (London),* 295, 553, 1982.
187. **Brown, I. R.,** Modification of gene expression in the mammalian brain following hypothermia, *J. Neurochem.,* Suppl. 41, S66, 1983.

Chapter 18

RETROGRADE TRANSPORT AND REGENERATION STUDIES

Mark A. Bisby

TABLE OF CONTENTS

I. INTRODUCTION

Axon injury is followed by a complex sequence of events: (1) cell body reaction, (2) axon sprouting and elongation, and (3) functional reinnervation of target. This sequence does not progress to completion for every injured neuron. The cell body reaction may result in death of the injured neuron; axon sprouting may be abortive, as in most instances of damage to the mature vertebrate central nervous system; regenerating axons may contact an inappropriate target. Nevertheless, in this essay I will review briefly the neuronal changes that occur at each stage of the successful regenerative sequence, with particular emphasis on the role of retrograde axonal transport in conveying to the cell body information about the status of the axon. Anterograde and retrograde transport are linked processes, since retrograde transport returns to the cell body materials originally reaching the axon by anterograde transport. Alterations in anterograde transport during regeneration will therefore be discussed where relevant to the issues of molecular information transfer within the neuron and the regulation of regeneration.

Before plunging into the retrograde control of neuronal cell body metabolism, a topic filled with speculations, I will present a few facts about the process of retrograde axonal transport.

II. RETROGRADE AXONAL TRANSPORT: KEY FACTS

A number of different phases of axonal transport exist,[1] characterized by their velocity range, by the molecules or organelles associated with each phase, and by their mechanisms (if we knew what they were). Retrograde axonal transport is a type of fast phase transport, with a velocity range varying enormously depending on the material under study and the procedure used. Velocities may approach those of fast-transport, (400 mm/day) but are often reported to be in the 100 to 300 mm/day range.[2,3] A slow component of retrograde transport has also been reported,[4-6] but the extent and nature of this component requires further study. Fast retrograde transport has many similarities with fast anterograde transport.[2] It can occur in isolated axons; it has the same sensitivity to temperature as fast-transport; and it is colchicine-sensitive, implicating microtubles in its mechanism. The major classes of fast anterograde transported molecules (proteins, glycoproteins, and phospholipids) are also retrograde transported.[7-14] For proteins, the same molecules are associated with both anterograde and retrograde transport,[7,15] raising the interesting problem of determination of transport direction. The answer seems to lie in the packaging of transported molecules. In both directions they are associated with membraneous organelles: smooth endoplasmic reticulum for anterograde transport[16-18] and various prelysosomal structures for retrograde transport.[17-19] Possibly the direction of transport is determined by the membrane properties of the two types of organelles.

Anterograde transported protein molecules, borne on the membranes of the smooth endoplasmic reticulum, are probably incorporated along with their companion lipid membrane components into the axonal membrane both at the axon terminal and all along the axon, since studies with labeled transported proteins show an exponential decline in transported radioactivity with distance from the cell body, consistent with an axonal drop-off.[20,21] Retrograde transport occurs following membrane retrieval by endocytosis.[22]

In keeping with this concept of retrograde transport as a recycling mechanism for used membrane, it has been shown that the activity of the enzyme dopamine-β-hydroxylase is reduced when it is undergoing retrograde transport,[23] so that modifications in protein structure accompany the reversal process. In the course of endocytosis, exogenous ligands bound to membrane receptors also become internalized and transferred within the prelysosomal particle back to the cell body. This feature of retrograde transport is used to advantage by neuroanatomists. They can apply to axon terminal regions fluorescent or enzymatic labels coupled to lectins which will bind to membrane surface glycoproteins and ultimately arrive in neuronal cell bodies, thus tracing the projection pathway of the axons back to the site of marker injection.[24] This is also the route by which some viruses gain access to the central nervous system.[25] From the standpoint of this essay, retrograde transport of exogenous molecules also allows hypothetical trophic substances, synthesized by target cells, or by Schwann cells associated with the axon, to exert an influence over the cell body.

In general terms retrograde axonal transport has two functions. First, the recycling of axonal constituents, particularly axonal membranes, and second, the transfer of putative information molecules from the environment of the axon to the cell body in order to elicit appropriate cell body response.[8]

III. EARLY CHANGES AFTER AXON INJURY: THE CELL BODY REACTION

Axon injury initiates a host of metabolic and structural changes in the neuron, its surrounding glial cells and the synapses of other neurons upon it. For details of these changes, please refer to comprehensive reviews.[26,27] Together these changes are referred to as the cell body reaction, though the term chromatolysis is sometimes used indiscriminately to mean the same.

In a thoughtful and stimulating review published in 1970, Cragg asked, "What is the signal for chromatolysis?"[28] The question, 14 years later has shifted to, "Are there multiple signals for chromatolysis, and if so, what aspects of the cell body reaction are controlled by each signal?" The evidence is mounting that axotomy triggers several signals. Two main categories of signals exist: target-disconnection signals and loss of axoplasm signals, with retrograde transport intimately involved in both categories.

A. Target-Disconnection Signals

Target-disconnection signals are negative and occur when the cell body is deprived of a target-derived trophic factor which is retrograde-transported to the cell body and maintains normal metabolism. Evidence for this type of signal is best for sympathetic nerve cells, where the trophic factor may be nerve growth factor (NGF).[29] Direct application of NGF to axotomized sympathetic neurons can protect them from some,[30-33] but not all,[34] the effects of axotomy.

There are also numerous in vitro experiments (for summary see Reference 35) that show improved survival of cholinergic neurons or spinal cord neurons in culture when co-cultured with muscle cells[38] or when the culture medium has been conditioned by the presence of muscle tissue within it prior to its addition to the neuron culture.[36,39] These studies show that a diffusible factor produced by muscle can influence cell body metabolism.

The dependency of immature neurons on target-derived factors is probably greater than in mature neurons since immature neurons usually die after axotomy or removal of their target,[27] while many mature neurons will survive indefinitely after axotomy even if their axons fail to contact the target.[40,41]

B. Loss of Axoplasm Signals

The existence of loss of axoplasm signals has been demonstrated by many investigators[42-46] who showed that a second axotomy, made before previously axotomized neurons had reinnervated their target cells, provoked a second cell body response which often differed in intensity, duration or character from the original response to axotomy, either because of the absence of the normally associated disconnection signal, or because of the pre-existing regenerating status of the neuron. Recently it has been shown that colchicine, applied to the axon between site of injury and cell body to block retrograde transport, retards the onset of a metabolic component of the cell body reaction, implicating some sort of signal arising from the site of injury and reaching the cell body by retrograde transport.[47]

There are at least three candidates for the loss of axoplasm signal:

1. Reduction in trophic factor supplied by Schwann or glial cells to the axon
2. Abnormal entry into the axon of exogenous materials at the site of injury
3. Abnormal retrograde transport developing at the site of injury

Candidate (1) is similar in nature to the disconnection signal, except that the factor is derived from supporting cells rather than the target. Assuming that this trophic material would enter the axon all along its length, axon shortening would result in less material reaching the cell body. There is little evidence to support or refute this hypothesis, beyond the demonstration of a trophic influence of Schwann-cell conditioned media on sympathetic ganglion and dorsal root ganglion neurones.[48]

Candidate (2) has direct experimental support, but is hard to test rigorously. The injured nerve suffers from a breakdown of normal permeability barriers, allowing exogenous tracers to enter the axon rapidly and to appear subsequently in the cell body after retrograde transport.[49,50]

Candidate (3) is alone in depending on purely endogenous materials to signal the onset of axotomy. For transported labeled protein[51,52] and for visible organelles[53] it has been demonstrated that the ability to reverse axonal transport develops within an hour after injuring an axon. Transported materials, including a variety of enzyme markers for different subcellular organelles,[54] normally destined for the distal axon accumulate proximal to the site of injury, but a proportion of the accumulated material subsequently returns to the cell body. The injury, by short circuiting the normal circulation of materials within the axon causes transported materials to return to the cell body earlier than usual. However, this is unlikely to be a signal: rather, changes in amount of material returned or composition may be significant. Increased amounts of transported protein are returned to the cell body after axotomy,[51] but a first attempt to demonstrate a difference in the composition of protein returning from an injury compared to protein returning from an intact axon was negative.[15]

A prediction of the increased return to the cell body hypothesis is that an increase in levels of transported protein in the cell body should occur prior to any metabolic reactions. Analysis of the acetylcholinesterase (AChE) content of individual frog motoneurons showed that AChE levels in the cell body did not rise transiently, as predicted, but that levels remained constant. Analysis of the total AChE content of the cell body and proximal axon showed that following axotomy there was a significant down-regulation of AChE synthesis so that some signal must have reached the cell body.[55] Thus, though reversal of AChE transport at a nerve injury occurs,[56,57] the arrival of a pulse of AChE at the cell body could not be detected. It could still be argued that the error signal for suppression of AChE synthesis is immeasurably small and the sensitivity of the regulatory process for AChE synthesis so high that increased return of AChE immediately reduced AChE synthesis so that changes in cell body content were not detected. In some cases where axons were cut very close to the cell bodies or transport was blocked by colchicine application, a transient rise in transported materials was observed,[58,59] perhaps because the cell body was unable to accommodate the massive short-circuiting of transported materials under these circumstances.

Rapid down-regulation of serotonin synthesis occurs in the giant cerebral neuron of *Aplysia* after axotomy of one of its two main axon branches. In this neuron the serotonin containing vesicles which reverse direction of transport at the site of injury are diverted into the remaining axon branch, and no increased return to the cell body could be detected.[60] In this situation, cell body synthesis does not seem to be regulated by a feedback inhibition of product returning from the axon following injury.

IV. AXON ELONGATION

After axotomy, axon sprouting can begin very rapidly, within a few minutes in vitro,[61] and from both ends of the severed axon,[62] suggesting that initial sprouting can occur without changes in cell body protein synthesis, but induced by purely local events at the site of injury. It has been suggested that calcium entry into the injured axon somehow initiates elongation.[63,64]

If sprouting can occur so readily we should perhaps be wondering why axons are not always growing, rather than why they grow when injured: an opposing hypothesis for regulation of axon growth[65] has just this emphasis. It has been proposed that outgrowth of axons occurs through the movement of cytoskeletal proteins in slow components of transport[66] and growth ceases at nerve terminals because of the activation of a cytoskeletal disassembly process by high terminal intracellular calcium concentrations. In growing or regenerating axons the disassembly process is inactivated and the advancing cytoskeleton propels the growth cone. This hypothesis would predict that the calcium concentration in the injured axon or in the growth cone would not be elevated above the level elsewhere in the axon, which seems unlikely given the existence of voltage dependent calcium channels at the tips of growing neurites[64,67,68] and the probable massive influx of calcium ions into the axoplasm adjacent to the site of nerve injury.

Whether or not regeneration can be sustained following initial sprouting in the absence of cell body changes is a moot point.[69] For example, in frogs maintained at 25°C, axotomy is followed by motoneuron chromatolysis and the rapid and complete regeneration of the axons. At 15°C, chromatolysis is not observed, and though regeneration begins, it does not proceed to completion.[70]

A. Changes in Amount of Transported Materials

The extent and function of changes in axonal transport following axotomy and during regeneration is debatable. It is clear that the synthesis and transport of transmitter related enzymes is immediately depressed, recovering more slowly as regeneration proceeds.[54,71-73] Transport of total fast-transported proteins is similarly depressed in regenerating peripheral nerves,[73,74] but is increased in regenerating optic nerve.[75]

In rat sciatic nerve motoneurons the return to normal amounts of fast-transport protein occurs mainly during the fist 10 days after crush injury,[74] that is, during the period of axon elongation but before the onset of reinnervation which does not occur until day 13 to 14.[76] Recovery of AChE transport in rat[77] and chicken[78] sciatic nerve also precedes reinnervation. In axons prevented from elongating, transport of protein and transmitter-related enzymes remains depressed.[71,72,74] The cell body somehow knows that its axon is elongating and modifies the synthesis and transport of fast-transported materials. Again, there are several candidates for the elongation signal: for example, contact with the regenerating axons could result in the resumption of the supply of Schwann cell trophic factor via retrograde transport to the cell body. Degenerating nerve has both a trophic and a tropic (directing) effect on axonal outgrowth. Regenerating axons grow preferentially towards degenerating nerve,[79] and fluid collected from chambers bridging proximal and distal nerve stumps contains a factor which has a tropic effect in vitro on all types of neurons found in peripheral nerve.[80] The amount of Schwann cell trophic factor reaching the cell body would be related to extent of axonal penetration into the distal stump.

B. Changes in Retrograde Transport

Retrograde transport of endogenous protein in the injured nerve also undergoes changes as the axon elongates.[74] Over the first 5 days following injury there is an increase in amount of retrograde transport, followed at later times by a shift in time-course of transport back to the normal time-course. The shift in time-course would be consistent with the growth-cone being the major site of transport reversal with longer transit time to the point on the nerve where transport is assessed accounting for the shift. In regenerating axons the growth cone is the principle destination of fast-transported material,[81-83] where the membrane constituents conveyed by fast-transport are inserted into the elongating axon.[84-86] It is possible that the progressive alterations in retrograde transport which occur during axon elongation inform the cell body of the progress of elongation, but the exact nature of the information conveyed by this process (amount or nature of material?) is unknown.

C. Changes in Composition of Transported Proteins

During regeneration changes also occur in the composition of the fast-transported proteins.[87-91] Relatively few of the hundred or more polypeptides conveyed by axonal transport increase or decrease in injured axons and the significance of these changes in transport are unknown. The changes could be totally unrelated to regeneration, being an incidental component of the cell body's reaction to injury. They could represent membrane components needed in greater or lesser amounts by regenerating axons, as compared to intact neurotransmitter-secreting axons. Finally, they might play a regulatory role in regeneration, for example, allowing regeneration to be sustained beyond the intrinsic sprouting behavior of the injured axon. In this respect it is intriguing that in optic nerves of rabbit[92] or rat[93] which do not regenerate after injury, the characteristic changes in composition of fast-transported protein associated with regeneration are not seen. This could be either cause or effect of the failure of these axons to regenerate.

If the switch for axon elongation is increased intracellular calcium levels,[63,64] as mentioned earlier, some means of raising intracellular calcium at the growth cone must persist following the initial traumatic entry of calcium following injury. The distal tips of growing or regenerating axons possess voltage-dependent calcium channels,[64,67,68] which might result in high intracellular calcium levels at the growth cone. Perhaps the alterations in transported proteins in regenerating axons are indicative of increased supply of calcium ionophores to the regenerating axon, or a decreased supply of intracellular calcium-buffering proteins. To carry this chain of speculation one weak link further, it is interesting that one of the fast-transported proteins which decreases in regenerating axons (which we call S_1)[94] is probably identical to a protein (Band A) which has been revealed, by peptide mapping, to have considerable homology with calmodulin, an important calcium-binding regulatory protein.[95] If S_1 has calcium-binding activity, reduced transport of S_1 into regenerating axons might permit elevated intracellular calcium concentrations.

Slow-transport also undergoes specific changes during regeneration, but the nature of these changes is more readily understandable than the obscure changes in fast-transport, mainly due to the work of Lasek's group. The synthesis and transport of the cytoskeletal elements conveyed by slow component b (SCb) at 3 to 4 mm/day increase relative to the elements carried by slow component a (SCa) at $\simeq$ 1 mm/day.[96] The cytoskeletal elements associated with SCa and SCb are the microtubule-neurofilament network and the actin-containing microfilament network respectively.[97] The correlation between SCb velocity and regeneration rate,[66] as well as the presence in SCb of actin and other constituents of the growth cone, strongly suggests that axonal growth occurs through the continued delivery of SCb materials which propel the growth cone forward. Changes in cell body synthesis of cytoskeletal proteins affect the site of axonal injury long after regeneration has begun, so the normal supply of these proteins to the axon must be sufficient to initiate and sustain regeneration. The increase in SCb transport which occurs in regenerating axons is not involved in initial axonal elongation, but may be related to a general increase in plasticity[96] of the axon in response to injury.

V. TARGET REINNERVATION

A. Regulation of Cell Body Metabolism by Interaction with Target

The time-course of changes in fast-transported protein composition very closely parallels events in regeneration and is quite different from the rapid recovery of amounts of transported materials. After crush of rat sciatic nerve, no change in composition of transported protein was detectable at day 1 but by day 3 significant change had occurred. Changes were maximal at approximately day 14 and by day 42 had returned to normal.[88] Electromyographic and behavioral studies show that functional reinnervation begins at day 14 and functional recovery does not improve much beyond day 30. Histological studies demonstrate the formation of neuromuscular junctions at days 13 to 14.[76] Thus, changes in composition of the transported proteins parallel the onset and termination of regeneration, rather than the elongation of the axon, which seems to regulate the amount of transported material. As might be expected if motor axons are prevented from reinnervating muscle, the composition of transported protein remains typical of the regenerating state for at least 100 days postinjury.[98] Recently we have made similar observations for transport in sensory axons,[94] with protein composition beginning to return towards normal at the time when the sensory axons reinnervate their skin fields,[99] but maintained in the regeneration state in axons prevented from reinnervation.

Functional contact of regenerating axons with their target causes cell body metabolism of fast-transported proteins to return to normal. Possibly the target-derived trophic factor, deprivation of which is one of the signals for the cell body reaction, is once again retrograde-transport to the cell body and turns off this aspect of the cell body reaction.

The conclusion that functional reinnervation is necessary to reverse completely the cell body reaction was reached in earlier experiments in which cell body nucleic acid metabolism[46] and ultrastructure[100] were examined. In particular, implantation of a sectioned motor nerve into a muscle already innervated by another nerve resulted in incomplete recovery with the neuron maintained in a "resting mode." However, when 84 days later the correct nerve was sectioned so that the implanted nerve could establish functional contact with the muscle, the motoneurons of the implanted nerve resumed a "seeking mode" and subsequently returned to a normal state or "secreting mode."[101]

Restoration of contact with muscle must also influence cell body synthesis of the structural proteins involved in slow-transport. Regenerating axons are considerably less in diameter than their parent axon, but motor axons which reinnervate muscle recover their normal dimensions, as shown by conduction parameters, while those that fail to reinnervate remain in the shrunken condition.[102] Axon diameter may be regulated by neurofilament and microtubule content,[103] and the reduction in transport of neurofilament proteins in slow component a following axotomy[96,104] probably accounts for the progressive proximo-distal reduction in diameter of the parent axons.[105] Later, during the process of axonal maturation that follows reinnervation[106] there should be a return to normal ratios of neurofilament to tubulin transport in parallel with the return to normal axon diameter. The increase in axonal diameter occurs in a proximal-distal fashion,[107] consistent with the increased transport, at a slow velocity, of materials causing axonal enlargement.

B. The Retrograde Signal for Reinnervation

The nature of the signal for reinnervation is not known but there are some important considerations. When growing motor axons reinnervate their target, the growth cones are converted within 1 day into functional terminals able to release acetylcholine, though the size of the terminals is considerably smaller than usual,[76] and the quantal content of endplate potentials is correspondingly reduced.[108] However, when contact has been made growth of the axon does not cease. A few microns proximal to the terminal, sprouting occurs and the sprouts continue to grow and innervate other muscle fibers, resulting in a polyneuronal innervation of the muscle fiber which peaks at day 21 after rat sciatic nerve crush[109] and is succeeded by elimination of polyneuronal innervation.[110] The signal which transforms growth cone into terminal must therefore be a local event since the parent axon continues growing. Most likely some contact inhibition process with the basal lamina at the site of the neuromuscular junction is responsible. Contact can be established with the basal lamina alone if the muscle fibers have been eliminated by irradiation.[111]

There are several possible candidates for the reinnervation signal which, exerted through some interaction between nerve and target, and conveyed by retrograde transport, causes cell body metabolism to return to normal. Returning to the idea that muscle produces a trophic factor, absence of which is one of the factors responsible for the cell body reaction to axotomy, it is possible that as more and more contacts are made between nerve and muscle, the supply of target-derived trophic factor taken up at the terminals and retrograde-transported to the cell body should increase, restoring normal cell body metabolism. In turn, restoration of normal fast-transported protein composition might lead to a general reduction in the tendency of axons to sprout and form further contacts with muscle. The muscle derived trophic factor would therefore have a generally suppressive effect on the growth of the axon.

Production of the muscle-derived trophic factor appears to be related to muscle activity. Botulinum toxin, which causes paralysis by blocking transmission, lead to a motoneuron cell body reaction[112] and caused intramuscular sprouting.[113] Blockade of motor nerve impulses (but not axonal transport) with tetrodotoxin or reduction in motoneuron activity by spinal transection also caused a cell body reaction.[114] Muscle hypoactivity induced by limb fixation caused considerable muscle atrophy and resulted in a reduction in diameter of both intact and regenerating motor axons, as occurs in motoneuron axons after unsuccessful attempts to regenerate.[115] Fixation of a muscle in either an extended position (which caused no muscle atrophy) or shortened position (which caused atrophy) was associated with no change or axotomy-type changes, respectively, in the motoneuron. It was concluded that production of the trophic factor is related to overall levels of muscle metabolism, which would be enhanced by nerve-stimulated activity.[116]

Alternatively, the trophic influence of muscle on nerve may be exerted only at a local, terminal, level. The tendency of axons to sprout, to form new connections or to restrict the number of nerve-muscle synapses may be influenced by local sprouting factors, with cell body changes consequent upon terminal changes: formation of new growth cones or the conversion of growth cones into secreting terminals would seem likely to alter the amount and/or composition of endogenous axonal materials returned to the cell body by retrograde transport and might in this indirect way influence cell body metabolism. After a careful review of factors responsible for inducing motor terminal sprouting, including changes in cell body metabolism, it was concluded that muscles undergoing denervation-type changes, whether produced by actual denervation or by experimental inactivity, generated a signal, possibly a diffusible factor, which stimulated terminal sprouting.[117] This sprouting factor might explain why intact motoneuron axons undergo terminal sprouting in response to partial denervation of muscle. The opposing view, that trophic effects of innervated muscle suppress the ability of axons to grow, does not explain sprouting in response to partial denervation because the intact axons that sprout have not lost any trophic support.

Much of the experimental evidence can support either view. Thus, botulinum toxin applied to muscles causes terminal sprouting and cell body changes.[112,113] Muscle inactivity could reduce production of the putative muscle trophic factor, to which cell bodies respond by undergoing metabolic changes supportive of axonal growth. Alternatively, muscle inactivity could increase production of sprouting factor, leading to sprouting and changes in the cell body secondary to sprout production. Denervated muscle has a greater growth-promotion effect on neurons in vitro than innervated muscle[118,119] which could also be due to production of growth-suppressing factors by innervated muscle, or growth promoting factors by denervated muscle.

It is possible that both types of factors work in concert, the sprouting factor of denervated muscle acting locally to produce intramuscular sprouting when regenerating axons regain the target muscle, and the trophic factor of innervated muscle acting through retrograde transport on the cell body to convert the metabolic state of the regenerating neuron to one emphasizing axonal maintenance and neurotransmission as neuromuscular junctions are being formed. Although, as previously mentioned, junction formation may be triggered by a surface recognition phenomenon with the basal lamina alone,[111] such contacts were not maintained, implicating the muscle cell as necessary for the long-term stabilization of motoneuron terminals.

VI. CONCLUSION

Retrograde axonal transport plays an important role in axon regeneration, primarily through the transfer of control signals from the axon to the cell body to elicit appropriate metabolic responses. While the initial onset of axonal sprouting and the conversion of growth cone into terminal upon contact with the target are probably controlled at the local axon level, it is likely that maintained axon elongation and the cessation of the sprouting ability of axons are dependent upon changes in cell body metabolism and subsequent axonal transport, both fast and slow, of materials to the axon. I have differentiated between two retrograde signals responsible for initiation of the cell body reaction to injury: one due to target disconnection, one due directly to damage to the axon, and I have argued that a similar duality exists during regeneration with separate signals for axon elongation and reconnection with the target.

This essay is rife with speculation. Doubtless some of the studies cited could be interpreted differently by those with other biases. It should be evident that a key area of research in understanding the regulation of regeneration must be the study of target-derived trophic factors. With the example of NGF as encouragement[29] it is necessary to purify these probably diverse factors, to demonstrate their retrograde transport and to show that their absence or presence has identical effects on cell body metabolism to those observed during the cell body reaction to injury and its recovery after successful regeneration.

ACKNOWLEDGMENTS

The author's research is supported by grants from the Medical Research Council of Canada (operating) and from the Alberta Heritage Foundation for Medical Research (equipment and research trainees).

REFERENCES

1. **Grafstein, B. and Forman, D. S.,** Intracellular transport in neurons, *Physiol. Rev.*, 60, 1167, 1980.
2. **Bisby, M. A.,** Retrograde axonal transport, in *Advances in Cellular Neurobiology*, Vol. 1, Federoff, S. and Hertz, L., Eds., Academic Press, New York, 1980, 69.
3. **Bisby, M. A. and Buchan, D. H.,** Velocity of labelled protein undergoing anterograde and retrograde transport, *Exp. Neurol.*, 74, 11, 1981.
4. **Fink, D. J. and Gainer, H.,** Axonal transport of proteins; a new view using *in vivo* covalent labeling, *J. Cell Biol.*, 85, 175, 1980.
5. **Gainer, H. and Fink, D. J.,** Evidence for slow retrograde transport of serum albumin in rat sciatic nerve, *Brain Res.*, 233, 414, 1982.
6. **Williams, L. R. and Agranoff, B. W.,** Retrograde transport of goldfish optic nerve proteins labeled by N-succinimidyl [^{3}H] propionate, *Brain Res.*, 259, 207, 1983.
7. **Abe, T., Haga, T., and Kurokawa, M.,** Retrograde axoplasmic transport: its continuation as anterograde transport, *FEBS Lett.*, 47, 272, 1974.

8. **Bisby, M. A.,** Orthograde and retrograde axonal transport of labeled protein in motoneurons, *Exp. Neurol.*, 50, 628, 1976.
9. **Bisby, M. A. and Guy, J. R.,** Anterograde and retrograde transport of phospholipid in rat sciatic nerve axons, *Can. J. Physiol. Pharmacol.*, 61, Aiii, 1983.
10. **Bray, J. J., Kon, C. M., and Breckenridge, B. Mc. L.,** Reversed polarity of rapid axonal transport in chicken motoneurons, *Brain Res.*, 33, 560, 1971.
11. **Edstrom, A. and Hanson, M.,** Retrograde axonal transport of proteins *in vitro* in frog sciatic nerves, *Brain Res.*, 61, 311, 1973.
12. **Frizell, M., McLean, W. G., and Sjostrand, J.,** Retrograde axonal transport of rapidly migrating labeled proteins and glycoproteins in regenerating peripheral nerves, *J. Neurochem.*, 27, 191, 1976.
13. **Sjostrand, J. and Frizell, M.,** Retrograde axonal transport of rapidly migrating proteins in peripheral nerves, *Brain Res.*, 85, 325, 1975.
14. **Whitnall, M. H., Currie, J. R., and Grafstein, B.,** Bidirectional axonal transport of glycoproteins in goldfish optic nerve, *Exp. Neurol.*, 75, 191, 1982.
15. **Bisby, M. A.,** Reversal of axonal transport: similarlity of proteins transported in anterograde and retrograde directions, *J. Neurochem.*, 36, 741, 1981.
16. **Rambourg, A. and Droz, B.,** Smooth endoplasmic reticulum and axonal transport, *J. Neurochem.*, 35, 16, 1980.
17. **Smith, R. S.,** The short term accumulation of axonally transported organelles in the region of localized lesions of single myelinated axons, *J. Neurocytol.*, 9, 39, 1980.
18. **Tsukita, S. and Ishikawa, H.,** The movement of membraneous organelles in axons. Electron microscopic identification of anterogradely and retrogradely transported organelles, *J. Cell. Biol.*, 84, 513, 1980.
19. **Breuer, A. C., Christian, C. N., Henkart, M., and Nelson, P. G.,** Computer analysis of organelle translocation in primary neuronal cultures and continuous cell lines, *J. Cell Biol.*, 65, 562, 1975.
20. **Gross, G. W. and Beidler, L. M.,** A quantitative analysis of isotope concentration profiles and rapid transport velocities in the C-fibres of the garfish olfactory nerve, *J. Neurobiol.*, 6, 213, 1975.
21. **Munoz-Martinez, E. J., Nunez, R., and Sanderson, A.,** Axonal transport: a quantitative study of retained and transported protein fraction in the cat, *J. Neurobiol.*, 12, 15, 1981.
22. **Holtzman, E. and Merucrio, A. M.,** Membrane circulation in neurons and photoreceptors: some unresolved issues, *Int. Rev. Cytol.*, 67, 1, 1980.
23. **Nagatsu, I., Kondo, Y., and Nagatsu, T.,** Retrograde axoplasmic transport of inactive dopamine-β-hydroxylase in sciatic nerves, *Brain Res.*, 116, 277, 1976.
24. **LaVail, J. H.,** A review of the retrograde transport technique, in *Neuroanatomical Research Techniques*, Robertson, R. T., Ed., Academic Press, New York, 1978, 356.
25. **Field, H. J. and Hill, T. J.,** The pathogenesis of pseudorabies in mice following peripheral inoculation, *J. Gen. Virol.*, 23, 145, 1974.
26. **Grafstein, B.,** The nerve cell body response to axotomy, *Exp. Neurol.*, 48, 32, 1975.
27. **Lieberman, A. R.,** Some factors affecting retrograde neuronal responses to axonal lesions, in *Essays on the Nervous System*, Bellairs, R. and Gray, E. G., Eds., Clarendon Press, Oxford, 1974, 71.
28. **Cragg, B. G.,** What is the signal for chromatolysis, *Brain Res.*, 23, 1, 1970.
29. **Thoenen, H. and Barde, Y. A.,** Physiology of nerve growth factor, *Physiol. Rev.*, 60, 1284, 1980.
30. **Banks, B. E. C. and Walter, S. J.,** The effects of postganglionic axotomy and nerve growth factor on the superior cervical ganglia of developing mice, *J. Neurocytol.*, 6, 287, 1977.
31. **Hamburger, V., Brunsobechtold, J. K., and Yip, J. W.,** Neuronal death in the spinal ganglia of the chick embryo and its reduction by nerve growth factor, *J. Neurosci.*, 1, 60, 1981.
32. **Hendry, I. A.,** The response of adrenergic neurons to axotomy and nerve growth factor, *Brain Res.*, 94, 87, 1975.
33. **Nja, A. and Purves, D.,** Effects of nerve growth factor and its antiserum on synapses in the superior cervical ganglion of the guinea pig, *J. Physiol.*, 227, 53, 1978.
34. **Hall, M. E. and Wilson, D. L.,** Nerve growth factor effects on protein synthesis after nerve damage, *Exp. Neurol.*, 77, 625, 1982.
35. **Barde, Y. A., Edgar, D., and Thoenen, H.,** New neurotrophic factors, *Ann. Rev. Physiol.*, 45, 601, 1983.
36. **Giller, E. L., Neale, J. H., Bulloch, P. N., Schrier, B. K., and Nelson, P. G.,** Choline acetyltransferase activity of spinal cord cell cultures increased by co-culture with muscle and by muscle-conditioned medium, *J. Cell Biol.*, 74, 16, 1977.
37. **Schnaar, R. L. and Schaffner, A. E.,** Separation of cell types from embryonic chicken and rat spinal cord: characterization of motoneuron-enriched fractions, *J. Neurosci.*, 1, 204, 1981.
38. **Hill, M. A. and Bennett, M. R.,** Cholinergic growth factor for skeletal muscle elevated following denervation, *Neurosci. Lett.*, 35, 31, 1983.
39. **Bennett. M. R., Lai, K., and Nurcombe, V.,** Identification of embryonic motoneurons *in vitro:* their survival is dependent on skeletal muscle, *Brain Res.*, 190, 537, 1980.

40. **Carlsen, J., Lais, A. C., and Dyck, P. J.,** Axonal atrophy from permanent peripheral axotomy in the adult cat, *J. Neuropath. Exp. Neurol.*, 38, 579, 1979.
41. **Davis, L. A., Gordon, T., Hoffer, J. A., Jhamandas, J., and Stein, R. B.,** Compound action potentials recorded from mammalian peripheral nerves following ligation or resuturing, *J. Physiol.*, 285, 543, 1978.
42. **Arvidsson, J. and Aldskogius, H.,** Effect of repeated hypoglossal nerve lesions on the number of neurons in the hypoglossal nucleus of adult rats, *Exp. Neurol.*, 75, 520, 1982.
43. **Brock, J. E. and Turner, J. E.,** Retinal ganglion cell body responses to repeated axotomy in the regenerating visual system of newt (Notophthalmus viridiscens), *Exp. Neurol.*, 78, 316, 1982.
44. **Sumner, B. E. H.,** Ultrastructural data, with special reference to bouton/glial relationships, from the hypoglossal nucleus after a second axotomy of the hypoglossal nerve, *Exp. Brain Res.*, 36, 107, 1979.
45. **Wells, M. R. and Bernstein, J. J.,** Amino-acid incorporation into rat spinal cord and brain after simultaneous and interval sciatic nerve lesions, *Brain Res.*, 139, 249, 1978.
46. **Watson, W. E.,** Some metabolic responses of axotomized neurons to contact between their axons and denervated muscle, *J. Physiol.*, 210, 321, 1970.
47. **Singer, P. A., Mehler, S., and Fernandez, H. L.,** Blockade of retrograde axonal transport delays the onset of metabolic and morphologic changes induced by axotomy, *J. Neurosci.*, 2, 1299, 1982.
48. **Varon, S., Skaper, S. D., and Manthorpe, M.,** Trophic activities for dorsal root and sympathetic ganglionic neurons in media conditioned by Schwann and other peripheral cells, *Dev. Brain Res.*, 1, 73, 1981.
49. **Anderson, P. N., Mitchell, J., and Mayor, D.,** On the mechanism of the uptake of horseradish peroxidase into the retrograde transport system of ligated postganglionic sympathetic nerves *in vitro*, *J. Anat.*, 133, 371, 1981.
50. **Kristensson, K. and Olsson, Y.,** Retrograde transport of horseradish peroxidase in transected axon. III. Entry into injured axons and subsequent localization in perikaryon, *Brain Res.*, 126, 154, 1977.
51. **Bisby, M. A. and Bulger, V. T.,** Reversal of axonal transport at a nerve crush, *J. Neurochem.*, 29, 313, 1977.
52. **O'Brien, D. W. and Snyder, R. E.,** Position-sensitive detector studies of the axonal transport of a pulse of radioisotope, *J. Neurobiol.*, 13, 435, 1982.
53. **Koles, Z. J., McLeod, K. D., and Smith, R. S.,** A study of the motion of organelles which undergo retrograde and anterograde rapid axonal transport in *Xenopus*, *J. Physiol.*, 328, 469, 1982.
54. **Schmidt, R. E. and McDougal, D. B., Jr.,** Axonal transport of selected particle-specific enzymes in rat sciatic nerve *in vivo* and its response to injury, *J. Neurochem.*, 30, 527, 1978.
55. **Sinicropi, D. V., Michels, K., and McIlwain, D. L.,** Acetylcholinesterase distribution in axotomized frog motoneurons, *J. Neurochem.*, 38, 1099, 1982.
56. **Schmidt, R. E., Yu, M. J. C., and McDougal, D. B.,** Turnaround of axoplasmic transport of selected particle-specific enzymes at an injury in control and diisopropylphosphorofluoridate-treated rats, *J. Neurochem.*, 35, 641, 1980.
57. **Couraud, J.-Y. and DiGiamberardino, L.,** Axonal transport of the molecular forms of acetylcholinesterase. Its reversal at a nerve transection, in *Axoplasmic Transport*, Weiss, D. G., Ed., Springer-Verlag, Basel, 1982, 144.
58. **Fonnum, F., Frizell, M., and Sjostrand, J.,** Transport, turnover and distribution of choline acetyltransferase and acetylcholinesterase in the vagus and hypoglossal nerves of the rabbit, *J. Neurochem.*, 21, 1109, 1973.
59. **Reis, D. J. and Ross, R. A.,** Dynamic changes in brain dopamine-β-hydroxylase activity during anterograde and retrograde reactions to injury of central noradrenergic neurons, *Brain Res.*, 57, 307, 1973.
60. **Aletta, J. M. and Goldberg, D. J.,** Rapid and precise down regulation of fast axonal transport of serotonin in an identified neuron, *Science*, 218, 913, 1982.
61. **Wessels, N. K., Johnson, S. R., and Nuttall, R. P.,** Axon initiation and growth-cone regeneration in cultured motor neurons, *Exp. Cell Res.*, 117, 335, 1978.
62. **Mason, A. and Muller, J.,** Axon segments sprout at both ends: tracking growth with fluorescent D-peptides, *Nature (London)*, 296, 655, 1982.
63. **Llinas, R.,** The role of calcium in neuronal function, in *The Neurosciences Fourth Study Programme*, Schmitt, F. O. and Worden, F. G., Eds., MIT Press, Cambridge, Mass., 1979, 555.
64. **Anglister, L., Farber, I. C., Shakar, A., and Grinvald, A.,** Localization of voltage-sensitive calcium channels along developing neurites: their possible role in regulating neurite elongation, *Dev. Biol.*, 94, 351, 1982.
65. **Lasek, R. J. and Hoffman, P. N.,** The neuronal cytoskeleton, axonal transport and axonal growth, in *Cell Motility, Book C, Microtubules and Related Proteins*, Pollard, T. and Rosenbaum, J., Eds., Cold Spring Harbor Laboratory, Cold Spring Harbor, N.Y., 1975, 1021.
66. **Wujeck, J. R. and Lasek, R. J.,** Correlation of axonal regeneration and slow component *B* in two branches of a single axon, *J. Neurosci.*, 3, 243, 1983.

67. **Grinvald, A. and Farber I. C.,** Optical recording of calcium action potentials from growth-cones of cultured neurons with a laser microbeam, *Science*, 212, 1164, 1981.
68. **Meiri, H., Spira, M. E., and Parnas, I.,** Membrane conductance and action potential of a regenerating axonal tip, *Science*, 211, 709, 1981.
69. **Duce, I. R. and Keen, P.,** The formation of axonal sprouts in organ culture and their relationship to sprouting *in vivo*, *Int. Rev. Cytol.*, 66, 211, 1980.
70. **Carlsen, R. C., Kiff, J., and Ryugo, K.,** Suppression of the cell body response in axotomized frog spinal neurons does not prevent initiation of nerve regeneration, *Brain Res.*, 234, 11, 1982.
71. **Heiwall, P. O., Dahlstrom, A., Larson, P. A., and Booj, S.,** The intra-axonal transport of acetylcholine and cholinergic enzymes in rat sciatic nerve after various types of axonal trauma, *J. Neurobiol.*, 10, 119, 1979.
72. **O'Brien, R. A. D.,** Axonal transport of acetylcholine, choline acetyltransferase and cholinesterase in regenerating peripheral nerve, *J. Physiol.*, 282, 91, 1978.
73. **Frizell, M. and Sjostrand, J.,** Transport of protein, glycoproteins and cholinergic enzymes in regenerating hypoglossal nerves, *J. Neurochem.*, 22, 845, 1974.
74. **Bulger, V. T. and Bisby, M. A.,** Reversal of axonal transport in regenerating nerves, *J. Neurochem.*, 31, 1411, 1978.
75. **McQuarrie, I. G. and Grafstein, B.,** Protein synthesis and fast axonal transport in regenerating goldfish retinal ganglion cells, *Brain Res.*, 235, 213, 1982.
76. **Carmignoto, G., Finesso, M., Siliprandi, R., and Gorio, A.,** Muscle re-innervation. I. Restoration of transmitter release mechanism, *Neuroscience*, 8, 393, 1983.
77. **DiGiamberardino, L., Couraud, J.-Y., Hassig, R., and Gorio, A.,** Recovery of axonal transport of acetylcholinesterase in regenerating sciatic nerve precedes muscle re-innervation, in *Axoplasmic Transport in Physiology and Pathology*, Weiss, D. G. and Gorio, A., Eds., Springer-Verlag, Basel, 1982, 77.
78. **Couraud, J. Y., DiGiamberardino, L., Hassig, R., and Mira, J. C.,** Axonal transport of the molecular forms of acetylcholinesterase in developing and regenerating peripheral nerve, *Exp. Neurol.*, 80, 94, 1983.
79. **Politis, M. J., Ederle, K., and Spencer, P. S.,** Tropism in nerve regeneration *in vivo*. Attraction of regenerating axons by diffusible factors derived from cells in distal nerve stumps of transected peripheral nerves, *Brain Res.*, 253, 1, 1982.
80. **Longo, F. M., Manthorpe, M., Skaper, S. D., Lundburg, G., and Varon, S.,** Neurotrophic activities accumulate *in vivo* within silicone nerve regeneration chambers, *Brain Res.*, 261, 109, 1983.
81. **Bisby, M. A.,** Fast axonal transport of labelled proteins in sensory axons during regeneration, *Exp. Neurol.*, 61, 281, 1978.
82. **Forman, D. and Berenberg, R. A.,** Regeneration of motor axons in the rat sciatic nerve studied by labelling with axonally transport radioactive proteins, *Brain Res.*, 156, 213, 1979.
83. **Griffin, J. W., Price, D. L., Drachman, D. B., and Morris, J.,** Incorporation of axonally transported glycoproteins into axolemma during nerve regeneration, *J. Cell Biol.*, 88, 205, 1981.
84. **Feldman, E. L., Axelrod, D., Schwartz, M., Heacock, A. M., and Agranoff, B. W.,** Studies on the localization of newly added membrane in growing neurites, *J. Neurobiol.*, 12, 591, 1981.
85. **Pfenniger, K. H. and Mayliepfenniger, M. F.,** Lectin labeling of sprouting neurons. II. Relative movement and appearance of glycoconjugates during plasmalemmal expansion, *J. Cell Biol.*, 89, 547, 1981.
86. **Tessler, A., Autiliogambetti, L., Gambetti, P.,** Axonal growth during regeneration — a quantitative autoradiograpic study, *J. Cell Biol.*, 87, 1197, 1980.
87. **Benowitz, L. I., Sashoua, V. E., and Yoon, M. G.,** Specific changes in rapidly transported proteins during regeneration of the goldfish optic nerve, *J. Neurosci.*, 1, 300, 1981.
88. **Bisby, M. A.,** Changes in the composition of labelled protein transported in motor axons during their regeneration, *J. Neurobiol.*, 11, 435, 1980.
89. **Levine, J., Skene, P., and Willard, M.,** GAPs and fodrin: novel axonally transported proteins, *Trends Neurosci.*, 4, 273, 1981.
90. **Perry, G. W. and Wilson, D. L.,** Protein synthesis and axonal transport during nerve regeneration, *J. Neurochem.*, 37, 1203, 1981.
91. **Theiler, R. F. and McClure, W. O.,** Rapid axoplasmic transport of proteins in regenerating sensory nerve fibres, *J. Neurochem.*, 31, 433, 1978.
92. **Skene, J. H. P. and Willard, M.,** Axonally transported proteins associated with axon growth in rabbit central and peripheral nervous systems, *J. Cell Biol.*, 89, 96, 1981.
93. **Redshaw, J. D. and Bisby, M. A.,** Fast axonal transport in central nervous system and peripheral nervous system axons following axotomy, *J. Neurobiol.*, 15, 109, 1984.
94. **Redshaw, J. D. and Bisby, M. A.,** Fast transported proteins within regenerating and non-regenerating peripheral sensory axons, *Proc. Can. Fed. Biol. Soc.*, 26, 102, 1983.
95. **Tashiro, T., Kasai, H., and Kurokawa, M.,** A calmodulin-related polypeptide rapidly migrates within the mammalian nerve, *Biomed. Res.*, 1, 292, 1980.

96. **Hoffman, P. N. and Lasek, R. J.,** Axonal transport of the cytoskeleton in regenerating motor neurons: constancy and change, *Brain Res.*, 202, 317, 1980.
97. **Lasek, R. J.,** The dynamic ordering of neuronal cytoskeletons, *Neurosci. Res. Prog. Bull.*, 19, 7, 1981.
98. **Bisby, M. A.,** Prolonged alterations in composition of fast-transported protein in axons prevented from regenerating after injury, *J. Neurobiol.*, 13, 377, 1982.
99. **Devor, M., Schonfeld, D., Seltzer, Z., and Wall, P. D.,** Two modes of cutaneous re-innervation following peripheral nerve injury, *J. Comp. Neurol.*, 185, 211, 1979.
100. **Sumner, B. E. H.,** Responses in hypoglossal nucleus to delayed regeneration of transected hypoglossal nerve, a quantitative ultrastructural study, *Exp. Brain Res.*, 29, 219, 1977.
101. **Watson, W. E.,** Cellular responses to axotomy and to related procedures, *Br. Med. Bull.*, 30, 112, 1974.
102. **Gordon, T. and Stein, R. B.,** Time-course and extent of recovery in re-innervated motor units of cat triceps surae muscles, *J. Physiol.*, 323, 307, 1982.
103. **Friede, R. L. and Samorajski, T.,** Axon calibre related to neurofilaments and microtubules in sciatic nerve fibres of rats and mice, *Anat. Rec.*, 167, 379, 1970.
104. **Sinicropi, D. V. and McIlwain, D. L.,** Changes in amount of cytoskeletal proteins within the perikarya and axons of regenerating frog motoneurons, *J. Cell Biol.*, 96, 240, 1983.
105. **Hoffman, P. N., Griffin, J. W., and Price, D. L.,** The role of slow component a in determining axon calibre: changes during regeneration, *Neurosci. Abstr.*, 6, 93, 1980.
106. **Gutmann, E. and Sanders, F. K.,** Recovery of fibre numbers and diameters in the regeneration of peripheral nerves, *J. Physiol.*, 101, 489, 1943.
107. **Devor, M. and Govrin-Lippimann, R.,** Maturation of axonal sprouts after nerve crush, *Exp. Neurol.*, 64, 260, 1979.
108. **Tonge, D. A.,** Physiological characteristics of re-innervation of skeletal muscle in the mouse, *J. Physiol.*, 241, 141, 1974.
109. **Gorio, A., Carmignoto, G., Finesso, M., Polato, P., and Nunzi, M. G.,** Muscle re-innervation. II. Sprouting, synapse formation and repression, *Neuroscience*, 8, 403, 1983.
110. **O'Brien, R. A. D., Ostberg, A. J. C., and Vrbova, G.,** Observations on the elimination of polyneuronal innervation in developing mammalian skeletal muscle, *J. Physiol.*, 282, 571, 1978.
111. **Sanes, J. R., Marshall, L. M., and McMahan, U. J.,** Re-innervation of muscle fibre basal lamina after removal of myofibres. Differentiation of regenerating axons at original synaptic sites, *J. Cell Biol.*, 78, 176, 1978.
112. **Watson, W. E.,** The response of motor neurons to intramuscular injection of botulinum toxin, *J. Physiol.*, 202, 611, 1969.
113. **Hopkins, W. G., Brown, M. C., and Keynes, R. J.,** Nerve growth form nodes of Ranvier in inactive muscle, *Brain Res.*, 222, 125, 1981.
114. **Czeh, G., Galego, R., Kudo, N., and Kuno, M.,** Evidence for the maintenance of motoneuron properties by muscle activity, *J. Physiol.*, 281, 239, 1978.
115. **Eisen, A. A., Carpenter, S., Karpati, G., and Bellavance, A.,** The effect of muscle hyper- and hypo-activity upon fibre diameters of intact and regenerating nerves, *J. Neurol. Sci.*, 20, 457, 1973.
116. **Gallego, R., Kuno, M., Nunez, R., and Snider, W. D.,** Dependence of motoneuron properties on the length of immobilized muscle, *J. Physiol.*, 219, 179, 1979.
117. **Brown, M. C., Holland, R. L., and Hopkins, W. G.,** Motor nerve sprouting, *Ann. Rev. Neurosci.*, 4, 17, 1981.
118. **Henderson, C. E., Huchet, M., and Changeux, J. P.,** Denervation increases a neurite promoting activity in extracts of skeletal muscle, *Nature (London)*, 302, 609, 1983.
119. **Slack, J. R. and Pockett, S.,** Motor neurotrophic factor in denervated adult skeletal muscle, *Brain Res.*, 247, 138, 1982.

Chapter 19

THE DESTINATION AND MODIFICATION OF FAST-TRANSPORTED PROTEINS IN NORMAL AND REGROWING NERVE

Bruce Tedeschi and David L. Wilson

TABLE OF CONTENTS

I. INTRODUCTION

The materials moving at different rates in axonal transport appear to consist of distinct subcellular entities.[1] In the case of the most rapidly transported materials (400 mm/day in mammals;[2] the rate varies with temperature in poikilothermic animals),[3] the components are membrane or membrane-bound materials.[4-9] There are two important functional implications for fast-transported materials. First, the materials conveyed in fast-transport probably play a major role in rapid adjustments to environmental changes. For example, some enzymes important for neurotransmitter synthesis are fast-transported[10,11] and the levels of such enzymes can be regulated by nerve cell activity.[12-14] Second, fast-transport probably provides the materials necessary for replenishment of axolemma. It is clear that these two functions are important when one considers the regenerative response of nerve. Successful regeneration requires a set of rapidly occurring events (signaling to nerve cell body, growth and extension of axolemma, and release of factors from axon to modulate the environment of the growing axon). In this review, we look at the role that fast axonal transport of protein might play in nerve regeneration, emphasizing the work in our lab which suggests changes in certain fast-transported proteins at local regions of nerve regeneration.

A. Possible Roles for Fast-Transported Materials in the Damaged, Regenerating Nerve

Following damage of the peripheral nerve and disconnection from its target tissue, several events occur in the first few weeks:[15—19]

1. The region of axon distal to the damage undergoes dissolution and is phagocytized by invading cells (Wallerian degeneration)
2. Glia proliferate and Schwann cell basement membranes may form a guidance path for the eventual regrowing tips
3. Growth cones emanate from the severed axon tips and the axon begins to regrow
4. As the regrowing axon advances it becomes myelinated by the surrounding Schwann cells

It is clear that extensive interactions must occur between the axon and its local nerve environment during such processes. Materials conveyed by fast-transport could have a number of diverse roles to play, both within the axon and between the axon and its extracellular nerve environment, during the early stages of nerve regeneration.

B. Response of Nerve Cell Body to Injury

It is thought that the cell body plays a major role in successful regeneration of nerve.[11,20] One idea has been that injury elicits (via the presence or absence of a signal) the synthesis of new materials, required for the regenerative response, for export down the axon.[21] This model seemed reasonable since major morphological responses have been observed in the soma subsequent to axotomy.[22,23] Some manifestations of this axon reaction (e.g., dispersal of Nissl bodies (chromatolysis), assymetric movement of nucleus, etc.) suggest modifications in RNA and protein synthesis. This model has been supported by several studies which have shown an increase in RNA and protein synthesis in the cell body following axotomy.[24-29] However, several observations have called into question whether changes in protein synthesis in the cell body play a key role in the regenerative response of axotomized nerve. First, the axon reaction of the soma does not occur in all cases of axotomy, and successful regeneration has been observed without chromatolysis.[20,28] Second, not all studies have found an increase in incorporation of amino acid precursors into protein following axotomy.[30-33]

Further, when individual proteins are examined in the cell body following axotomy, using two-dimensional gel electrophoresis, the results have shown, at least in some nerves, that:

(1) no synthesis of new protein is observed, (2) a minority of the detected proteins change their rate of synthesis (some increase, some decrease), and (3) the effects are minor in most of those proteins that do change their synthesis rates.[30,31] These results do not eliminate a role for the cell body in the regenerative response. It is taken for granted that most of the materials required for axon regrowth are supplied by the cell body, but most of these materials may be supplied on a continuing basis, even in undamaged nerve. Since radical morphological reorganization occurs at growing axon tips, one might suspect that biochemical events occurring after protein translation might also play an important role in mediating and defining successful regeneration.

C. Response of Fast-Transported Proteins to Axotomy

The fast-transported proteins represent minor species synthesized in the cell body.[34] In fact, it is difficult to identify most of them in the cell body when newly synthesized proteins are examined on two-dimensional gels. Other protein species are synthesized in much greater abundance. It is possible, therefore, that the proteins destined for fast-transport could undergo a dramatic response following axotomy without being detected in the cell body. However, the results, using two-dimensional gel electrophoresis (2D-PAGE), do not show a dramatic change in most proteins in fast-transport following axotomy but do show quantitative differences in a few species.

Skene and Willard have found a small group of fast-transported polypeptides which are transported in elevated amounts following axotomy.[35,36] One of these growth associated proteins (called GAP-43) is elevated in transport in crushed toad optic nerve following a brief lag period.[35] It remains elevated for a period of time and then returns to control levels at about the time visual function is restored. A fast-transported protein with the same molecular weight (43,000 daltons) and isoelectric point (pI = 4.3) as GAP-43 has been found to increase in the rabbit regenerating peripheral hypoglossal nerve but not in the nonregenerating rabbit optic nerve (except the immature, growing optic nerve).[36] These results suggest an association of this fast-transported protein with axon growth.

Perry and Wilson have done a quantitative analysis of fast-transported proteins in the frog sciatic nerve following axotomy.[30] They did not detect any large changes in the total amount of protein destined for fast-transport but did find abundance changes in some species in response to axotomy. These changes, however, might not be necessary for axon regrowth since no such quantitative responses were observed in the dorsal root ganglion (DRG) or sciatic nerve following damage to and regrowth of the dorsal root axons.[37]

The results of these and other studies suggest that some quantitative changes in a few fast-transported proteins are observed following axotomy. At least some axons can regrow without any detected changes. Thus, axon regrowth does not appear to require changes in the major materials being supplied by fast-transport. Thus, it is possible that most of the important changes during regeneration are taking place locally at the sites of axotomy and regrowth. Such changes might involve reorganization of existing materials. Some of these materials might require modifications. We have conducted a series of experiments which has attempted to detect protein modifications in the axotomized frog sciatic nerve.[38,39] We have detected the dramatic appearance of a fast-transported protein solely at regions of regeneration. The time course of its appearance suggests that it might be uniquely related to regeneration. We have also found the release of a small subset of fast-transported proteins from regions of regenerating nerve.[48] These proteins could be important in modulating the activities of the regrowing axons' environment during regeneration. These results suggest that some of the activities required for regeneration may necessitate the modification of materials or of their destinations locally, at the site of regrowth.

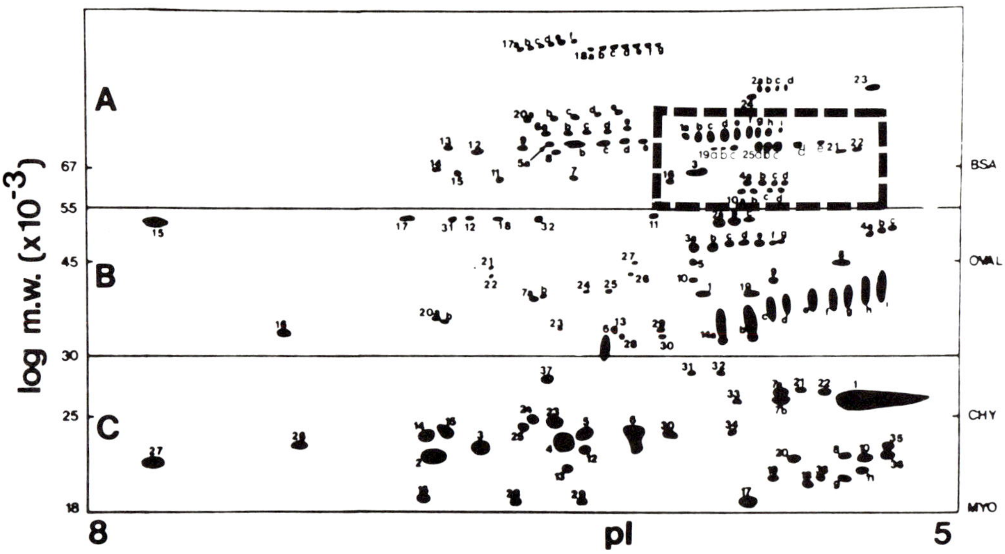

FIGURE 1. Map of the most abundant and consistently labeled (^{35}S-methionine) spots of fast-transported in frog sciatic nerve after their arrival at a crush site. The coordinate and numbering system has been discussed in previous publications.[30,34,37,39] Polypeptides are separated by isoelectric point (horizontally) from pI = 8 (left) to pI = 4.5 (right) and by molecular weight (vertically) from 200,000 (top) to 15,000 (bottom). The boxed area refers to that region where the protein. A25, is found. All other figures are enlargements of this region.

II. MODIFICATION AND DESTINATION OF FAST-TRANSPORTED PROTEINS AT LOCAL REGIONS OF NERVE REGENERATION

A. Post-Translational Modification of a Fast-Transported Protein at Regions of Nerve Regrowth

In our experiments, the bullfrog *(Rana catesbeiana)* sciatic nerve is crushed *in situ* and time is allowed for regeneration to take place. The sciatic nerves (crushed and undamaged, contralateral control), with the associated DRG, are removed for in vitro labeling of DRG proteins with radioactive amino acids. Following in vitro axonal transport, the labeled fast-transported proteins in nerve are separated by 2D-PAGE and examined by fluorography. The procedures have been previously described in detail.[30-34, 37-41]

In order to analyze some of the local biochemical responses to axotomy, we have allowed the fast-transported proteins to reach the sites of damage and regrowth. In the regenerating frog sciatic nerve, we have analyzed the ^{35}S-methionine-labeled, fast-transported proteins using two-dimensional gel electrophoresis. We have found a multiple-spotting protein on the gels which appears dramatically only at regrowing regions of nerve. Figure 1 shows a coordinate map of the most abundant fast-transported proteins in frog sciatic nerve. The polypeptide that we have found to increase at growing nerve regions is arbitrarily designated A25. A25 has a molecular weight of approximately 70,000 daltons and a pI of 5 to 6. Figure 2 gives a close-up view of A25 and its neighboring polypeptides on gels from a crushed (A) and control (B) nerve. The pattern of A25 is characterized by a triplet of spots directly above A4 and under the most acidic spots of A1. In many cases, one or two fainter spots which are more acidic than the triplet can be seen. It is difficult to evaluate the additional pair of basic spots which can be seen since this overlaps with the region where another polypeptide (A19) is located. A19 is usually seen in fluorographs of gels from control nerves but is not abundantly labeled. The multiple-spotting characteristic of A25 suggests it may be a glycosylated moiety. However, we have been unable to confirm this by testing with several radiolabeled sugars. Al, which has a somewhat larger apparent molecular weight than A25, is glycosylated.

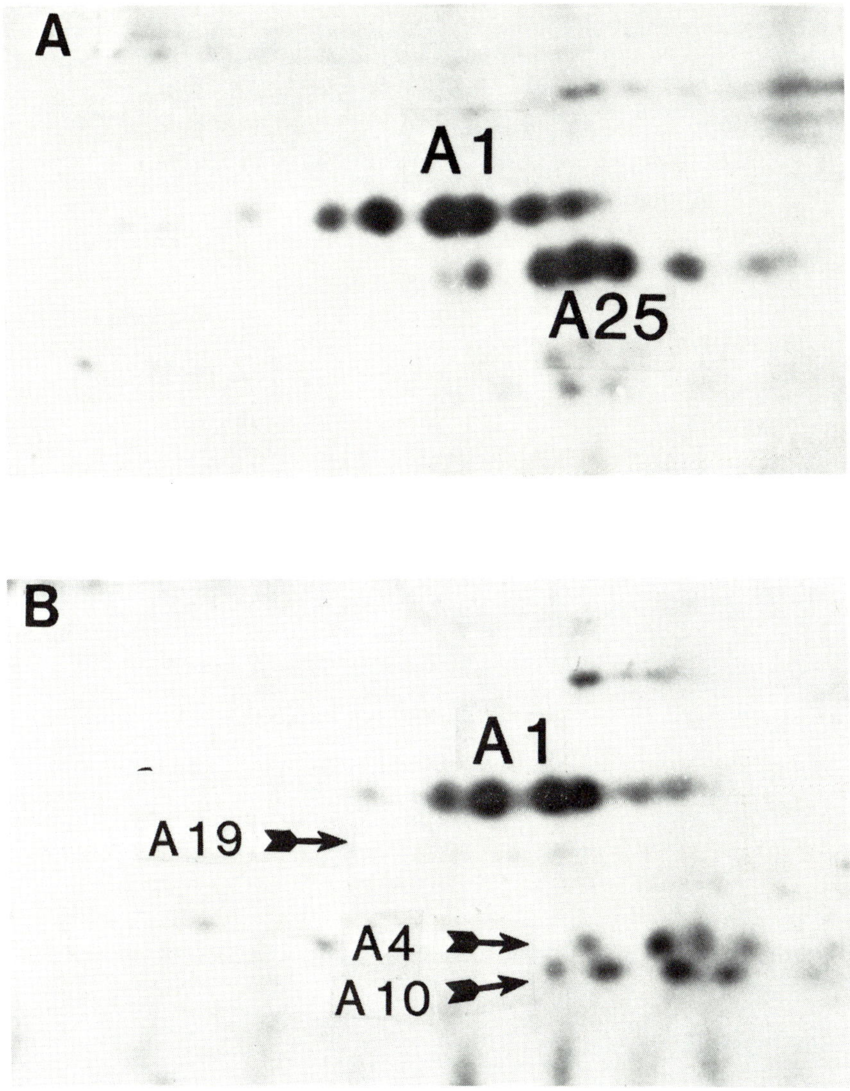

FIGURE 2. Fluorographs of the A25 gel region showing labeled tranported proteins from (A) nerve crush site and (B) ligated segment in contralateral control nerve. Frog was sacrificed 14 days after axotomy.

1. Distribution of A25 Along the Length of the Regenerating Nerve

A25 shows distinct differences in relative abundance when different points along the length of the regenerating nerve are examined (Figure 3). At regions of nerve proximal to a crush site, there is little or no detectable A25. A25 only appears at the crush site and regenerated regions distal to the crush site. It appears, therefore, that the presence of A25 is specific to regrowing nerve regions. We have made ligations at various distances proximal to crush sites and found either barely detectable or no A25 (Figure 4). We now have preliminary evidence that A25 may be concentrated at the growing tips themselves.[49]

2. Time of Appearance of A25

We initially determined that A25 only becomes dramatically present at regenerating tips between 5 and 6 days post-crush.[39] However, in recent studies using a new solubilization

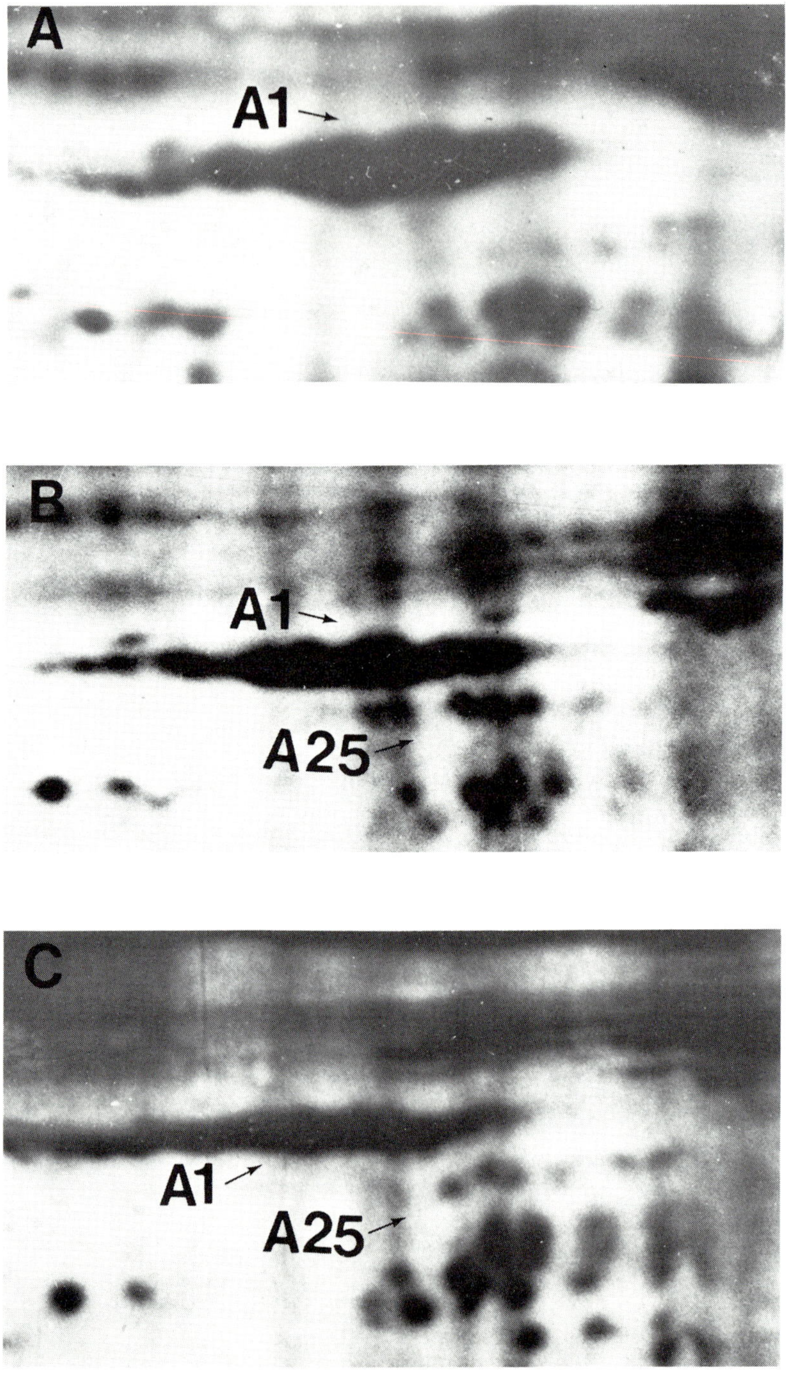

FIGURE 3. Distribution of A25 along the length of the regenerating nerve. In order to obtain significant regeneration for a spatial profile, frog was killed 24 days following nerve injury. (A) Nerve region immediately proximal to crush site. (B) Crush site. (C) Nerve region distal to crush site.

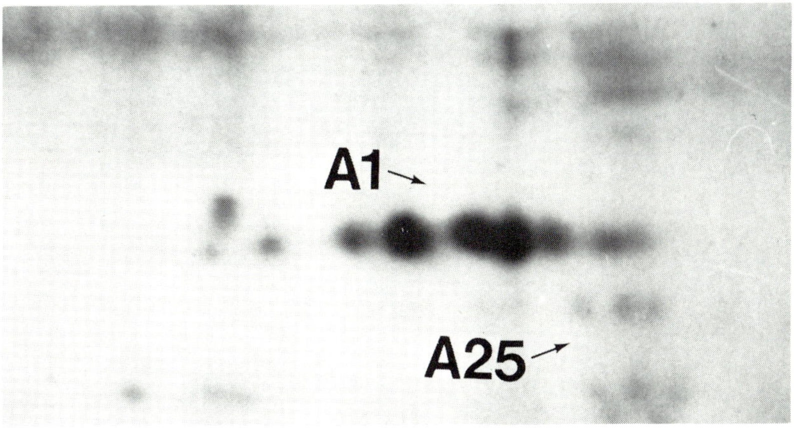

FIGURE 4. Fluorograph of A25 region after 2D-PAGE of proteins from proximal ligature on crushed nerve. Frog was sacrificed 14 days after crush. A ligature was placed on the nerve approximately 15 mm proximal to the crush site before labeling and transport. A small amount of labeling in A25 can be seen in the orthogradely fast-transported proteins at the ligature, before their arrival at the crush site. In two other similar experiments, no labeling of A25 was seen.

procedure we have shown that A25 appears much earlier than 5 days.[49] This early response of A25 at regenerating nerve tips may suggest a possible role in signalling to the cell body.

3. Retrograde Transport of A25

An interesting speculation is that biochemical responses occurring locally at sites of nerve injury or regrowth might provide information to the cell body about the state of the nerve. Assuming A25 to be a local response of the nerve to axotomy, we asked whether A25 is retrogradely transported to the cell body.[39] When the fast-transport wave of labeled materials reached a crush site, we made a ligation 15 mm proximal, to collect the retrograde flow of this material. Figure 5 shows that we collected significant amounts of retrogradely transported A25 at the proximal ligation.

The most likely explanation for the dramatic appearance of A25 at regions of regeneration is the post-translational modification of a precursor fast-transported protein. We have done several experiments to eliminate the possibilities of local labeling, local breakdown of labeled protein and reutilization of label, artifactual modification of a transported protein during homogenization/preparation for electrophoresis, and differential solubility.[39]

B. Release of a Subset of Fast-Transported Proteins from Regrowing Nerve

The growing axon appears to have some control over its immediate environment. One mechanism by which it could accomplish this is the secretion of materials from the growing axon. The presence or absence of such molecules could then modify the extracellular matrix and/or initiate some of the activities of the Schwann cells (proliferation, ensheathment, myelination) subsequent to axotomy.[15,42-45] Some of the fast-transported materials appear good candidates for such functions because they include membrane-bound entities.

A massive release of fast-transported proteins from intact frog sciatic nerve was previously reported,[46] but this observation was not confirmed when strict controls were used to avoid local incorporation of radiolabel.[47] However, when the release of radiolabeled fast-transported proteins from a nerve crush site was examined, we were able to obtain significant release of labeled protein into the surrounding bath.[48] We examined the transported proteins released into the bath using 2D-PAGE. We found that a small set of the fast-transported proteins are released into the surrounding bath. These released proteins do not represent the

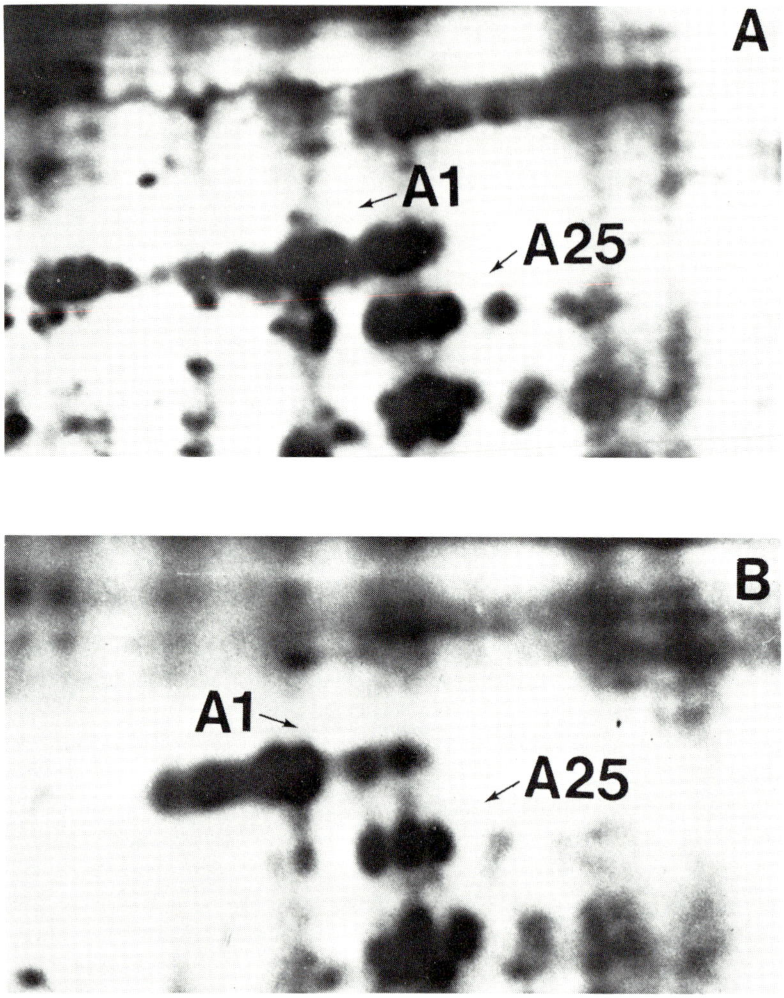

FIGURE 5. Retrograde transport of A25. Frog was sacrificed 16 days following nerve crush. Following the normal 24 hr of labeling and transport, a ligature was placed on the nerve approximately 15 mm proximal to the crush site. Transport was allowed to proceed for an additional 7 hr. The build-up in retrogradely transported material at the proximal ligature is shown in (A). In (B) is shown the pattern at the crush site.

most abundant of the fast-transported group. We determined from sugar-labeling experiments that many of these released proteins are glycosylated.

It is too early to evaluate the relevance of these findings to the regenerative process. It does not appear that these proteins are released as a general consequence of membrane or cell damage due to crush since we see the release at later post-crush times (3 to 21 days) but not in nerves soon after crush (1 to 2 days), nor in control nerves. Since many of the fast-transported proteins are probably the source of membrane for the maintenance and rebuilding of the axolemma, the insertion of some of these into growing axolemma could result in the release of soluble proteins carried within the inserted membrane vectors. Thus, the appearance of these proteins in bath could be just a consequence of other fast-transported protein being inserted into new membrane. On the other hand, knowing the importance of the axon for the organization of its local environment, some of these released proteins could function as important extracellular or trophic agents.

III. SUMMARY

The results of experiments in our laboratory and others are beginning to identify events occurring at local regions of nerve damage and regrowth that could play important roles in regeneration. This does not exclude a role for the cell body or more proximal regions of axon but suggests a division of functions. For example, the cell body likely provides the necessary precursor materials which are then specifically modified or reorganized at local regions of nerve to accomplish a given task. This allows greater diversification in the expression of genetic information in response to environmental change.

The regenerative response of nerve to injury involves a complex interplay of reorganizational events within the axon and between the axon and its nerve environment. The selective release and modifications of some fast-transported proteins could play a role in the reorganization.

ACKNOWLEDGMENTS

We gratefully acknowledge the support of NSF (BNS-8117817) and NIH (NS 18263) grants.

REFERENCES

1. **Lasek, R. J.,** Axonal transport: a dynamic view of neuronal structures, *Trends Neurosci.*, 3, 87, 1980.
2. **Ochs, S.,** Rate of fast axoplasmic transport in mammalian nerve fibers, *J. Physiol. (London)*, 226, 627, 1972.
3. **Edström, A. and Hanson, M.,** Temperature effects on fast axonal transport of proteins *in vitro* in frog sciatic nerves, *Brain Res.*, 58, 345, 1973.
4. **Droz, B., Koenig, H. L., and DiGiamberardino, L.,** Axonal migration of protein and glycoprotein to nerve endings. I. Radioautographic analysis of the renewal of protein in nerve endings of chicken ciliary ganglion after intracerebral injection of (^{3}H)lysine, *Brain Res.*, 60, 93, 1973.
5. **Bennett, G., DiGiamberardino, L., Koenig, H. L., and Droz, B.,** Axonal migration of protein and glycoprotein to nerve endings. II. Radio-autographic analysis of the renewal of glycoproteins in nerve endings of chicken ciliary ganglion after intracerebral injection of (^{3}H) fucose and (^{3}H)glucosamine, *Brain Res.*, 60, 129, 1973.
6. **DiGiamberardino, L., Bennett, G., Koenig, H. L., and Droz, B.,** Axonal migration of protein and glucoprotein to nerve endings. III. Cell fraction analysis of chicken ciliary gangion after intracerebral injection of labeled precursors of proteins and glucoproteins, *Brain Res.*, 60, 147, 1973.
7. **Cancalon, P. and Beidler, L. M.,** Distribution along the axon and into various subcellular fractions of molecules labeled with (^{3}H)leucine and rapidly transported in the garfish olfactory nerve, *Brain Res.*, 89, 225, 1975.
8. **Lorenz, T. and Willard, M.,** Subcellular fractionation of the intra-axonally transported polypeptides in the rabbit visual system, *Proc. Natl. Acad. Sci. U.S.A.*, 75, 505, 1978.
9. **Hammerschlag, R. and Stone, G. C.,** Membrane delivery by fast axonal transport, *Trends Neurosci.*, 5, 12, 1982.
10. **Brimijoin, S. and Wiermaa, M. H.,** Rapid axonal transport of tyrosine hydroxylase in rabbit sciatic nerves, *Brain Res.*, 121, 77, 1977.
11. **Grafstein, B. and Forman, D. S.,** Intracellular transport in neurons, *Physiol. Rev.*, 60, 1167, 1980.
12. **Zigmond, R. E.,** The long term regulation of ganglionic tyrosine hydroxylase by preganglionic nerve activity, *Fed. Proc. Fed. Am. Soc. Exp. Biol.*, 39, 3003, 1980.
13. **Thoenen, H., Mueller, R. A., and Axelrod, J.,** Trans-synaptic induction of adrenal tyrosine hydroxylase, *J. Pharmacol. Exp. Ther.*, 169, 249, 1969.
14. **Reis, D. J., Ross, R. A., Gilad, G., and Joh, T. H.,** Reaction of central catecholaminergic neurons to injury: model systems for studying the neurobiology of central regeneration and sprouting, in *Neuronal Plasticity*, Cotman, C. W., Ed., Raven Press, New York, 1978, 197.
15. **Lubínska, L.,** Early course of Wallerian degeneration in myelinated fibres of the rat phrenic nerve, *Brain Res.*, 130, 47, 1977.

16. **Spencer, P. S. and Weinberg, H. J.,** Axonal specification of Schwann cell expression and myelination, in *Physiology and Pathobiology of Axons,* Waxman, S. G., Ed., Raven Press, New York, 1978, 389.
17. **Bunge, R. P. and Bunge, M. B.,** Cues and constraints in Schwann cell development, in *Studies in Developmental Neurobiology,* Cowan, W. M., Ed., Oxford Univ. Press, New York, 1981, 322.
18. **Pfenniger, K. H. and Bunge, R. P.,** Freeze-fracturing of nerve growth cones and young fibers. A study of developing plasma membrane, *J. Cell. Biol.,* 63, 180, 1974.
19. **Pfenniger, K. H. and Rees, R. P.,** From the growth cone to the synapse: Properties of membranes involved in synapse formation, in *Neuronal Recognition,* Barondes, S. H., Ed., Plenum Press, New York, 1976, 131.
20. **Grafstein, B. and McQuarrie, I. G.,** Role of the nerve cell body in axonal regeneration, in *Neuronal Plasticity,* Cotman, C. W. Ed., Raven Press, New York, 1978, 155.
21. **Cragg, B. G.,** What is the signal for chromatolysis?, *Brain Res.,* 23, 1, 1970.
22. **Nissl, F.,** Ueber die Veranderungen der Ganglienzellen am Facialiskern des Kaninchens nach Ausreissung der Nerven, *Allg. Z. Psychiatr.,* 48, 197, 1982.
23. **Lieberman, A. R.,** The axon reaction: a review of the principal features of perikaryal responses to axon injury, *Int. Rev. Neurobiol.,* 14, 49, 1971.
24. **Murray, M.,** ^{3}H-uridine incorporation by regenerating retinal ganglion cells of goldfish, *Exp. Neurol.,* 39, 489, 1973.
25. **Watson, W. E.,** Observations on the nucleolar and total cell body nucleic acid of injured nerve cells, *J. Physiol. (London),* 196, 655, 1968.
26. **Brattgärd, S. O., Hydén, H., and Sjöstrand, J.,** Incorporation of orotic acid-^{14}C and lysine-^{14}C in regenerating single nerve cells, *Nature (London),* 182, 801, 1958.
27. **Miani, N., Rizzoli, A., and Bucciante, G.,** Metabolic and chemical changes in regenerating neurons. II. *In vitro* rate of incorporation of amino acids into proteins of the nerve cell perikaryon of the C. 8 spinal ganglion of rabbit, *J. Neurochem.,* 7, 161, 119.
28. **Murray, M. and Grafstein, B.,** Changes in the morphology and amino acid incorporation of regenerating goldfish optic neurons, *Exp. Neurol.,* 23, 544, 1969.
29. **Heacock, A. M. and Agranoff, B. W.,** Enhanced labelling of a retinal protein during regeneration of optic nerve in goldfish, *Proc. Natl. Acad. Sci. U.S.A.,* 73, 828, 1976.
30. **Perry, G. W. and Wilson, D. L.,** Protein synthesis and axonal transport during nerve regeneration, *J. Neurochem.,* 37, 1203, 1981.
31. **Hall, M. E., Wilson, D. L., and Stone, G. C.,** Change in synthesis of specific proteins following axotomy: detection with two-dimensional gel electrophoresis, *J. Neurobiol.,* 9, 353, 1978.
32. **Hall, M. E. and Wilson, D. L.,** Modification of protein synthesis by deafferentation of rat sympathetic ganglia, *Brain Res.,* 168, 414, 1979.
33. **Hall, M. E.,** Changes in synthesis of specific proteins in axotomized dorsal root ganglia, *Exp. Neurol.,* 76, 83, 1982.
34. **Stone, G. C., Wilson, D. L., and Hall, M. E.,** Two-dimensional gel electrophoresis of proteins in rapid axonal transport, *Brain Res.,* 144, 287, 1978.
35. **Skene, J. H. P. and Willard, M.,** Changes in axonally transported proteins during regeneration in toad retinal ganglion cells, *J. Cell Biol.,* 89, 86, 1981.
36. **Skene, J. H. P. and Willard, M.,** Axonally transported proteins associated with growth in rabbit central and peripheral nervous systems, *J. Cell Biol.,* 89, 96, 1981.
37. **Perry, G. W., Krayanek, S. R., and Wilson, D. L.,** Protein synthesis and rapid axonal transport during regrowth of dorsal root axons, *J. Neurochem.,* 40, 1590, 1983.
38. **Tedeschi, B. and Wilson, D. L.,** Modification of a rapidly transported protein in regenerating nerves, *Am. Soc. Neurochem. Abstr.,* 13, 225, 1982.
39. **Tedeschi, B. and Wilson, D. L.,** Modification of a rapidly transported protein in regenerating nerve, *J. Neurosci.,* 3, 1728, 1983.
40. **Wilson, D. L., Hall, M. E., Stone, G. C., and Rubin, R.W.,** Some improvements in two-dimensional gel electrophoresis of proteins: protein mapping of eurcaryotic tissue extracts, *Anal. Biochem.,* 83, 33, 1977.
41. **Wilson, D. L.,** Two dimensional polyacrylamide gel electrophoresis of proteins, in *Handbook of Neurochemistry,* Vol. 2, 2nd ed., Lajtha, A., Ed., Plenum Press, New York, 1982, 133.
42. **Weinberg, H. and Spencer, P.,** Studies on the control of myelinogenesis. I. Myelination of regenerating axons after entry into a foreign unmyelinated nerve, *J. Neurocytol.,* 5, 395, 1975.
43. **Wood, P. M. and Bunge, R. P.,** Evidence that sensory axons are mitogenic for Schwann cells, *Nature (London),* 256, 662, 1975.
44. **Salzer, J. L. and Bunge, R. P.,** Studies of Schwann cell proliferation . I. An analysis in tissue culture of proliferation during development, Wallerian degeneration, and direct injury, *J. Cell Biol.,* 84, 739, 1980.
45. **Bray, G. M., Rasminsky, M., and Aguayo, A. J.,** Interactions between axons and their sheath cells, *Ann. Rev. Neurosci.,* 4, 127, 1981.

46. **Hines, J. F. and Garwood, M. M.,** Release of proteins from axons during axonal transport: an *in vitro* preparation, *Brain Res.*, 125, 141, 1977.
47. **Tedeschi, B., Wilson, D. L., Zimmerman, A., and Perry, G. W.,** Are axonally transported proteins released from sciatic nerves?, *Brain Res.*, 211, 175, 1981.
48. **Tedeschi, D. and Wilson, D. L.,** Release of fast-transported proteins from regenerating nerve, *Soc. Neurosci. Abstr.*, 8, 48, 1983.
49. **Perry, G. W., Tedeschi, B., and Wilson, D. L.,** Early appearance of A24 (a modified, rapidly transported polypeptide) in frog sciatic nerve following damage, and the effects of a conditioning lesion, *Soc. Neurosci. Abstr.*, 10, 420, 1985.

Chapter 20

THE MECHANISM OF AXOPLASMIC TRANSPORT

Dieter G. Weiss

TABLE OF CONTENTS

I. INTRODUCTION

This chapter is meant to summarize and to discuss our present knowledge on the general questions with regard to the axoplasmic transport mechanism:

1. How is the transport in axons organized?
2. Which properties are acceptable on the basis of experimental data?
3. What are the possible modes of force generation?

Lacking detailed knowledge of the mechanisms underlying the phenomenon of axoplasmic transport, we shall call axoplasmic transport all movement of molecules and particles inside nerve cell processes, which cannot be related to diffusion or Brownian movement. Whether one, unitary, mechanism can account for all aspects of this phenomenon, or several mechanisms are involved, remains to be clarified. The term mechanism wherever used in this article is therefore meant to include as many different mechanisms as might be needed ultimately to account for all of the phenomena.

In order to evaluate the feasibility of the various models of axoplasmic transport we need to discuss to what extent they are compatible with the experimentally defined properties of axoplasmic transport. The properties necessary for such an evaluation have been reviewed in the various chapters of this book as well as in earlier books and articles.[1-4]

The question of whether material is moved through a stationary axoplasm or transported as part of a streaming medium poses a general problem to our understanding of the transport phenomenon. A theoretical choice of one of these two alternatives should be of substantial help in data interpretation and experimental strategy.

Most of the topics in this article can be discussed irrespectively of the type of mechanism involved. However, when treating the energy requirement, the identity of the mobile compartment, and the type of force generation, we have to consider the two possible modes of transport separately.

In an attempt to discriminate between these two main transport modes and perhaps other possibilities, we employed well documented experimental results and the limitations imposed by the laws of physics. Wherever the experimental data are insufficient we have tried to close the gaps by theoretical considerations.

II. VISCOSITY

The choice between various types of transport modes can be based on energetic feasibility considerations. Energy limitations may be imposed by the energy required to bring about and to maintain movement in viscous systems. We therefore, start with a discussion of the problems related to viscosity and energy consumption.

As discussed by Rubinson and Baker,[5,6] viscosity is a property of fluid, the quantitative description of which depends greatly on the size of the probe used for measurement. Viscosity values in cytoplasm have been measured by a number of investigators. A comparison of the values[7] clearly demonstrated the effect of measurement techniques on the values obtained. The small molecular weight probes (determination of microviscosity), used to electron spin resonance and diffusion measurements, consistently yield viscosity values close to that of water (1 cP), whereas the determination of macroviscosity employing measurement of particulate components or capillary measurement of the whole axoplasm gives remarkably high values of viscosity. The discrepancy between the macroviscosity obtained by Biondi et al.[8] and by Rubinson and Baker[5] is probably caused by the use of different measurement techniques as well as by the different systems studied. Although both laboratories employed the capillary flow method, only the latter study attempted to provide the proper ionic environment for the extruded axoplasm. The value of 14,600 cP may therefore be considered a good approximation for the bulk axoplasm viscosity in giant axons, although smaller axons may display much higher macroviscosities. The macroviscosities are the result of the density, degree of cross-linking, and rigidity of the cytoskeletal elements.

On the whole, amino acids encounter much lower viscosities than proteins moving through the same subcellular region, because the former are hindered to a lesser extent by the local microstructure of the cytoplasm. Moreover, the microstructure may not be uniform throughout the axon. Such regional differences would exert a substantial influence on the mobility of molecules. Furthermore, the microstructure at a specific locus may change with time due to a variety of physical and chemical factors such as organelle movement, pH changes, variation in the ionic milieu, or enzymatic activity. Viscosity is therefore difficult to assess in discrete subcellular regions, which makes the exact theoretical analysis of the axoplasmic transport phenomenon rather difficult.

Organelles in transit are affected only by the macroviscosity of axoplasm. The measurement of macroviscosities along the longitudinal axis of the axon does not provide information about the regional (or radial) macroviscosities.[5] However, the inhomogeneous arrangement of cytoskeletal elements, especially the existence of microtubule and neurofilament domains,[9-13] implies that the concept of regional viscosity differences requires a serious consideration (Reference 7).

III. ENERGY

Axoplasmic transport is an active process driven by ATP which is derived from oxidative phosphorylation[14,15] (for review see Reference 3).Diffusion does not contribute significantly to the transport of material.[16] This also applies to small molecules.[17] In order to decide which of a number of modes of force generation proposed are energetically feasible, we have to compare their energy requirement with the energy available.

A. Transport of Organelles Through Stationary Cytoplasm

Almost all hypothetical transport mechanisms suggest that the transported material is moved through a stationary cytoplasm.[18-25] Cytoplasm is, however, known to be a gel of high macroviscosity as already discussed above. It is, therefore, reasonable to first consider the amount of particulate material that can be moved through stationary axoplasm on the basis of estimates of the energy available.

The energy available to the transport mechanism in the form of ATP in a nerve fiber was also estimated. An upper limit of 68,800 available ATP molecules was calculated for the 4.3 sec it takes a vesicle to saltate a distance of 10 μm at a velocity of 300 mm/day.[26]

The thermodynamic efficiency of this process is unknown, but since that of muscle contraction, another motile phenomenon, ranges from near 0 to a maximum of 50%,[27] these authors suggested an average value of 25% as a first estimate.[26] As the outcome of a study of this kind a nomograph which gives the energy values needed to move one or more model vesicles with various velocities for a distance of 10 μm through cytoplasm of various viscosities has been published (see Figure 1 in Reference 26).

The energy available to the transport mechanism in the form of ATP in a nerve giver was also estimated. An upper limit of 68,800 available ATP molecules was calculated for the 4.3 sec it takes a vesicle to saltate a distance of 10 μm at a velocity of 300 mm/day.[26]

At the present stage of analysis it can be stated that these calculations do not rule out transport through viscous axoplasm if only one 80 nm vesicle were to be moved. However, it should be recognized that a mitochondrion may use the energy equivalent of over twenty 80 nm vesicles. Such a movement would not be possible through a medium of a macroviscosity of 14,600 cP and the simultaneous movement of other material would also be unrealistic. It is therefore apparent that viscosities above 1,000 cP stretch the limits of the available energy and make transport through such a medium energetically very unattractive and probably even impossible. This suggests that axoplasmic transport has to occur in microregions that have a viscosity below that of gelled axoplasm.

B. Carrier Stream Movement

Other models for the transport mechanism[12,28,29] propose that part of the axoplasm is in a streaming state and thus conveys the organelles as well as the other material which is to be transported. Estimates of the amount of cytoplasm streaming and the dimensions of such hypothetical streams have been presented on the basis of theoretical considerations and morphological data.[7,12,26,30] For reasons discussed later in this article, such microstreams can only be envisaged to exist in annular regions of about 100 nm diameter around single or clustered microtubules. The energy required to maintain streaming in these regions has been estimated and compared with the energy available in axons.[30]

The viscosity of the carrier stream is unknown. In the study discussed here[30] a nearly total solation of the cytoplasm in the carrier streams is assumed. Moreover, the zones of exclusion around microtubules (see below), where the streaming has been hypothesized to occur, are devoid of neurofilaments. The authors contended that the viscosity in these regions approaches that of the microviscosities measured. As an approximation they have therefore chosen 10 cP as the average viscosity of the carrier stream.[30]

The energy required to move fluid over a distance of 1 μm in one circumtubular annulus, as described above, was estimated as 20.3×10^{-12} erg. To obtain the energy required per second, one has to multiply by the velocity of 2.55 μm/sec which yields 51.8×10^{-12} erg/sec compared to the 53.4×10^{-12} erg/sec of available energy as calculated from ATP levels.[30] It should also be noted that it may not be necessary that *all* circumtubular regions are streaming all the time, as was assumed in these calculations.

Therefore, carrier stream movement is also feasible if the streaming portion of the axoplasm is of low viscosity and/or if only a fraction of the regions is moving at a given time. The correspondence of the orders of magnitude between energy required and energy available, especially in view of the rather conservative assumptions made, forces us to consider the streaming mode of transport seriously and compare it to the other transport modes.

IV. REDUCTION OF VISCOSITY

As shown above there is a severe shortage of energy for both the movement of particles through viscous axoplasm and for the maintenance of streaming in high viscosity regions. Therefore, we should search for possibilities to lower the viscosity and/or for modes of force generation which avoid the viscosity problem to a degree where the already well documented movement of larger organelles is energetically feasible.

The high intracellular macroviscosity is apparently due to the extensive network of cross-linked cytoskeletal elements such as microtubules (25 nm diameter), neurofilaments (9 to 10 nm diameter), and perhaps microfilaments (6 to 7 nm diameter) present in axons.[11,12,31-33]

Macroviscosities could be much lower if measured parallel to the cytoskeletal fibers as compared to measurements made across these filamentous arrays. The longitudinal macroviscosity, e.g., in bundles of microtubules may be as low as to approach microviscosity values. In this case there would no longer be a need for a reduction of viscosity to allow transport. The determination of macroviscosities by the capillary method as employed by Rubinson and Baker[5] is a procedure which measures mainly in a longitudinal direction. However, since the entire axoplasm is squeezed through the capillary, no direct information can be obtained from these experiments on viscosities in subregions, such as microtubule bundles, within the axon. Until direct measurements are available, we have to use these high values which force us to consider mechanisms of viscosity reduction.

Three-dimensional gel networks are formed by the cross-linking of proteins that have more than two reactive groups and are therefore well suited for gel formation. On the basis of volume change during gelation Freundlich defined three classes.[35] Type 1 undergoes no volume change and has no temperature coefficient; type 2 undergoes an exothermic volume decrease; and type 3 increases in volume and absorbs heat.

Proteins in the cytoplasm form gels.[36-38] However, the determination of the type of gel formed has presented serious experimental difficulties. Furthermore, it is not at all clear whether the simple classification mentioned above is sufficient to explain the behavior of all cellular gels. Nevertheless, a type 3 behavior has been identified.[36,39] Type 1 gels have also been found in the cytoplasm of animal cells,[40] slime mold,[41] synovial fluid,[42] solutions of egg albumin,[43] and cortical gel of *Amoeba proteus*.[44]

A type 1 gel is also called a thixotropic gel, the characteristics of which are of special importance to our search for low viscosity regions. Thixotropic gels decrease in viscosity with time despite fixed rates of shear and form a low viscosity sol for as long as the shearing force is maintained.[45] The network structure is reformed when the agitation is stopped. Actually, most biogels show some thixotropic behavior[34] with the degree of thixotropy depending on protein concentration and electrolyte composition. Baker et al.[46] observed thixotropic behavior in squid axoplasm. Agitation, i.e., movement of or in thixotropic cytoplasm, acts to produce low viscosity regions as an inevitable byproduct of local shear forces. The movement itself, therefore, can reduce the viscosity along the tracks of particle or fluid transport.

In addition it is possible that the microtubules situated in a neurofilament array may weaken the cytoskeletal cross-linkage, since the microtubules appear to have fewer or weaker cross-linking sites. Tighter linkage is expected where interfilament distances are smaller, which is the case for neurofilaments (see Figure 3 in Reference 26).

Microtubule regions may therefore weaken the lattice and can be compared to a defect in a crystal where structural breakage is most likely to occur. In other words, organelle movement in regions of reduced viscosity — regardless of how it is powered — might be expected to be easier in the vicinity of microtubules.

Finally, thixotropy or a similar process is not the only way to achieve low viscosity regions. Controlled enzymatic reactions and solation factors could also play an important role. Calcium-activated proteases, which can break down neurofilaments, have been found in axoplasm.[47,48] If elevated calcium levels can be generated in specific regions of the axon, neurofilaments can be enzymatically degraded or at least destabilized to a degree that local pressure or shear forces can accomplish a gel-sol transition. Gelation and solation factors have been shown to exist in many types of cells affecting cytoskeletal cross-linking with a concomitant effect on viscosity (e.g., References 37, 49 to 52). Similar mechanisms can be expected to influence the axonal microtubules' state and interactions.

V. LOCUS OF LOW VISCOSITY

The physical principles already discussed should allow us to choose a likely locus for the low viscosity regions where rapid transport is suggested to take place. There are only four subcellular loci that can be considered: the plasma membrane with the submembraneous region, the neurofilament domains, the microtubule domains, and the tubular structures of the endoplasmic reticulum. The fact that microtubules are often excluded from the neurofilament regions (or vice versa) has only recently been generally recognized (see References 7, 12, and 53), although it was already apparent in a number of electron micrographs published over the years. This has been shown for myelinated axons[9-11,31,54-58] as well as for unmyelinated fibers.[59,60] An example of an axon displaying a clear-cut separation of microtubule domains from neurofilament domains is shown in Reference 26. In this particular axonal cross section, the microtubule domains take up 40% of the area and the neurofilament region 60%. The average distance between microtubules is 107 ± 24 nm (halo regions) whereas the neurofilaments are separated by 49 ± 12 nm. Since the distance between microtubules is greater and the degree of possible cross-linkage by side arms appears to be less than in the case of the neurofilament domains, it should be the microtubule domains where a low viscosity region can be expected. This is supported by recent electron microscopic studies on negatively stained material where the author concludes that the projections on neuronal microtubules may have no cross-linking function.[61] A similar view is presented by Sasaki et al.[62] who suggest that the microtubules with their halos are not part of the cross-linked cytoplasmic lattice but represent a separate cytoskeleton with the high molecular weight microtubule-associated proteins serving as noncross-linking spacers.

Microtubules are frequently surrounded by clear zones with a diameter of about 100 nm called halos by Ledbetter and Porter.[63] These apparently represent one form of zones of exclusion, i.e., cellular regions of cytoplasm free of identifiable cell components.[64] These regions seem to be especially prominent in axons and in other cells that display active intracellular transport[59,60,62,65,66] and can also be seen in freeze fracture replicas.[67] In invertebrates[68] as well as in vertebrates[57,69,70] microtubules have been found to traverse cisternae of smooth endoplasmic reticulum passing through holes of a size comparable to that of the halos. Furthermore, in the constricted axonal segments at the nodes of Ranvier, the discontinuity of neurofilaments is a common feature,[11,71] whereas microtubules are present and continuous. Therefore, morphological considerations alone strongly suggest that the microtubular halos are the best and possibly the only region, where a rapid movement of cytoplasm and organelles can occur.

Several light microscopic studies[72-76] show essentially independent movement of organelles along invisible tracks or channels within the axoplasm which are sometimes spiraled, thus causing organelle movement in and out of the focus of the microscope. The plasma membrane and the region underlying it which is lined with actin filaments[12] can therefore be eliminated as a locus for such movement.

The endoplasmic reticulum can also be ruled out as a channel since it cannot carry organelles within its lumen and is too irregular to be considered as the seat of a force generating mechanism (see Reference 29). It should be mentioned that these tubulovesicular organelles are believed to constitute the majority of the rapidly transported material.[10,77-80] Note, however, that Ellisman described a smooth cisternal system apparently not undergoing axoplasmic transport.[81]

In light of all these considerations, we are again led to the conclusion that all or part of the microtubule domains represent low viscosity transport regions embedded in a more viscous neurofilament gel.

We cannot expect the low viscosity domains to be static in time or with distance. Domains may merge, shift in position, or change shape. Large moving organelles, such as mitochondria, may have great influence on the position of microtubules and on the viscosity within the domains. This has already been directly shown for the cytoplasmic transport in protozoa, where Allen et al.[82] using highest resolution light microscopy were able to see vigorous microtubule translocations accompanying the passage of organelles. The close association of numerous microtubules with mitochondria[11,57,83-86] supports this concept and suggests that the microtubules can be shifted from their positions when a large organelle is squeezed between them (see Reference 62). Furthermore, the creation of single superdomains of microtubules is possible without destroying the force generating mechanism.[83,87,88]

Finally, the definitive proof for the microtubular location of organelle transport was achieved by using AVEC-DIC microscopy.[89-91] These studies clearly showed transport along single microtubules both inside cells and on single native microtubules derived from squid giant axons.

VI. LOCUS AND STATE OF TRANSPORTED MATERIAL

In this chapter our knowledge of the state and of the location of the transported material in the axon is reviewed. The term moving compartment will be used for this material. This moving compartment has been suggested to be a moving organelle,[20-24] the transport organelle or transport package,[18] a transport filament with attached material,[3,19] or a streaming fraction of the cytoplasm,[28,29] depending on the various models.

It is known from several theoretical as well as experimental studies that the conveyed material is reversibly associated with the transport system, that it is able to exist in a mobile and in a stationary state, and that part of this material is deposited during its passage through the axon.[92-97] The terms mobile phase and stationary phase will subsequently be used for the two phases between which the material partitions. It should be noted that the terms moving compartment and mobile phase are synonyms for the microstream model[28,29] whereas this is not necessarily the case for the other models.

A. The Mobile Phase

Autoradiographic studies attempting to directly demonstrate the axonal location of rapidly transported material have so far yielded a variety of results.[98-101] This is probably due to the fact that in some of these studies substantial amounts of material may have already left the rapid transport system to become associated with stationary or slowly moving phases by the process of axonal deposition or retention.[83,92-94] To obtain autoradiographs showing a maximal amount of material in the transport compartment one must sample the axon at the peak of the moving isotope distribution. This has only recently been attempted for rapid transport.[80,88,102]

In axons treated with the drug IDPN (β,β′-iminodipropionitrile) a very large microtubule superdomain incorporating all the individual microtubule domains which are normally scattered in the axoplasm is formed.[83,86-88] In these altered axons, rapid transport continues virtually unimpaired.[87] In autoradiographs of axons containing rapidly transported radioactive material, the label is found in the microtubule superdomain.[88,102] These studies strongly support all concepts of circumtubular flow and/or of transport along microtubules (see References 12, 28, 73, 103 to 105).

Moreover, the careful studies correlating light microscopic observations with electron microscopic analyses[10,75,104] clearly demonstrate the location of the transported organelles along microtubules or microtubule regions. It can be seen from the micrographs of Smith[10] that all organelles accumulating at a constriction are retained in channels equivalent to microtubule domains, whereas the neurofilament domains appear free of organelles. Previous work by Smith et al.[57] had also shown vesicles which have accumulated and are apparently queued up in rows along microtubules before entering terminal segments of lamprey axons. The work on single microtubules from extruded squid axoplasm is, however, the most convincing proof in this regard.[89,91] Intracellular movement in other cells seems to be very similarly organized since it has been shown, that also in keratocytes and fibroblasts the paths of the moving organelles are associated with microtubules and not with other filaments.[90,106,107] The membrane bounded organelles in the circumtubular regions are therefore clearly identified as being at least part of the moving compartment.

The microstream concept[12,28,29,105] goes further and proposes that the halos around individual microtubules are regions where low viscosity axoplasm is in a streaming state. In this model, the hydrodynamic coalescence of these regions into microtubule domains produces larger streams that are capable of carrying organelles up to the size of mitochondria through the cooperative effect of individual microstreams.[28,29] These authors also consider it very likely that the number and diameters of such streaming regions can vary with temperature or the physiological state of the neuron. In this model it is also not necessary that all microtubules be involved in the maintenance of streaming regions. This may be the explanation for the reported spare capacity (up to 300%) for rapid transport[108] and of the fact that transport could in some studies be impaired only when a high percentage of microtubules is destroyed by colchicine or low temperature.[109,110]

B. The Stationary Phase

Tracer studies[111-114] and microscopical observations[76,125] have shown that, compared to the velocities of rapid transport or particle saltation, the movement of the major part of the cytoplasmic material is very slow and can, therefore, be approximated by a stationary phase when dealing with rapid transport. On the basis of theoretical considerations outlined above it was concluded that the stationary phase outside the microtubule domains can be regarded as a high viscosity gel.[30] Ultrastructural studies on constricted axons suggest that these gelled regions are apparently so viscous that transported organelles, even if jammed in front of the constriction, apparently cannot penetrate into the neurofilament regions but queue up in the microtubule domains.[10] Furthermore, it is generally observed that membraneous organelles are excluded from regions where neurofilaments are concentrated.[11-13,71,83,88]

Actin and other proteins may also play a structural or cytoskeletal role in the axon[116,117] by interconnecting neurofilaments or forming a gel themselves.[37,116] Actin filaments appear to form random filamentous arrays associated with the axolemma,[12,117-119] but may otherwise be absent or very rare in the axoplasm proper.[120]

C. Exchange Between the Mobile and the Stationary Phases

Deposition or retention of the transported material along the axon is well documented[83,92-94] and is interpreted as the result of exchange processes between the mobile and the stationary phases. Theoretical treatments of this aspect have been published in association with the transport filament model[94] as well as the microstream model.[95]

The circumtubular flow model treated theoretically by Gross and Weiss[28-30] proposes a transport stream which together with diffusion of molecules or of supramolecular aggregates between the two phases will result in a spectrum of transport velocities, since the outer boundary of the microstream annulus is porous and permits some of the transported material to leave the carrier stream. Such molecules will stop migrating until their random walk will again bring them into the domain of a microstream.

This exchange of material between a moving and a stationary phase is, of course, the molecular dynamics of partition chromatography. Furthermore, these authors expect the exchange to be size dependent because the cross-linked neurofilaments probably allow easy penetration only of molecules or molecular complexes that are smaller than 20 to 40 nm. This cutoff is suggested by scanning electron micrographs of freeze-etched axons.[12,121,122] Consequently, phenomena akin to size exclusion chromatography can be expected to occur during axoplasmic transport. Within the stationary phase, molecules may also combine with specific binding sites, a process which is analogous to affinity chromatography. The complexity of the biophysical parameters of the flow field may also contribute to molecular separation similar to the phenomenon of "field flow fractionation".[123] Therefore, the circumtubular flow together with the structural properties of the gelled neurofilament domains, as proposed by the microstream concept, should result in chromatographic phenomena.[28,29,95,96,124]

The characteristic changes in isotope distributions obtained during transport in various nerve systems[92-94, 113] are the subject of successful computer simulation studies based on chromatographic theory.[95,96] The separation of material in the rapid transport isotope profiles,[124] the increasing number of observations of velocity groups for specific proteins[125] and the well documented difference between slowly and rapidly transported material[1,126] have been suggested to be direct consequences of chromatographic mechanisms.[28,29,96,124]

VII. LOCUS OF FORCE GENERATION

Chemomechanical energy conversion requires a structural arrangement by virtue of which a chemical reaction is vectorially directed and establishes a gradient.[127] In neurons this chemical reaction of the force generating mechanism is generally assumed to be catalyzed by ATPases since it has been shown that axoplasmic transport requires ATP.[14,15] Here they are called more generally force generating enzymes, since their molecular identity is still unclear. All force generating enzymes must have very similar orientation and polarity of action relative to the longitudinal axis of the axon, since a random force generation would merely intensify random motion.

Theoretically, force generating enzymes may be situated either on the transported material or on the guide element structure(s). It would be difficult, however, to orient force generating enzymes in or on the membrane of the moving organelle. Furthermore, Smith[25] observed, that during saltatory movement particles may spin and tumble, and concluded that organelles themselves cannot possess directional polarity based on vectorial moieties at their surface. Ideally, force generating enzymes should be attached to a polar structure that can absorb the recoil of the force generation without changing their vectorial arrangement. This was proposed in several models.[18,20,28,128]

There is a variety of axonal structures to which such force generating enzymes can be attached, but the orientation and recoil criteria are not easily satisfied. The placement of force generating enzymes in or on membranes is only meaningful with regard to transmembrane transport where the force must be applied perpendicularly to the membrane. This vector would not change, despite lateral movement and rotation within the plane of the membrane. However, the generation of a force tangential to the plasma membrane in the direction of the long axis of the axon would not be possible unless the force generating enzymes were prevented from thermal rotation. In addition, the question arises of how the initial orientation of membrane-attached force generating enzymes can be achieved. The submembrane microfilamentous meshwork is randomly oriented and seems, therefore, to be incapable of orienting intramembrane components, although it could produce and restrict their movement (see Reference 116 for review).

Thus, only the following three filamentous structures can be envisaged as orienting structures since only these are or could be vectorial.

1. Microfilaments are difficult to detect in axons except if one interferes with the g-actin/f-actin system by adding heavy meromyosin.[118,119] Their existence in the axoplasm proper is, therefore, still equivocal.[120] Submembraneous microfilaments are abundant in axons but we have to assume that they are, similarly to other cells,[129] arrayed in a highly ordered but not longitudinally aligned or vectorial manner.
2. Neurofilaments are, as discussed above, excluded from regions where rapid transport occurs.[83,88]
3. Only microtubules are long and massive[11,130] rigid,[131] and present in all axons.

Furthermore, various types of ATPases or ATPase-binding elements (e.g., actin) have been shown or suggested to bind to microtubules,[132-137] although the importance of these ATPases for transport is not known. Orientation of the force generating enzymes can be accomplished by collision kinetics and may not change until the microtubule is disassembled.

The above considerations point to the microtubule as by far the most suitable candidate for the primary backbone of the force generating system.

VIII. TYPES OF FORCE GENERATION

The chemomechanical energy transformation producing directional movement in biological systems requires vectorial structures which add directionality to a chemical reaction such as cleavage of ATP.[7,28,138,139] One may ask whether the chemical energy utilized is directly converted into kinetic energy or must first assume the form of potential, i.e., stored, energy (as for example in the form of a conformational change of a protein or of a chemical or electrochemical gradient across a membrane).[127] Chemical, electrochemical, or pressure gradients acting over distances seem to be ruled out for axoplasmic transport since the material is moving with various velocities and in opposite directions simultaneously and independently, and since such movement is intrinsic even to short axon segments.[26] Otherwise, however, both possibilities, direct and indirect energy transformation, are under discussion for the various types of chemomechanical conversions including axoplasmic transport.[26,138,140-144]

A. Modes Involving Specific Interactions

The utilization of potential energy of conformational changes which is mechanically transferred specifically to the structure to be moved, is generally believed to be realized in muscle contraction and motion of cilia. Most models of axoplasmic transport also suggest that movement is brought about by specific interactions between the motor and the transported material[18,20-24,73,103,128,145] Such models must involve steps for recognition of the transported material, selection of the transported material, binding of the transported material, sorting into different velocity groups, regulation of each more or less saltatory jump, and finally release of the transported material from the mechanism. This requires a number of rather complicated events to ensure the proper transport direction, velocity group, or structure of destination. As structural moieties involved in the recognition steps one often envisions peptide sequences, in analogy to the hypothesis of Blobel[146] for protein topogenesis. This type of ratchet-like force generation should result in interrupted or saltatory particle movement.[25,147,148]

These models are widely accepted because they provide an easy structuralistic explanation and are in line with the general impression, nourished by modern ultrastructural research, of a rigid, structured cytoplasm. There is little direct experimental support for specific interaction steps and for recognition except the study on transport of synaptic vesicles in sympathetic neurons by Schwab and Thoenen.[149] They present evidence that vesicles are specifically recognized and most probably hooked onto the mechanism since vesicles that distally in axons undergo membrane modifications without changes in ultrastructure move retrogradely instead of anterogradely as they did previously. Whether subtle chemical modifications or changes in gross physical parameters such as surface charge are responsible for such behavior is unknown.

In favor of this type of mechanism one can also infer certain microscopically observable aspects of organelle movement. Allen et al. first observed with AVEC-DIC microscopy[89,91,106,150] that in dispersed squid axoplasm or in intact keratocytes small organelles undergo irregular Brownian-like movements until they come near microtubules. Once they are close to these they stop Brownian motion and eventually travel along the microtubules in a directed and ordered manner. The fact that in very narrow axons of 0.2 μm diameter, which contain only 4 microtubules,[60] organelles of diameters up to 0.2 μm move in both directions[151,152] is also more easily explained by a ratchet-type mechanism, than by a streaming mechanism.

However, all these models explain only the transport of special recognizable particles but cannot explain the fact that a great variety of material, if not all axonal constituents[2] are transported. This includes soluble compounds of high and low molecular weight which are transported with various velocities and which may or may not be stored in carrier organelles.[17,124,153-164] These models require a very high amount of molecular information (genetically coded) and a large number of assumptions. Energetically all hypotheses are only acceptable if transport occurs in low viscosity microtubule regions and if the single steps of the process are not so energy expensive that the efficiency falls below 25%.

All the above models imply step-wise force transduction modes similar to that in muscle contraction, which are known to occur in nature in systems where interaction is between only two invariable partner molecules or structures (muscle, cilia). To suggest this type of force transduction in axoplasmic transport, where a variety of partner molecules or structures would have to interact, could be logical if one unitary transport organelle was to be moved.[18] However, at least for the more rapid velocity groups this is not the case, since they contain various types of organelles.[1,124,165]

Furthermore, the wide spectrum of velocities is difficult to explain by such ratchet-type mechanisms, since the generally observed maximal velocity suggests a rather invariable basic speed of the ratchet or conveyor belt. This can be overcome in some of these models, if additional controls governing the variable duration of association to the mechanism of each type of organelle are involved (average velocity, population velocity).[92,97,128,166] The explanation of the transport of soluble material which appears to be present together with particles in rapid-transport is especially difficult. Since the transport dynamics of soluble material differs significantly from that of particulate material,[17,163] these models would have to provide for mechanisms whereby soluble material, including macromolecules, could be continuously released from the carrier organelles in transit.

All models involving specific interactions require specific recognition and interaction sites for both the motor and the transported material. The entire variety of at least the particular contents of the axon would be expected to bear such recognition and interaction molecules, which would have to be transported with the same, generally rapid, velocity as the particle. However, no such molecules have been found so far. The many studies involving tracer application and characterization of the radioactive proteins of various transported organelles have not yet shown a common protein component which could serve this role (see References 1 and 156 for review). All the known potential candidates, especially the proteins of filamentous and contractile elements investigated so far, have been found to be transported exclusively in the two slowest transport groups, and therefore cannot be constituents of rapidly moving organelles. These include actin,[167,168] myosin-like actin binding proteins,[169] clathrin,[170] tubulins, and neurofilament proteins.[171] One axonal protein, fodrin, appears in several velocity groups and could, therefore, function as glue between the motor and the organelles.[172] It is, however, not part of any of the rapidly transported organelles and could only be involved in transport of slow (group IV) components.[172] Similar arguments can be applied to models proposing a general carrier to which a variety of transported material binds.[103] No moving carrier material of this type has yet been found. Such material, if present, ought to be transported rapidly and in large amounts so that it should have been possible to demonstrate it.

B. Nonspecific Modes

The potential energy of conformational changes could theoretically be transferred nonspecifically to the surrounding medium, mainly water molecules, resulting in streaming. This case is apparently realized in actin-based streaming models and probably also in muscle contraction.[143] In the case of microtubule-based motility such a mechanism could be realized by microtubule-associated proteins, if these would go through a cycle of conformational changes of proper geometry to result in rowing, paddling, or microciliary movement on the surface of the microtubule as was proposed by Gross and Weiss.[26,28-30,105] When this kind of force generation is tested, one should search for possible involvement of microtubule associated ATPases, or of soluble and only intermittently associated ATPases (this is probably the case in the actin-based streaming models[143]) or of HMW-MAPs which could go through a phosphorylation cycle.

The other possible type of chemomechanical energy conversion which can be coupled to the medium resulting in nonspecific force generation is the direct conversion of chemical energy to kinetic energy. This has been first proposed for actin-based force generation by Oplatka and his collaborators[139,142,173,174] and as one possibility of force generation for axoplasmic transport by Gross and Weiss.[26,28-30,105] In actin based motility it is probable that one or both of these mechanisms are utilized,[139,143] whereas the situation is far from being clear in the case of microtubule based motility which is discussed here.

If charged products arise from the vectorially oriented chemical reaction they are vectorially released by repulsive forces. If occurring often enough, these forces could be coupled to move larger structures or part of the cytoplasm containing these structures. This does not involve specific interactions between the motor and special molecules or organelles. Instead, the movement of the reaction products is coupled to the surrounding medium in a nonspecific manner by viscous coupling or electroosmosis.[29] This type of nonspecific force generation would result in streaming domains within the cytoplasm, a phenomenon which is widely represented in cells.[36,148] Storing the chemical energy in the form of potential energy, e.g., conformational change or membrane potential which itself is then vectorially set free in a nonspecific but directed manner, would have the same effect.[26,28]

There are only a few models for axoplasmic transport which fall into this group (see Reference 26 for discussion).[28,175] They have the disadvantage of being less easily understood since they are based not as much on the ultrastructural as on the biophysical and hydrodynamic aspects of cytoplasm. However, if axoplasmic transport is a complex and not very selective phenomenon, as it appears to be, it may be more adequately described by models for nonspecific force generation which usually allow bulk flow of all types of cytoplasmic constituents. Generally, such nonspecific mechanisms work more simply and coherently (on a hydrodynamic basis), thus providing for greater transport continuity at loci of discontinuity of the linear guide-elements, i.e., microtubules.

Since the processes involved are characterized by very low Reynolds numbers[176-178] of the order of $Re = 10^{-9}$ to 10^{-6} it is evident that all movement of particles or fluid is absolutely laminar and inertialess.[176] Movement would, therefore, stop at the same moment as the motive force. Furthermore, it is possible that movement on two- (or almost one-) dimensional surfaces would have kinetics much more favorable than in three-dimensional space.[179] Hydrodynamic models have to cope with the additional difficulty that instead of macroscopic, i.e., intuitive, hydrodynamics very-small-Reynolds-number, i.e., counterintuitive, hydrodynamics have to be applied. The question whether streams of fluid can be brought about by cytoskeletal filaments can already be answered, since Shimizu and Yano have been successful to prove experimentally that streaming with considerable velocities is produced along actin filaments by the actin-activated ATPase myosin.[143] It is, therefore, reasonable to discuss the possibility of a similar process based on microtubules, as suggested by the microstream hypothesis.[28,29]

This hypothesis could explain the variety of transported material and the wide spectrum of velocities observed by suggesting a nonspecific force generation. In this way most of the drawbacks of the models suggesting specific interactions are avoided. Saltatory movement is permitted and an explanation of its origin is also provided.[30] Furthermore, this model is, besides the hypotheses of Ochs[103] and Schwartz et al.,[128] the only one which accounts for the characteristic radioactivity profiles and their variation with time.[95,96] This model can also explain the experimentally observed size dependent molecule separation during transport.[124,180] The energy requirement seems acceptable, since bulky particles move as part of a carrier stream. (See Section III.A.)

The microstream hypothesis[28] is probably the simplest way to explain intracellular movement and is based on only three basic assumptions and their consequences.[29] These form the foundation of an unitary model which is compatible with the majority of the experimental and theoretical evidence on rapid axoplasmic transport gathered so far.[29]

Evidence supporting the view that the transported organelles are not in contact but in the vicinity of the microtubules and that, part of the axoplasm may be streaming has been published by Schnapp and Reese[12]. The impression of coherent and continuous instead of saltatory movement of very small organelles which can be visualized now with AVEC-DIC microscopy[150] also fits into this picture of an underlying smooth carrier stream. Hayden et al.,[106] using this technique to study organelle movement in fibroblasts, also suggest a model where transport occurs in low viscosity circumtubular regions. The microstream hypothesis proposes[30] that the larger organelles more frequently experience channel obstructions or less regular microtubule arrangements which could cause the saltatory aspect of their movement. The most direct support in favor of a nonspecific, streaming mechanism comes from the observation of Adams and Bray[181] that microinjected artificial particles of various types are transported in crustacean giant axons.

There is one feature which distinguishes the mode of force generation of the microstream hypothesis from all other force generation modes: the activity of the force-generating enzyme is not necessarily dependent on the presence of material, i.e., organelles, to be transported. The mechanism can therefore be expected to work with plain native microtubules and to produce streaming. Among the many predictions which were derived from this hypothesis[26,29] is the prediction that microtubules would require anchoring filaments to prevent their movement in the direction opposite to the bulk water flow due to the action-reaction-principle. Liberation of the microtubules from the cytoskeletal meshwork is therefore expected to result in ATP-dependent movement of microtubules.

IX. FORCE GENERATING ENZYMES

The question of which ATPase may be the force generating enzyme of axoplasmic transport remains unsolved. The requirement for a vectorial structure with associated ATPase can in principle be fulfilled by at least two well-known systems, the actin-myosin and the tubulin-dynein system.

The impressive body of evidence in favor of microtubule involvement led early to the suggestion that a microtubule-associated ATPase may generate the force.[18,19,28] There is indeed a wealth of reports on microtubule-associated ATPases (for review see Reference 135).[134,136] Dynein-like ATPases have been reported to exist not only in cilia but also in cytoplasm and to be involved in chromosome movement.[182,183] Ciliary dynein can bind to brain microtubules,[133] however, the isolation of a dynein-like ATPase from brain tissue has not yet been achieved. ATPases co-purifying with microtubules have been characterized recently as microtubule-associated[137] whereas other workers concluded that these ATPases are part of a contaminating membrane fraction.[184,185]

The pharmacology of axoplasmic transport not only closely parallels the pharmacology of microtubule inhibitors,[186] there is also close resemblance between the effects of a variety of sulfhydryl reagents on microtubule associated Ca^{2+}, Mg^{2+}-ATPase,[187] and on axoplasmic transport.[188] The same behavior towards such reagents is as well characteristic for dynein ATPase.[189] Recently two inhibitors of ciliary dynein have been used in axoplasmic transport studies, orthovanadate and EHNA (erythro-9-(3,2-(hydroxynonyl))adenine). Both have been shown to inhibit transport in axons[190,191] and in other cell types.[192-194] An earlier study[195] had shown transport in the presence of microinjected vanadate, but due to the large volumes injected (up to the volume of the cell body itself) hydrodynamic convection caused by the pressure injection could not be excluded. If, as Schliwa[148] pointed out, the specificity of the anti-dynein drugs can be convincingly defended, then the present situation is clearly in favor of a dynein-like ATPase being the force generating enzyme of axoplasmic transport.

Actin and myosin are present in axons as well.[196] Are they involved in force generation or not? Earlier studies with cytochalasin B which interferes with actin assembly or the actin network itself, showed no inhibition of transport.[197-200] More recently a microinjection study reported inhibition of transport with this drug.[195] DNAse I which binds to g-actin and even depolymerizes f-actin, and phalloidin which has the opposite effect were also reported to block axoplasmic transport in leech and *Aplysia* giant axons.[195,201] These findings led many to accept actin as the force generating moiety.[202] However, by injecting the interfering molecules into the cell body, one could have affected the process of initiation of axonal transport[148] which has a pharmacology different from that of axonal transport itself.[203] The large volumes of fluid microinjected into leech neurons[195] not only seem to have disrupted the plasma membrane, as can be inferred from the labeling of surrounding cells, but may have also caused hydrodynamic effects, whose evaluation would require statistical treatment of the data. In a subsequent study Goldberg microinjected cytochalasin B, DNAse I, and the dynein inhibitor EHNA into the axon proper[191] and found no inhibition by cytochalasin-B but by the latter two substances. In addition, injection of NEM-HMM (*N*-ethyl-maleimide-coupled heavy meromyosin), which inhibits the actin-myosin interaction did not block transport. This was interpreted as evidence that actin is not involved in force generation, but a dynein-like ATPase may be, and that an actin network only maintains the necessary structure of the cytoplasm.[191]

In conclusion, there seems to be not much evidence left for actin-based force generation. For the very related movement in chromatophores such a role for actin has been excluded[204,205] and Pollard, although he had recently proposed an actin-based mechanism for axoplasmic transport,[145] showed that blocking of myosin II by injected antibody does inhibit locomotion, but not the intracellular particle movement in *Acanthamoeba*.[206]

The intriguing reports of α-actinin being associated with chromaffin granules[207,208] and actin being bound to cholinergic vesicles[209] are to be confronted with the findings that such vesicles are known to be transported rapidly, while actin and α-actinin are both transported only slowly.[167,168] This would make all proposed mechanisms for rapid transport which suggest that rapid transport is brought about by the pulling action of actin filaments very difficult to accept.[20,21,106,145] Another strong argument against actomyosin involvement is the recent finding that Ca^{2+} is not needed to maintain axoplasmic transport[15,25,210] although it is required for its initiation.[203]

X. DIRECTIONALITY AND REVERSAL

It is clear from a number of studies on organelles accumulated at constrictions[10,71,79,211] that anterogradely and retrogradely transported material is different.

Although there is less information available on retrograde as compared to anterograde transport, there appear to be only few differences with regard to the dynamics. Various population velocities, sometimes as rapid as the anterograde rapid transport rate, have been measured,[1,212] the rapid retrograde phase being also mainly particulate in nature (e.g., References 10, 79, and 213). Slow retrograde transport has also been reported.[214] The turnaround of material from one direction to the other, although rarely observed in the axon proper,[151,215] seems to occur readily at constrictions.[10,216-218] There are few studies where the same type of organelle or label could be used for both directions. For pike olfactory nerve axons the same maximal velocities were found for both directions for horseradish peroxidase[152] and for free amino acids.[219] Lysosome-like organelles showed identical average velocities in both directions, whereas mitochondria moved in the retrograde direction twice as rapid as in the anterograde direction, although both populations had the same size distribution.[151]

The direction of movement can theoretically be determined by the force generating enzymes, by the linear guide elements (microtubules) or by the moving particle. Most models propose that particles be coded for transport in one of the two directions (e.g., References 24 and 104), although this is not yet well-documented experimentally. The only convincing finding in this respect is by Schwab and Thoenen[149] who showed that in adrenergic axons dense core vesicles, which after arrival at the synaptic sites through anterograde transport, undergo a modification and are subsequently transported retrogradely.

It has been proposed that two subsets of microtubules with opposite polarity could produce transport in the two opposite directions.[28] However, it was shown recently that the growth vector of microtubules in axons is uniformly directed distally, thus the (+) end of microtubules points away from the perikaryon.[220-222] Since this is also true for cilia, chromatophores, and mitotic spindles,[223-225] all being examples where transport in both directions is well documented, it appears that the direction of intracellular movement can be determined independently of the growth vector of microtubules or that a microtubule with a given growth vector can generate transport in both directions simultaneously.

The possibility that two different types of force generating enzymes specific for the two opposite directions can account for the directionality seems difficult to envisage since this would require a special mechanism distributing them properly among microtubules of the same type or even on the same microtubule in some kind of order. Opposite vectors at the surface of different microtubules could be created theoretically by two distinct modes of attachment of a microtubule-associated ATPase.[26]

It was pointed out in the case of the microstream hypothesis where bulk water flow occurs, that a near equilibrium recirculation of water is assumed to be brought about by anterograde and retrograde streams.[26,29] This means that water moved a certain distance would not locally create pressure but is proposed to enter or create and stabilize neighboring microstreams in the opposite direction. These microstreams of opposite direction were proposed to be realized in neighboring microtubule domains,[26,29] but it may as well turn out that the local laminar eddies recirculate in a larger (center vs. periphery of the axon) or even much smaller scale (at different distances on the surface of the same microtubule). The latter case would explain the otherwise puzzling finding that even single microtubules are capable of transporting organelles in both directions.[89-91] This question may, perhaps, be solved by a quantitative treatment of the phenomenon applying the principles of low Reynolds number hydrodynamics.[176]

XI. SLOW-TRANSPORT OF SOLUBLE PROTEINS

Material passing through the axon with population velocities of between 0.1 and approximately 20 mm/day is found to fall mainly into three velocity classes. Class III contains myosin-like material moving with about 8 mm/day.[169] Class IV contains many soluble and cytoskeletal proteins such as actin, clathrin, calmodulin, enolase, creatine phosphokinase, and proteins of the glycolytic pathway[114,165,168,170,226,227] which are transported at maximum velocities between 1 and 2 mm/day (for review see References 125, 165, and 227). The slowest material, class V, moves with a maximum population velocity of 0.1 to 1 mm/day (for review see References 2, 165, 227, and 228) and is comprised of the neurofilament proteins, the two tubulins and the spectrin-like protein fodrin, which is also present in the other slow-transport classes. Biochemically the slowly transported material can be divided into three classes:[1,2,112,126,165,227] soluble proteins, cytoskeletal components, and low molecular weight molecules (after they have left the rapid transport phase with which they initially moved).[124,180,229]

The term "soluble" proteins originates from the biochemical literature and is operationally defined there to describe the polypeptides, which appear in the supernatant after a series of procedures including homogenization and high speed-ultracentrifugation. From such work it cannot be definitely concluded that the soluble form is the native state of such polypeptides in the living, undisrupted cell, although this view was not challenged until recently. From the electronmicroscopical literature, however, originated the alternative view of the microtrabecular lattice which is operationally defined as the presumably proteinaceous material of the cytosol which appears to form a coating on the surface of organelles and thin filaments interconnecting them. This microtrabecular lattice can be observed by high-voltage electronmicroscopy after dehydration of the tissues.[230] Whether proteins can exist in a free, soluble form in cells, or whether all cytoplasmic proteins are structured, i.e., associated or assembled to supramolecular structures, is an unresolved issue in cell biology.[231] This leads to two opposing views on the axonal transport of soluble proteins, namely the structural hypothesis and the microstream hypothesis.

The dynamic events, leading to the relatively cohesive slow movement of soluble proteins[114,226] are explained by the structural hypothesis as follows: "If, as we propose, each group of proteins corresponds to a specific structure, then the movement of the proteins in each group will be determined by the motile properties of that structure".[228] The mechanism for the movement of the various slow structures is rarely addressed. The structural hypothesis would require most probably different force-generating mechanisms for the various fast and slow structures. The subaxolemmal matrix containing the spectrin-like protein fodrin and actin filaments was proposed to move the material of classes IV and V by a mechanism involving actin and myosin.[171,232,233]

The cohesive dynamics of slowly transported soluble proteins is explained by the microstream hypothesis as being the result of chromatographic distribution of soluble material between a rapidly moving fraction of cytoplasm, the microstreams, and the stationary domains.[28-30, 96, 105, 124] In terms of chromatography: material not privileged by its physical properties to stay effectively in the mobile phase is expected to move slowly.[234,235] The theory of chromatography provides us with two general effects which can be responsible for retardation. First the ability of a soluble substance to freely pass in and out of the mobile phase, as well as its diffusion into retrograde channels, are expected to cause a slowing of velocity (analogous to counter current and gel permeation chromatography). Slow-transport was attributed to counter-current movement by Lubinska already in 1964.[237] Second, the presence of numerous and/or high affinity binding sites for a given molecule in the stationary phase will also slow the movement considerably (affinity chromatography). The latter has been especially suggested for soluble subunits of cytoskeletal elements which, after short periods of movement and diffusion, will assemble on the cytoskeleton again (see below). Slow transport may, therefore, result from the combined effects of affinity for stationary components and retardation by frequently leaving the mobile phase.[29,236]

These soluble macromolecules have a slow transport velocity in common, which is proposed to be due to their common physical properties. One such property could be that these proteins are indeed soluble in their native state and therefore smaller than the more rapidly moving membrane-bounded organelles. This would allow them to leave the mobile phase of the cytoplasm freely and to penetrate stationary or even retrograde compartments.[29,96,124] Due to the greater "solubility" these proteins may have in common, the first of the two above-mentioned chromatographic effects is suggested to dominate their transport behavior. This view is supported by the finding that the proteins in velocity class IV show similar but independent transport behavior and may, therefore, not be associated during their migration.[238] According to the microstream hypothesis slow-transport is regarded as an epiphenomenon of rapid transport which does not require additional or different force generating mechanisms.[29]

There is little direct experimental evidence on the native state of slowly transported polypeptides in the living axon, which would help in deciding between the two hypotheses.

XII. SLOW-TRANSPORT OF CYTOSKELETAL PROTEINS

The material of velocity class V, namely the microtubule and neurofilament proteins, and to a certain extent also actin moving with group IV, differ from the soluble proteins already discussed since the latter proteins have been shown to exist in an assembled filamentous and in a disassembled diffusable state.[239-241] There is a characteristic equilibrium for each of the three filaments between the assembled state and its subunits.

The structural hypotheses claim that the cross-linked cytoskeletal structures move as a whole while the free subunits are stationary and merely subject to local diffusion.[58,228,240,242,243] The microstream hypothesis, on the other hand, proposes that the material is conveyed in soluble subunit form or as small fragments whereas the large skeletal arrays onto which these molecules can assemble are stationary.[28,29,236] Thus, the latter hypothesis proposes a very similar type of movement for the cytoskeletal proteins (group V) as was described for the soluble proteins (group IV) in the preceding section.

According to the structural hypotheses the transport behavior of the various cytoskeletal proteins should be identical. It has been shown first by Hoffman and Lasek[171] that the neurofilament and microtubule proteins move with the same velocities. The similar behavior of the proteins in velocity group V[114,165,228] is most easily explained by the structural hypothesis. There are however reports that neurofilament proteins and tubulins move independently, i.e., with different velocities.[240,244-247] This is also found if colchicine, IDPN, or 3,4-dimethyl-2,5-hexanedione is applied, which block predominantly the transport of microtubule or neurofilament proteins respectively.[87,248,249]

According to the structural hypotheses it can be expected that radioactively labeled subunits once inserted into the cytoskeleton move as a cytoskeletal region which does not broaden with time. This was first described by Hoffman and Lasek.[171] However, a later study analyzing their data quantitatively[236,250] showed that peak spreading does occur in slow-transport and that the values for apparent diffusion coefficients are equivalent to those for diffusion of soluble proteins in water. This was interpreted to mean that the radioactive material may well have been subject to diffusion or an active exchange process during transport. This view is also supported experimentally by a study on the oculomotor nerve, where proteins of class V are found to be dispersed during the course of their transit along the nerve.[251] Cancalon[113,252] showed for olfactory nerves only one almost symmetrical peak of radioactivity which apparently includes the skeletal and soluble proteins (group IV and V). This suggests that either soluble proteins move as part of the cytoskeleton or that cytoskeletal subunits move as "soluble" proteins. In addition it has been shown that part of the radioactive material is not transported down to the synapse but remains stationary in the axon.[48,113,114,167,251,253]

According to the structural hypotheses microtubules and neurofilaments would have to be cross-linked. There is good biochemical and ultrastructural evidence for this,[254,255] even though they quite often form well separated domains.[9,11,12,26] However, in contrast to the microtubules, the neurofilaments are discontinuous at the nodes of Ranvier.[11,71,122] This finding is very difficult to reconcile with the models proposing movement of the whole cytoskeletal scaffold. It would mean that the cytoskeleton is not the transport form or that at least at the nodes the neurofilaments have to disassemble to subunits or short pieces in order to pass through the nodal constriction.

Synthesis and assembly of subunits into the cytoskeleton should according to this hypothesis occur only in the perikaryon. This is supported in the case of the neurofilaments by the fact that there is virtually no free pool of subunits.[33,233,241] There are, however, also reports that neurofilaments differ from intermediate filaments of other cells insofar as neurofilaments disassemble reversibly.[233,256,257] The neurofilament proteins have also been reported to be synthesized as a soluble product.[258] On the other hand, in various cells ribosomes have been shown to be attached to the cytoskeleton so that cytoskeletal proteins could be inserted directly into the existing filamentous arrays.[259] In the case of microtubules insertion at the perikaryon would require assembly at the biochemically less favored end (i.e., the slowly growing (−) end).[13] Transport of subunits which are available in considerable amounts[241] and their insertion at the distal (+) ends seem to be the more feasible alternative. This shows that the locus and the mode of cytoskeletal assembly remains not very well understood.

The same is true for disassembly and degradation of neurofilaments and microtubules. According to the structural hypotheses the degradation should occur in the synaptic terminals.[260] There is however, evidence for degradation in the axon, since the slowly moving radioactivity decreases during transport and radioactive amino acids become available to glial cells along the axon.[113,114,167,168,171,261] The experimental evidence for neurofilament degradation at synaptic sites is regarded to be only indirect.[48]

The structural hypotheses would predict that in tracer experiments the specific radioactivity (dpm/mg protein) for the individual cytoskeletal proteins should remain constant during their transport through the axon. This has been shown not to be the case.[262] Local supply of radioactively labeled and nonlabeled subunits to a stationary cytoskeleton could explain changes of the specific activity.

Slow-transport has been reported to stop within 1 or 2 hr after the axon is severed[263] or protein synthesis is blocked,[243] whereas rapid transport is known to continue for up to 24 hr. The microstream hypothesis would predict that slow-transport as an epiphenomenon of rapid transport should continue for the time rapid transport is maintained. Slow-transport has, however, also been reported to continue in severed optic nerve for 24 hr[243] and in severed olfactory nerve for several days.[264]

A number of other findings pose difficult questions to the structural hypotheses. Energetic difficulties exceeding those mentioned for small vesicles[7] by orders of magnitude would be expected if large skeletal arrays were to be moved *in toto*. It is especially difficult to imagine how two cytoskeletal structures, at least one of them laterally cross-linked, could move with different velocities through one another and through the axon at the same time as was proposed by the structural hypothesis.[114,240] The question of the generation of the motive force is rarely addressed in connection with the structural hypothesis.[171,228] Moreover, the demonstration of slow-transport of ^{3}H-colchicine over a certain distance[265,266] has to be interpreted to mean that tubulin subunits which bind the drug[267] are moving.

The structural hypothesis incorporates a great number of very detailed requirements, which evidently are not all met by the experimental findings. It would, therefore, be premature to consider this model as proven. The much more general view of the microstream hypothesis[28,29] can encompass most of the findings easily, but is far from being proven as well. This hypothesis proposes that free subunits or short pieces of the cytoskeletal filaments are subject to transport, deposition, chromatographic distribution, and degradation, either in groups or independently from each other, depending on their degree of aggregation in the living cytoplasm.[26,28,29] The driving force has been suggested to be the drag flow from rapid transport.[28,29]

XIII. TRANSPORT OF LOW MOLECULAR WEIGHT MATERIAL

Small molecules, mainly transmitter substances, have been shown to be transported in storage vesicles.[1,268] For many other low molecular weight substances it is not known whether they can be stored in vesicles, but axoplasmic transport has been shown to convey a wide variety of such compounds including amino acids,[17,124,153,156,157,160,163,269-271] sugars,[158,163] UDP-glucose,[155] and amines.[272,273] Axoplasmic transport of such molecules not covalently bound and not being neurotransmitters is discussed here.

Early studies in which labeled compounds were injected into the perikaryal region (eye) and only the label arriving in the synaptic region was measured, failed to show axoplasmic transport of these compounds.[111,274-276] If, however, the distribution of the label along the axon was monitored, several researchers were able to show transport of free amino acids, i.e., amino acids not covalently bound to proteins and not precipitable in trichloroacetic acid.[17,153,156,157,160,163,269,277]

Several studies have shown that transport of low molecular weight material and of proteins are very similar, since both depend on oxidative phosphorylation,[17,163] both classes of molecules are transported in the form of radioactivity peaks,[17,163,269] and both can be inhibited by microtubule inhibitors.[17,160] These substances are transported rapidly[17,163,269,277] and no somal delay time as seen with protein transport[93] can be observed.[17,163,269] The maximal velocity has been shown to be, at least in the olfactory nerve axons, identical with the maximal velocity of proteins.[17,163,269] Proteolysis is excluded as a source of the rapidly transported amino acids.[17,269,277]

It was also reported that the amount of low molecular weight material decreases during transport.[17,156,163] In protein transport a similar linear reduction (2 to 3% per cm) of the rapidly transported material had been shown,[93,94] but for amino acids the losses from the rapidly moving compartment are much more dramatic.[17,163] In the case of leucine transport it has been shown that the rapidly moving peak loses 20% of its contents per hour.[17] This leads to a situation where a rapidly moving radioactive peak is no longer detectable after 5 hr or a transport distance of 4 to 5 cm (19°C). The authors[17] concluded that this can explain the frequent failures to detect transport of low molecular weight material[111,274-276] to the terminals of long nerves.

That diffusion cannot account for this movement has only recently been shown.[17] In this study diffusion has been calculated to be much slower than the observed displacement of radioactive amino acids.The velocity of diffusion decreases with time, whereas the observed velocity was constant. Microtubule inhibitors did block this movement (see also Reference 160) and the same is true if the oxygen supply is interrupted. The finding that transport of various small molecules occurs in the form of moving peaks[17,163,269] also excludes diffusion.

The state of this low molecular weight material during transport is unknown. It could move either free in solution, associated with macromolecules, or stored in organelles. Transport by carrier proteins appears, however, unlikely since free proteins are in general regarded to be transported only slowly.[1,165] Furthermore, the amount of all the soluble protein in axons would not be enough to carry the various small molecules transported.[17,163]

Transport of the small molecules stored in vesicles would be compatible with most proposed models for axoplasmic transport, including the microstream hypothesis. The hypotheses involving specific force generation (see Section VIII.A.) and the structural hypotheses would require that these molecules were packaged in order to undergo transport. In the well-studied olfactory nerve this is, however, not easily compatible with the fact that all low molecular weight substances studied so far have been shown to exhibit the same behavior. These substances include leucine and proline, but also exogenous and metabolically inert substances such as β-alanine and 3-0-methyl-D-glucose.[17,163,269] The existence of mechanisms for uptake of all of these substances into vesicles seems rather unlikely.

Transport in free solution is likewise unproven, but it has been shown that a considerable amount of low molecular weight compounds is indeed present in free form in axons.[278] Transport of these substances while in free form would easily fit the microstream hypothesis and was even predicted by this hypothesis.[26,28-30,124,163,269] All features of the transport of these substances, especially the decrease in peak height, are compatible with chromatographic dynamics as proposed by the microstream hypothesis.[17,163]

XIV. REGULATION OF AXOPLASMIC TRANSPORT

Even in extreme physiological states of the neuron such as in situations of blocked or increased electrical and exocytotic activity at the synapse there seems to be no significant influence on axoplasmic transport. This is especially true for the transport velocity which cannot be influenced by altering the physiological activity of the neuron.[1] The amount of transported material can be increased in certain situations of physiological stimulation, especially in the hypothalamo-neurohypophysial system[1,279,280] This may be explained, however, by increased perikaryal synthesis of the transported material, since it is known that transport has a considerable excess capacity and can in some cases accommodate four times the amount of material transported in the resting state.[108] This situation seems to be the same in various neuronal and nonneuronal secretory cells, which have been shown to have two separate pathways for synthesis, transport, and exocytosis of the protein constituents of membrane-bounded vesicles. The assembly line for the maintenance of the plasma membrane is not regulated (constitutive pathway) whereas the pathway for secretory vesicles and their contents, e.g., neurohormones, is regulated according to the physiological situation.[281,282]

That axoplasmic transport cannot be regulated very effectively is also evident from the fact that by electrical stimulation synapses can be depleted of vesicles and transmission be abolished.[283,284] Elevated velocities could be observed only if the amount of organelles was increased.[51,285] Davison[286] had already speculated that due to the lack of regulation the neuron has always to convey a surplus of material, thus necessitating retrograde transport (which itself seems also not to be regulated).[287]

There are many proposed models of how axonal transport can be regulated intracellularly[3,22-24,232,288,289] although there seems to be no need that the neuron makes use of any of these.

XV. CONCLUSION

Reviewing the mechanisms of intracellular organelle transport in various cell types, Schliwa[148] tried to divide them in two classes. In the first class streaming is the predominant aspect, such as in plant cells, and in a second one saltation is the main feature. He discusses if streaming is actomyosin-based and if saltation is driven by a tubulin-dynein mechanism. It seems, however, that in some cases these two types cannot be readily distinguished. He points out examples for microtubule-based streaming in pigment cells, plant cells, and axons, as well as for microfilament based saltatory movement in plants. Already Rebhun, who did some of the pioneering work on saltatory movement,[147] did not exclude microstreams as a possible basis for such movements.[290]

This ambivalence seems to be general in cell motility research. Even the seemingly established cross-bridge model of muscle contraction which is the paradigm for force generation, involving a specific, direct interaction between two types of filaments has been challenged by the results of in-vitro studies on the interaction of actin filaments with free myosin heads.[139,142,143,173,174] These authors claim that the vectorial translocation of soluble reaction products and/or the cycle of conformational changes in the myosin head bring about movement and contraction by a hydrodynamic, i.e., streaming, mechanism. Furthermore, Shimizu and collaborators showed that actin based motility including muscle contraction would not be possible without the dynamic cooperativity introduced by the microstreams along the actin filaments.[143,291] Recent reviews on the cross-bridge model are refreshingly critical in this respect.[140,292]

A similar situation is encountered in the field of axoplasmic transport research. The ratchet-type models generally imply specific force generation and movement of organelles through a stationary cytoplasm, whereas the models proposing nonspecific force generation imply movement of a certain percentage of the cytoplasm itself. At present none of the two modes can be ruled out on the basis of available experimental data.

It should be pointed out, however, that any mechanism moving cytoplasmic particles by specific interaction, such as the movement of myosin coated particles along actin cables in algal cells,[293] will result in streaming of cytoplasm mediated by viscous coupling. Viscous coupling is treated theoretically by Nothnagel and Webb[294] and is suggested by these authors to be the basis of cytoplasmic streaming in plant cells.

Reports on the movement of exogenous particles such as latex and colloidal gold do add considerable support to the nonspecific, streaming aspect of intracellular particle movement.[181,295] Moreover, in some of these experiments the particles have been coated with serum albumin and polyethylene glycol to prevent the binding of specific recognition and/or transport proteins such as actin and myosin.[181,295] This treatment, however, did not influence the particles' motile behavior. It has also been reported that very small particles move in a smooth, streaming manner instead of saltating.[89,91,150] Even if this streaming-like movement of large numbers of particles along microtubules is brought about by a ratchet-type mechanism with specific force transduction,[106] nonspecific viscous coupling would give rise to the formation of streams in the surrounding cytoplasm which could then carry larger organelles. This may lead the way to a possible explanation of the saltatory and streaming aspects through a combination of both specific and nonspecific force generation. Clarification in this regard may be expected from work using the new technique of AVEC-DIC microscopy[89,150,296,297] which permits one to directly observe the motile behavior of the organelles and microtubules.

It should also be borne in mind that several physical effects may produce part of the many aspects of the phenomenon of intracellular transport. Vigorous and rapid movements occur under certain conditions in mixtures of liquids, emulsions, and especially in multi-component polymer solutions.[298-301] They can be ascribed to charge repulsion effects, to surface tension and to structured flows,[299,301] the latter being a phenomenon which is brought about by self-organization in nonequilibrium systems.[143,291]

Concerning slow transport it can be concluded that many of its features fit the idea that the cytoskeleton and other supramolecular structures move in assembled form through a stationary cytoplasm. This would however require additional mechanisms to propel these structures. However, at least as many aspects of slow-transport are compatible with the idea that soluble proteins, subunits, and short segments of the cytoskeletal elements are conveyed by microstreams while the cytoskeleton remains stationary. Both hypotheses are not yet supported by experimental evidence to a degree which would allow to accept one of them as proven. The history of science tells us, that the large body of evidence supporting each of the two concepts may be indicative for a mechanism which in some way, as yet hidden to us, combines the two concepts which still seem to be mutually excluding.

XVI. NOTE ADDED IN PROOF

Since this chapter went to press the field has made extraordinary progress (References 302-316).

Video microscopy (AVEC-DIC)[82] has been used successfully to directly observe the complex motile properties of microtubules and organelles.[91,302-306] These studies definitely proved that it is indeed the microtubule that supports organelle movement and that the individual native microtubule supports bidirectional organelle movement.[89-91,302,306] Electron microscopy[306] and pharmacological work[303] did not provide any evidence for the involvement of actin filaments, nor for a role of Ca^{2+} in organelle movement. Furthermore, these studies clearly supported the earlier prediction of the microstream hypothesis, namely the active movement of microtubules freed from the cytoskeletal lattice.[302] Organelle and microtubule movements were shown to be closely correlated to each other so that it seems that anterograde and retrograde organelle movements as well as the microtubule movements (which were shown to be retrograde in direction[302]) are the result of one molecular mechanism.

Cross-bridges and firm linkages between microtubules and organelles have been reported [307-310] which seem to be more specific and are especially prominent in the absence of ATP.[307,309,310] These findings support the view that force generation involves specific interactions. However, new reports which show that such minute connections can be created artifactually[311,312] may force us to be very cautious in interpreting such studies (see Reference 313). Further support for cross-bridge models of force generation comes from the studies showing that nonhydrolyzable ATP-analogues attach organelles to microtubules,[309,310] This can be interpreted to represent a situation analogous to the rigor state in muscle contraction. It remains unknown which components bind the organelles and how they may do so. However, proteins have already been purified which are part of the microtubule-vesicle complex formed in the absence of ATP[310,314] and which at least in one case possess ATPase activity.[310]

These new and until quite recently unexpected findings[302-314] had a major impact on our view of the transport mechanism. The possible involvement of treadmilling, the microtrabecular lattice, the smooth endoplasmic reticulum, actin filaments, and of the neurofilamentous domains can now be regarded as less important or negligible (see References 302, 313). The models which locate the force-generating enzymes on organelles and other surfaces[308,315] and suggest that such enzymes may be dissociable factors[315] are thought-provoking alternatives to those models which propose a microtubular location.[91,302,313] Only the latter models together with a nonspecific mode of force generation were reported to be capable of explaining the bidirectional organelle *and* the unidirectional microtubule movements by a unitary mechanism,[315,316] while the new hypotheses[308,315] would require two or three different mechanisms. It is especially the gliding and serpentine movements of microtubules[302] that can hardly be reconciled with the location of force generation on structures other than the microtubule itself.[313] Although there is no direct proof for the existence of a cross-bridge mechanism, this is supported by circumstantial evidence,[308-310,314,315] while only little additional support for a hydrodynamic mechanism was reported.[316] This means that the hydrodynamic aspects that were amply discussed in this chapter may have to be regarded as secondary and as being the result of the organelle translocation.[313]

Thus, the new findings clarified several old questions and made axoplasmic transport of organelles a directly observable phenomenon. Nevertheless its many facets cannot at present be reconciled with any of the classical hypotheses of force generation. However, axoplasmic transport has attracted a great number of active researchers studying intracellular transport so that more major achievements are soon to be expected.

REFERENCES

1. **Grafstein, B. and Forman, D. S.,** Intracellular transport in neurons, *Physiol. Rev.*, 60, 1167, 1980.
2. **Weiss, D. G.,** General properties of axoplasmic transport, in *Axoplasmic Transport in Physiology and Pathology*, Weiss, D. G. and Gorio, A., Eds., Springer-Verlag, Basel, 1982, 1.
3. **Ochs, S.,** *Axoplasmic Transport and its Relation to Other Nerve Functions*, John Wiley & Sons, New York, 1982.
4. **Weiss, D. G.,** *Axoplasmic Transport*, Springer-Verlag, Basel, 1982.
5. **Rubinson, K. A. and Baker, P. F.,** The flow properties of axoplasm in a defined chemical environment: influence of anions and calcium, *Proc. R. Soc. London, Ser. B*, 205, 323, 1979.
6. **Baker, P. F.,** The viscosity of axoplasm: its meaning, measurement and sensitivity to calcium, in *Axoplasmic Transport*, Weiss, D. G., Ed., Springer-Verlag, Basel, 1982, 99.

7. **Gross, G. W. and Weiss, D. G.,** Theoretical considerations on rapid transport in low viscosity axonal regions, in *Axoplasmic Transport,* Weiss, D. G., Ed., Springer-Verlag, Basel, 1982, 330.
8. **Biondi, R. J., Levy, M. J., and Weiss, P. A.,** An engineering study of the peristaltic drive of axonal flow, *Proc. Natl. Acad. Sci. U.S.A.,* 69, 1732, 1972.
9. **Krammer, E. B. and Zenker, W.,** Effekt von Zinkionen auf Struktur und Verteilung der Neurotubuli, *Acta Neuropathol.,* 31, 59, 1975.
10. **Smith, R. S.,** The short term accumulation of axonally transported organelles in the region of localized lesions of single myelinated axons, *J. Neurocytol.,* 9, 39, 1980.
11. **Tsukita, S. and Ishikawa, H.,** The cytoskeleton in myelinated axons: serial section study, *Biomed. Res.,* 2, 424, 1981.
12. **Schnapp, B. J. and Reese, T. S.,** Cytoplasmic structure in rapid-frozen axons, *J. Cell Biol.,* 94, 667, 1982.
13. **Soifer, D.,** Assembly and disposition of microtubule proteins: some boundary conditions for possible role of microtubules in axoplasmic transport, in *Axoplasmic Transport,* Weiss, D. G., Ed., Springer-Verlag, Basel, 1982, 81.
14. **Ochs, S.,** Energy metabolism and supply of -P to the fast axoplasmic transport mechanism in nerve, *Fed. Proc. Fed. Am. Soc. Exp. Biol.,* 33, 1049, 1974.
15. **Adams, R. J.,** Organelle movement in axons depends on ATP, *Nature (London),* 297, 327, 1982.
16. **Copeland, A. R.,** Axonal transport. I. Diffusion, *Bull. Math. Biol.,* 38, 425, 1976.
17. **Schmid, G., Wagner, L., and Weiss, D. G.,** Rapid axoplasmic transport of free leucine, *J. Neurobiol.,* 14, 133, 1983.
18. **Schmitt, F. O.,** Fibrous proteins — neuronal organelles, *Proc. Natl. Acad. Sci. U.S.A.,* 60, 1092, 1968.
19. **Ochs, S.,** Characteristics and a model for fast axoplasmic transport in nerve, *J. Neurobiol.,* 2, 331, 1971.
20. **Heslop, J. P.,** Fast transport along nerves, in *Transport at the Cellular Level,* Soc. Exp. Biol. Symp. XXVII, Univ. Press, Cambridge, U. K., 1974, 209.
21. **Kerkut, G. A.,** Mini-review: axoplasmic transport, *Comp. Biochem. Physiol. A.,* 51, 701, 1975.
22. **Porter, K. R., Byers, H. R., and Ellisman, M. H.,** The cytoskeleton, in *The Neurosciences,* Schmitt, F. O. and Worden, F. G., Eds., MIT Press, Cambridge, Mass., 1979, 703.
23. **Stearns, M. E.,** Lattice-dependent regulations of axonal transport by calcium ions, in *Microtubules and Microtubule Inhibitors 1980,* De Brabander, M. and De Mey, J., Eds., Elsevier/North Holland, Amsterdam, 1980, 17.
24. **Ellisman, M. H.,** A hypothesis for rapid axoplasmic transport based upon focal interactions between axonal membrane systems and the microtrabecular crossbridges of the axoplasmic matrix, in *Axoplasmic Transport,* Weiss, D. G., Ed., Springer-Verlag, Basel, 1982, 390.
25. **Smith, R. S.,** Axonal transport of optically detectable particulate organelles, in *Axoplasmic Transport,* Weiss, D. G., Ed., Springer-Verlag, Basel, 1982, 181.
26. **Weiss, D. G. and Gross, G. W.,** Intracellular transport in axonal microtubular domains, I. Theoretical consideration on the essential properties of a force generating mechanism, *Protoplasma,* 114, 179, 1983.
27. **Zierler, K. L.,** Mechanism of muscle contraction and its energetics, in *Medical Physiology,* Vol. 2, 12th ed., Mountcastle, V. B., Ed., C. V. Mosby, St. Louis, 1968, 1128.
28. **Gross, G. W.,** The microstream concept of axoplasmic and dendritic transport, *Adv. Neurol.,* 12, 283, 1975.
29. **Weiss, D. G. and Gross, G. W.,** The microstream hypothesis of axoplasmic transport: characteristics, predictions and compatibility with data, in *Axoplasmic Transport,* Weiss, D. G., Ed., Springer-Verlag, Basel, 1982, 362.
30. **Gross, G. W. and Weiss, D. G.,** Intracellular transport in axonal microtubular domains. II. Velocity profile and energetics of circumtubular flow, *Protoplasma,* 114, 198, 1983.
31. **Wuerker, R. B. and Kirkpatrick, J. B.,** Neuronal microtubules, neurofilaments and microfilaments, *Int. Rev. Cytol.,* 33, 45, 1972.
32. **Ellisman, M. H. and Porter, K. R.,** Microtrabecular structure of the axoplasmic matrix: visualization of cross-linking structures and their distribution, *J. Cell Biol.,* 87, 464, 1980.
33. **Lasek, R. J. and Shelanski, M. L., Eds.,** Cytoskeletons and the architecture of nervous sytems, *Neurosci. Res. Prog. Bull.,* 19, 1, 1981.
34. **Bull, H. B.,** *An Introduction to Physical Biochemistry,* Davis, Philadelphia, 1964, 391.
35. **Freundlich, H. and Röder, H. L.,** Dilatancy and its relation to thixotropy, *Trans. Faraday Soc.,* 34, 308, 1938.
36. **Kamiya, N.,** Protoplasmic streaming, *Protoplasmatologia,* Vol. 8, 3a, Springer-Verlag, Basel, 1959.
37. **Taylor, D. L. and Condeelis, J. S.,** Cytoplasmic structure and contractility in amoeboid cells, *Int. Rev. Cytol.,* 56, 57, 1979.
38. **Pollard, T. D., Aebi, U., Cooper, J. A., Fowler, W. E., and Tseng, P.,** Actin structure, polymerization, and gelation, in *Cold Spring Harbor Symp. Quant. Biol.,* 46, 513, 1982.

39. **Marsland, D. A. and Brown, D. E. S.,** The effect of pressure on sol-gel equilibria with special reference to myosin and other protoplasmic gels, *J. Cell. Comp. Physiol.*, 20, 295, 1942.
40. **Scott-Blair, G. W.,** *Elementary Rheology*, Academic Press, New York, 1969, 74.
41. **Abe, S.,** The viscoelasticity of slime mold protoplasm, in *Symp. Biorheology, Proc. 4th Int. Congr. Rheol.*, Part 4, Copley, A. L., Ed., Interscience, New York, 1965, 147.
42. **Dintenfass, L.,** Rheology of synovial fluid; its role in joint lubrication, in *Symp. Biorhelogy, Proc. 4th Int. Congr. Rheol.*, Part 4, Copley, A. L., Ed., Interscience, New York, 1965, 487.
43. **Myers, W. G. and France, W. G.,** A study of some properties of thixotropic gels containing egg albumin as the disperse phase, *J. Phys. Chem.*, 44, 1113, 1940.
44. **Heilbrunn, L. V.,** *The Dynamics of Living Protoplasm*, Academic Press, New York, 1956.
45. **Bauer, W. H. and Collins, E. A.,** Thixotropy and dilatancy, in *Rheology — Theory and Application*, Eirich, F. R., Ed., Academic Press, New York, 1967, 446.
46. **Baker, P. F., Ladds, M., and Rubinson, K. A.,** Measurement of the flow properties of isolated axoplasm in a defined chemical environment, *J. Physiol. (London)*, 269, 10, 1977.
47. **Gilbert, D. G., Newby, B., and Anderton, B. M.,** Neurofilament disguise, destruction and discipline, *Nature (London)*, 256, 586, 1975.
48. **Nixon, R. A.,** Proteolysis of neurofilaments, in *Neurofilaments*, Marotta, C. A., Ed., University of Minnesota Press, Minneapolis, 1983, 117.
49. **Moore, P. B. and Carraway, K. L.,** Proteolytic enhancement of gelation of ascites tumor cell extracts. Relationship to actin binding protein, *Biochem. Biophys. Res. Commun.*, 80, 560, 1978.
50. **Isenberg, G., Aebi, U., and Pollard, D. T.,** An actin-binding protein from *Acanthamoeba* regulates actin filament polymerization and interactions, *Nature (London)*, 288, 455, 1980.
51. **Schwartz, J. H. and Goldberg, D. J.,** Studies on the mechanism of fast axoplasmic transport in single identified neurons, in *Axoplasmic Transport*, Weiss, D. G., Ed., Springer-Verlag, Basel, 1982, 351.
52. **Schliwa, M.,** Microtubule-dependent intracellular transport in chromatophores, in *International Cell Biology 1980-1981*, Schweiger, H. G., Ed., Springer-Verlag, Basel, 1981, 275.
53. **Soifer, D. Czosnek, H. H., Mack, K., and Wisniewski, H. M.,** Properties and dynamics of neurofilament proteins, in *Axoplasmic Transport*, Weiss, D. G., Ed., Springer-Verlag, Basel, 1982, 64.
54. **Berthold, C.-H.,** Morphology of normal peripheral axons, in *Physiology and Pathobiology of Axons*, Waxman, S. G., Ed., Raven Press, New York, 1978, 3.
55. **Hammond, G. R. and Smith, R. S.,** A statistical approach to examination of axonal microtubule distribution, *Microsc. Soc. Can.*, 3, 78, 1976.
56. **Raine, C. S.,** Neurocellular anatomy, in *Basic Neurochemistry*, Siegel, G. J., Albers, R. W., Katzman, R., and Agranoff, B. W., Eds., Little Brown, Boston, 1976, 5.
57. **Smith, D. S., Järlfors, U., and Cameron, B. F.,** Morphological evidence for the participation of microtubules in axonal transport, *Ann. N.Y. Acad. Sci.*, 253, 472, 1975.
58. **Weiss, P. A. and Mayr, R.,** Organelles in neuroplasmic ("axonal") flow: neurofilaments, *Proc. Natl. Acad. Sci. U.S.A.*, 68, 846, 1971.
59. **Green, L. S., Donoso, J. A., Heller-Bettinger, I. E., and Samson, F. E.,** Axonal transport disturbances in vincristine-induced peripheral neuropathy, *Ann. Neurol.*, 1, 255, 1977.
60. **Kreutzberg, G. W. and Gross, G. W.,** General morphology and axonal ultrastructure of the olfactory nerve of the pike, *Esox lucius*, *Cell Tissue Res.*, 181, 443, 1977.
61. **Langford, G. M.,** Length and appearance of projections on neuronal microtubules in vitro after negative staining: evidence against a crosslinking function for MAPs, *J. Ultrastruct. Res.*, 85, 1, 1983.
62. **Sasaki, S., Stevens, J. K., and Bodick, N.,** Serial reconstruction of microtubular arrays within dendrites of the cat retinal ganglion cell: the cytoskeleton of a vertebrate dendrite, *Brain Res.*, 259, 193, 1983.
63. **Ledbetter, M. C. and Porter, K. R.,** A "microtubule" in a plant cell fine structure, *J. Cell Biol.*, 19, 239, 1963.
64. **Mollenhauer, H. H. and Morré, D. J.,** Structural compartmentation of the cytosol: zones of exclusion, zones of adhesion, cytoskeletal and intercisternal elements, *Subcell. Biochem.*, 11, 327, 1978.
65. **Hall, S. M. and Williams, P. L.,** Studies on the "incisures" of Schmidt and Lanterman, *J. Cell Sci.*, 6, 767, 1970.
66. **Hyams, J. S. and Stebbings, H.,** The distribution and function of microtubules in nutritive tubes, *Tissue Cell*, 9, 537, 1977.
67. **Stebbings, H. and Willison, J. H. M.,** Structure of microtubules: a study of freeze-etched and negatively stained microtubules from the ovaries of *Notonecta*, *Z. Zellforsch.*, 138, 387, 1973.
68. **Peracchia, C.,** A system of parallel septa in crayfish nerve fibers, *J. Cell Biol.*, 44, 125, 1970.
69. **Holz, A. and Weber, W.,** Periodisch auftretende Querstrukturen in Nervenfasern des Bulbus olfactorius der Elritze *Phoxinus laevis*, *Experientia*, 26, 1349, 1970.
70. **Lieberman, A. R.,** Microtubule-associated smooth endoplasmic reticulum in the frog's brain, *Z. Zellforsch.*, 116, 564, 1971.

71. **Berthold, C.-H.,** Some aspects of the ultrastructural organization of peripheral myelinated axons in the cat, in *Axoplasmic Transport,* Weiss, D. G., Ed., Springer-Verlag, Basel, 1982, 40.
72. **Kirkpatrick, J. B., Bray, J. J., and Palmer, S. M.,** Visualization of axoplasmic flow in vitro by Nomarksi microscopy: comparison to rapid flow of radioactive proteins, *Brain Res.,* 43, 1, 1972.
73. **Cooper, P. D. and Smith, R. S.,** The movement of optically detectable organelles in myelinated axons of *Xenopus laevis, J. Physiol. (London),* 242, 77, 1974.
74. **Leestma, J. E.,** Velocity measurements of particulate neuroplasmic flow in organized mammalian CNS tissue cultures, *J. Neurobiol.,* 7, 173, 1976.
75. **Leestma, J. F. and Freeman, S.,** Computer-assisted analysis of particulate axoplasmic flow in organized CNS tissue cultures, *J. Neurobiol.,* 8, 453, 1977.
76. **Forman, D. S., Padjen, A. L., and Siggins, G. R.,** Axonal transport of organelles visualized by light microscopy: cinemicrographic and computer analysis, *Brain Res.,* 136, 197, 1977.
77. **Droz, B.,** Synthése et transfer des protéines cellulaires dans les neurones ganglionnaires. Étude radioautographique quantitative en microscopie électronique, *J. Microsc.,* 6, 201, 1967.
78. **Droz, B.,** Protein metabolism in nerve cells, *Int. Rev. Cytol.,* 25, 363, 1969.
79. **Tsukita, S. and Ishikawa, H.,** The movement of membranous organelles in axons. Electron microscopic identification of anterogradely and retrogradely transported organelles, *J. Cell Biol.,* 84, 513, 1980.
80. **Papasozomenos, S. Ch., Autilio-Gambetti, L., and Gambetti, P.,** Distribution of proteins migrating with fast axonal transport. Their relationship to smooth endoplasmic reticulum, *Brain Res.,* 278, 232, 1983.
81. **Ellisman, M. H. and Lindsay, J. D.,** The axoplasmic reticulum within myelinated axons is not transported rapidly, *J. Neurocytol.,* 12, 393, 1983.
82. **Allen, R. D., Allen, N. S., and Travis, J. L.,** Video-enhanced contrast, differential interference contrast (AVEC-DIC) microscopy: a new method capable of analyzing microtubule-related motility in the reticulopodial network of *Allogromia laticollaris, Cell Motility,* 1, 291, 1981.
83. **Griffin, J. W., Hoffman, P. N., and Price, D. L.,** Axonal transport in β,β′-iminodipropionitrile neuropathy, in *Axoplasmic Transport in Physiology and Pathology,* Weiss, D. G. and Gorio, A., Eds., Springer-Verlag, Basel, 1982, 109.
84. **Raine, C. S., Ghetti, B., and Shelanski, M. L.,** On the association between microtubules and mitochondria within axons, *Brain Res.,* 34, 389, 1071.
85. **Smith, D. S., Järlfors, U., and Beránek, R.,** The organization of synaptic axoplasm in the lamprey *(Petromyzon marinus)* central nervous system, *J. Cell Biol.,* 46, 199, 1970.
86. **Spencer, P. S. and Griffin, J. W.,** Disruption of axoplasmic transport by neurotoxic agents. The 2,5-hexanedione model, in *Axoplasmic Transport in Physiology and Pathology,* Weiss, D. G. and Gorio, A., Eds., Springer-Verlag, Basel, 1982, 92.
87. **Yokoyama, K., Tsukita, S., Ishikawa, H., and Kurokawa, M.,** Early changes in the neuronal cytoskeleton caused by β-β′-iminodipropionitrile: Selective impairment of neurofilament polypeptides, *Biomed. Res.,* 1, 537, 1980.
88. **Papasozomenos, S. Ch., Autilio-Gambetti, L., and Gambetti, P.,** The IDPN axon: Rearrangement of axonal cytoskeleton and organelles following β-β′-iminodipropionitrile (IDPN) intoxication, in *Axoplasmic Transport,* Weiss, D. G., Ed., Springer-Verlag, Basel, 1982, 241.
89. **Allen, R. D., Brown, D. T., Gilbert, S. P., and Fujiwake, H.,** Transport of vesicles along filaments dissociated from squid axoplasm, *Biol. Bull.,* 165, 523, 1983.
90. **Hayden, J. H. and Allen, R. D.,** Deletion of single microtubules in living cells: particle transport can occur in both directions along the same microtubule, *J. Cell Biol.,* 1984, in press.
91. **Allen, R. D. and Weiss, D. G.,** An experimental analysis of the mechanisms of fast axonal transport in the squid giant axon, in *Proc. II Yamada Conf. Cell Motility 1984,* Ishikawa, H., Sato, H., and Hatano, S., Eds., Univ. Tokyo Press, 1986, 342.
92. **Ochs, S.,** Retention and redistribution of proteins in mammalian nerve fibres by axoplasmic transport, *J. Physiol. (London),* 253, 459, 1975.
93. **Gross, G. W. and Beidler, L. M.,** A quantitative analysis of isotope concentration profiles and rapid transport velocities in the C-fibers of the garfish olfactory nerve, *J. Neurobiol.,* 6, 213, 1975.
94. **Muñoz-Martínez, E. J., Núñez, R., and Sanderson, A.,** Axonal transport: a quantitative study of retained and transported protein fraction in the cat, *J. Neurobiol.,* 12, 15, 1981.
95. **Gross, G. W., Stewart, G. H., and Horwitz, B.,** Computer generation of typical axoplasmic transport isotope distribution using theoretical principles of liquid chromatography, *Soc. Neurosci. Abstr.,* 6, 503, 1980.
96. **Stewart, G. H., Horwitz, B., and Gross, G. W.,** A chromatographic model of axoplasmic transport, in *Axoplasmic Transport,* Weiss, D. G., Ed., Springer-Verlag, Basel, 1982, 414.
97. **Mackey, S., Schuessler, G., Goldberg, D. J., and Schwartz, J. H.,** Dependence of fast axonal transport on the local concentration of organelles, *Biophys. J.,* 36, 455, 1981.

98. **Droz, B., Koenig, H. L., and DiGiamberardino, L.,** Axonal migration of protein and glycoprotein to nerve endings. I. Radioautographic analysis of the renewal of protein in nerve endings of chicken ciliary ganglion after intracerebral injection of ^{3}H-lysine, *Brain Res.*, 60, 93, 1973.
99. **Hendrickson, A. E.,** Electron microscopic distribution of axoplasmic transport, *J. Comp. Neurol.*, 144, 381, 1972.
100. **Schonbach, J., Schonbach, C., and Cuénod, M.,** Rapid phase of axoplasmic flow and synaptic proteins: an electron microscopical autoradiographic study, *J. Comp. Neurol.*, 141, 485, 1971.
101. **Goldman, J. E., Kim, K. S., and Schwartz, J. H.,** Axonal transport of (^{3}H)serotonin in an identified neuron of *Aplysia californica*, *J. Cell Biol.*, 70, 304, 1976.
102. **Papasozomenos, S. Ch., Yoon, M., Crane, R., Autilio-Gambetti, L., and Gambetti, P.,** Redistribution of proteins of fast axonal transport following administration of β,β′-iminodipropionitrile: a quantitative autoradiographic study, *J. Cell Biol.*, 95, 672, 1982.
103. **Ochs, S.,** Fast transport of materials in mammalian nerve fibers, *Science*, 176, 252, 1972.
104. **Breuer, A. C., Christian, C. N., Henkart, M., and Nelson, P. G.,** Computer analysis of organelle translocation in primary neuronal cultures and continuous cell lines, *J. Cell Biol.*, 65, 562, 1975.
105. **Gross, G. W.,** A Quantitative Characterization of Rapid Axoplasmic Transport in the C-fibers of the Garfish Olfactory Nerve, Ph.D. dissertation, Florida State University, Tallahassee, 1973.
106. **Hayden, J. H., Allen, R. D., and Goldman, R. D.,** Cytoplasmic transport in keratocytes: direct visualization of particle translocation along microtubules, *Cell Motility*, 3, 1, 1983.
107. **Collot, M., Louvard, D., and Singer, S. J.,** Lysosomes are associated with microtubules and not with intermediate filaments in cultured fibroblasts, *Proc. Natl. Acad. Sci. U.S.A.*, 81, 788, 1984.
108. **Brimijoin, S.,** On the kinetics and maximal capacity of the system for rapid axonal transport in mammalian neurones, *J. Physiol. (London)*, 292, 325, 1979.
109. **Byers, M. R.,** Structural correlates of rapid axonal transport: evidence that microtubules may not be directly involved, *Brain Res.*, 75, 97, 1974.
110. **Brimijoin, S., Olsen, J., and Rosenson, R.,** Comparison of the temperature-dependence of rapid axonal transport and microtubules in nerves of the rabbit and bullfrog, *J. Physiol.*, 287, 303, 1979.
111. **McEwen, B. S. and Grafstein, B.,** Fast and slow components in axonal transport of protein, *J. Cell Biol.*, 38, 494, 1968.
112. **Karlsson, J.-O. and Sjöstrand, J.,** Synthesis, migration and turnover of protein in retinal ganglion cells, *J. Neurochem.*, 18, 749, 1971.
113. **Cancalon, P.,** Influence of temperature on the velocity and on the isotope profile of slowly transported labeled proteins, *J. Neurochem.*, 32, 997, 1979.
114. **Black, M. N. and Lasek, R. J.,** Slow components of axonal transport: two cytoskeletal networks, *J. Cell Biol.*, 86, 616, 1980.
115. **Smith, R. S. and Koles, Z. J.,** Mean velocity of optically detected intraaxonal particles measured by a cross-correlation method, *Can. J. Physiol. Pharmacol.*, 54, 859, 1976.
116. **Condeelis, J. S.,** Reciprocal interactions between the actin lattice and cell membrane, *Neurosci. Res. Prog. Bull.*, 19, 83, 1981.
117. **Goldman, R. D., Chojnacki, B., Goldman, A. E., Starger, J., Steinert, P., Talian, J., Whitman, M., and Zackroff, R.,** Aspects of the cytoskeleton and cytomusculature of nonmuscle cells, *Neurosci. Res. Prog. Bull.*, 19, 59, 1981.
118. **LeBeux, Y. J. and Willemot, J.,** An ultrastructural study of the microfilaments in rat brain by means of heavy meromyosin labeling. I. The perikaryon, dendrites, and the axon, *Cell Tissue Res.*, 160, 1, 1975.
119. **Alonso, G., Gabrion, J., Travers, F., and Assenmacher, I.,** Ultrastructural organization of actin filaments in neurosecretory axons of the rat, *Cell Tissue Res.*, 214, 323, 1981.
120. **Taylor, E. W.,** Cytoskeletons and the architecture of nervous systems. Retrospective, *Neurosci. Res. Prog. Bull.*, 19, 136, 1981.
121. **Heuser, J. E. and Kirschner, M. W.,** Filament organization revealed in platinum replicas of freeze-dried cytoskeletons, *J. Cell Biol.*, 86, 212, 1980.
122. **Ishikawa, H. and Tsukita, S.,** Morphological and functional correlates of axoplasmic transport, in *Axoplasmic Transport*, Weiss, D. G., Ed., Springer-Verlag, Basel, 1982, 251.
123. **Giddings, J. C., Fisher, S. R., and Myers, M. N.,** Field-flow fractionation. One-phase chromatography for macromolecules and particles, *Int. Lab.*, May/June, 15, 1978.
124. **Gross, G. W. and Weiss, D. G.,** Subcellular fractionation of rapidly transported axonal material in olfactory nerve: evidence for a size-dependent molecule separation during transport, *Neurosci. Lett.*, 5, 15, 1977.
125. **Willard, M., Cowan, W. M., and Vagelos, P. R.,** The polypeptide composition of intra-axonally transported proteins: evidence for four transport velocities, *Proc. Natl. Acad. Sci. U.S.A.*, 71, 2183, 1974.
126. **Schwartz, J. H.,** Axonal transport: components, mechanisms and specificity, *Ann. Rev. Neurosci.*, 2, 467, 1979.
127. **Oplatka, A.,** Foreword, in *Biological Structures and Coupled Flows*, Oplatka, A. and Balaban, M., Eds., Academic Press, New York, 1983, xi.

128. **Schwartz, J. H., Goldman, J. E., Ambron, R. T., and Goldberg, D. J.,** Axonal transport of vesicles carrying serotonin in the metacerebral neuron of *Aplysia californica*, *Cold Spring Harbor Symp. Quant. Biol.*, 40, 83, 1975.
129. **Small, J. V., Rinnerthaler, G., and Hinssen, H.,** Organization of actin meshworks in cultured cells: the leading edge, *Cold Spring Harbor Symp. Quant. Biol.*, 46, 599, 1982.
130. **Bray, D. and Bunge, M. B.,** Serial analysis of microtubules in cultured rat sensory axons, *J. Neurocytol.*, 10, 589, 1981.
131. **Bereiter-Hahn, J.,** A model for microtubular rigidity, *Cytobiologie*, 17, 298, 1978.
132. **Griffith, L. M. and Pollard, T. D.,** Evidence for actin filament-microtubule interaction mediated by microtubule-associated proteins, *J. Cell Biol.*, 78, 958, 1978.
133. **Haimo, L. T., Telzer, B. R., and Rosenbaum, J. L.,** Dynein, binds to and cross-bridges cytoplasmic microtubules, *Proc. Natl. Acad. Sci. U.S.A.*, 76, 5759, 1979.
134. **Larsson, H., Wallin, M., Edström, A.,** Some characteristics of ATPase activity in a brain microtubule protein preparation, *J. Neurochem.*, 33, 1249, 1979.
135. **Prus, K. and Wallin, M.,** Microtubule-associated ATPase: fact or artifact?, in *Axoplasmic Transport*, Weiss, D. G., Ed., Springer-Verlag, Basel, 1982, 91.
136. **Sharp, G. A., Fitzsimmons, J. T. R., and Kerkut, G. A.,** Electron microscopic localization of an ATPase associated with microtubules in the cerebral cortex of the rat, *Comp. Biochem. Physiol.*, 66A, 415, 1980.
137. **Tominaga, S. and Kaziro, Y.,** Adenosine triphosphatases associated with bovine brain microtubules. I. Presence of two distinct ATPases and their purification, *J. Biochem.*, 93, 1085, 1983.
138. **Katchalsky, A. and Oplatka, A.,** Mechano-chemical conversion, in *Handbook of Sensory Physiology. I. Principles of Receptor Physiology*, Autrum, H., Jung, R., Loewenstein, W. R., MacKay, D. M., and Teuber, H. L., Eds., Springer-Verlag, Basel, 1971, 1.
139. **Oplatka, A., Levy, R., and Friedman, J. E.,** The elementary mechanochemical unit of the actomyosin system, in *Biological Structures and Coupled Flows*, Oplatka, A. and Balaban, M., Eds., Academic Press, New York, 1983, 207.
140. **Pollack, G. H.,** The cross-bridge theory, *Physiol. Rev.*, 63, 1049, 1983.
141. **Oplatka, A. and Balaban, M., Eds.,** *Biological Structures and Coupled Flows*, Academic Press, New York, 1983.
142. **Tirosh, R., Liron, N., and Oplatka, A.,** A proposal for the molecular basis of cytoplasmic streaming, in *Cell Motility: Molecules and Organization*, Hatano, S., Ishikawa, H., and Sato, H., Eds., University of Tokyo Press, Tokyo, 1979, 133.
143. **Shimizu, H. and Yano, M.,** Studies of the chemo-mechanical conversion in artificially produced streamings. III. Dynamic cooperativity — a new cooperativity in actomyosin systems with a polarized arrangement of F-actin, *J. Biochem.*, 84, 1093, 1978.
144. **Jaffe, L.,** An electrophoretic model of axonal organelle transport, *Biol. Bull.*, 167, 502, 1984.
145. **Pollard, T. D.,** Which organelles are necessary for fast neuronal transport?, *Neurosci. Res. Prog. Bull.*, 20, 92, 1981.
146. **Blobel, G.,** Intracellular protein topogenesis, *Proc. Natl. Acad. Sci. U.S.A.*, 77, 1492, 1980.
147. **Rebhun, L. I.,** Polarized intracellular parlticle transport: saltatory movements and cytoplasmic streaming, *Int. Rev. Cytol.*, 32, 93, 1972.
148. **Schliwa, M.,** Mechanisms of intracellular organelle transport, in *Cell and Muscle Motility*, Vol. 5, Shay, J. W., Ed., Plenum Press, New York, 1984, 1.
149. **Schwab, M. E. and Thoenen, H.,** Mechanism of uptake and retrograde axonal transport of noradrenaline in sympathetic neurons in culture: reserpine-resistant large dense-core vesicles as transport vehicles, *J. Cell Biol.*, 96, 1538, 1983.
150. **Allen, R. D., Metuzals, J., Tasaki, I., Brady, S. T., and Gilbert, S. P.,** Fast axonal transport in squid giant axons, *Science*, 218, 1127, 1982.
151. **Buchner, K., Gulden, J., and Weiss, D. G.,** Bidirectional cytoplasmic transport of organelles in olfactory nerve axons: an AVEC-DIC study, in *International Cell Biology 1984—1985*, Seno, S. and Okada, Y., Eds., Academic Press, New York, 1984, 491.
152. **Buchner, K. and Weiss, D. G.,** Anterograde and retrograde axoplasmic transport of exogenous proteins in C-fibers, *J. Neurochem.*, 41, S96, 1983.
153. **Kidwai, A. M. and Ochs, S.,** Components of fast and slow phases of axoplasmic flow, *J. Neurochem.*, 16, 1105, 1969.
154. **Sabri, M. I. and Ochs, S.,** Characterization of fast and slow transported proteins in dorsal root and sciatic nerve of cat, *J. Neurobiol.*, 4, 145, 1972.
155. **Wolburg, H.,** Intraaxonaler Transport von Ethidium-Bromid-sensitiven RNS-und niedermolekularen ^{3}H-Uridin-Verbindungen im Tractus opticus von Teleosteern, *Exp. Brain, Res.*, 15, 348, 1972.
156. **Csanyi, V. Gervai, J., and Lajtha, A.,** Axoplasmic transport of free amino acids, *Brain Res.*, 56, 271, 1973.

157. **Roberts, P. J., Keen, P., and Mitchell, J. F.,** The distribution and axonal transport of free amino acids and related compounds in the dorsal sensory neuron of the rat, as determined by the dansyl reaction, *J. Neurochem.*, 21, 199, 1973.

158. **Breer, H. and Rahmann, H.,** Axonal transport of ^{3}H-glucose radioactivity in the optic system of *Scardinius erythrophthalmus*, *J. Neurochem.*, 22, 245, 1974.

159. **Cancalon, P. and Beidler, L. M.,** Distribution along with the axon and into various subcellular fractions of molecules labeled with (^{3}H)leucine and rapidly transported in the garfish olfactory nerve, *Brain Res.*, 89, 225, 1975.

160. **Price, C. H., McAdoo, D. J., Farr, W., and Okuda, R.,** Bidirectional axonal transport of free glycine in identified neurons R3-R14 of *Aplysia*, *J. Neurobiol.*, 10, 551, 1979.

161. **Starkey, R. R and Brimijoin, S.,** Stop flow analysis of the axonal transport of DOPA decarboxylase (EC 4.1.1.26) in rabbit sciatic nerves, *J. Neurochem.*, 32, 437, 1979.

162. **Elam, J.,** Compositon, subcellular distribution and turnover of axonally transported glycoproteins, in *Axoplasmic Transport*, Weiss, D. G., Ed., Springer-Verlag, Basel, 1982, 125.

163. **Weiss, D. G.,** 3-O-methyl-D-glucose and β-alanine: rapid axoplasmic transport of metabolically inert low molecular weight substances, *Neurosci. Lett.*, 31, 241, 1982.

164. **Hawlitschek, G.,** Schneller Axonaler Transport löslicher und Partikulärer Proteine, Ph.D. thesis, Universität München, München, 1984.

165. **Baitinger, C., Levine, J., Lorenz, T., Simon, C., Skene, P., and Willard, M.,** Characteristics of axonally transported proteins, in *Axoplasmic Transport*, Weiss, D. G., Ed., Springer-Verlag, Basel, 1982, 110.

166. **Rubinow, S. I. and Blum, J. J.,** A theoretical approach to the analysis of axonal transport, *Biophys. J.*, 30, 135, 1980.

167. **Black, M. M. and Lasek, R. J.,** Axonal transport of actin: slow component b is the principal source of actin for the axon, *Brain Res.*, 171, 401, 1979.

168. **Willard, M., Wiseman, M., Levine, J., and Skene, P.,** Axonal transport of actin in rabbit retinal ganglion cells, *J. Cell. Biol.*, 81, 581, 1979.

169. **Willard, M.,** The identification of two intra-axonally transported polypeptides resembling myosin in some respects in the rabbit visual system, *J. Cell Biol.*, 75, 1, 1977.

170. **Garner, J. A. and Lasek, R. J.,** Clathrin is axonally transported as part of slow component b: the microfilament complex, *J. Cell. Biol.*, 88, 172, 1981.

171. **Hoffman, P. N. and Lasek, R. J.,** The slow component of axonal transport. Identification of major structural polypeptides of the axon and their generality among mammalian neurons, *J. Cell Biol.*, 66, 351, 1975.

172. **Levine, J., Skene, P., and Willard, M.,** GAPs and fodrin. Novel axonally transported proteins, *Trends Neurosci.*, 4, 273, 1981.

173. **Oplatka, A. and Tirosh, R.,** Active streaming in actomyosin solutions, *Biochim. Biophys. Acta*, 305, 684, 1973.

174. **Tirosh, R. and Oplatka, A.,** Active streaming against gravity in glass microcapillaries of solutions containing acto-heavy meromyosin and native tropomyosin, *J. Biochem.*, 91, 1435, 1982.

175. **Odell, G. M.,** A new mathematical continuum theory of axoplasmic transport, *J. Theor. Biol.*, 60, 223, 1976.

176. **Happel. J. and Brenner, H.,** *Low Reynolds Number Hydrodynamics*, Prentice-Hall, Englewood Cliffs, N.J., 1965.

177. **Purcell, E. M.,.** Life at low Reynolds number, *Am. J. Phys.*, 45, 3, 1977.

178. **Koles, K. J., McLeod, K. D., and Smith, R. S.,** The determination of the instantaneous velocity of axonally transported organelles from filmed records of their motion, *Can. J. Physiol. Pharmacol.*, 60, 670, 1982.

179. **Adam, G. and Delbrück, M.,** Reduction of dimensionality in biological diffusion processes, in *Structural Chemistry and Molecular Biology: A Volume Dedicated to Linus Pauling by his Students, Colleagues, and Friends*, Rich, A. and Davidson, N., Eds., W. H. Freeman, New York, 1968.

180. **Weiss, D. G., Gross, G. W., and Kreutzberg, G. W.,** Subcellular fractionation of rapidly transported axonal constituents, *Proc. Int. Soc. Neurochem.*, 6, 114, 1977.

181. **Adams, R. J. and Bray, D.,** Rapid transport of foreign particles microinjected into crab axons, *Nature (London)*, 303, 718, 1983.

182. **Pratt, M. M., Otter, T., and Salmon, E. D.,** Dynein-like Mg^{2+}-ATPase in mitotic spindles isolated from sea urchin embryos (Strongloc-entrutus droebuchiensis), *J. cell Biol.*, 86, 738, 1980.

183. **Sakai, H., Abuchi, I. M., Shimoda, S., Kuriyama, R., Ogawa, K., and Mohri, H.,** Induction of chromosome motion in the glycerol isolated mitotic apparatus. Nucleotide specificity and effects of anti-dynein and myosin sera on the motion, *Dev. Growth Differ.*, 18, 211, 1976.

184. **Murphy, D. B., Hiebsch, R. R., and Wallis, K. T.,** Identity and origin of the ATPase activity associated with neuronal microtubules. I. The ATPase activity is associated with membrane vesicles, *J. Cell Biol.*, 96, 1298, 1983.
185. **Prus, K. and Wallin, M.,** The ATPase activity in brain microtubule preparations is membrane-associated, *Histochemistry*, 78, 181, 1983.
186. **Hanson, M. and Edström, A.,** Mitosis inhibitors and axonal transport, *Int. Rev. Cytol.*, Suppl. 7, 373, 1978.
187. **Wallin, M., Larsson, H., and Edström, A.,** Effects of sulfhydryl reagents on brain microtubule-associated ATPase activity in vitro, *J. Neurochem.*, 33, 1095, 1979.
188. **Edström, A. and Mattsson, H.,** Inhibition and stimulation of rapid axonal transport in vitro by sulfhydryl blockers, *Brain Res.*, 108, 381, 1976.
189. **Mabuchi, I.,** Biochemistry of dynein and its role in cell motility, *Horizons Biochem. Biophys.*, 5, 1, 1978.
190. **Forman, D. S., Brown, K. J., and Livengood, D. R.,** Fast axonal transport in permeabilized lobster giant axons is inhibited by vanadate, *J. Neurosci.*, 3, 1279, 1983.
191. **Goldberg, D. J.,** Microinjection into an identified axon to study the mechanism of fast axonal transport, *Proc. Natl. Acad. Sci. U.S.A.*, 79, 4818, 1982.
192. **Forman, D. S.,** Vanadate inhibits saltatory organelle movement in a permeabilized cell model, *Exp. Cell Res.*, 141, 139, 1982.
193. **Clark, T. G. and Rosenbaum, J. L.,** Pigment particle translocation in detergent-permeabilized melanophores of *Fundulus heteroclitus*, *Proc. Natl. Acad. Sci. U.S.A.*, 79, 4655, 1982.
194. **Beckerle, M. C. and Porter, K. R.,** Inhibitors of dynein-activity block intracellular transport in erythrophores, *Nature (London)*, 295, 701, 1982.
195. **Isenberg, G., Schubert, P., and Kreutzberg, G. W.,** Experimental approach to test the role of actin in axonal transport, *Brain Res.*, 194, 588, 1980.
196. **Goldberg, D.,** Actin and myosin in nerve cells, in *Axoplasmic Transport*, Weiss, D. G., Ed., Springer-Verlag, Basel, 1982, 73.
197. **Abe, T., Haga, T., and Kurokawa, M.,** Rapid transport of phosphatidylcholine occuring simultaneously with protein transport in the frog sciatic nerve, *Biochem. J.*, 136, 731, 1973.
198. **Anderson, K. E., Edström, A., and Mattsson, H.,** Effects of cytochalasin B on uptake of glucosamine, leucine, and sulfate into nerve cells: incorporation into glycoproteins and rapid axonal transport, *Brain Res.*, 43, 343, 1972.
199. **Banks, P., Mayor, D., and Mraz, P.,** Cytochalasin B and the intra-axonal movement of noradrenaline storage vesicles, *Brain Res.*, 49, 417, 1973.
200. **Edström, A. and Hanson, M.,** Retrograde axonal transport of proteins *in vitro* in frog sciatic nerves, *Brain Res.*, 61, 311, 1973.
201. **Goldberg, D. J., Harris, D. A., Lubit, B. W., and Schwartz, J. H.,** Analysis of the mechanism of fast axonal transport by intracellular injection of potentially inhibitory macromolecules: evidence for a possible role for actin filaments, *Proc. Natl. Acad. Sci. U.S.A.*, 77, 7448, 1980.
202. **Thoenen, H. and Kreutzberg, G. W., Ed.,** The role of fast transport in the nervous system, *Neurosci. Res., Prog. Bull.*, 20, 1, 1981.
203. **Hammerschlag, R. and Stone, G. C.,** Fast axonal transport as endomembrane flow, in *Axoplasmic Transport*, Weiss, D. G., Ed., Springer-Verlag, Basel, New York, 1982, 406.
204. **Beckerle, M. C. and Porter, K. R.,** Analysis of the role of microtubules and actin in erythrophore intracellular motility, *J. Cell Biol.*, 96, 354, 1983.
205. **Beckerle, M. C. and Porter, K. R.,** Intracellular motility in erythrophores examined by microinjection, *J. Cell Biol.*, 91, 302a, 1981.
206. **Pollard, T. D., Aebi, U., Cooper, J. A., Fowler, W. E., Kiehart, D. P., Smith, P. R., and Tseng, P. C.,** Actin and myosin function in *Acanthameoba, Phil. Trans. R. Soc. London Ser B.*, 299, 237, 1982.
207. **Jockusch, B. M., Burger, M. M., Da Prada, M., Richards, J. G., Chaponnier, C., and Gabbiani, G.,** α-Actinin attached to membrane secretory vesicles, *Nature (London)*, 270, 629, 1977.
208. **Aunis, D., Guerold, B., Bader, M. F., and Cieselski-Treska, J.,** Immuno-cytochemical and biochemical demonstration of contractile proteins in chromaffin cells in culture, *Neuroscience*, 5, 2261, 1980.
209. **Stadler, H. and Dove, G. H. C.,** Identification of a heparan sulfate containing proteoglycan in cholinergic synaptic vesicles of *Torpedo marmorata*, *EMBO J.*, 1, 1982.
210. **Forman, D. S.,** Saltatory organelle movement and the mechanism of fast axonal transport, in *Axoplasmic Transport*, Weiss, D. G., Ed., Springer-Verlag, Basel, 1982, 234.
211. **Zelená, J., Lubińska, L., and Gutmann, E.,** Accumulation of organelles at the ends of interrupted axons, *Z. Zellforsch.*, 91, 200, 1968.
212. **Shield, L. K., Griffin, J. W., Drachman, D. B., and Price, D. L.,** Retrograde axonal transport: a direct method for measurement of rate, *Neurology*, 27, 393, 1977.

213. **La Vail, J. H., Rapisardi, S., and Sugino, I. K.,** Evidence against the smooth endoplasmic reticulum as a continuous channel for the retrograde axonal transport of horseradish peroxidase, *Brain Res.*, 191, 3, 1980.
214. **Gainer, H. and Fink, D. J.,** Covalent labeling techniques and axonal transport, in *Axoplasmic Transport*, Weiss, D. G., Ed., Springer-Verlag, Basel, 1982, 464.
215. **Forman, D. S., Padjen, A. L., and Siggins, G. R.,** *Movement of Organelles in Living Nerve Fibers* (scientific film), National Audio Visual Center, Washington, 1975.
216. **Bisby, M. A. and Bulger, V. T.,** Reversal of axonal transport at a nerve crush, *J. Neurochem.*, 29, 313, 1977.
217. **Bisby, M. A.,** Reversal of axonal transport: similarity of proteins transported in anterograde and retrograde directions, *J. Neurochem.*, 36, 741, 1981.
218. **Snyder, R. E. and Smith, R. S.,** Application of position-sensitive detectors to the study of the axonal transport of β-emitting isotopes, in *Axoplasmic Transport*, Weiss, D. G., Ed., Springer-Verlag, Basel, 1982, 442.
219. **Wannenwetsch, E., Buchner, K., and Weiss, D. G.,** Anterograder und retrograder Transport von freiem (^{3}H)Leucin entlang des Nervus olfactorius, *Hoppe-Seyler's Z. Physiol. Chem.*, 364, 1283, 1983.
220. **Heidemann, R. S.,** Visualization of microtubule polarity, in *Microtubules and Microtubule Inhibitors 1980*, De Brabander, M. and De Mey, J., Eds., Elsevier/North Holland, Amsterdam, 1980, 341.
221. **Burton, P. R. and Paige, J. L.,** Polarity of axoplasmic microtubules in the olfactory nerve of the frog, *Proc. Natl Acad. Sci. U.S.A.*, 78, 3269, 1981.
222. **Filliatreau, G. and DiGiamberardino, L.,** In vivo reassembly of axonal microtubules: a fast undirectional process, *Biol. Cell*, 41, 63, 1981.
223. **Heidemann, R. S. and McIntosh, J. R.,** Visualization of the structural polarity of microtubules, *Nature (London)*, 286, 517, 1980.
224. **Euteneuer, U. and McIntosh, J. R.,** Structural polarity of kinetochore microtubules in PtK1 cells, *J. Cell Biol.*, 89, 338, 1981.
225. **Telzer, B. R. and Haimo, L. T.,** Decoration of spindle microtubules with dynein: evidence for uniform polarity, *J. Cell Biol.*, 89, 373, 1981.
226. **Garner, J. A. and Lasek, R. J.,** Cohesive axonal transport of the slow component b complex of polypeptides, *J. Neurosci.*, 2, 1824, 1982.
227. **Brady, S. T. and Lasek, R. J.,** The slow components of axonal transport: movements, compositions, and organization, in *Axoplasmic Transport*, Weiss, D. G., Ed., Springer-Verlag, Basel, 1982, 206.
228. **Tytell, M., Black, M. M., Garner, J., and Lasek, R. J.,** Axonal transport: each of the major rate components consist of distinct macromolecular complexes, *Science*, 214, 179, 1981.
229. **Weiss, D. G., Schmid, G., and Wagner, L.,** Influence of microtubule inhibitors on axoplasmic transport of free amino acids. Implications for the hypothetical transport mechanism, in *Microtubules and Microtubule Inhibitors 1980*, De Brabander, M. and De Mey, J., Eds.,Elsevier/North Holland, Amsterdam, 1980, 31.
230. **Wolosewick, J. J. and Porter, K. R.,** Microtrabecular lattice of the cytoplasmic ground substance. Artifact or reality?, *J. Cell Biol.*, 82, 114, 1979.
231. Organization of the cytoplasm, *Cold Spring Harbor Symp. Quant. Biol.*, Vol. 46, 1982.
232. **Durham, A. C. H.,** A unified theory of the control of actin and myosin in nonmuscle movements, *Cell*, 2, 123, 1974.
233. **Willard, M.,** Neurofilaments and axonal transport, in *Neurofilaments*, Marotta, C. A., Ed., Univerity of Minnesota Press, Minneapolis, 1983, 86.
234. **Giddings, J. C.,** *Dynamics of Chromatography, Part I: Principles and Theory*, Marcel Dekker, New York, 1965, chap. 2.
235. **Stewart, G. W.,** in *Encyclopedia of Chemical Technology*, Vol. 5, 2nd ed., Stanton, A., Ed., John Wiley & Sons, New York, 1964, 413.
236. **Weiss, D. G. and Gross, G. W.,** Intracellular transport in nerve cell processes: the chromatographic dynamics of axoplasmic transport, in *Biological Structures and Coupled Flows*, Oplatka, A. and Balaban, M., Eds., Academic Press, New York, 1983, 387.
237. **Lubińska, L.,** Axoplasmic streaming in regenerating and in normal nerve fibres, in *Mechanisms of Neural Regeneration, Progr. Brain Res.*, Vol. 13, Singer, M. and Schadé, J. P., Eds., Elsevier/North Holland, Amsterdam, 1964, 1.
238. **Mori, H. and Kurokawa, M.,** Intra-axonal tranport of clathrin and actin, *Biomedical Res.*, 2, 677, 1981.
239. **Pipeleers, D. G., Pipeleers-Marichal, M. A., and Kipnis, D. M.,** Physiological regulation of total tubulin and polymerized tubulin in tissues, *J. Cell Biol.*, 74, 351, 1977.
240. **Lasek, R. J., McQuarrie, I. G., and Brady, S. T.,** Transport of cytoskeletal and soluble proteins in neurons, in *Biological Structures and Coupled Flows*, Oplatka, A. and Balaban, M., Eds., Academic Press, New York, 1983, 329.
241. **Morris, J. R. and Lasek, R. J.,** Monomer-polymer equilibria in the axon: direct measurement of tubulin and actin as polymer and monomer in axoplasm, *J. Cell Biol.*, 98, 2064, 1984.

242. **Lasek, R. J. and Brady, S. T.,** The structural hypothesis of axonal tranport: two classes of moving elements, in *Axoplasmic Transport,* Weiss, D. G., Ed., Springer-Verlag, Basel, 1982, 397.
243. **Grafstein, B. and Alpert, R. M.,** Properties of slow axonal transport: studies in goldfish optic axons, in *Axoplasmic Transport,* Weiss, D. G., Ed., Springer-Verlag, Basel, 1982, 226.
244. **Tashiro, T. and Kurokawa, M.,** Rapid transport of a calmodulin-related polypeptide in the vagal nerve, in *Axoplasmic Transport,* Weiss, D. G., Ed., Springer-Verlag, Basel, 1982, 153.
245. **McQuarrie, I. G., Brady, S. T., and Lasek, R. J.,** Polypeptide composition and kinetics of SCa and SCb in sciatic nerve motor axons and optic axons of the rat, *Soc. Neurosci. Abstr.,* 6, 501, 1980.
246. **Mori, H., Komiya, Y., and Kurokawa, M.,** Slowly migrating axonal polypeptides. Inequalities in their rate and amount of transport between two branches of bifurcarting axons, *J. Cell Biol.,* 82, 174, 1979.
247. **Tashiro, T., Kurokawa, M., and Komiya, Y.,** Two populations of axonally transported tubulin differentiated by their interactions with neurofilaments, *J. Neurochem.,* 1984, 43, 1220, 1984.
248. **Komiya, Y. and Kurokawa, M.,** Preferential blockade of the tubulin transport by colchicine, *Brain Res.,* 190, 505, 1980.
249. **Griffin, J. W., Anthony, D. C., Fahnestock, K. E., Hoffman, P. N., and Graham, D. G.,** 3,4-Dimethyl-2,5-hexanedione impairs the axonal transport of neurofilament proteins, *J. Neurosci.,* 4, 1516, 1984.
250. **Gross, G. W., Stewart, G. H., and Horwitz, B.,** Molecular diffusion cannot account for spreading of isotope distribution peaks during rapid axoplasmic transport, *Brain Res.,* 216, 215, 1981.
251. **Filliatreau, G. and Di Giamberardino, L.,** Quantitative analysis of axonal transport of cytoskeletal proteins in chicken oculomotor nerve, *J. Neurochem.,* 39, 1033, 1982.
252. **Cancalon, P.,** Subcellular and polypeptide distributions of slowly transported proteins in the garfish olfactory nerve, *Brain Res.,* 161, 115, 1979.
253. **Stromska, D. P. and Ochs, S.,** Patterns of slow transport in sensory nerves, *J. Neurobiol.,* 12, 441, 1981.
254. **Hirokawa, N.,** Cross-linker system between neurofilaments, microtubules, and membranous organelles in frog axons revealed by the quick-freeze, deep-etching method, *J. Cell Biol.,* 94, 129, 1982.
255. **Runge, M. S. and Williams, R. C., Jr.,** Formation of an ATP-dependent microtubule-neurofilament complex *in vitro, Cold Spring Harbor Symp. Quant. Biol.,* 46, 483, 1982.
256. **Zackroff, R. V., Idler, W. W., Steinert, P. M., and Goldman, R. D.,** *In vitro* reconstitution of intermediate filaments from mammalian triplet polypeptides, *Proc. Natl. Acad. Sci. U.S.A.,* 79, 754, 1982.
257. **Zackroff, R. V. and Goldman, R. D.,** *In vitro* reassembly of squid brain intermediate filaments (neurofilaments); purification by assembly-disassembly, *Science,* 208, 1152, 1980.
258. **Czosnek, H., Soifer, D., and Wisniewski, H. M.,** Studies on the biosynthesis of neurofilament proteins, *J. Cell Biol.,* 85, 726, 1980.
259. **Penman, S., Fulton, A., Capco, D., Ben Ze'ev, A., Wittelsburger, S., and Tse, C. F.,** Cytoplasmic and nuclear architecture in cells and tissue: form, functions, and mode of assembly, *Cold Spring Harbor Symp. Quant. Biol.,* 46, 1013, 1982.
260. **Lasek, R. J. and Hoffman, P. N.,** The neuronal cytoskeleton, axonal transport and axonal growth, in *Cell Motility,* Vol. 3, Goldman, R., Pollard, T., and Rosenbaum, J., Eds., Cold Spring Harbor Laboratory, Cold Spring Harbor, New York, 1976, 1021.
261. **Nixon, R. A.,** Protein degradation in the mouse visual system. I. Degradation of axonally transported and retinal proteins, *Brain Res.,* 200, 69, 1980.
262. **Davison, P. F.,** Neuronal fibrillar proteins and axoplasmic transport, *Brain Res.,* 100, 73, 1975.
263. **Frizell, M., McLean, W. G., and Sjöstrand, J.,** Slow axonal transport of proteins; blockade by interruption of contact between cell body and axon, *Brain Res.,* 86, 67, 1975.
264. **Cancalon, P.,** Proximodistal degeneration of C-fibers detached from their perykarya, *J. Cell Biol.,* 97, 6, 1983.
265. **Grafstein, B., McEwen, B. S., and Shelanski, M. L.,** Axonal transport of neurotubule protein, *Nature (London),* 227, 289, 1970.
266. **Karlsson, J.-O. and Sjöstrand, J.,** Transport of microtubular protein in axons of retinal ganglion cells, *J. Neurochem.,* 18, 975, 1971.
267. **Wilson, L., Bamburg, J. R., Mizel, S. B., Grisham, L. M., and Creswell, K. M.,** Interaction of drugs with microtubule proteins, *Fed. Proc. Fed. Am. Soc. Exp. Biol.,* 33, 158, 1974.
268. **Dahlström, A.,** Axoplasmic transport (with particular respect to adrenergic neurons), *Phil. Trans. R. Soc. London Ser. B.,* 261, 325, 1971.
269. **Gross, G. W. and Kreutzberg, G. W.,** Rapid axoplasmic transport in the olfactory nerve of the pike. I. Basic transport parameters for proteins and amino acids, *Brain Res.,* 139, 65, 1978.
270. **Cuénod, M. and Streit, P.,** Amino acid transmitters and local circuitry in optic tectum, in *Neurosciences. Fourth Study Program,* Schmitt, F. O. and Worden, F. G., Eds., MIT Press, Cambridge, Mass., 1979, 989.
271. **Streit, P., Knecht, E., and Cuénod, M.,** Transmitter related retrograde labeling in the pigeon optic lobe; a high resolution autoradiographic study, *Brain Res.,* 187, 59, 1980.

272. **Ingoglia, N. A., Sturman, J. A., Lindquist, T. D., and Gaull, G. E.,** Axonal migration of taurine in the goldfish visual system, *Brain Res.*, 115, 535, 1976.
273. **Ingoglia, N. A., Sturman, J. A., and Eisner, R. A.,** Axonal transport of putrescine, spermidine and spermine in normal and regenerating goldfish optic nerves, *Brain Res.*, 130, 433, 1977.
274. **Elam, J. S. and Agranoff, B. W.,** Rapid transport of protein in the optic system of the goldfish, *J. Neurochem.*, 18, 375, 1971.
275. **Neale, J. H., Elam, J. S., Neale, E. A., and Agranoff, B. W.,** Axonal transport and turnover of proline- and leucine-labeled protein in the goldfish visual system, *J. Neurochem.*, 23, 1045, 1974.
276. **Karlsson, J.-O.,** Is there an axonal transport of amino acids?, *J. Neurochem.*, 29, 615, 1977.
277. **Di Giamberardino, L.,** Independence of the rapid axonal transport of protein from the flow of free amino acids, *Acta Neuropathol.*, Suppl. V, 132, 1971.
278. **Koike, H. and Nagata, Y.,** Intra-axonal diffusion of (^{3}H)-acetylcholine and (^{3}H)-γ-aminobutyric acid in a neuron of *Aplysia*, *J. Physiol. (London)*, 295, 397, 1979.
279. **Norström, A.,** Axonal transport and turnover of neurohypophysial protein in the rat, *Ann. N.Y. Acad. Sci.*, 248, 46, 1975.
280. **Ikeda, H. and Pessoa, V. F.,** Axonal transport in demyelinating optic nerve of mice infected with Semliki Forest virus, *J. Physiol. (London)*, 339, 16P, 1983.
281. **Moore, H. P., Walker, M. D., Lee, F., and Kelly, R. B.,** Expressing a human proinsulin cDNA in a mouse ACTH secreting cell. Intracellular storage, proteolytic processing, and secretion on stimulation, *Cell*, 35, 531, 1983.
282. **Moore, H. P., Gumbiner, B., and Kelly, R. B.,** Chloroquine diverts ACTH from a regulated to a constitutive pathway in AtT-20 cells, *Nature (London)*, 302, 434, 1983.
283. **Heuser, J. E. and Reese, T. S.,** Evidence for recycling of synaptic vesicle membrane during transmitter release at the frog neuromuscular junction, *J. Cell Biol.*, 57, 315, 1973.
284. **Zimmermann, H. and Whittaker, V. P.,** Effect of electrical stimulation on the yield and composition of synaptic vesicles from the cholinergic synapses of the electric organ of *Torpedo:* A combined biochemical, electrophysiological and morphological study, *J. Neurochem.*, 22, 435, 1974.
285. **Goldberg, D. J., Schwartz, J. H., and Sherbany, A. A.,** Kinetic properties of normal and perturbed axonal transport of serotonin in a single identified axon, *J. Physiol. (London)*, 281, 559, 1978.
286. **Davison, P. F.,** Axoplasmic transport: physical and chemical aspects, in *The Neurosciences: Second Study Program*, Schmitt, F. O., Ed., Rockefeller Univ. Pres., New York, 1970, 851.
287. **Stöckel, K., Dumas, M., and Thoenen, H.,** Uptake and subsequent retrograde axonal transport of nerve growth factor (NGF) are not influenced by neuronal activity, *Neurosci. Lett.*, 10, 61, 1978.
288. **Friede, R. L. and Ho, K.-C.,** The relation of axonal transport of mitochondria with microtubule and other axoplasmic organelles, *J. Physiol. (London)*, 265, 507, 1977.
289. **Henkart, M. P., Reese, T. S., and Brinley, F. J.,** Endoplasmic reticulum sequesters calcium in the squid giant axon, *Science*, 202, 1300, 1978.
290. **Rebhun, L. I. and Sander, G.,** Electron microscope studies of frozen-substituted marine eggs. III. Structure of the mitotic apparatus of the first meiotic division, *Am. J. Anat.*, 130, 35, 1971.
291. **Shimizu, H. and Haken, H.,** Co-operative dynamics in organelles, *J. Theor. Biol.*, 104, 261, 1983.
292. **Hoyle, G.,** *Muscles and Their Neural Control*, John Wiley & Sons, New York, 1983.
293. **Sheetz, M. P. and Spudich, J. A.,** Movement of myosin-coated fluorescent beads along actin cables in vitro, *Nature (London)*, 303, 31, 1983.
294. **Nothnagel, E. A. and Webb, W. W.,** Hydrodynamic models of viscous coupling between motile myosin and endoplasm in Characean algae, *J. Cell Biol.*, 94, 444, 1982.
295. **Beckerle, M. C.,** Microinjected fluorescent polystyrene beads exhibit saltatory motion in tissue culture cells, *J. Cell Biol.*, 98, 2126, 1984.
296. **Allen, R. D. and Allen, N. S.,** Video-enhanced microscopy with a computer frame memory, *J. Microsc.*, 129, 3, 1983.
297. **Martz, D., Lasek, R. J., Brady, S. T., and Allen, R. D.,** Mitochondrial motility in axons: Membranous organelles may interact with the force generating system through multiple surface binding sites, *Cell Motil.*, 4, 89, 1984.
298. **Bütschli, O.,** Untersuchunger über mikroskopische Schäume und das Protoplasma, Leipzig, 1892, translation in *Investigation of Microscopic Foams and of Protoplasm* , Minchin, E. A., Ed., Adam & Charles, London, 1894.
299. **Preston, B. N., Laurent, T. C., Comper, W. D., and Checkley, G. J.,** Rapid polymer transport in concentrated solutions through the formation of ordered structures, *Nature (London)*, 287, 499, 1980.
300. **Jebsen-Marwedel, H.,** Spontane Mischbewegungen in Flüssigkeiten als Folge "dynaktiven" Materialtransportes durch Grenzflächenkräfte, *Umschau.*, 6, 165, 1957.
301. **Comper, W. D., Barry, N. P., and Austin, L.,** A paradigm for axonal transport, *Neurochem. Res.*, 8, 943, 1983.

REFERENCES ADDED IN PROOF

302. **Allen, R. D.,Weiss, D. G., Hayden, J. H., Brown, D. T., Fujiwake, H., and Simpson, M.,** Gliding movement of and bidirectional transport along single native microtubules from squid axoplasm: evidence for an active role of microtubules in cytoplasmic transport, *Cell Biol.*, 100, 1736, 1985.

303. **Brady, S. T., Lasek, R. J., and Allen, R. D.,** Video microscopy of fast axonal transport in extruded axoplasm: a new model for study of molecular mechanisms, *Cell Motil.*, 5, 81, 1985.

304. **Gilbert, S. P., Allen, R. D., and Sloboda, R. D.,** Translocation of vesicles from squid axoplasm on flagellar microtubules, *Nature (London)*, 315, 245, 1985.

305. **Vale, R. D., Schnapp, B. J., Reese, T. S., and Sheetz, M. P.,** Movement of organelles along filaments dissociated from the axoplasm of the squid giant axon, *Cell*, 40, 455, 1985.

306. **Schnapp, B. J., Vale, R. D., Sheetz, M. P., and Reese, T. S.,** Single microtubules from squid axoplasm support bidirectional movement of organelles, *Cell*, 40, 449, 1985.

307. **Suprenant, K. A. and Dentler, W. L.,** Association between endocrine pancreatic secretory granules and in vitro-assembled microtubules is dependent upon microtubule-associated proteins, *J. Cell Biol.*, 93, 164, 1982.

308. **Miller, R. H. and Lasek, R. J.,** Cross-bridges mediate anterograde and retrograde transport along microtubules in squid axoplasm, *J. Cell Biol.*, 101, 2181, 1985.

309. **Lasek, R. J. and Brady, S. T.,** Attachment of transported vesicles to microtubules in axoplasm is facilitated by AMP-PNP, *Nature (London)*, 316, 645, 1985.

310. **Brady, S. T.,** A novel brain ATPase with properties expected for the fast axonal transport motor, *Nature (London)*, 317, 73, 1985.

311. **Ris, H.,** The cytoplasmic filament system in critical point-dried whole mounts and plastic-embedded sections, *J. Cell Biol.*, 100, 1474, 1985.

312. **Kondo, H.,** Reexamination of the reality or artifacts of the microtrabeculae, *J. Ultrastruc. Res.*, 87, 124, 1985.

313. **Weiss, D. G. and Allen, R. D.,** The organization of force generation in microtubule-based motility, im *Microtubules and Microtubule Inhibitors 1985*, De Brabander, M. and De Mey, J., Eds., Elsevier/North-Holland, Amsterdam, 1986, in press.

314. **Vale, R. D., Reese, T. s., and Sheetz, M. P.,** Identification of a novel force-generating protein, kinesin, involved in microtubule-based motility, *Cell*, 42, 39, 1985.

315. **Vale, R. D., Schnapp, B. J., Reese, T. S., and Sheetz, M. P.,** Organelle, bead, and microtubule translocations promoted by soluble factors from the squid giant axon, *Cell*, 40, 559, 1985.

316. **Weiss, D. G., Keller, F., Gulden, J., and Maile, W.,** Towards a new classification of intracellular particle movements based on quatitative analyses, *Cell Motil. Cytoskel.*, 6, 1986, in press.

INDEX

A

D

H

I

K

L

M

N

O

P

R

S

T

U

V

W

X

Z